Advances in
Usability Evaluation
Part II

Advances in Human Factors and Ergonomics Series

Series Editors

Gavriel Salvendy
Professor Emeritus
School of Industrial Engineering
Purdue University

Chair Professor & Head
Dept. of Industrial Engineering
Tsinghua Univ., P.R. China

Waldemar Karwowski
Professor & Chair
Industrial Engineering and
Management Systems
University of Central Florida
Orlando, Florida, U.S.A.

3rd International Conference on Applied Human Factors and Ergonomics (AHFE) 2010

Advances in Applied Digital Human Modeling
Vincent G. Duffy

Advances in Cognitive Ergonomics
David Kaber and Guy Boy

Advances in Cross-Cultural Decision Making
Dylan D. Schmorrow and Denise M. Nicholson

Advances in Ergonomics Modeling and Usability Evaluation
Halimahtun Khalid, Alan Hedge, and Tareq Z. Ahram

Advances in Human Factors and Ergonomics in Healthcare
Vincent G. Duffy

Advances in Human Factors, Ergonomics, and Safety in Manufacturing and Service Industries
Waldemar Karwowski and Gavriel Salvendy

Advances in Occupational, Social, and Organizational Ergonomics
Peter Vink and Jussi Kantola

Advances in Understanding Human Performance: Neuroergonomics, Human Factors Design, and Special Populations
Tadeusz Marek, Waldemar Karwowski, and Valerie Rice

4th International Conference on Applied Human Factors and Ergonomics (AHFE) 2012

Advances in Affective and Pleasurable Design
Yong Gu Ji

Advances in Applied Human Modeling and Simulation
Vincent G. Duffy

Advances in Cognitive Engineering and Neuroergonomics
Kay M. Stanney and Kelly S. Hale

Advances in Design for Cross-Cultural Activities Part I
Dylan D. Schmorrow and Denise M. Nicholson

Advances in Design for Cross-Cultural Activities Part II
Denise M. Nicholson and Dylan D. Schmorrow

Advances in Ergonomics in Manufacturing
Stefan Trzcielinski and Waldemar Karwowski

Advances in Human Aspects of Aviation
Steven J. Landry

Advances in Human Aspects of Healthcare
Vincent G. Duffy

Advances in Human Aspects of Road and Rail Transportation
Neville A. Stanton

Advances in Human Factors and Ergonomics, 2012-14 Volume Set:
Proceedings of the 4th AHFE Conference 21-25 July 2012
Gavriel Salvendy and Waldemar Karwowski

Advances in the Human Side of Service Engineering
James C. Spohrer and Louis E. Freund

Advances in Physical Ergonomics and Safety
Tareq Z. Ahram and Waldemar Karwowski

Advances in Social and Organizational Factors
Peter Vink

Advances in Usability Evaluation Part I
Marcelo M. Soares and Francisco Rebelo

Advances in Usability Evaluation Part II
Francisco Rebelo and Marcelo M. Soares

Advances in
Usability Evaluation
Part II

Edited by
Francisco Rebelo
and
Marcelo M. Soares

CRC Press
Taylor & Francis Group
Boca Raton London New York

CRC Press is an imprint of the
Taylor & Francis Group, an **informa** business

CRC Press
Taylor & Francis Group
6000 Broken Sound Parkway NW, Suite 300
Boca Raton, FL 33487-2742

First issued in paperback 2019

© 2013 by Taylor & Francis Group, LLC
CRC Press is an imprint of Taylor & Francis Group, an Informa business

No claim to original U.S. Government works

ISBN-13: 978-1-4665-6054-3 (hbk)
ISBN-13: 978-0-367-38108-0 (pbk)

Visit the Taylor & Francis Web site at
http://www.taylorandfrancis.com

and the CRC Press Web site at
http://www.crcpress.com

Table of Contents

Section II: Theoretical Issues in Usability

Section III: Usability in Web Environment

Section IV: Miscellaneous

Preface

Successful interaction with products, tools and technologies depends on usable designs, accommodating the needs of potential users and does not require costly training. In this context, this book is concerned about emerging concepts, theories and applications of human factors knowledge focusing on the discovery and understanding of human interaction with products and systems for their improvement.

The book is organized into four sections that focus on the following subject matters:

- Usability Methods and Tools
- Theoretical Issues in Usability
- Usability in Web Environment
- Miscellaneous

In the section "Usability Methods and Tools," studies related with new and improved methods and tools for the advancement in efficiency of the usability studies are reported. In this context, this book provides studies, which cover everything from checklists and heuristics development to kaizen and biometrics measurement techniques. Also, the use of tools, like eye tracker, virtual reality and augmented reality is discussed.

The section "Theoretical Issues in Usability" concentrates on theorical approaches of usability that allow justifying the impact of usability in our lives. Review studies about the importance of usability and connections between ergonomics and virtual reality were reported. General approaches raised the concepts of modeling and simulation to explain changes in human performance and accidents.

The section "Usability in Web Environment" concentrates on studies associated with the use of the Internet environment and mainly discusses the development of new services and creates social communities.

The section "Miscellaneous" shows various studies that focus on aesthetic, affective and emotional design, corporate and inclusive design.

We would like to thank the Editorial Board members for their contributions.

B. Amaba,USA
D. Feathers, USA
W. Friesdorf, Germany
S. Hignett, UK
W. Hwang, S. Korea
Y. Ji, Korea
B. Jiang,Taiwan
S. Landry,USA

Z. Li, PR China
A. Moallem, USA
F. Rebelo, Portugal
V. Rice, USA
C. Stephanidis, Greece
A. Yeo, Malaysia
W. Zhang, PR China

This book will be of special value to a large variety of professionals, mainly researchers and students in the broad field of usability who are interested in the development and application of methods and tools to improve the design of products and systems. We hope this book is informative, but even more - that it is thought provoking. We hope it inspires, leading the reader to contemplate other questions, applications, and potential solutions in creating good designs for all.

April 2012

Francisco Rebelo
Ergonomics Laboratory
Interdisciplinary Centre for the Study of Human Performance
Technical University of Lisbon
Portugal

Marcelo Soares
Federal University of Pernambuco
Brazil

Editors

Section I

Usability Methods and Tools

Using Space Exploration Matrices to Evaluate Interaction with Virtual Environments

Luís Teixeira[1,2], Emília Duarte[2,3], Júlia Teles[2,4], Mariana Vital[1], Francisco Rebelo[1,2] and Fernando Moreira da Silva[5]

[1] Ergonomics Laboratory – FMH – Technical University of Lisbon, Cruz Quebrada-Dafundo, Portugal
[2] CIPER – Interdisciplinary Center for the Study of Human Performance, Technical University of Lisbon, Cruz Quebrada-Dafundo, Portugal
[3] UNIDCOM/IADE – Institute of Arts, Design and Marketing, Lisbon, Portugal
[4] Mathematics Unit – FMH – Technical University of Lisbon, Cruz Quebrada-Dafundo, Portugal
[5] CIAUD - Research Centre in Architecture, Urban Planning and Design – FA – Technical University of Lisbon, Lisbon, Portugal
{lmteixeira@fmh.utl.pt; emilia.duarte@iade.pt; jteles@fmh.utl.pt; marianavital@fmh.utl.pt; frebelo@fmh.utl.pt; fms.fautl@gmail.com}

ABSTRACT

The purpose of this study was to evaluate human interaction with Virtual Environments (VEs) through the use of matrices of space exploration. The matrices are the outputs of the participants' exploration of a VE in five experimental conditions, defined by two levels of visual pollution (i.e., more visually polluted and less visually polluted) containing two types of warnings (i.e., static and dynamic), plus a control condition (i.e., less visually polluted without warnings). One hundred and fifty individuals participated in this study, thirty in each experimental condition. They were asked to perform a work-related task, which required the search of some button switches that were placed in different areas of the VE and that were signalized by the warnings. Admitting that the more salient warnings (dynamic) would be detected more easily, the hypothesis is that there would be differences in

3

the space exploration between the experimental conditions and that these differences would be reflected on the matrices. The data required for the matrices generation (i.e., the position of the participant at all times, the time spent and distances travelled in the VE) were automatically recorded, resulting in information about the cells stepped by the participants and the frequency that they were stepped. The "agreement" of the space exploration matrices was evaluated through the application of the Concordance Correlation Coefficient over the frequencies that the cells were stepped on. The obtained results show high values of "agreement" between the matrices, which suggests that the space exploration in the VE was not strongly influenced by the experimental condition.

Keywords: *Virtual Reality, warnings, salience, space exploration matrices*

1. INTRODUCTION

There might be situations where not giving or maintaining attention on given information might not be severe, but that is not the case of safety information (e.g., warnings). Furthermore, several studies showed the existence of a relationship between the warnings' salience and their ability to promote compliant behavior (e.g., Wogalter, Kalsher, & Racicot, 1993). However, in part, the importance of the salience level is also dependent on the individual's previous experiences and knowledge. For example, repeated exposure a warning leads to the formation of a memory that, in turn, leads to habituation, resulting in a warning becoming less salient to the receiver (Wogalter and Vigilante 2006).

Commonly, the tendency is to pay attention (e.g., looking, hearing) to the stimuli that are more conspicuous. In this regard, recent studies suggest that dynamic warnings, sometimes multimodal, occasionally technology-based, are more efficient than the traditional equivalent counterparts (i.e., static), since they possess characteristics that make them more salient and more resistant to habituation (e.g., Wogalter and Mayhorn 2006; Mayhorn and Wogalter 2003; Smith-Jackson and Wogalter 2004). Attention, as with other stages of the information processing, cannot be evaluated directly. Thus, its evaluation is made through variables that, potentially, reflect the phenomenon (operational definition) such as, for example, detection/response time, observation behavior, eye fixations and several subjective evaluations (e.g., classification of salience of different alternatives), among others.

Virtual Reality (VR) has been used, as a research tool, in the field of warnings, for the evaluation of the influence of the safety information during egress situations (e.g., Glover and Wogalter 1997; Ren, Chen, and Luo 2008; Gamberini et al. 2003; Tang, Wu, and Lin 2009). Its potentialities for this field of research are diverse, emphasizing the potential to provide realistic contexts of use, very interactive and with high levels of immersion, benefiting external or ecological validity of the studies. VR can also provide access to all kinds of contexts, especially those that are difficult to access with relatively low costs. With VR, the studies can be repeated and replicated, in a systematic and rigorous manner, benefiting the internal validity of the studies. Additionally, VR can help to overcome some of the constraints related with ethical and participants' safety concerns, which limit many of these studies (Duarte, Rebelo, and Wogalter 2010).

In this context, this study's main objective was the evaluation of human interaction with Virtual Environments (VEs), through matrices of space exploration. The analyzed matrices were created with the data collected automatically, during the VR simulation, by the ErgoVR system (Teixeira, Rebelo, and Filgueiras 2010). Assuming that the most salient warnings (dynamic) would be detected more easily than the static counterparts, it was hypothesized that such differences would be manifested in the spatial exploration of the VE. More dispersion of the participant's position in the VE, resulting in a higher diffusion of the stepped cells (an indicator of greater search, in vaster areas), as well as higher frequencies of "steps" in some of the cells (indicator of hesitation and/or need of a more detailed search in that particular area), are expected in the experimental conditions where the warnings are less salient (static) and the environment is more visually polluted.

2. METHODOLOGY

2.1. Apparatus

The apparatus used in this study was: two magnetic motion sensors from Ascension-Tech®, model Flock of Birds, with 6 DOF, to capture the movements of the head and the participant's left hand; one joystick from Thrustmaster® to navigate in the VE; one Head-Mounted Display (HMD) from Sony®, model PLM-S700E, which presented the VE in a 800×600 pixels resolution with an approximate diagonal field of view of 30°; wireless headphones from Sony®, model MDR-RF800RK. It was also used one monitor that presented a copy of the VE, as seen by the participant, for the researcher.

2.2. Scenario and Virtual Environment

The VE was an interior space, ground floor, with typical characteristics of an office building, composed by two major areas: Area 1 – Rooms and Area 2 – Escape routes (see Figure 1). Area 1 consists in four rooms (Meeting Room, Laboratory, Cafeteria and Warehouse), each one with 144 m², in which the participants should enter to fulfill some assigned tasks. Area 2 consists in a sequence of six "T"-shaped corridors, 2 m wide, with one dead-end hallway. The environment was provided with diverse equipment (e.g., tables, chairs, cabinets) according with the tasks that were going to take place in each room. Also, there was informational material posted on the walls (e.g., message boards, posters, signs and safety information). Some of the physical properties of the VE, such as lights, shadows, textures and sounds, were manipulated in order to obtain an acceptable level of realism but without compromising the correct operation of the ErgoVR system.

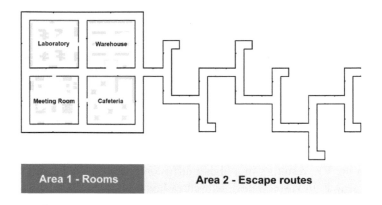

Figure 1 – Top view of the Virtual Environment

The scenario created for this study consisted in an end-of-day routine check task that at a given point is interrupted by an explosion followed by a fire. Participants were told that they had been selected to replace a coworker that had to leave early due to a health issue. The participants were therefore, entering in a section of the building for the first time. Before leaving, the coworker left some written instructions about what had to be done, to facilitate the task completion. As such, participants should look for the instructions, comply with them and only then, move to the next room. The first room of the sequence was the Meeting Room. The instructions, written on boards placed on the walls, informed the participants that they should press specific buttons (e.g., cut energy to the machine room), identified by warning/safety signs and positioned in diverse places of the VE. The explosion and fire were not mentioned in the pre-simulation dialogues with the participants. The simulation would stop when participants reached the exit of the building (final corridor of the Escape Routes), if the established maximum duration of the simulation was reached (i.e., 20 minutes) or if they manifested will to stop the simulation.

2.3. Experimental conditions and stimuli

Five independent samples indexed to each of the experimental conditions were used in this study. Four of them resulted from the combination of the two types of warnings (static and dynamic) and two types of environment (more visually polluted and less visually polluted). There was also a neutral condition (without warnings and the less visually polluted environment). The experimental conditions were therefore designated as: SNP (static not polluted), DNP (dynamic not polluted), SP (static polluted), DP (dynamic polluted) and Neutral (NE).

In the VE there were seven buttons, with their corresponding warnings/safety signs. These consisted in plates (30×40cm) with two components, text and image consistent with ISO 3864-1 (ISO 2002) recommendations.

The static and dynamic versions differ in modality and state. The static ones were only visual while the dynamic were visual and auditory (beep), backlit (two times brighter than the background), supplemented with five orange flashing lights (four flashes per second, with equal duration in the on and off state) that had 4 cm in diameter and were positioned at the top and bottom of the plate. The dynamic version also had two different states, on and off, being activated by proximity sensors. Safety signs and warnings were similar in what refers to their design variables (e.g., lettering, organization, dimensions) but differed on their content and purpose. The signs indicated the presence of an equipment, and were referred explicitly in the instructions being, therefore, expectedly searched in an active way by the participants. The warnings alerted for a potential hazard and required a specific action to ensure safety. These were never mentioned in the instructions although they were coherent with the type of activity of each room.

The two types of environment, more visually polluted and less visually polluted, were characterized by the amount of visual information present in the VE (more or less information), by the degree of legibility offered by the warnings (degradation was visible or not) and by the level of environmental lighting (i.e., adequate for tasks with low visual demands or adequate for task with high visual demands). The type of environment was thought of to facilitate, or to difficult, the efficiency of the participants' performance, evaluated by the behavioral compliance with the signs/warnings, time to fulfill the task and distance travelled in the VE.

2.4. Sample

One hundred and fifty individuals participated in this study, aged between 18 and 35 years (mean = 21.20, SD = 2.85). Participants were assigned to one of the five experimental conditions. They were equally distributed in gender and number between the conditions. All the participants were university students from different areas and they declared not having any previous experience with an immersive VR system.

2.5. Protocol

The procedure was divided into three main stages: 1) Preliminary activities: after receiving the initial explanations regarding the study and the equipment, participants answered a demographic questionnaire, filled the Free and Informed Consent form, and were screened for color blindness with the Ishihara test (Ishihara 1988). After that, they completed a training session in a VE created for that purpose in which the calibration of the equipment was included; 2) Experimental session: participants were indexed to one of the experimental conditions and, before starting the simulation, heard the instructions. When they declared that were ready, the simulation started. For safety reasons, the researcher remained throughout the simulation in the same room with the participants who were dully warned that they could abandon the procedure at any time; 3) Post-hoc Questionnaire: after the experimental session, participants answered a questionnaire to evaluate their experience with the VE.

3. RESULTS AND DISCUSSION

For this study, each log entry was composed by the position of the "actor" in the VE and by instant of time when the log entry was saved. With that data it was possible to generate the paths taken, as well as the matrices of space exploration of the VE, for the regions of interest (with visual representation and numerically).

The matrices of space exploration of the VE consist in a grid (with configurable cell dimensions) with a color and an associated number to each cell. The number represents the frequency that each cell was stepped on during the simulation. In this study, a "stepped on cell" is one that was intersected by the point that represents the middle axis of the "actor", perpendicular to the floor. Each time that the "actor" enters the cell, the frequency value of that cell increases by one unit. Since this study used groups of participants (one for each experimental condition), the frequency value of each cell represents the sum of the frequencies of every individual in that group. The color of each cell is assigned from a 11 pseudo-color scale (Ware 2004) to indicate the "warm" zones (warm colors = more stepped on) e "cold" zones (colder colors = less stepped on). The highest value from all the cells (the most stepped on cell), from all five experimental conditions, was used as a base to the normalization of the showed colors.

In Figure 2 it is possible to visualize the five matrices (from the complete VE), one for each experimental condition, with cell size of 50×50 cm. This size was adopted since it was closer to the 95th percentile for the male shoulders' width (51 cm) (Pheasant and Haslegrave 2006).

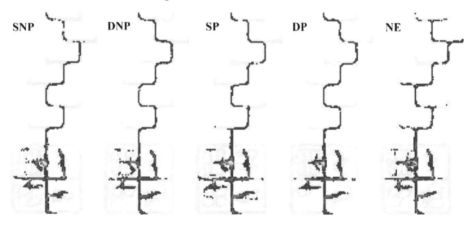

SNP   DNP   SP   DP   NE

Figure 2 – Space exploration matrices of the VE, per experimental condition

The log viewer allows creating delimiting areas where data should be analyzed (called inclusion areas). These areas correspond to each of the four rooms, including the adjacent corridor, the one with the door of the room and lastly there is an inclusion area on the escape routes area.

The "agreement" of the space exploration matrices, for the areas of interest, was evaluated through the application of the Concordance Correlation Coefficient

(CCC) (L. Lin 1989; Barnhart, Haber, and Song 2002; L. Lin et al. 2002) considering the frequency that the cells were stepped on as the variable of interest.

The evaluation of the "agreement" of the space exploration of the VE in the five experimental conditions, was made through the use of the Overall Concordance Correlation Coefficient (OCCC) (Barnhart, Haber, and Song 2002) over the frequencies that the cells were stepped on. The OCCC values obtained are presented on the first row of Table 1. Taking into account that this coefficient takes values in the interval [-1,1], it is possible to verify that the "agreement" is very high, in every analyzed area, suggesting that the space exploration was not strongly influenced by the experimental condition. Nonetheless, it is to be noted that the lowest values of CCC were found in the Escape Routes and in the Warehouse, with the highest values in the Cafeteria and in the Meeting Room.

Table 1 – OCCC values per area analyzed

	Warehouse	Cafeteria	Laboratory	Meeting Room	Escape Routes
50 x 50 cm	0.843	0.948	0.918	0.946	0.838
75 x 75 cm	0.856	0.960	0.927	0.961	0.841
100 x 100 cm	0.858	0.972	0.940	0.966	0.860

Table 2 – Partial CCC values for the Warehouse and the Escape Routes

	Warehouse				Escape Routes			
	DNP	SP	DP	NE	DNP	SP	DP	NE
SNP	0.762	0.909	0.877	0.926	0.895	0.889	0.887	0.825
DNP		0.717	0.838	0.658		0.873	0.909	0.751
SP			0.884	0.925			0.852	0.823
DP				0.833				0.720
	Laboratory				Cafeteria			
	DNP	SP	DP	NE	DNP	SP	DP	NE
SNP	0.945	0.921	0.917	0.952	0.941	0.960	0.951	0.959
DNP		0.892	0.927	0.945		0.937	0.942	0.944
SP			0.871	0.903			0.955	0.947
DP				0.909				0.942
	Meeting Room							
	DNP	SP	DP	NE				
SNP	0.929	0.935	0.942	0.950				
DNP		0.947	0.955	0.943				
SP			0.952	0.950				
DP				0.959				

The CCC partial values were also calculated, that is, the CCC values considering the experimental conditions in pairs. In the Warehouse, the highest CCC values were obtained between the SNP/NE, SP/NE and SNP/SP pairs of experimental conditions and the lowest CCC values were obtained between the pairs DNP/NE and DNP/SP. In the Escape Routes the highest values were registered in the DNP/DP and SNP/DNP pairs of experimental conditions, with the lowest in the DP/NE and DNP/NE pairs. The CCC partial values obtained for the Laboratory, Cafeteria and Meeting Room were very similar. These data can be seen in Table 2.

Since the ErgoVR's log viewer allows the generation of matrices with different cell dimensions, the OCCC was also applied to the obtained matrices with cell sizes of 75 cm and 100 cm. The change in the size of the cell did not originate relevant changes in the OCCC values. It should be noted that the OCCC values decrease with the decreasing of the size of the cell. A summary of the OCCC values, for all the cell dimensions can be found in Table 1.

4. CONCLUSION

It was hypothesized that the way how the participants explore the space, during a VR experience, could be used to evaluate their interaction with VEs. In this context, this study evaluated the interaction with VEs, in different experimental conditions, using matrices of space exploration in VEs, generated from data collected automatically by the ErgoVR system.

The matrices contained the number of times each cell was stepped on and reflected how the participants explored the VE in the five experimental conditions. Assuming that the more salient warnings (dynamic) would be detected more easily, it was expected to find these differences reflected in the space exploration of the VE. Greater dispersion of the participant's path in the VE, resulting in a greater dispersion on the stepped on cells was expected in the experimental conditions where the warnings were less salient (static) and the environment was more visually polluted.

The Concordance Correlation Coefficient was used to determine the degree of "agreement" between the experimental conditions, regarding the frequency of the stepped on cells. The obtained results reveal high values of "agreement" between the different matrices, which allows concluding that there was not a significant influence of the experimental condition on the space exploration of the VE.

In this study it was not considered the individual behavior of the participants in the VR experience, since that the coefficient was used to evaluate the correlation of the frequency of the stepped cells by the thirty participants that were assigned to one of the five experimental conditions. One limitation of this study it that for each one of the experimental conditions it was not taken into account if the registered frequency for a cell was due to only one individual that stepped on that cell repeatedly or several different individuals. In future studies it is intended to evaluate the individual paths, to compare the paths taken with reference paths (shorter, faster and easier paths) and, by assigning scores to the distance to the reference path, to gather some performance values. Additional information, such as the orientation of the body of the "actor" and its gaze orientation, in each instant of the interaction, may complement the analysis.

The use of VR, in this type of studies, allows an easy collection of data regarding the paths taken by the participants as well as times and distances travelled. This type of data would be more difficult to collect in the field. VR provides, simultaneously, an optimal relationship between the strict control of variables and the necessary degree of ecological validity of the study.

ACKNOWLEDGMENTS

This study was developed in the scope of a research project (PTDC/PSI-PCO/100148/2008) funded by the Portuguese Science Foundation (FCT). It was also sponsored by the Human Factors Group of the Interdisciplinary Centre for the Study of Human Performance (CIPER).

REFERENCES

Barnhart, Huiman X, Michael Haber, and Jingli Song. 2002. "Overall concordance correlation coefficient for evaluating agreement among multiple observers." *Biometrics* 58 (4): 1020-1027.

Duarte, E., F. Rebelo, and M. S. Wogalter. 2010. "Virtual Reality and Its Potential for Evaluating Warning Compliance." *Human Factors and Ergonomics in Manufacturing & Service Industries* 20 (6): 526-537. doi:10.1002/hfm.20242.

Gamberini, L, P Cottone, A Spagnolli, D Varotto, and G Mantovani. 2003. "Responding to a fire emergency in a virtual environment: different patterns of action for different situations." *Ergonomics* (731736530).

Glover, B.L., and M. S. Wogalter. 1997. Using a computer simulated world to study behavioral compliance with warnings: Effects of salience and gender. In *Human Factors and Ergonomics Society Annual Meeting Proceedings*, 41:1283–1287. Santa Monica: Human Factors and Ergonomics Society.

ISO. 2002. "ISO 3864-1 Graphical symbols — Safety colours and safety signs." *Journal of the Acoustical Society of America* 130 (4): 2525. doi:10.1121/1.3655080.

Ishihara, S. 1988. *Tests for colour-blindness: 38 plates edition*. Tokyo, Japan: Kanehara & Co. LTD.

Lin, L. 1989. "A concordance correlation coefficient to evaluate reproducibility." *Biometrics* 45 (1): 255-268.

Lin, L., A S Hedayat, Bikas Sinha, and Min Yang. 2002. "Statistical methods in assessing agreement: Models, issues, and tools." *Journal of the American Statistical Association* 97 (457): 257-270.

Mayhorn, C.B., and M.S. Wogalter. 2003. "Technology-based warnings: Improvising safety through increased cognitive support to users." *15th International Ergonomics Association Congress* 4: 504-507.

Pheasant, S, and C M Haslegrave. 2006. *Bodyspace: Anthropometry, Ergonomics, and the Design of Work. Health San Francisco*. Taylor & Francis.

Ren, A, C Chen, and Y Luo. 2008. "Simulation of Emergency Evacuation in Virtual Reality." *Tsinghua Science Technology* 13 (5): 674-680. doi:10.1016/S1007-0214(08)70107-X.

Smith-Jackson, T. L., and M. S. Wogalter. 2004. "Potential Uses of Technology to Communicate Risk in Manufacturing." *Human Factors and Ergonomics in Manufacturing* 14 (1): 1?14. doi:10.1002/hfm.10049.

Tang, Chieh-Hsin, Wu-Tai Wu, and Ching-Yuan Lin. 2009. "Using virtual reality to determine how emergency signs facilitate way-finding." *Applied Ergonomics* 40 (4): 722-730.

Teixeira, Luís, Francisco Rebelo, and Ernesto Filgueiras. 2010. Human interaction data acquisition software for virtual reality: A user-centered design approach. In *Advances in Cognitive Ergonomics. Advances in Human Factors and Ergonomics Series*, ed. D.B. Kaber and G. Boy, 793-801. Miami, Florida, USA: CRC Press/Taylor & Francis, Ltd.

Ware, Colin. 2004. *Information Visualization: Perception for Design*. Ed. S K Card, J Grudin, J Nielsen, and T Skelly. *Information Visualization*. Vol. 22. Morgan Kaufmann.

Wogalter, M. S., Michael J. Kalsher, and Bernadette M. Racicot. 1993. "Behavioral compliance with warnings: effects of voice, context, and location." *Safety Science* 16 (5–6): 637-654. doi:10.1016/0925-7535(93)90028-C.

Wogalter, M. S., and C. B. Mayhorn. 2006. The Future of Risk Communication: Technology-Based Warning Systems. In *Handbook of Warnings*, 783-793. Lawrence Erlbaum Associates.

Wogalter, M. S., and W. J. Vigilante. 2006. Attention Switch and Maintenance. In *Handbook of Warnings*, 245-265.

CHAPTER 2

Measuring and Managing the Consistency and Effectiveness of Checklists Used to Assess Jobs at Risk for Musculoskeletal Disorders

Thomas J. Albin

High Plains Engineering Services
Minneapolis, USA
Talbinus@comcast.net

ABSTRACT

The use of checklists and similar tools to assess jobs with regard to risk of musculoskeletal symptoms is fairly common; jobs with some level of risk factors present are identified as problem jobs. However, knowledge of the reliability and validity of such tools is less common. This paper briefly describes a method of measuring and managing the reliability of checklist users' scores. It also describes more completely a means of calculating the probability that a checklist correctly identifies a problem job

Keywords: checklist, reliability, validity, musculoskeletal disorder

1   INTRODUCTION

At some time or another, nearly every ergonomics practitioner uses a checklist or a similar assessment tool, either to identify problem jobs, e.g., jobs where MSD risk factors are present, or to characterize the risk factors once they have been identified - this many repetitive movements, that much flexion or extension. Further, these checklists are often scored in such a way as to indicate greater or lesser need for intervention to correct identified risk factors; jobs with scores above some criterion are given first priority, those with scores below the criterion are given lesser priority for remediation.

A tool with a poor ability to discriminate between problem jobs and non-problem jobs has poor validity and will create credibility problems for the ergonomics practitioner. It will fail to identify issues that need to be corrected and it will waste resources fixing things that don't need to be fixed. Similarly, a tool that gives inconsistent results from time to time or when different individuals use it has poor reliability and will also lead to wastage. A good tool must be effective and consistent, or, in other words, it must be both valid and reliable.

While the use of such tools is common, knowledge of their validity and reliability is much less common. When validity and reliability data is available, it is often historical and static, e.g., a study done by someone else in different circumstances than those in which the practitioner practices. While this is of some utility, it behooves an individual practitioner to develop a dynamic and current understanding of the validity and reliability of the tool or tools that he or she uses.

A second concept presented in this paper is that checklists need not be static in form. While it may be of historical interest that a checklist had a certain content and format when it was developed, the current context of use may be so different as to warrant modification. Similarly, experience with a checklist may suggest that something is missing that should be assessed, or that some information gathered is redundant.

Just as a cutting tool may be sharpened or re-shaped to suit the purpose at hand, so should checklists be sharpened and reshaped to suit the context of use. Consequently, the practitioner needs a dynamic means of assessing checklist validity and validity that can respond dynamically to changes in the checklist.

This paper describes simple methods for assessing both the reliability and validity of checklists used by practitioners, although the validity measure is more fully described.

2   A BRIEF NOTE REGARDING CONSISTENCY

There are two types of reliability that are of concern to the practicing ergonomist and that must be managed. The first is intra-rater reliability. Suppose that an ergonomist uses a checklist to evaluate the same job at two different times, perhaps a couple of weeks apart. If nothing about the job changes, we would expect the checklist results to be very similar. The second type of consistency is inter-rater reliability.   In this case, two or more ergonomists concurrently use the same checklist to evaluate the same job. Again, we would expect the checklist results to be very similar. For both these cases, we might calculate a correlation coefficient or a similar measure to quantitatively assess how similar the ratings are.

However, this assessment of consistency or reliability should be an on-going process. Consider the analogy of a sharp drill bit used to bore a hole that is exactly 5 mm in diameter in a hard steel plate. We know that the bit will begin to wear and dull when it is used, so we carefully monitor the diameter of the hole with a go-no go gauge to determine when the bit needs to be re-sharpened.

Checklist users are the same. We need to closely monitor the consistency with which evaluators use a checklist over time and should have some criterion, or a go-no go gauge to determine if and when the evaluators' observational skills need to be re-sharpened.

1.2   Managing reliability

One way to do this would be to periodically provide a test job to each evaluator. The test job, perhaps a video, would present known values for each quantity assessed on the checklist of interest. For example, if one of the things that the checklist assessed were wrist flexion, there would be instances of wrist flexion on the video where the angle is known, although not to the evaluator. The evaluator would receive feedback on the correctness of his or her performance.

Here is the management part. So long as the performance met the go – no go criterion, e.g. 90 percent of items on the checklist were correctly assessed; all concerned would be satisfied that the consistency or reliability of the checklist was satisfactory. What happens if it is less than the go-no go criterion? A careful analysis of performance on the test job or jobs might indicate weak points in the evaluators' performance. For example, a checklist might assess force, posture and repetition at a number of body joints. Performance in assessing force and repetition might be acceptable (greater than 90 percent correct), but posture is much less, say 65 percent. Clearly, we need to re-sharpen.

A process control chart is a tool that might be effectively used to monitor evaluator reliability. Briefly, a control chart would be used to plot the average percent correct. Control limits, often plus or minus 3 standard errors of the mean above or below the mean, would also be shown on the chart. In general, so long as the data points stay within the control limits, variation is assumed to be random and acceptable. However, if data points fall outside the control limits, or give evidence of developing trends, some cause other than random variation is assumed to be present, and must be identified and corrected. Of course, the mean should also be equal to or greater than the go – no go criterion. This is not meant to be a complete discussion of control charts and statistical process control; there are many resources available where more complete descriptions can be found. [Wikipedia, 2012]

There are two essential points regarding reliability. First, without good reliability checklists are of limited worth. Second, reliability should be managed as a dynamic quantity.

3   A MEASURE OF CHECKLIST VALIDITY

The utility of a checklist is that it can be used to identify problem jobs. It has validity if it is a good predictor of problem jobs. So, to begin to assess the validity of a checklist, we have to know what it is that we want it to identify, that is, how is a problem job defined? It may be simple, e.g. any job with a history of a lost-time musculoskeletal disorder within the past two years, or more complex, e.g. any job with a history of lost-time, restricted-time, high turnover rate, and employee complaints within the past six months.

Once we define problem jobs, we need to know the prevalence of problem jobs as a proportion of all the jobs of interest, say within a factory that has 1,253 total jobs. We may know that there are 132 problem jobs within the factory (population prevalence), or we may estimate the prevalence based on a sample of jobs picked at random (sample prevalence).

Suppose that we take the latter course and randomly pick 100 jobs. We determine that 10 of those randomly picked jobs are problem jobs: they have a history of at least one lost-time musculoskeletal disorder within the past two years. The sample prevalence is then 10 percent or 0.10.

3.1   Guessing

If we want to identify the problem jobs among the remaining 1,153 jobs in the factory, we might simply guess, based on the prevalence of problem jobs in the sample, that one in every ten jobs is a problem job.

What is the probability that a guess correctly identifies a problem job? Based on the sample prevalence of 10 percent, we would expect that there are 115 more problem jobs in the factory. By guessing, we identify one in ten of these (11.5) as problem jobs. However, we also guess and identify one in ten of the 1,038 non-problem jobs as problem jobs (103.8). The probability of correctly guessing that a job is a problem job is 11.5 correct guesses out of a total of 115.3 guesses, or ten percent. Not too surprising, but it gives us something with which to compare our checklist results.

3.2   Estimating the probability a checklist correctly identifies a problem job

We next use our checklist to analyze the 100 sampled jobs. Suppose that the checklist is designed so that a score of 6 or higher indicates a problem job and that the checklist results identify 16 jobs as problem jobs. Checking against the list of known problem jobs, we determine that the checklist has correctly identified 7. The complete results are summarized in Table 1.

Table 1. Results of using the checklist to analyze a sample of 100 jobs

	Problem Job Yes	Problem Job No	Row Sum
Checklist Yes	7	9	16
Checklist No	3	81	84
Column Sum	10	90	100

The checklist correctly identified 7 jobs as problem jobs, however it incorrectly identified an additional 9 jobs as problem jobs. Consequently a positive checklist result was correct 7 out of 16 times, or about 44 percent of the time. Similarly, the sample data suggest that a non-problem job was incorrectly identified as a problem job 9 out of 90 times, or about 10 percent of the time.

We can now use the sample results to estimate the probability that the checklist will correctly identify problem jobs for the remaining 1,153 jobs in the factory.

Based on the sample prevalence, we would expect 115 of the remaining jobs in the factory to be problem jobs (0.10 x 1,153) and 1,038 to be non-problem jobs. The sample data suggests that about 44 percent of the problem jobs (50.6) will be correctly identified.   However, 10 percent of the non-problem jobs will be incorrectly identified as problem jobs (103.8). The probability that the checklist correctly identifies a problem job is 0.327 (50.6)/(50.6+103.8).

We argue that the checklist is valid with regard to identifying problem jobs, as it is more effective than guessing by a factor of about 3.3 times. If one wished to evaluate whether the difference was statistically significant, a t-test of proportions could be used. [Walpole et al, 1978]

3.3   Tuning the checklist

Although the checklist is more efficient than guessing, we might wish to try to improve it. Suppose that we want to see the effect of changing the criterion score on the checklist from 6 to 8. We re-classify the sample according to the new criterion score and obtain the following results.

Table 2. Results of modifying the checklist used to analyze the same sample of 100 jobs

	Problem Job Yes	Problem Job No	Row Sum
Checklist Yes	9	4	13
Checklist No	1	86	87
Column Sum	10	90	100

The checklist now correctly identifies the problem jobs 69 percent of the time (9 of 13) and incorrectly identifies non-problem jobs as problem jobs 4.4 percent of the time (4 of 90).

Again, based on the sample prevalence, we would expect 115 of the remaining jobs in the factory to be problem jobs (0.10 x 1,153) and 1,038 to be non-problem jobs. The sample data suggests that about 69 percent of the problem jobs (79.4) will be correctly identified and that 4.4 percent of the non-problem jobs will be incorrectly identified as problem jobs (45.7). The probability that the checklist correctly identifies a problem job is now 0.635 (79.4)/(79.4+45.7). Clearly the efficiency of the checklist has been improved by changing the criterion score.

3 SUMMARY AND CONCLUSIONS

While checklists are commonly used to assess jobs with regard to musculoskeletal risk factors, information regarding the reliability and validity of the checklists is less common. Even when available, the ability to generalize reliability and validity data from the setting where that data was generated to a novel situation is an open question. In addition, those data, particularly the reliability data, are nearly always static and do not reflect any "drift" in the users' performance during repetitive use of the checklist.

This paper suggests that reliability and validity of checklists must be managed as dynamic processes. One suggested means of evaluating reliability is to periodically assess users' performance in using a checklist against known standard jobs. The resulting data can be used to manage the reliability, e.g. by training evaluators, simplifying the checklist, etc., to ensure that it is of an acceptable level.

Similarly, one way of looking at the validity of a checklist is to determine the probability that a positive result on the checklist will correctly identify a problem job. This paper describes a method to calculate that probability.

REFERENCES

Walpole, R.E., Myers, R.H. (1978) Probability and Statistics for Engineers and Scientists 2
ed. Macmillan, New York

Wikipedia (2012) Control Charts. http://en.wikipedia.org/wiki/Control_chart Boyd, J. and S.
Banzhaf. 2007. What are ecosystem services? *Ecological Economics* 63: 616–626.

.

CHAPTER 3

Can Virtual Reality Methodologies Improve the Quality of Life of People with Disabilities?

Lígia M Presumido Braccialli*, Francisco Rebelo**, Leonor M Pereira**

*UNESP – Univ Estadual Paulista
São Paulo, Br
bracci@marilia.unesp.br
**CIPER – Ergonomics Laboratory FMH-Technical University of Lisbon

ABSTRACT

Despite of the importance of assistive technology devices for the independence and quality of life of people with disabilities large part of the prescription is abandoned after some time. The abandonment of assistive technology (AT) resources has been justified by the difficulties of user interaction / technology, mainly due to problems in the aesthetics of the product, difficulty in learning to use and lack of assessment of use in situations of stress or imminent risk. This article discusses the use of virtual reality (VR) as a tool for the evaluation of assistive technology products during the development process, which when fitted into appropriate methodologies, can be trusted, viable and present some advantages such as being more comprehensive have ecological validity, keeping track of variables, besides enabling the application to situations of danger and stress. Were analyzed the possibilities of use of VR in different situations for evaluation of device AT: (a) product aesthetics; (b) of the learning process for use of the product; (c) of viability of resource usage in risk situations imminent.

Keywords: disability, virtual reality, assistive technology

1 ASSISTIVE TECHNOLOGY AND QUALITY OF LIFE OF PEOPLE WITH DISABILITIES

The quality of life of people with disabilities depend on interventions during rehabilitation that aim to better mobility, self care and functionality, but also the possibility of participation and social interaction, independence and choice of favorite activities.

In order to compensate for sensory and functional deficits of individuals with some limitation in order to enable it to achieve maximum independence and life satisfaction are routinely prescribed assistive technology devices (Verza et al., 2006). Assistive Technology (AT) can be defined therefore as a wide range of equipment, services, strategies and practices designed and implemented to reduce the functional problems encountered by individuals with disabilities (Cook and Hussey, 2001). The use of a device assistive technology reduces the difficulty that the individual has to perform the activity and dependence that the same has in relation to another (Brummel-Smith and Dangiolo, 2009).

Although it is indisputable that the contribution of AT brings a device to its user, access to the resource often becomes difficult due to fragmentation of services necessary for the prescription of different pieces of equipment, lack of information regarding the user services offered, the lack of eligibility criteria for access to these services and inequalities in provision within and between geographical areas within the same country (Cowan and Khan, 2005).

Several studies have shown that Over 30% of all devices acquired are abandoned by Medals between the first and fifth year of use, and some not arrive nor even to be used (Goodman et al., 2002; Huang et al., 2009; B. Phillips and Zhao, 1993; Scherer, 2002; Verza et al., 2006). These authors reported a number of reasons that lead the user to abandon the use of prescribed and purchased. Among them, 1) lack of user participation during the selection of the device, 2) ineffective performance of the device, 3) changes in user needs, 4) lack of user training; 5) device inappropriate to user needs; 6) devices use complicated; 7) social acceptance of the device; 8) lack of motivation to use the device; 9) lack of knowledge of the device; 10) devices with appearance, weight and size aesthetically not approved.

The user of AT, regardless of disability you have, considers it important that the action has prescribed and used some attributes: enable communication with others, improve mobility, improve physical security, enabling autonomy, confidence, competence and independence personnel, and improving skills for entering the labor market and community (Lupton and Seymour, 2000).

When the technology has positive attributes, it provides social interaction and improves self-esteem and quality of life of the user. When choosing a technology, the user finds what meaning friends and society in general attribute that device (Louise-Bender et al., 2002). The authors reported also that the device will have a greater chance of being abandoned by the user if it has a negative connotation by society or is stigmatizing, even if there was better functional performance, ie the higher the invisibility of the appeal and acceptance this community are less likely to be abandoned.

For prescription and acquisition of a resource of assistive technology to succeed, four steps must be considered: 1) evaluating the user to make certain that there is a deficiency and that the device can bring benefits, 2) assessment of user motivation to make use of an assistive technology device, 3) request the user collaboration and sharing of decisions taken in relation to the acquisition of a device, 4) implementation of user training for the use of the device indicated and acquired (Brummel-Smith and Dangiolo, 2009).

Thus, to implement a program using assistive technology to succeed, it is necessary to plan and support based on individual needs and abilities of each user (Donegan et al., 2009).

2 TRADITIONAL METHODOLOGIES AND TOOLS USED TO ASSESS AT RESOURCES

The literature indicates that the abandonment of the AT is due to problems related to: interactive user / device and the aesthetics of the product, learning to use, and difficulty when using the device in situations of stress and risk (Goodman et al., 2002; Huang et al., 2009; B. Phillips and Zhao, 1993; Scherer, 2002; Verza et al., 2006). From these findings it is possible to assume that the AT abandonment may not be related to the prescription or purchase of the product, but to the way these products have been evaluated during the development process.

Most current investigations has focused on Product Assessment by user's perspective after the device is already available on the market. However, at this point it is difficult to make any change on the device because of financial costs corresponding to the development of a new product. Traditionally biomechanical and social psychology tools have been used to assess the use of TA resources and to obtain information related to the difficulties of using the product.

Studies on resource assessment of AT, with the biomechanical approach, have especially been investigating the efficacy of wheelchairs, orthotics and furniture adapted for people with physical disabilities (Troy, 2011).

These investigations have numerous advantages as they are grounded on experimental research and, within laboratory environment, they accurately assess and control variables isolated by techniques for measuring physical quantities and with high internal validity. The methodological rigor of these studies also allows to establish cause-effect relationships, in addition to allowing its replicability.

Studies on the evaluation of AT with the use of social psychology tools are designed to evaluate the perception and user's satisfaction regarding the device as well as its functionality in the natural environment (M. Phillips et al., 2011; J. K. Martin et al., 2011; Murchland and Parkyn, 2010; Ryan et al., 2009). To achieve these objectives a more qualitative approach using questionnaires, interviews and direct and indirect observations have been used. The advantage of such approach is that it enables comprehensive studies in an ecological environment that allows to study the interaction between person / environment, person / person, person / equipment, data collection instruments with relatively low cost, however without complete control of the variables.

Both methods seem to have disadvantages for assessing AT resources, studies in the field of biomechanics are not considered ecological, while the ones on the area of social psychology apply to the product already developed, and do not allow tests in situations involving danger and stress for the participant because of ethical factors.

Given these difficulties, Virtual Reality (VR) is presented as a tool that fitted into appropriate methodologies, can overcome some of the problems presented in addition to presenting some advantages such as the possibility of simulating a wide range of contexts, a good ecological validity, the possibility to easily replicate experimental conditions, have a good internal validity and a high control of variables and possible use in dangerous situations. The use of virtual AT prototypes may also overcome the difficulties of using conventional methodologies and tools, which require a physical prototype to be applied.

3 VIRTUAL REALITY (VR): ADVANTAGES, USE AND LIMITATIONS

Virtual Reality (VR) can be defined as a computer-user interface approach that involves real-time simulation of an environment, scenery, or activity that allows user interaction via multiple sensorial channels. (Adamovich et al., 2009). While real-world knowledge about the environment is obtained through the senses, hearing, sight, touch, proprioception, and smell, in the virtual world the same senses are used to obtain information about the virtual world through man-machine interface (Holden, 2005).

VR offers the opportunity to bring the complexity of the physical world to the controlled environment of the laboratory, allows measurement of natural movement within natural complex environments, ie, is the creation of a synthetic environment with precise control over a large number of variables influencing the physical behavior during the generation of physiological and kinematics responses. (Keshner, 2004).

In recent decades VR has been widely used as a tool for research in rehabilitation and education of people with disabilities. Researchers have found that the use of VR in this context brings several advantages: (a) possibility of adapting the virtual environment (VE) relatively easily to meet the needs and physical abilities of the user; (b) the activities are conducted in a safe environment, without risk of injury; (c) easy performance feedback in real time; (d) independent motivational training; (e) possibility of increasing the complexity of tasks (P. L. T. Weiss et al., 2003; Jannink et al., 2008; Laver et al., 2011); (f) activities can be developed in a collaborative environment (Craig and Sherman, 2009); (g) accurate record of the user's performance in a database, (h) capacity to produce applications to test and evaluate the user in relevant daily tasks (Kisony R, Katzand and Tamar, 2003; Rizzo et al., 2006; Gourlay et al., 2000); (i) enrichment of environment in which activities are carried out (Gourlay et al., 2000; Laver et al., 2011); (j) ecological validity (Standen and Brown, 2006); (l) ability to intervene during the practice to provide more education or counseling (Kisony, Katzand and Tamar,

2003); (m) sensory output control and consistency; (n) full control over the content (Gourlay et al., 2000; Standen and Brown, 2006; Laver et al., 2011); (o) the possibility of extended practice in the user's own home or some other community environment; (Viau et al., 2004; Laver et al., 2011); (p) flexible learning environment (Standen and Brown, 2006); (q) improves residual capacity without causing fatigue and frustration. (R Kizony et al., 2004; Adelola et al., 2009).

The assessment in a virtual environment still presents some limitations such as the presence of wires that limit the user's movements, depth perception and insufficient tactile feedback, inadequate association between vision and action that may lead to performances different from those performed in physical reality, possibility of side effects such as dizziness, stomachache, headache, eyestrain, dizziness, and nausea.

However, few studies have addressed the use of this tool for assessing assistive technology devices from the moment of its conception to the impact and effectiveness of the devices on the user's daily routine.

4 VIRTUAL REALITY AS A TOOL TO ASSESS AT

The most important reason why AT devices are not used by consumers is the lack of user involvement in selecting the device, and the solution proposed by the author is to conduct a comprehensive evaluation process of the product with the active participation of the disabled person (Scherer, 2002).

From this perspective, when designing an AT resource, one must consider the features of the product in terms of interaction with the environment and the user versus the capabilities and limitations of the disabled person from the perspective of the disabled person.

According to ISO 13407 there are four required activities during the development process of a product: (1) understand and specify context of use, (2) specify user and organizational requirements, (3) to produce design solutions, and (4) to assess projects in relation to user's requirements. (Standard et al., 1994).

Because VR allows the creation of a scenario that permits user interaction with objects through multiple sensory channels, it can be used as an effective tool to assess AT development projects.

Watanuki (2010) points out two advantages in the development of product prototypes within immersive VR environment: (1) the representation of three-dimensional image helps the user create a mental three-dimensional image; (2) the image size produced in the virtual environment facilitates understanding the volume of the object, regardless the cognitive ability of the user.

Under this perspective, during the process of product development, VR can be used as a tool to assess an AT device in different situations: (a) the product aesthetics; (b) the learning process for using the device; (c) use feasibility in situations of imminent risk

The assessment the aesthetic aspect of the product is crucial for future acceptance. In this context, VR presents advantages for assessing the product in relation to other methodologies due to its good ecological validity (Standen and

Brown, 2006). In a VR environment it is possible to create objects with different levels of complexity, (Weiss et al., 2003; Jannink et al.,, 2008; Laver et al. 2011); in real size with stereoscopic image (Watanuki, 2010), that can be manipulated by users so that they can give their opinions in real-time and immersive environment (Laver et al., 2011).

Another advantage of assessing an AT prototype in a VR environment is that the team developing the project is able to modify the characteristics of the resource and test other projects in the VR environment considering the feedback provided by the users in a shorter period of time and with affordable cost compared to traditional methodologies. Performing this procedure during the conception stage of a product can contribute to providing the market with products at higher rates of acceptance among users and thus increase the usage time.

Evaluating the learning process of an AT resource during the conception stage is crucial once factors such as inefficient device performance; complicated use devices, excessive time spent by caregivers to make adjustments and inadequate to user's needs in different contexts are mentioned as factors contributing to the abandonment of these technologies.(Phillips and Zhao, 1993).

Assessments of the learning process in a virtual environment offers advantages over those made in the laboratory or field, because it provides objective information, repeatable and quantitative results, in addition to standardized instructions and tasks performed in a controlled environment (Sveistrup, 2004) and with good ecological validity, which increases possibility of transferring or generalizing the skill learned in the virtual environment to the real world (Standen and Brown, 2006).

The virtual environment prevents the person with learning disabilities who is making mistakes from suffering real consequences of their mistakes; virtual worlds can be manipulated while the physical world cannot. The abstract concepts can be taught without the use of language or other symbols, ie, the qualities of objects can be discovered by direct interaction with them and not through verbal instructions (Standen and Brown, 2006).

Due to ethical issues, the AT devices currently available on the market are not usually tested in situations of danger to the user. Studies have shown that the virtual environment is a powerful tool for assessing or safely teaching tasks that are potentially dangerous for people with different disabilities (Craig and Sherman, 2009; Coles et al., 2007; Padgett et al., 2006). The simulation of hazardous contexts through VR provides greater ecological validity in comparison with most other methods except, of course, if executed in real world (Duarte et al., 2010).

In this context, Virtual Reality constitutes a powerful method to investigate the feasibility of the product in an emergency or high stress level situations with advantages in relation to assessment in the real world, since it allows the user to engage in potential risky situations without being exposed to real risk; the environment can be programmed to a predictable and systematic distribution of stimulus that is customized to the skill level and learning strategy of the person

In these situations, it is possible to analyze the variable error committed during the interaction with the AT device through the collection of measures on behavioral compliance, mistakes made by the user, attention span and level of success.

The children with disabilities have difficulties in behavioral consonance and failure to follow instructions may result from deficits in some skill or motivational factors, or a combination of both, so it is important to assess behavioral consonance in relation to warnings or safety instructions on AT equipments (Smith and Lerman, 1999). The warning available on devices or in an environment aims to convince the user to adopt security measures to prevent damage or actions hazardous to their physical integrity. Wogalter, Conzola and Jackson-Smith propose some guidelines for developing effective warnings: (1) they need to be salient, ie, must be noticed and attended, (2) they must clearly and concisely write four messages: a signal word to attract attention, hazard identification, explanation of the consequences if exposed to danger, and guidance on how to avoid the danger, (3) layout and location of the warning, (4) inclusion of pictorial symbols, (5) use of auditory warnings, (6) they must consider individual factors such as age, gender, demographic and cultural differences, familiarity with the product and task, training (Wogalter et al., 2002) .

In the virtual environment it is possible to develop 3D virtual models that allow the manipulation of types and forms of security warnings, visual, audible, kinesthetic, available in AT equipment to select the most suitable model according to user's ability and the most motivating one to ensure effectiveness. It is also possible to perform behavior assessment through behavioral and cognitive task analysis since every interaction is recorded and fragmented analysis of the information collected can be done.

The behavioral task analysis involves breaking the task into task component, activities and sequences. The cognitive task analysis involves breaking mental processes into components, such as decision points and information that must be remembered.(Wogalter et al., 2002).

In virtual environment, it is also possible to monitor and assess the maintenance or change of attention focus attention in risky situations. The head or eye tracking device can collect measures on visual fields movement, attention span and feedback time to the generated stimulus (Schwebel and McClure, 2010; Duarte et al., 2010), ou or still through reports for subjective data collection regarding the attention (Duarte et al., 2010). The subjective data collection can be performed using a semi-structured interview with questions about what and where they were searching, why and the sequence of actions performed, among others.

VR may still, in situations of emergency and stress, assess mistakes made during the interaction with AT products with advantages over other methods, for the continuous record allows the researcher to identify factors that might influence the user to make mistakes and fail the actions taken.

5 CONCLUSIONS

This study aimed to reflect on the methodologies that have been used to evaluate AT devices especially the possibilities of using other approaches that may be more appropriate to AT developing process focusing on potential users. A solution may involve the use of VR, which when integrated into appropriate methodologies, can constitute a reliable and feasible way to evaluate AT devices during the process of product development. The active participation of future users and people who live with them may contribute to the production of more efficient equipment, and meet the wishes and needs of people with disabilities and their families, both in the technological aspect of the product and in relation to social acceptance. The use of virtual prototypes integrated into the VR, framed in appropriate methodologies, may also overcome the difficulties of using conventional methodologies and tools which require a physical prototype to be applied. In this context, VR can solve some of the problems presented while bringing some advantages such as the possibility of simulating a wide variety of contexts, having a good ecological validity, the possibility of easily replicating the experimental conditions, having a good internal validity, the accurate control of variables and the possibility of use in dangerous situations. It should also be noted that navigation in a virtual environment may be impaired or impossible depending on the degree of motor or sensory difficulties presented by the user with disabilities.

ACKNOWLEDGMENTS

The authors would like to acknowledge UNESP – Univ Estadual Paulista and (PTDC/PSI-PCO/100148/2008) Portuguese Science Foundation (FCT).

REFERENCES

Adamovich, S.V. et al., 2009. Sensorimotor training in virtual reality: a review. *NeuroRehabilitation*, 25(1), pp.29–44.

Adelola, I.A., Cox, S.L. & Rahman, A., 2009. Virtual environments for powered wheelchair learner drivers: Case studies. *Technology and Disability*, 21(3), pp.97–106.

Brummel-Smith, K. & Dangiolo, M., 2009. Assistive technologies in the home. *Clinics in geriatric medicine*, 25(1), pp.61-77, vi.

Coles, C D et al., 2007. Games that "work": using computer games to teach alcohol-affected children about fire and street safety. *Research in Developmental Disabilities*, 28, pp.518-530.

Cook, A.M. & Hussey, S.M. 1995. Assistive Technologies: Principles and Practices.St. Louis, Missouri. *Mosby - Year Book Inc.*

Craig, A. & Sherman, W., 2009. Developing virtual reality applications: Foundations of effective design. , p.382.

Donegan, M. et al., 2009. Understanding users and their needs. *Universal Access in the Information Society*, 8(4), pp.259-275.

Duarte, E., Rebelo, F. & Wogalter, M.S., 2010. Virtual Reality and its potential for evaluating warning compliance. *Human Factors and Ergonomics in Manufacturing & Service Industries*, 20(6), pp.526–537.

Goodman, G., Tiene, D. & Luft, P., 2002. Adoption of assistive technology for computer access among college students with disabilities. *Disability and rehabilitation*, 24(1-3), pp.80-92.

Gourlay, D., Lun, K. & Lee, Y., 2000. Virtual reality for relearning daily living skills. *International Journal of Medical*, 60, pp.255-261.

Holden, M.K., 2005. Virtual environments for motor rehabilitation: review. *Cyberpsychology & behavior*, 8(3), pp.187–211.

Huang, I.-C., Sugden, D. & Beveridge, S., 2009. Assistive devices and cerebral palsy: factors influencing the use of assistive devices at home by children with cerebral palsy. *Child: care, health and development*, 35(1), pp.130-9.

Jannink, M.J. a et al., 2008. A low-cost video game applied for training of upper extremity function in children with cerebral palsy: a pilot study. *Cyberpsychology & behavior*, 11(1), pp.27-32.

Keshner, E.A., 2004. Virtual reality and physical rehabilitation: a new toy or a new research and rehabilitation tool ? *Journal of NeuroEngineering and Rehabilitation*, 2, pp.1-2.

Kisony R, Katzand, N. & Tamar, P.L., 2003. Adapting an immersive virtual reality system for rehabilitation. *The Journal of Visualization and Computer Animation*, 268, pp.261-268.

Kizony, R, Katz, N. & Weiss, P.L., 2004. Virtual reality based intervention in rehabilitation: relationship between motor and cognitive abilities and performance within virtual environments for patients with stroke. *Proc 5° Intl conf. Disability, virtual reality, & assoc tech.*

Laver, K. et al., 2011. Virtual reality stroke rehabilitation–hype or hope? *Australian Occupational Therapy Journal*, 58, pp.215-219.

Louise-Bender, P.T., Kim, J. & Weiner, B., 2002. The shaping of individual meanings assigned to assistive technology: a review of personal factors. *Disability and rehabilitation*, 24(1-3), pp.5-20.

Lupton, D. & Seymour, W., 2000. Technology, selfhood and physical disability. *Social Science & Medicine*, 50(12), pp.1851–1862.

Martin, J.K. et al., 2011. The impact of consumer involvement on satisfaction with and use of assistive technology. *Disability and rehabilitation Assistive technology*, 6(3), pp.225-242.

Murchland, S. & Parkyn, H., 2010. Using assistive technology for schoolwork: the experience of children with physical disabilities. *Disability and rehabilitation Assistive technology*, 5(6), pp.438-447.

Padgett, L.S., Strickland, D. & Coles, C.D., 2006. Case study: Using a virtual reality computer game to teach fire safety skills to children diagnosed with fetal alcohol syndrome. *Journal of pediatric psychology*, 31(1), p.65.

Phillips, B. & Zhao, H., 1993. Predictors of assistive technology abandonment. *Assistive Technology*, 5(1), pp.36–45.

Phillips, M., Radford, K. & Wills, A., 2011. Ankle foot orthoses for people with Charcot Marie Tooth disease--views of users and orthotists on important aspects of use. *Disability and rehabilitation Assistive technology*, 6(6), pp.491-499.

Rizzo, A. a et al., 2006. A virtual reality scenario for all seasons: the virtual classroom. *CNS spectrums*, 11(1), pp.35-44..

Ryan, S.E. et al., 2009. The impact of adaptive seating devices on the lives of young children with cerebral palsy and their families. *Archives of physical medicine and rehabilitation*, 90(1), pp.27-33.

Scherer, M.J., 2002. The change in emphasis from people to person: introduction to the special issue on assistive technology. *Disability and rehabilitation*, 24(1-3), pp.1-4.

Schwebel, D.C. & McClure, L.A., 2010. Using virtual reality to train children in safe street-crossing skills. *Injury Prevention*, 16(1), pp.1-5.

Smith, M.R. & Lerman, D.C., 1999. A preliminary comparison of guided compliance and high-probability instructional sequences as treatment for noncompliance in children with developmental disabilities. *Research in developmental disabilities*, 20(3), pp.183–195. .

Standard, T. et al., 1994. ISO 13407 : Human Centred Design Process for Interactive Systems The Standard describes : *Human Factors*.

Standen, P. & Brown, D., 2006. Virtual reality and its role in removing the barriers that turn cognitive impairments into intellectual disability. *Virtual Reality*, 10(3), pp.241–252.

Sveistrup, H., 2004. Motor rehabilitation using virtual reality. *Journal of neuroengineering and rehabilitation*, 1(1), p.10.

Troy, K.L., 2011. Biomechanical validation of upper extremity exercise in wheelchair users: design considerations and improvements in a prototype device. *Disability and rehabilitation Assistive technology*, 6(1), pp.22-28.

Verza, R. et al., 2006. An interdisciplinary approach to evaluating the need for assistive technology reduces equipment abandonment. *Multiple sclerosis (Houndmills, Basingstoke, England)*, 12(1), pp.88-93.

Viau, A. et al., 2004. Reaching in reality and virtual reality: a comparison of movement kinematics in healthy subjects and in adults with hemiparesis. *Journal of neuroengineering and rehabilitation*, 1(1), p.11.

Watanuki, K., 2010. Development of virtual reality-based universal design review system. *Journal of Mechanical Science and Technology*, 24(1), pp.257-262.

Weiss, P.L.T., Bialik, P. & Kizony, Rachel, 2003. Virtual reality provides leisure time opportunities for young adults with physical and intellectual disabilities. *Cyberpsychology & behavior : the impact of the Internet, multimedia and virtual reality on behavior and society*, 6(3), pp.335-42.

Wogalter, M.S., Conzola, V.C. & Smith-Jackson, T.L., 2002. Research-based guidelines for warning design and evaluation. *Applied Ergonomics*, 33(3), pp.219–230.

CHAPTER 4

A Proposal of Usability Task Analysis for User Requirements and Evaluation

Toshiki Yamaoka

Wakayama University
Wakayama, Japan
Tyamaoka6@gmail.com

ABSTRACT

Usability task analysis is used for collecting user requirements and usability problems of products, GUI and so on. The process of usability task analysis is as follows. (1) The test participants check and evaluate every task of products or GUI. Every task of products or GUI is evaluated from view point of the good and bad points of the task. Test participants are asked to fill out appropriate evaluation words in the blanks in Sentence Completion Test (SCT) after operating the products. (2) The test participants check all tasks of products as the synthetic evaluation and grade using the Likert scaling. (3) The collected user requirements and problems are structured using card sorting. (4) The collected user requirements are analyzed using DEMATEL method and FCA . For an example, four staplers are examined using usability task analysis for the validation. As results, it is concluded that usability task analysis is useful method for collceting user requirements and evaluation.

Keywords: usability task analysis, DEMATEL, Formal Concept Analysis(FCA), Sentence Completion Test(SCT)

1   BACKGROUND AND PURPOSE

Usability task analysis is used for collecting user requirements and usability problems of products, GUI and so on. It is constructed based on task analysis which is a famous method in Ergonomics. The results created by the usability task analysis can be analysed using quantitative methods such as DEMATEL method, FCA (Formal Concept Analysis), the quantification 1(Multiple regression analysis) and the Quine-McCluskey method.

Usually engineers or designers must master methods respectively for collecting user requirements and usability evaluation. Interview, observation and task analysis are useful for collecting user requirements, and protocol analysis as user test and heuristic method as inspection method are used frequently for usability evaluation. If engineers and designers can master only the usability task analysis, they can collect a lot of user requirements and evaluate the usability of product, GUI and so on. This paper describes the method and effectiveness of the usability task analysis.

2   THE PROPOSED METHOD

The usability task analysis is developed based on a kind of the task analysis.

The procedure of collecting user requirements is as follows.

(1) The test participants check and evaluate every task of products or GUI.

Every task of products or GUI is evaluated from view point of the good and bad points of the task. Test participants are asked to fill out appropriate evaluation words in the blanks in Sentence Completion Test (SCT) after operating the products. SCT is aimed at collecting the causal relationship of usability of the task.

SCT

As the stapler is [(a)   ], I feel that [(b)   ].

So, the stapler has [good or poor] usability.

The every task of products is graded according to the Likert scaling: strongly agree(5), agree(4), neutral(3), disagree(2), strongly disagree(1). The good points become user requirements while the bad points are changed into good user requirements. For example, bad point "heavy" is changed into user requirement "light" which is an antonym of "heavy". The common user requirements of products for every task are very important. The user requirements are structured. A design concept is constructed based on the user requirements structured.

Table 1 The bad point(c) is changed into user requirement (c')

	Product-n	
	Point [2]	
	good point	user requirements
Task-n	[(a)]→[(b)]	(a),(c') (c is changed into c')
	bad point	evaluation
	[(c)]→[(d)]	(c) +(d)

(2) The test participants check all tasks of products as the synthetic evaluation and grade using the Likert scaling.

Table 2 Usability task analysis

	Product-1	Product-n	
	Point [ ]	Point [ ]	
	good point	good point	user requirements
task(1)	[(a)]→[(b)]	[(a)]→[(b)]	
	bad point	bad point	evaluation
	[(c)]→[(d)]	[(c)]→[(d)]	
	Point [ ]	Point [ ]	
	good point	good point	user requirements
task(n)	[(a)]→[(b)]	[(a)]→[(b)]	
	bad point	bad point	evaluation
	[(c)]→[(d)]	[(c)]→[(d)]	
	Point [ ]	Point [ ]	
Synthetic evaluation	good point	good point	user requirements
	[(a)]→[(b)]	[(a)]→[(b)]	
	bad point	bad point	evaluation
	[(c)]→[(d)]	[(c)]→[(d)]	

(3) The collected user requirements are structured using card sorting.

Especially, when items of bad point are changed into user requirements which are the same meaning of items of good point, these items changed are important .

(4) The collected user requirements are analyzed using DEMATEL method and FCA , especially,the quantification 1(Multiple regression analysis) and the Quine-McCluskey method.

The procedure of evaluations as follows.

(1) Bad items of each task are collected.

(2) The collected items are structured using card sorting.

The tasks of low points (1 or 2) should be analyzed in order to examine the cause.

(3)If possible, the quantification 1(Multiple regression analysis) is done.

The multiple regression analysis is done using the points of each task as independent variables and the point of synthetic evaluation as dependent variable .The analysis identified some tasks which influence the point of synthetic

3   APPLIED THE METHOD TO STAPLERS

The usability task analysis was applied to staplers.

3.1   Method

(1) Participants: 16 university students (Male & Female,20—22 old years)
(2) Object: Four staplers

Product-1

Product-3

Product-2

Product-4

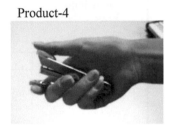

Figure 1 The four staplers

(3) The test participants check staplers and evaluate the tasks. The participants were asked to fill out appropriate evaluation words in the blanks in SCT in good point and bad point after operating them.

3.2   Results and consideration

(1) The results of usability task analysis
 Four tasks, "easy to lift", "easy to hold", "easy to push", "easy to insert lead", were analyzed. The table3 shows items of good point and bad point.
(2) The data of usability task analysis (table3) were structured by card sorting and analyzed based on DEMATEL method.
   The following items were extracted for DEMATEL method.
   1.easy to lift, 2.easy to hold, 3.easy to push, 4.easy to insert lead, 5.large form,
   6.light weight, 7.form show the direction to use, 8.easy to understand,
   9.round form at the rear, 10.fit hand, 11.heavy, 12.sense of stability,
   13.wide width,14.support by the palm of hand, 15.push concave button by finger,
   16.fit finger,17.form fit palm, 18.don't touch sides of body, 19.push tightly,
   20.smooth operation, 21.push by palm, 22.light feeling to push,

23.push concave button by finger, 24.soft pushing, 25.good feeling of pushing , 26.don't touch the sides of body,27.light cover, 28.understanding operation, 29.open cover widely, 30.insert lead tightly, 31.close cover easily
(3) The relationship between four tasks and four products were analyzed based on FCA(fig3).

Table 3 The results of usability task analysis of four staplers

task	Product-1	Product-2	Product-3	Product-4
easy to lift	Point [2]	Point [1]	Point [4]	Point [4]
	good point	good point	good point	good point
	large form →easy to hold		light →easy to hold	light →easy to hold, form shows direction to use→easy to understand
	bad point	bad point	bad point	bad point
	heavy→bad to lift, no direction to use→hard to understand	no direction to use→hard to understand		
easy to hold	Point [4]	Point [3]	Point [4]	Point [2]
	good point	good point	good point	good point
	round form of the rear→fit hand, heavy →sense of stability	wide width→ support by the palm of hand, and easy to hold	push concave button by finger →fit finger, form fit palm→ easy to hold	push concave button by finger →fit finger
	bad point	bad point	bad point	bad point
	heavy→bad to hold			touch side of body→painful
easy to push	Point [2]	Point [4]	Point [2]	Point [2]
	good point	good point	good point	good point
	push tightly →easy to push	push by palm →easy to push	push concave button by finger →easy to push	light weight →easy to push, good feeling of pushing→easy to push
	bad point	bad point	bad point	bad point
	not smooth operation→bad to push	bad feeling to push→bad to push	hard pushing →bad to push	touch side of body→painful

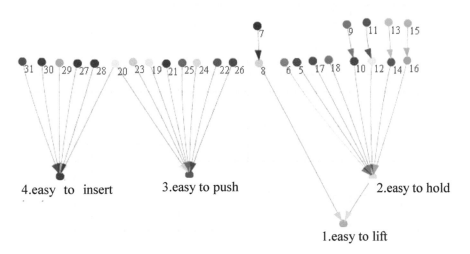

Figure 2 The results of DEMATEL

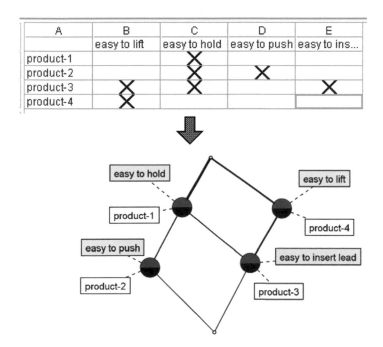

Figure 3 The results of FCA

(4) The figure2 shows the relationship and structure between user requirements.

As " 20.smooth operation" connects "3.easy to push" and "4.easy to insert lead", it is important item. The figure2 also shows "2.easy to hold" connects "1.easy to lift", and if users can hold staplers easily, they can also lift it easily.

(5) According to the figure3, "1.easy to lift" and "2.easy to hold" are common items for four products. This means staplers which are lifted and held easily have developed well, but staplers which are pushed and inserted lead easily need to be improved.

(6) Any problems of usability were collected from view point of the bad points.

The collected problems were structured and the structured information shows structural problems of usability for each product.

4   CONCLUSION

The usability task analysis is useful method to extract a lot of user requirements and structural problems of usability for each product.

The characteristics of the usability task analysis are as follows.

(1) The good point and bad point of the usability task analysis are efficient for collecting requirements and problems. Bad items can be changed into good items which mean user requirements. Sometimes the requirements become good design items for developed products. Because they are not usually noticed. The collected user requirements and problems are structured using card sorting.

(2) SCT also is useful for applying the data to DEMATEL method.

(3)The relationship between user requirements can be shown by DEMATEL method and FCA. DEMATEL method also can make requirements structured.

(4) SCT is very convenient method to collect participant's idea. The data answered by SCT can be analyzed using DEMATEL method and FCA.

REFERENCES

Yamaoka,T.,2001, Human Design Technology as a new product design method, First
    International Conference on Planning and Design, Taipei, CD JP003-F 01-10
Ganter,B. and Wille,R.,1999. Formal Concept Analysis, Springer
Yamaoka, T., Muramatsu, A.,Hamatani,Y.,2005, A propasal on two new usability
    evaluation methods using the Boolean algebra and so on, International Design
    Congress - IASDR 2005, 10 pages-CDROM,

CHAPTER 5

An Application of the Objects' Layout Optimization Models in the Ergonomic Analysis — A Case Study

Joanna Koszela, Jerzy Grobelny, Rafał Michalski

Institute of Organization and Management (I23),
Faculty of Computer Science and Management (W8),
Wrocław University of Technology,
27 Wybrzeże Wyspiańskiego,
50-370 Wrocław, Poland

ABSTRACT

The problem of objects' arrangement appears often during the design process of the workplace environment. The quality of the solution may significantly influence the effectiveness and efficiency of manufacturing. Therefore, this issue gains much attention of building planners (architects, industrial engineers) as well as operations research scientists and practitioners. The facility layout problem is generally well-researched one. Since the 1960s, there have been also many attempts to create computer systems that were supposed to support the application of the developed algorithms and heuristics in practice. In this paper, the computerized Decision Support System for solving facility layout problem is described and applied to a practical problem. The presented solution is a menu-driven program with a graphical user interface that utilizes various optimization heuristics including Virtual force, DISCON, Simulated Annealing, and CRAFT. The case study deals with the ergonomic problem of designing the workplace in the laboratory conducting tests for the road building purposes.

Keywords: Facility Layout Problem, Decision Support Systems, heuristic algorithms

1   INTRODUCTION

Facility Layout Problems (FLP) are family of design problems concerned with creating suitable physical environment to support different activities such as manufacturing, administration, health care, education and, many others. This issue has a significant impact on many things including operating costs, the flexibility of the system, task completion times, equipment utilization, general productivity, business effectiveness etc. It has also considerable influence on quality of work and health of employees. Generally the aim of FLP is to assign objects to proper locations on planar site, minimize or maximize objective function of the problem so that to find an appropriate arrangement.

The FLP is one of the traditional quadratic assignment problem (QAP) originally presented by Koopmans and Beckmann (1957). QAP can be written as follows:

$$\min_{p \in \Pi N} \Sigma \Sigma f_{ij} d_{p(i)p(j)}$$ where f_{ij} is the flow between the facility i and facility j, and d_{ij} denotes the distance between locations of facility i and j,

From the mathematical point of view, the FLP belongs to at least NP-complete problems which was proved by Sahni and Gonzales in 1976. It means that there is not any method that allows for finding the optimal solution within a reasonable time for a large size problem (number of objects over a dozen). Hence, for many years researchers were interested in developing heuristic algorithms that would find reasonable suboptimal solutions. Nowadays, there are many such approaches available in the prestigious literature. Extensive surveys about classifications of algorithms and/or optimization systems are presented in e.g. Hergau and Kusiak (1987), Meller and Gau (1996), Singh and Sharma (2006), Drira et al. (2007), and Schouman et al. (2001).

Because of the complicated nature of producing effective facilities layouts, scientists are forced to create and apply Decision Support Systems (DSS). Various ideas have been proposed since 1960s in the field of FLP. The list of 35 programs utilizing different heuristic approaches was demonstrated by Singh and Sharma (2006). DSS includes various methods among which one can distinguish two main domains: classic algorithms and solutions based on artificial intelligence. The first group consists of optimal and suboptimal algorithms commonly known as heuristics. Optimal algorithms always provide the best possible solution nevertheless they have the generic defect resulting from the nature of NP hard problems, therefore heuristics have greater importance in practice. This group is usually composed of construction, improvement (neighborhood search) and metaheuristics approaches. The characteristic feature of construction algorithms is to create solutions *ab initio*, i.e. step by step, each element is being assigned individually to a system. Multiple examples of the construction approaches can be identified, among them the best known are: ALDEP (Seehof and Evans, 1967), CORELAP (Lee and Moore, 1967), HC66 (Hillier and Connors, 1966). Another group of heuristics includes improvement algorithms. In contrast to the previously mentioned, they produce an initial solution randomly or by construction methods and then make improvements by applying the strategy of neighborhoods searching. The most widely used improvement algorithms are CRAFT (Armour and Buffa, 1963), FRAT (Khalil, 1973), COFAD (Tompkins and Reed, 1976) and many others.

Unlike the improvement and construction approaches, metaheuristics seek for a solution by supervising other heuristic procedures to explore the solution space beyond the local optimum. There are plenty of effective metaheuristics e.g. the taboo search, simulated annealing, ant and genetic algorithms, beam search and many others (Silver, 2004).

Artificial intelligence (AI) appeared in the area of facility layout design in early 1980s. It usually contains four different methods: expert systems (ES), artificial neural networks (ANN), genetic algorithms (GA), and the fuzzy logic. Each of them imitates some unique processes occurring in nature. Expert systems are based on knowledge and experience of the human experts and thus can solve complex, real-world problems in different scientific fields. ANN, in turn, are inspired by the study of biological neural processing. It can be considered as a parallel-distributed processing structure which mathematical model consists of two components: neurons and weighted synaptic links (Cox et al., 1995). Genetic algorithms are included in the class of AI methodologies because they emulate the evolutionary processes especially natural selection and genetics. The solutions for optimization problems are accomplished by means of techniques inspired by mutation, selection, and crossing-over processes. The membership of fuzzy sets in the AI is not as obvious as in the above-mentioned examples. In the center of fuzziness lies the technique of linguistic variables that reflects natural human language. The idea is to employ linguistic expressions which although less precise, are more convenient for people than numbers (Grobelny, 1988).

The next sections of this research are organized as follows: the Decision Support System for solving the facility layout problem (FLP) called Alinks is presented in Section 2 along with its basic description, main characteristics as well as an overview of two implemented algorithms. The case study showing the practical usage of the Alinks carried out on the real-life example is presented in Section 3. The conclusions are discussed in Section 4.

2 ALINKS SOFTWARE CHARACTERISTICS

2.1 General description

The Alinks software was created in order to find solutions for Facility Layout Problems and optimize the arrangement of objects on a plane. It is the menu-driven computer program with a graphical user interface working in Microsoft Windows environment and written in a Delphi programming language. In this information system, the designer operates on models of real objects presented in a form of rectangular outlines. The utilization of various types of analyses helps the decision makers to choose the correct solution for a given layout problem. During the software operation values assessing the quality of solutions are being continuously updated and displayed by the program. Optimizations issues may be applied in different areas, such as the work organization in enterprises, the creation of efficient facilities' layouts, the machines' placement, the ergonomic design of control panels, creation of computer interfaces, and many others.

2.2   Main features

As it was mentioned earlier, the most important and distinctive feature of Alinks is the ability to apply several algorithmic approaches in one working environment. Among them, one can distinguish methods based on a regular grid and a concept of scatter plots. The vast majority of common, classical algorithms are based on a regular gird, which can be considered as a model of available area or space composed of potential locations for facilities. Such an approach facilitates the development of various algorithms and allows for creating their computerized implementation. Methods associated with rearranging objects by changing the location require this methodology to define the obtainable position e.g. CRAFT, COFAD, FRAT, COL, metaheuristics etc. Program supports two algorithms of this type: CRAFT and Simulated Annealing that are well known in the theory of operational research. The second category is represented by the Virtual force algorithm and the Drezner's method developed in 1987. Both of them are based on the unique idea of scatter plots that enables the investigator to resign from the regular gird and search for solutions in a more unconstrained way.

The last two aforementioned algorithms available in Alinks were employed in the case study presented in this research, therefore only they are presented in the next section. The Virtual force and Drezner's algorithms are accessible respectively in *AlgorytmLinks* and *Eigen* tabs of the software. The verification of the Virtual force solution quality is possible due to objective functions f, ff, ID which are displayed in text boxes. The objective function f is defined as the sum of links product (c) between each pair of facilities (i, j) and the distances between each pair of facilities (d): $f = Q1 = \Sigma(c[i, j] \times d[i, j])$. The unit depends on how in a particular case connections between analyzed objects are characterized e.g. if it is the number of transitions between pairs of objects then Q is the whole distance traveled in a given system. The ff is a value independent of project units and described as quotient of the objective function f and the total distance traveled in a given solution: $ff = Q2 = Q1/\Sigma(d[i, j])$.

The mean index of difficulty (ID) is one of the key measures of the solution quality for a facilities' layout at the workplace and is associated with the Fitts' law. Originally, this idea relates the difficulty of visually controlled hand movement with the distance to the target and its width in the following manner:

$$ID = \log_2\left(\frac{2A}{W}\right)$$, where A - the amplitude of a movement
W - the dimension of a target

In the described software, *ID* is computed as an arithmetical average of IDs between all pairs of the examined layout.

In the *Eigen* tab of the system one can obtain solutions computed according to the Drezner's proposals. After entering data, the *Find layout* button should be press in order to get an arrangement in the form of a scatter plot. There is also a possibility to adjust a scale to a screen size by pressing the *Scale layout* button.

2.3   Description of applied algorithms

In this Section detailed description of algorithms employed during the case study is given. Drezner's and Virtual force algorithms are popular and commonly used in the theory of operational research. Both of them are based on a unique idea of scatter plots.

Drezner

A flexible approach to the solution of the facilities layout problem on the plane was presented by Zvi Drezner in 1987. This procedure was based on the DISCON algorithm that was proposed by the same author in 1980. DISCON is an analytical approach based on Lagrangian Differential Gradient method and can be divided into two stages called DISpersion and CONcentracion. Firstly, all facilities are placed at one point and then they *explode* and are being scattered on a plane. The purpose of the Dispersion phase is to find good initial conditions, in the form of the so-called scatter plot, for the Concentration phase. In the Concentration phase, facilities are accumulated in a cluster without overlapping.

In the subsequent algorithm, Drezner changed the method for obtaining scatter plots (the DIS-phase) by modifying the formula of the objective function and thus applying properties of eigenvectors and eigenvalues. Optimal solutions on a plane are obtained by assigning the eigenvector associated with the second smallest eigenvalue of the matrix S (in which $s_{ij} = c_{ij}$ for $i = j$ and $s_{ii} = -\Sigma_j\ c_{ij}$ for all i) to the x-coordinates of the analyzed objects, and treating the the eigenvector associated with the third smallest eigenvalue as y-coordinates.

Virtual force

The two-steps idea, similar to the DISCON, appeared also in the so-called Virtual Forces algorithm (VF) developed by Grobelny (1999). The Drezner's analytical approach was abandoned and replaced by a heuristic procedure of a similar nature. In the Alinks approach, each facility is represented by a material point. In order to scatter facilities on a plane, all of them are initially located at the center of the plane and than are gradually moved in a randomly chosen direction and distance. Such a scattering corresponds to a random layout of facilities with no interconnections. Physically it can be obtained by the pushing out force acting on each object from the center of the analyzed area. The force, called the scattering one, decreases when moving away from the center. In the next stage, after a given period the attractive force, proportional to interconnections between pairs of objects, locates them in a proper arrangement. The interconnections between facilities act inversely proportional to the scattering force - the bigger is the connection between facilities, the stronger they are attracted.

3   CASE STUDY

The case study deals with the ergonomic problem of designing a workplace in a laboratory conducting tests for the road building purposes. The main idea was to optimize arrangement of objects used during laboratory tests to increase efficiency and to reduce the employee's workload. The next subsections provide details about the applied methodology and obtained results.

3.1   Methods

The main objective of this research was to optimize the arrangement of laboratory equipment used in the tests. After the analysis, three, the most frequently carried out, laboratory tests were chosen for further examination. They included: (1) the compressive strength of cement-stabilized soil, (2) the sand equivalent test, (3) the recipe designation of cement-stabilized soil. Additionally, the analysis taking into account objects employed in all 12 available laboratory procedures was conducted. The list of the remaining tests includes the Proctor method, sieving method, determination of frost resistance, determination of frost resistance with soil, natural moisture, determination of water absorption, compressive strength of cement, tensile splitting strength of test specimens, and tests for properties of aggregates. The total number of analyzed objects equaled 39. The location preferences were assigned to eight objects, which resulted in keeping their coordinates unmodified during simulations. Both the laboratory space and the objects' dimensions corresponded to their real values in an appropriate scale. Afterwards, the data about functional connections between objects were entered in the *Sequence* tab of the software. The degree of relationship was specified as the frequency of objects usage registered during three years of the laboratory functioning. The collected data were separately investigated by means of the Virtual force and Drezner's algorithms. Six hundred simulations were performed for the Virtual force approach. Projects with the best objective function were selected and improved in order to achieve solutions acceptable from a practical viewpoint. In this study, irrespectively of Alinks program, the Drezner's heuristic based on eigenvectors and eigenvalues was carried out as well. This procedure was executed only for the option including all 12 laboratory tests. The analysis was conducted in the Matlab mathematical package instead of the Alinks computer system because the lack of links between some of the investigating objects made it impossible to get any meaningful results. The necessary for further computations matrix S was obtained by modifying the array of connections between objects made available by the Alinks program.

3.2   Results and discussion

Methods applied in this study aimed at developing a safe and efficient working space. For optimizing the objects layout in the lab space two different algorithms were utilized. The obtained results were meant to serve as guidance for the further inference about the optimal arrangement of objects that should also take into consideration the practical aspect of the proposed solutions.

Simulation results and analyses for the Drezner's algorithm

The application of the Eigen algorithm based on assumptions of the Drezner method proved to be impossible. Due to the absence of links between space limiters and other objects, many of the eigenvectors and eigenvalues were equaled zeros. Because the algorithm treated the eigenvector associated with the second smallest eigenvalue as the x-coordinates of the solution, and the eigenvector associated with the third smallest eigenvalue as y-coordinates. In such circumstances, the Alinks software placed all objects in one point of the plane. For this reason, it was necessary to conduct the entire procedure required for the Drezner method independently of Alinks program. Calculations were carried out in the Matlab software and they confirmed the results obtained in the Alinks software. The elimination from the S matrix the space limiters that were not connected with other objects, allowed for obtaining results in the form of proper objects' coordinates, which were used for constructing the scatter plots.

Outcomes observed for 33 elements showed a very strong concentration of 31 items with two objects (the styrofoam form and the press attachment) situated quite far away from this group. This was probably the case because of the weak relationship of those objects with other elements. Coordinates obtained for 31 items indicated that the place of their locations should be near the entrance to the laboratory and almost in one place. The application of such proposals in practice was impossible. Execution of the entire procedure excluding the styrofoam form and the press attachment gave another set of coordinates. The illustration of the obtained results for the 31 items is presented in Figure 1. From this figure, one can easily observe a very strong concentration of 21 objects located between the lab table and the case. Four of the remaining items i.e. the freezer, tray, shelf case, and magnifier, had almost identical coordinates and were placed in the original location of the desk. The stopwatch, funnel, buret and prourer were scattered at a location of a laboratory table. The cork and piston had the smallest degree of relationship with other objects and therefore were located on the outskirts of the laboratory. As it was the case during analyzing 33 objects, a relatively high degree of concentration made it impossible to fully apply the solution in practice. The exclusion from the S matrix the styrofoam form and the press attachment resulted in changes in the objects arrangement. The styrofoam form was previously associated with a buret and the value of relationships was three. After the removal of this element, the burette and a few other objects used in the sand equivalent test were moved away from the point of concentration.

44

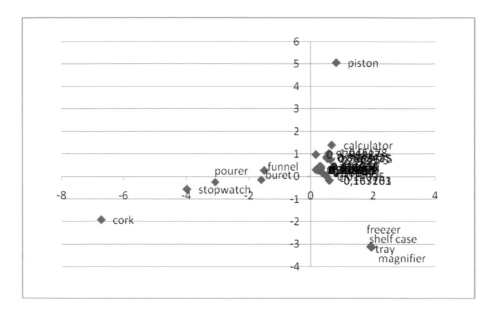

Figure 1 Illustration of the application Drezner's algorithm to the 31 of the examined items.

The press attachment was associated with the large press. After excluding this element from a simulation, there were no visible changes in the location of the large press. This probably resulted from very high relationships with many other items.

Simulation results an analyses for the Virtual force algorithm

A series of simulations conducted by means of the Virtual force algorithm resulted in 40 projects: 30 for the three specific laboratory tests and 10 including all 12 procedures. Each arrangement proposal was manually refined e.g. when a tool or a group of tools were placed at the entrance to the laboratory they were moved away from this location. In this way, projects acceptable from the practical point of view were obtained. The best solutions in terms of the objective functions are presented in Figure 2.

In the compressive strength of cement stabilized soil layout (C), tools tended to form one or two aggregations: near presses and on the left hand side of the laboratory table. Close to presses often appeared the notebook, pen and computer. Other items were located in the second accumulation place. This group of tools either formed a tight cluster near the shelf case or were more scattered to the left of the laboratory table. The best objective function was obtained for the first configuration and any attempts to scatter elements significantly worsened the results. The fact that objects formed a tight group is undoubtedly due to the high degree of linkage between them. Similar conclusions can be drawn about locations of such objects as the notebook, pen or computer, because their relationship to the place selected by the algorithm directly results from their relations with presses.

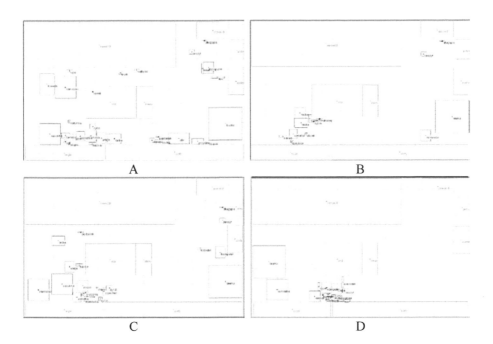

Figure 2 Illustration of the application Alinks's algorithm for all of the procedures (A), sand equivalent test (B), comprehensive strength of cement stabilized soil (C), recipe designation of cement stabilized soil (D).

Solutions found for the sand equivalent test (B) in many aspects were convergent with previous results. Objects always tended to create tight group along the shelf case and in a straight line from the shelf case to the row of cases. All obtained solutions for this test were very similar to each other. Uncorrected versions obtained immediately after running the algorithm were characterized by a large shift of objects with preference more or less in the same places. The excessive dispersal of objects with preferences seemed to have no significant impact on projects quality. Reducing the degree of preference from 1 to 0.99 resulted in solutions where objects without preferences were more scattered and did not have such a homogeneous character. Objective functions obtained during simulations for these parameters were comparable with previous ones (the best outcome for the original simulation: $f = 27512$; the worst $f = 44135$; results obtained with preferences 0.99 $f = 39106$). For other tests, the similar slight change in preferences did not affect the results.

In proposals received for the recipe designation of cement stabilized soil (D) elements such as the notebook and pen in each project constantly appeared in the same area - near presses. The calculator and the computer were placed around the case and presses. Other objects in a quite compact group were placed between laboratory table and shelf case or slightly to the left of the table.

Of course, all of the above-described layouts clearly arise from the links between examined objects. The notebook, pen, calculator and computer were functionally associated with presses, and therefore were put in their neighborhood. Other objects in more or less compact cluster were set near the shelf case due to the presence of functional connections between the shelf case and some of these elements. The most innovative proposal seemed to be moving the freezer and the dryer from their original position near the presses to the vicinity of the shelf case.

Because of large number of objects used in projects combining 12 laboratory procedures (A), it was difficult to drawn reasonable conclusions from the simulations' outcomes. In most cases, compact groups of objects appeared consistently e.g. close to the shelf case or near presses. Some of the laboratory equipment was scattered on the laboratory surface. The obtained locations of the notebook, pen and computer were very similar to the results obtained for individual procedures. In each project, there was the aggregation of elements connected with the shelf case, sometimes forming a tight point or stretched in a straight line. Among items that most frequently were associated with this element, the tray, trowel, spray, spatula, brush, and towel may be mentioned. Objects that in majority of projects were scattered included the buret, calipers, calculator, and funnel.

Generally, it can be concluded that in relation to the solutions proposed by the system for all the procedures it was very difficult to make clear inferences. When carrying out simulations with such a large number of objects in the Alinks program one should take into consideration consequences described above. Much easier interpretable solutions were registered for the smaller number of elements. Those projects were clearer and to less degree unambiguous and it were easier to draw conclusions about the objects' layout in a real lab environment.

4   CONCLUSIONS

This paper presents computerized Decision Support System called Alinks and its application to the practical problem. The proposed solution is based on the menu-driven program with a graphical user interface. It was created in order to find solutions for Facility Layout Problem and optimize the objects arrangement on a plane. The most characteristic and distinguishing feature of Alinks is the use of various heuristic approaches to problem solving available in single software. The system supports algorithms that are widely used in the theory of operational research including CRAFT, Simulated Annealing, the Drezner's heuristic and the Virtual Forces algorithm. These approaches enable the researcher to seek for solutions taking advantage of a regular gird as well as more flexible methods that use the idea of scatter plots.

The described system was later applied to solve the practical problem of designing the workplace in the laboratory conducting tests for the road building purposes. The outcomes of the case study show that the application of some of these methods results in solutions that may serve as guidelines for practitioners in designing a suitable and efficient work environment.

Despite the usefulness of the software in a practical situation, some of the existing problems related with this application were identified. The most important one arises from the relatively big number of items involved in the examination. Probably in future studies researchers or practitioners should consider dividing such a sophisticated problem into smaller subproblems at the cost of decreasing the possibility of finding the globally optimal solution for the analyzed set of objects. In this specific study, probably conducting a layout analysis separately for the lab table would give additional insight into the analyzed problem. The argument for this approach is the fact that many of tools used in analyses are normally stored in this place. Moreover, in subsequent experiments it would be advisable to consider the possibility of dividing objects into two categories: tools and objects. Among the tools there ought to be relatively small items such as a spatula, bowl, balance, buret, etc. Analyses of their deployment should not be made on the plan of the laboratory (or other considered spaces), but on a surface of an object on which they are normally stored in relation to links with other elements. In this way, one can avoid scattering of small objects on a plane that from the practical point of view is difficult to accept. In the objects' category, one can place larger elements such as presses, lab table, shelf cases, dryer, freezer, or desk, which optimization should be performed using only the links between them independently of their links with tools.

Simulations carried out using the Eigen algorithm showed that the utilization of space limiters with zero links prevented the proper functioning of the algorithm. In subsequent studies, projects without such objects should be designed. Another solution may consist in modifying the program so that similar problems would not occur.

REFERENCES

Al-Ghanim A. M., L. D. Cox and D. E. Culler 1995. A neural network-based methodology for machining knowledge acquisition. *Computers and Engineering* 29(1-4): 217-220.

Backmann M. and T. Koopmans 1957. Assignment problems and the location of economic activities, *Econometrica* 25(1): 53-76.

Baum S. and R. Clarson 1979. On solutions that are better than most. *Omega* 7: 249-255.

Connors M. and F. S. Hillier 1966. Quadratic assignment problem algorithms and the location of indivisible facilities. *Management Science* 13: 42-57.

Drezner Z. 1980. DISCON: A new method for layout problem. *Operations Research* 28(6): 1373-1384.

Drezner Z. 1987. A heuristic procedure for the layout of a large number of facilities. *Management Science* 33(7): 907-915.

Drira A., S. Hajri-Gabouj, and H. Pierreval 2007. Facility layout problems: A survey. *Annual Reviews in Control* 31: 255–267.

Evans W. O. and J. M. Seehof 1967. Automated layout design program. *The Journal of Industrial Engineering* 18(2): 690-695.

Eldrandaly K. A., G. M. Nawara, M.A. Shouman, and A. H. Reyad 2001, Facility Layout Problem and Intelligent Techniques: A Survey, *7th International Conference on Production Engineering, Design and Control*, PEDAC, 2001, Alex, Egypt, pp.: 409-422.

Evans W. O. and J. M. Seehof 1967. Automated layout design program. *The Journal of Industrial Engineering* 18(2): 690-695.

Gau K. Y. and R. D. Meller 1996. The facility layout problem: recent and emerging trends and perspectives. *Journal of Manufacturing Systems* 15(5): 351-366.

Gonzales T. and S. Sahni 1976. P-complete approximation problem. *Journal of Associated Computing Machinery* 23(3): 555-565.

Grobelny J. 1988. The 'linguistic pattern' method for a workstation layout analysis. *International Journal of Production Research* 26(11): 1779-1798.

Grobelny J. 1999. Some remarks on scatter plots generation procedures for facility layout. *International Journal of Production Research* 37(5): 1119-1135.

Hergau S. and A. Kusiak 1986. A construction algorithm for facility layout problem. draft 14/86, Department of Mechanical and Industrial Engineering, University of Manitoba, Winnipeg, Manitoba, Canada.

Khalil T. M. 1973. Facilities relative allocation technique (FRAT). *International Journal of Production Research* 11(2): 183-194.

Lee R. and J. M. Moore 1967. CORELAP – computerized relationship layout planning. *The Journal of Industrial Engineering* 18: 195-200.

Silver E. 2004. An overview of heuristic solutions methods. *Journal of the Operation Research Society* 55: 936-956.

Singh S. P. and R. R. K. Sharma 2006. A review of different approaches to the facility layout problems, A solution to the facility layout problem using simulated annealing, *Advanced Manufacturing Technology* 30: 425-433.

Tompkins J. A. and R. Jr. Reed 1976. An applied model for the facilities design problem. *International Journal of Production Research* 14(5): 583-595.

CHAPTER 6

Eye Movement and Environmental Affordances in an Emergency Egress Task: A Pilot Study

Paulo Noriega[1,2], Elisângela Vilar[1,2], Francisco Rebelo[1,2], Leonor M. Pereira[1,2], Inês P. Santos[2]

[1] Ergonomics Laboratory, Faculty of Human Kinetics - Technical University of Lisbon, Estrada da Costa 1499-002, Cruz Quebrada - Dafundo, Portugal
[2] CIPER – Interdisciplinary Center for the Study of Human Performance, Faculty of Human Kinetics - Technical University of Lisbon, Estrada da Costa,1499-002 Cruz Quebrada - Dafundo, Portugal

ABSTRACT

Understanding and predicting people's displacement movement is particularly important to avoid wayfinding problems as well as to improve egress in emergency situation within complex buildings (e.g., hospitals, convention centers, subway stations and university campus). Environmental cues (i.e., affordances) can act as attractiveness factor which can influence some decisions taken by the visitors while choosing what route to follow. The study of eye movement can be an interesting approach to examine the influence of environmental cues about people´s decision-taking. Therefore, this pilot study aims to investigate the use of eye movement analysis to understand the association between decision-taking during an emergency egress and environmental affordances. For this, a non-immersive virtual-reality (VR)-based methodology was adopted, involving the projection of images in a wall-screen. To collect the users' responses, a constant stimulus method combined with a two-forced choices method was used. The results obtained, allow us to conclude that eye movement analysis may be used to investigate the association between decision-taking in an egress task and environmental affordances.

Keywords: Eye movements, decision-taking, affordances, wayfinding, virtual reality

1   INTRODUCTION

When complex buildings (e.g., hospitals, convention centers and train stations) are the focus of intervention by professionals involved in planning these structures (e.g., architects, designers, managers), the ability to predict people's movement is particularly important to avoid wayfinding problems within these built environments. According to Conroy (2001), architects and designers are often unable to determine how people navigate through complex buildings, so they either neglect this information altogether or make judgments based solely on intuition. Additionally, as settings grow larger and more complex, emergency evacuation emerges as a key problem, and wayfinding becomes a matter of life and death (Arthur and Passini 2002).

The importance of signage for wayfinding problem solving is generally recognized among researchers; however, Arthur and Passini (2002) point out that sometimes people are often as lost with the signs present as they are without them. Thus, an important issue to consider when studying human wayfinding is the influence of the external information when it is presented in a lower level of awareness, which sometimes is implicit in the overall configuration of the building. Some studies suggest that in ordinary conditions (e.g., where relatively symmetrical intersections have the same environmental characteristics on both sides) there is a tendency for the public to bear to the right (Robinson 1933; Scharine and McBeath 2002; Vilar et al. 2012). The influence of lighting on people's decision about what path to choose in an everyday situation was also examined (Taylor and Socov 1974; Vilar et al. 2012). They found that when paths had the same intensity of light on both sides, about 60% of the subjects chose the path on the right side. However, when there was a noticeable difference in the light intensity on one of the sides, subjects preferred to follow the brighter path. An approach to study this implicit external information can be based on the affordances that the environment furnishes or affords the observer (Gibson 1986). Some authors state that affordances are environmental properties that have some meaning to guide the observer's behavior (Turvey 1992; Stoffregen 2003). The conscious use of this concept in architecture is based on the definition of a set of variables, which express the different priorities and capabilities of users under various conditions, creating congruence between what people realize they can do and the activities they can really perform.

According to Wiener, Büchner, and Hölscher (2009), eye-tracking studies have primarily investigated the role of gaze for the control of locomotors behavior, considering the context of navigation and wayfinding. They also stated that very few eye-tracking studies investigated the processes related to spatial memory, path planning, and decision making at choice points, and this approach can be very useful for answering questions related to human wayfinding.

In this way, the main objective of this pilot study is to investigate the use of eye movement analysis to understand the association between decision-taking during an emergency egress and environmental affordances. For this, a non-immersive VR-based methodology was adopted, involving the projection of images in a wall-screen. To collect the users' responses, a constant stimulus method combined with a two-forced choices method was used.

2   METHODOLOGY

For this study, a set of static images of virtual indoor hallways was presented to participants using a projector. Eye-movement data was acquired through the use of an eye-tracker system. To collect the participants' responses, the images' sequence was presented using a constant stimulus method combined with a two-forced choice method.

The participant's decision-taking (i.e., path selection) at the corridors' intersection points along simulated corridors (i.e., escape routes) was the study's main focus. As such, these concerns have conditioned the architecture of the experimental VE designed for this study.

For this pilot study, data were collect from a female participant, with 24 years old, who participated as volunteers in this study. She was right-handed and had normal sight or had corrective lenses and no color vision deficiencies. Color blindness was screened by the Ishihara Test (Ishihara 1988). She also reported no physical or mental conditions that would prevent her from participating in a VR simulation.

2.1   Design of the experiment

In order to assess the influence of the independent variables – corridor width and lighting – on the path's selection, by participants in a simulated emergency egress, three situations were considered (i.e., corridors width with same bright, lighting in wider corridors and lighting in narrower corridors). Considering this, nine experimental conditions representing indoor situations formed by two corridors linked by a "T-type" intersection were designed, providing two alternative arms or directional choices, (Figure1). Thus, the participant was assigned to a setting which was composed of a main corridor that ends in a perpendicular corridor with two side corridors (i.e., an alternative hallway to where the user could turn), offering two alternative paths (i.e., to turn left or to turn right).

The width of the side corridor was manipulated to test the influence of the corridor's width on the route decisions. Thus, the main corridor's width was fixed at 2 m wide, but the perpendicular alternative corridors (left and right) had their width varying between 2m and 3.5m wide each, corresponding to the corridors C2 and C3 in Figure 1.

The existence of lighting was also manipulated to test the influence of this variable on participants' decision about the path to follow. Thus, lighting was added in the wider corridors and two stimuli were obtained through this (i.e., corridors C5 and C6 in Figure 1). In order to verify which variable had more influence in participant's decision, two stimuli were created with lighting in narrower corridors,

52

designing a situation that confronts narrower and brighter *vs.* wider and darker corridors. These stimuli can be seen in Figure 1 as corridors C8 and C9. Neutral condition was defined as corridor C1 in which all corridors have the same width (2m) and bright.

Considering light positioning, corridor C4 and C7 were designed with lighting at the corridors of the left and right sides.

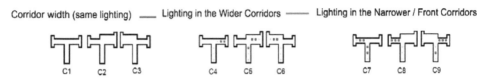

Figure 1 Nine different corridors according to the studies' independent variables.

An example of the virtual environments developed from those conditions can be seen in Figure 2.

The stimuli were organized according to the method of constant stimuli, and a method of forced choice, between two alternatives, was used to collect answers. All the participants were exposed to 2 blocks of 135 trials, in a randomized sequence, in which all of the stimuli were repeated 15 times. The second block had the trials organized in the inversed order used in the first block. To exclude an eventual sequence effect, each half of the sample was assigned, in first place, to one of the blocks.

The inter-stimulus interval varied from 800 to 1000 ms and the stimulus maximum duration was 1400 ms, but it could be less, because in the moment that the participant pushed the button to select an answer (i.e., a direction), the corridor's image disappeared and an inter-stimulus screen was presented. The inter-stimulus screen was a gray screen with the image of a black cube in the center.

Figure 2 Examples of the images of corridors presented to the participants. In the left image there is an example of the narrower corridor with more lighting vs. the wider and darker corridor. The middle image has an example of a situation in which the lighting is in the wider corridor. The right image show the situation where left and right corridors have the same width but the right corridor has more lighting.

The participant was unaware of the real objective of the experiment and was asked to act in a realistic/natural manner in order to evaluate a new system for VR simulation. She was told that she should choose one of the available paths as fast as possible, since she was in an emergency situation.

2.2   Virtual Environment

The VE used in this study was a simplified version of those used for Vilar et al. (2012). Later, a free plugin (OgreMax v1.6.23), exported the environment which were presented by the ErgoVR system (Teixeira et al. 2010), developed by the Ergonomics Laboratory at FMH – Technical University of Lisbon.

2.3   Experimental settings

A Lightspeed DepthQ 3D video projector and a MacNaughton Inc's APG6000 active glasses comprised the VR system used for the experimental tests. A Thrustmaster FireStorm Dual Analogue 3 Gamepad was used as an input device, in order to collect the participants' answers. Participants were asked to press the Gamepad's functional buttons on the right, according to the chosen direction (i.e., left and right).

The projected image size was 1.72 m (horizontal) by 0.95 m (vertical) with an aspect ratio of 16:9. The observation distance (i.e., the distance between the observers' eyes and the screen) was 1.50 m resulting in a 35.2° of vertical field-of-view (FOV) and 59.7° of horizontal FOV. All participants remained standing during the experimental session at the same location (marked on the floor) to ensure the same observation distance.

Data for eye- movement analysis was collected using the eye-tracker system Mobile Eye v. 1.33 from Applied Science Laboratories (ASL).

2.4   Procedure

The participant was asked to sign a form of consent and advised she could stop the experimental session at any time. The duration of the experimental session was 15 minutes.

The experimental session started with a training stage, in which some explanations about the experiment and the equipment involved were given to the participant. She also saw images of the intersection type and received instructions regarding to the task she was requested to fulfill. Participant received task-related instructions such as "You are in an emergency situation, the building is on fire. You have to choose between two alternative paths to find the building egress". The training stage also comprised of two trials using a sequence of images like those used in the experimental test. In the first trial, the participant was asked to point, with her hands, to each alternative corridor that she could see in the image. This procedure intended to ensure that the participant realized the alternative paths that she had in front of her. The second trial intended to make participants familiar with: i) the command buttons in order to choose the direction and, ii) the time available for the answers.

After the training stage, the calibration of the eye-tracker system was done asking the participant to look at five points projected.

The experimental test started after participant had given the required answers, in the time available, and had declared she felt confident and comfortable with the command buttons. For the experimental test, participant was assigned to the first sequence of 135 trials. When the first sequence was fulfilled, and after a five minute break, in the absence of simulator sickness symptoms, participant was assigned to the second sequence of 135 trials. At the end of the experimental test, a demographic questionnaire was applied to collect information such as age, gender, occupation and dominant hand. The participant was also asked to answer questions related to the experimental test.

3   RESULTS AND DISCUSSION

The main objective of this pilot study is to investigate the use of eye movement analysis to understand the association between decision-taking during an emergency egress and environmental affordances.

All data obtained with the eye tracker were analyzed using the ASL Results and the Captiv L-2100.

Table 1 Percentages of participants' directional choices and percentage of first eye-gaze coincident with the choice for the experimental conditions.

Characteristic	Condition	% Choice	% First Eye Gaze
Same Bright Same Width	Neutral	66.7 (%Left)	76.7
Same Bright Different width	Width	98.3 (%Wider)	38.9
Different Bright Same Width	Bright	100 (%Brighter)	98.3
Different Bright Different Width	Bright Vs. Width	95 (%Brighter)	100
Different Bright Different Width	Bright + Widht	100 (%Both)	88.3

The participant's choice in what concerns the direction, for each experimental condition, considering that the subject was escaping from a building in an emergency situation, is presented in Table 1, as well as the first eye-gaze for each directional choice. The results encompass the participant's directional choice, by experimental condition, for all trials, (30 observations for each stimulus). All missing answers were considered invalid.

Regarding to the neutral condition, where both alternative corridors had the same width and bright (C1), 66.7% of choices were favoring the left corridor. When this choice is made, the analysis of the eye-gaze show that the first eye-gaze was coincident with the directional choice in 76,7% of the trials related to this stimulus (30 trials).

The situation where the corridors had the same bright but different widths, 98,3% of choices were favoring the wider path. The first eye-gaze was coincident with the wider corridor only in 38.9% of the cases.

Others conditions had the majority of the directional choices favoring the brighter corridors. In these situations also the first eye-gaze was near 100% favoring chose option.

Nine categories related to the strategies of image exploration were defined considering the three main areas of interest, left, center and right, as shown in Figure 3. Exploration patterns were analyzed only based on the presence of the eye-gaze in each area. Thus, up, center and down eye-gazes into the same area were classified as same (i.e., they were not discriminated).

Figure 3 The three main areas of interest defined for strategies macro-analysis.

In this way, the image exploration categories defined were:
1. Choice area: the participant looked only to the area which is the same of her directional choice;
2. Choice area – Opposite area: the participant looked first to the area coincident with her choice and after she looked to the opposite area of her directional choice;

3. Choice area – Center area- Choice area: the participant looked first to the area coincident with her choice, after she looked to the center area and returned to look at the choice area;
4. Choice area - Opposite area - Choice area: the participant looked first to the area coincident with her choice, after she looked to the opposite area of her directional choice and looked again for the choice area;
5. Opposite area: the participant looked only to the area which is the opposite of her directional choice;
6. Opposite area - Choice area: the participant looked first to the opposite area of her choice and after she looked to the Choice area;
7. Opposite area - Center area - Opposite area: the participant looked first to the opposite area of her choice, after she looked to the center area and looked again for the opposite area;
8. Opposite area - Choice area - Opposite area: the participant looked first to the opposite area of her choice, after she looked to the choice area and looked again for the opposite area;
9. Opposite area - Center area - Choice area: the participant looked first to the opposite area of her choice, after she looked to the center area and looked to the area which is coincident of her choice.

Table 2 Percentages for each eye-movement category according to the participants' directional choices .

Eye-movement categories	C1 %Neutral	C4 + C7 %Light	C2 + C3 %Width	C5 + C6 %Bright + Width	C8 + C9 %Bright Vs. Width
1	51,7	76,3	36,2	75,0	93,0
2	17,2	20,3	1,7	5,0	1,8
3	3,4	1,7	0,0	5,0	3,5
4	6,9	0,0	1,7	3,3	1,8
5	3,4	0,0	13,8	0,0	0,0
6	10,3	1,7	39,7	8,3	0,0
7	3,4	0,0	0,0	0,0	0,0
8	3,4	0,0	0,0	1,7	0,0
9	0,0	0,0	6,9	1,7	0,0

When bright is higher in one of the alternative corridors, the exploration pattern is more concentrated in the areas of choice, with few variations among other areas. Nonetheless the condition bright *vs.* width had opposite cues, the eye-gaze exploration was very focused in the brighter areas.

In Neutral and Width situations, there were more variability in the eye-gaze strategies as can be seen in the Table 2.

4   CONCLUSION

The main objective of this pilot study was to investigate the gaze behavior during spatial decision taking to assess the influence of cues, which may increase the affordance of interior hallways in complex buildings during an emergency situation. In this paper we present an eye-tracking experiment investigating the relation of gaze behavior and spatial decision-taking.

Generalizations about eye movement's strategies in emergency egress task, are not possible since we used only one participant. Nonetheless, results regarding the choice of the path in the intersection in our participant, was similar to previous studies (Vilar et al. 2012).

Regarding eye movement analysis, results are according to previous studies that show that light is a factor of eye-gaze attraction (Taylor and Socov 1974) . In our study, in conditions where one of the intersections was brighter, first gaze was in majority of cases directed to that intersection. In the corridors that light was equal (C4+C7 "neutral" and C2+C3 "width") there was no such clear eye-gaze pattern.

Conjointly the results obtained, allow us to conclude that eye movement analysis may be used to investigate the association between decision-taking in an egress task and environmental affordances.

ACKNOWLEDGMENTS

This research was supported by the Portuguese Science and Technology Foundation (FCT) grants (PTDC/PSI-PCO/100148/2008 and SFRH/BD/38927/2007). It was also sponsored by the Interdisciplinary Centre for the Study of Human Performance (CIPER).

REFERENCES

Arthur, Paul, and Romedi Passini. 2002. *Wayfinding: People, Signs, and Architecture*. McGraw-Hill Companies.

Conroy, Ruth. 2001. Spatial navigation in immersive virtual environments. *Faculty of Built Environment*. London: University of London.

Gibson, James J. 1986. *The Ecological Approach to Visual Perception*. Boston: Lawrence Erlbaum Associates.

Ishihara, S. 1988. *Test for Colour-Blindness*. 38th ed. Tokyo: Kanehara & Co., Ltd.

Robinson, Edward S. 1933. "The Psychology of Public Educations." *American Journal of Public Health* 23 (2): 123-128.

Scharine, Angelique A, and Michael K McBeath. 2002. "Right-Handers and Americans Favor Turning to the Right." *Human Factors: The Journal of the Human Factors and Ergonomics Society* 44 (2): 248-256. doi:10.1518/0018720024497916.

Stoffregen, Thomas A. 2003. "Affordances as Properties of the Animal-Environment System." *Ecological Psychology* 15 (2): 115-134. doi:10.1207/S15326969ECO1502_2.

Taylor, Lyle H, and Eugene W Socov. 1974. "The movement of people toward lights." *Journal of the Illuminating Engineering Society* 3 (3): 237-241.

Teixeira, Luís, Elisângela Vilar, Emília Duarte, and Francisco Rebelo. 2010. ErgoVR – Uma abordagem para recolha automática de dados para estudos de ergonomia no design. In *Proceedings of SHO2010 International Symposium on Occupational Safety and Hygiene 1112 February 2010 Guimarães*, ed. P Arezes, J S Baptista, M P Barroso, P Carneiro, P Cordeiro, N Costa, R Melo, A S Miguel, and G P Perestrelo, 505-509. Sociedade Portuguesa de Segurança e Higiene Ocupacionais - SPOSHO.

Turvey, M T. 1992. "Affordances and Prospective Control: An Outline of the Ontology." *Ecological Psychology* 4 (3): 173-187.

Vilar, Elisângela, Luís Teixeira, Francisco Rebelo, Paulo Noriega, and Júlia Teles. 2012. "Using environmental affordances to direct people natural movement indoors." *Work (Reading, Mass.)* 41 (0): 1149-56.

Wiener, Jan M, Simon J Büchner, and Christoph Hölscher. 2009. "Taxonomy of Human Wayfinding Tasks: A Knowledge-Based Approach." *Spatial Cognition & Computation: An Interdisciplinary Journal* 9 (2): 152-165.

CHAPTER 7

Comparing Likert and Pictorial Scales to Assess User Feedback

Fabio Campos, Rui M. Belfort, Walter Franklin Correia

Universidade Federal de Pernambuco - UFPE
Recife, Brazil
ffcc@ieee.org, ruibelfort@gmail.com, ergonomia@terra.com.br

ABSTRACT

This paper deals with the problem of eliciting user feedback. Information acquired from the feedback of users is crucial in fields like ergonomics, design and statistics. To be able to obtain such feedback ones usually use some kind of questionnaire with questions and a scale where the users answer their opinion about the regarded subject. It is believed that in questions non-directly correlated with numeric quantities a pictorial scale would allow the users to better express their feedback. This work compares the performance of a pictorial scale against a 5-point Likert scale, and shows that a well-designed numeric scale can be as efficient as a pictorial scale when measuring the feedback of users.

Keywords: Likert Scale; Pictorial Scale; User feedback.

1   INTRODUCTION

Eliciting information from users is essential in ergonomics as well in several others fields like design, sociology, statistics and so on. To collect this information is usual the employment of questionnaires with scales in which the users can declare their opinion. The way of asking the questions and the options provided for answering are both important. This work deals with the answering part of the problem of eliciting user feedback, in particular, the kind of scale chosen.

Regarding the scale used in the questionnaires, it is believed (TWYMAN, 1985) that in questions non-directly correlated with numeric quantities a pictorial scale would allows the users to better express their feedback, having an hedonic factor, an intrinsic linguistic universality and focusing on particularities, rather than generalities (allowing to better represent specific kinds of user experience).

The main aim of this work is compare the performance of a pictorial scale against a well-designed numeric scale, in our case study a 5-point Likert scale (KASS et al.,1999), and to show that it can be as efficient as a pictorial scale when measuring the feedback of users.

To accomplish this a case to study based on an online game was executed and the performance of the scales were compared with the feedback of more than a hundred real-users.

2   CASE STUDY

2.1   Introduction to the Case Study

In this case study was chosen an online game as the object upon which the users would give feedback about their experience with the game and its usability.

This online game was chosen because of the feasibility of presenting the different scales randomly to the also random users.

As the object chosen was a game, some precautions were taken to minimize interference to the flow of it.

The "flow state" (CSÍKSZENTMIHÁLYI, 1997) is an experience of intrinsic motivation, where the user is completely immersed in what he or she is doing. In this state, it is important to assess the user experience and feedback in real-time, in a real environment, and directly from them. Thus, the challenge is to access user data causing minimal interference to the flow.

2.2   Process of the Case Study

The digital artifact used in the experiment was a casual game called Racing Against Time, Figure 1, developed by Jynx Playware (PLAYWARE, 2011) for the Contest of Games and Education, Joy Street (STREET, 2011).

In this game the user controls Raíza (character) and must help her arrive at school before the end of the school games. The time taken to reach the end is crucial, as the score is higher with faster times.

When the potential flow is low, that is, the challenges were too simple and few skills required (BELFORT, 2011), the 5-point Likert, Figure 2, or the Pictorial scale, Figure 3, was shown to the user.

Each time the game was used, only one kind of scale would appear to the user; this alternance configures an A/B test (EISENBERG, 2008).

In each game session, the scales were shown to the users 3 times per game, in fixed points of low game flow.

It should be noted that both scales are 5 points scales.

To normalize both scales in the analyses, "points" were assigned to each corresponding segments, that is, to the first segment of both scales ("-2" in the 5-point Likert scale and the first pictogram in the pictorial scale) was assigned 2 points, to the second one ("-1" in the Likert and second pictogram in the pictorial) was assigned 4, and so on till 10 points to the fifth segment ("2" in the Likert and fifth pictogram in the pictorial). The data collected from the all users were segmented by scale kind and averaged.

Also, the time that each user spent, after the question was posed, to answer it was recorded, segmented by scale kind and averaged.

In the start of the experiment, the link of the game was made available through the Twitter of one of the authors, which used to have 200 followers, as if he were doing marketing. The post text was "Racing Against Time: game developed by @jynxplayware, to @oje_news. Check it out! www.jynx.com.br/corridacontrao tempo/". That is, no judgment value was used, in order to not influence opinions and, consequently, the user experience.

The game was available to play for 30 days, while data were being collected remotely, using a custom software combined to Google Analytics to track the number of visitors, average time spent, number of registered participants, the values assigned by the users in each segment evaluated and the time spent to answer. After all, results by each type of scale were compiled and compared.

Corrida Contra o Tempo

Figure 1: Screen of the game Racing Against Time (used with permission of Jynx Playware and Joy Street)

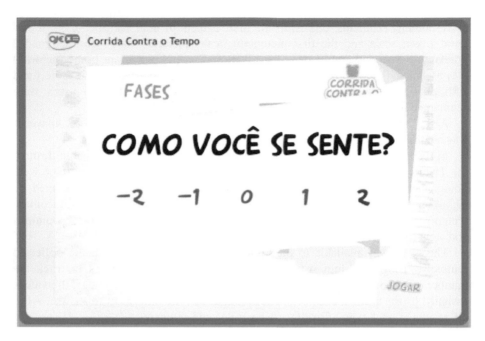

Figure 2: 5-point Likert scale implemented in the game (used with permission of JynxPlayware and Joy

Figure 3: Pictorial scale implemented in the game (used with permission of JynxPlayware and Joy Street)

2.3   Data collected

During the 30 days of availability of the game 163 visitors were registered. These visitors were from Brazil, Canada, Finland, Norway and Switzerland.

The average time spent in each visit was 2 minutes 45 seconds.

These 163 "gamers" accessed the game 396 times (measured by the page hits). From these 396 accesses, 294 (74.24%) were converted into effective participation in the experiment (measured by number of users reaching the final steps of a game play).

The data collected from the answers of all users were segmented by scale kind, normalized and averaged. The plot of this data is shown in Figure 5. In this chart the horizontal axis exhibits the 3 moments along the segments of the game play where the questions and scales were shown to the users, and the vertical axis exhibits average of the normalized score assigned by them using both scales.

From the plot in Figure 5 is clear the convergent behavior of the scales. That is, both scales present similar performance. If one averages the results of the 3 segments where the questions were presented, would obtain 6.87 for the 5-point Likert scale and 6.84 for the Pictorial scale, again evidencing the convergence of the scales.

The data from the time spent to answer the questions is plotted in Figure 6. The vertical axis in this chart is the average of the time in seconds spent to answer the questions, and the horizontal axis the segment of the game where the questions were presented (as in the earlier chart). The comparison of the behavior of the scales also presented a big degree of convergence, but with a better overall performance from users who provided feedback through 5-point Likert scale. The average of the time performance among the segments was 3.15s for the 5-point Likert scale and 3.52s for the Pictorial scale.

3   CONCLUSIONS AND FINAL THOUGHTS

Although there is a long time belief (Twyman, 1985) that pictorial scales can better represent the feedback of users when the questions asked are not directly correlated to numeric quantities, the experiment done in this work showed that a well designed numeric scale, like the 5-point Likert scale, is able to produce similar results. In fact, the cognitive load of a pictorial scale seems to be higher than a numeric one, which can be evidenced by the lower time-to-answer exhibited by it.

Surely other experiments are needed to isolate other variables and confirm these findings in a statistical significant way, but the evidence posed by this case study is striking anyway.

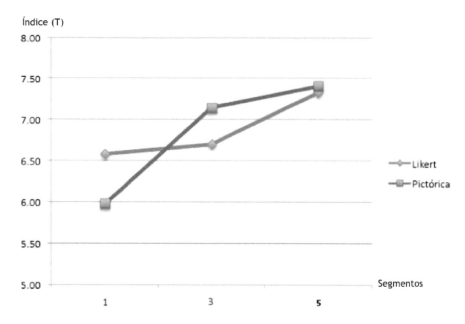

Fig 5: Averaged scores from the answers using the 5-point Likert and the Pictorial scales (authors).

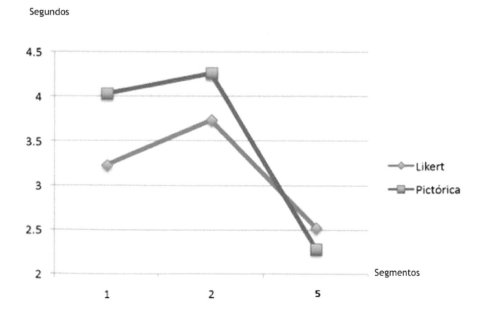

Fig 6: Time spent to answer the questions in each scale (authors).

ACKNOWLEDGMENTS

We thank Jynx Playware and Joy Street, for allowing us to utilize the "Racing Against Time" game as the object of the experiment described here. Without this it would have been impossible to obtain these results. ·

REFERENCES

Belfort, Rui. Investigações Teórico-Metodológicas sobre Experiência de Usuário no Âmbito do Design. Recife: UFPE, 2011;

Csíkszentmihályi, Mihály. Finding Flow. The Psychology of Engagement With Every Day Life. New York: HarperCollins Publishers, 1997;

Eisenberg, Bryan. Always Be Testing: The Complete Guide to Google Website Optimizer. Indianapolis: Wiley Publishing, 2008;

Kasser, Virginia et. al. The Relation of Psychological Needs for Autonomy and Relatedness to Vitality, Well-Being, and Mortality in a Nursing Home. Columbia: Journal of Applied Social Psychology, 1999;

Playware, Jynx. Jynx Playware. Disponível em: www.jynx.com.br/. Acesso em: 01/04/2011;

Street, Joy. Joy Street. Disponível em: www.joystreet.com.br/. Acesso em: 01/04/2011;

Twyman, Michael. Using pictorial language: a discussion of the dimensions. In: DUFTY, M. et. al. Designing usable text. Orlando: Academic Press, 1985.

CHAPTER 8

Evaluation of Safety and Usability in Projectual Methodologies with Focus on Design of Physical Artifacts

Walter Correia, Fábio Campos, Marcelo Soares, Marina Barros

Universidade Federal de Pernambuco, Design Department, CAC, Recife-PE, Brazil

ergonomia@terra.com.br

ABSTRACT

Actual changing in technology and the advances in the science are always surprisingly users in a global and expanded market. What can be seen in many market shelves are some products with a number of information and devices which require a high level of knowledge and discernment by users. Apparently the society is not prepared to these advances. The lack of knowledge, understanding and attention are causes of many accidents with consumer products inside the residential environment. Consumer products are considered in this study as those used inside the domestic environment. A strong input from ergonomics is used to obtain subsidies and support to some case studies and analysis carried out in this dissertation. Some methods to the analysis of accidents are introduced and used in this study, however an emphasis in the analysis of some selected accidents is given. It is introduced a methodology used to collect and apply data to a sample of accidents, and to carry out simulations of accidents.

Keywords: Ergonomics, Consumer Products, Safety, Design

1   INTRODUCTION

According to the National Safety Council of the United States, "*1 in every 13 Americans will suffer a serious accident in the home, which could be avoided if there is prevention.*" This is an alarming fact, if we think that are not included in this statement, the lesions in smaller scale.

The products are developed and designed to meet needs, or at least its primary function should be this. A product makes life easier for its users aiding in actions that could not be performed by them without the aid of a mechanical device or material. In theory, this presents itself as a major reason for consumer products exist. The text above helps to provide an area that has given rise to much discussion around the world: the safety of the product. Consumer products often cause some sort of harm to its users, that is a fact, and more frequently, its usability is weak. The reality is that very little has been done in Brazil to change that. It is a reality extremely difficult to be handled, considering that the legal framework in which it protects the consumer is very idle with respect to these aspects.

Denari (2000) clarifies that any technological advance is to contribute in some way so that there is multiplication of risks arising from defects in industrial products. This opens the way for it to be adopted, with regard to liability for accidents consumption, other types of responsibilities in order to safeguard the rights of injured consumers. The entrepreneurs, designers and planners have a key role with respect to this matter, for they are the main issues involved in legal proceedings that will lead to financial losses and a permanent stain on the image of their products. Although it does not have an apparatus which accounts for the Brazilian population how careless people are with respect to home accidents of nature, and also in relation to household products (consumer products), North American research sources report that only in 1995, approximately 5.5 million people were victims of some type of accident in his own house, and more than 30,500 deaths were recorded, and these numbers are growing according to the same sources. Thus, the worry is that Brazil has to reveal.

Chiozzotti (2002) affirms that case study is a comprehensive characterization to describe a variety of surveys that collect and record data in a particular case or several cases of individuals to organize an orderly and critical report of an experience, or evaluate it analytically, aiming to make decisions about it or propose a transformation. Thus, for the submission of this article, was performed a case study with a sample taken from the two accidents questionnaires submitted by Correia (2002) in a universe of eight accidents.

The main objectives of this case study were: a) detailed analysis of each accident, b) to analyze the activity through a flowchart of task c) perform a careful analysis of the causes of accidents and d) assess the level of usability of products. The products that were part of the study and their accidents were (i) a common drill, leading to a cut and bruise, and (ii) a pressure cooker, causing concussion. The case studies involved the users of accident victims too.

2   CASE STUDY PROCEDURE

With the evolution and, in some way with the "revolution" of the technologies, the way to think and act has intensely changed. It has been created a new culture.

And especially, when analyzing the chronological point of view, with the advent of electrification, the machines have changed so much that even your identity or signage, somehow, been changed. According to Guimarães (2001 b), in this respect, one should mention the example of the use of washing machine, that among the household was one that has undergone great changes.

During the interview stage with each of the users, the victim of some kind of incident, followed the roadmap proposed byWeegles (2001), and thus took into account several aspects of the accident such as how this happened. The accidents simulations were made from the reports sent by users, since it is aimed to reconstruct the situation for future analysis. They were asked to describe the accident, repeatedly, to obtain a recovery as close as possible to the fact. The considerations were based on (i) the "User's voice," which were recorded their descriptions and opinions about the accident, and (ii) the author's own insight to replenish the same. The evaluation of each of the SUS questionnaire was made subjectively, it depends on the User's own opinion, and its quantification and calculations were made based on parameters provided by the SUS.

Almost half of study participants (47%, n = 149) said and affirmed they had suffered some kind of accident with any consumer product. To these respondents were asked to identify the product type and briefly describe the accident. The responses are recorded in Table 1. 39 respondents did not indicate the product that caused the accident and / or described the type of accident suffered.

Table 1 – Some of products that have caused accidents with their frequencies and kind of consequence

Product (frequency)	Type of Accident	Product (frequency)	Type of Accident
Iron (16)	Shock and burn	Pressure cooker (4)	Stain by Explosion
Electric shower (13)	Shock	Plastic Chair (3)	Injury (fall)
Knife (12)	Deep cuts	Computer (3)	Shock
Stove (9)	Deep cut / stain	Refrig. Bottle (3)	Deep cut
Blender (8)	Shock	Monitor (3)	Shock
Refrigerators (5)	Shock	Radio (3)	Shock
Tin (5)	Deep cut	Television (3)	Shock
Fan (5)	Cutting and stain	Vacuum Cleaner (3)	Shock
Can opener (4)	Deep cut	Shavers (2)	Deep cut
Screwdriver (4)	Deep cut	Coffee Machine (2)	Burn
Wash Machine (4)	Injury and cut	Air Debugger (1)	Shock and injury
Microwave (4)	Injury and cut	Electric drill (1)	Deep cut / Stain

Survey participants were practically divided when asked whether he thought the accident was his fault or the product itself: 42% (n = 63) stated that the product was to blame and 41% (n = 61) said that the blame was his own. 17% of respondents (n = 25) said they did not know.

2.1 Accident Analysis

For this simulation, from the eight volunteers the one who had suffered an accident with the pressure cooker was selected. In this situation the researcher requested that the artifact to be used in the simulation be similar to the product that was involved in the accident.

The Figures 1 to 6 show a simulation done for the accident with the pressure cooker. It should be pointed out that the actual product involved in the accident was destroyed and, therefore, another of the same brand and model

Figures 1, 2 and 3 – Images of the pressure cooker closed, opened, and in use by user.

Figures 4, 5 and 6 – Simulation of the accident with the pressure cooker

The user reported that the pressure cooker exploded shortly after being placed on the stove for cooking beans. She said she did not hear the noise of the exhaust valve. There was no physical damage, because the user was in another room. However, there was damage to the stove, ceiling and the pot. It should be mentioned that during the simulation, the user took some time (about 30 seconds) impatiently trying to put the cover of the pan in place. The simulation was held without any food inside the pot.

The sequence of activities, described by the user is those that are usually made for the use of an artifact like that which was presented during the simulation.

It was required that during the face to face interview the user demonstrate the following steps in the use of the pressure cooker:

- Place the beans into the pan;
- Secure the top;

- Check to see if the top is secured correctly,
- Take the pan to the stove,
- Turn on the stove,

The user reported that after a few minutes she heard an explosion and the cover of the pot had hit the ceiling.

The user also noted that she knew of other cases in which pressure cookers had exploded while in use. In this study, four users say they had suffered the same type of accidents with the use of pressure cookers. Two of these reported no physical damage to the user, and the other two did not respond to the question.

It was found that the pressure cooker used in the simulation was difficult to seal completely due to the cover of the pot having a side opening between it and the pot. It was necessary to make several attempts at closing the pot to have a secure seal. The Figures 7 and 8 show this defect in the product design.

Figures 7 and 8 Images of the pressure cooker - details of the gap in the cover

To reach conclusions about the real cause of the accident, there would be a need for a technical report. However, it can be assumed that failures in the sealing system and / or failures in the exhaust valve may have been the cause of the accident.

The fault tree (Figure 9) was developed based on information gathered from the face-to-face interview. These had been developed taking into account the steps that culminated in the accident.

It is perceived that the lack of reading the instruction manual and inadequate closure of the pressure cooker lid are the main factors for the accident (steps 5 and 6). However, this latter fact is due to some inadequate design of this pressure cooker. Step 4 in the tree demonstrates the insecurity and reluctance of the user regarding the use of the pressure cooker.

In this Fault Tree the steps flow from the bottom up. Solid lines represent the direct flow of activity to get to the accident. The dashed lines represent factors involved indirectly with the accident.

However, these faults are to be related since all the accidents could have been avoided if the products had been designed with an emphasis on the caution and safety items, moreover had been paid particular attention to the prevention of misuse of them. It can also be seen that the products analyzed did not show characteristics of a "friendly" product. This fact must consider (i) physical composition (ii) components, (iii) manual of instruction and (iv) warnings.

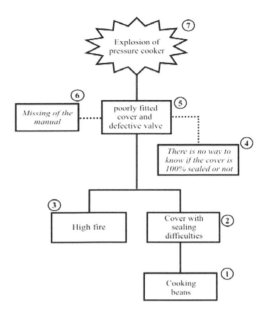

Figure 9 Fault tree for the accident with the pressure cooker

In the case of the drill's accident, this occurred when the user tried to make a hole in the wall for placement of a thing using a automatic drill (Figure 10). The photos, shown in Figures 11 and 12, are simulating the use of the drill at the time of the accident. The drill is the same with which the User had his accident.

Figures 10, 11 and 12 Pictures of the drill and its use for accident's simulation

The related accident happened due to an excessive resistance from the wall, and then the User decided to tilt the drill during operation. At this point, the drill broke and hit the face of the User. There was a bruise and a small cut. Part of the drill remained attached to drill.

The actions that caused the accident were described as follows by user:

- Choice of equipment;
- Search the drill bit ideal for the bush in question;
- Using the reference manual for the drill bit;
- Start the activity using both hands;
- Slightly inclination to try to break the barrier (strongest part - wall);

- The drill latch and part of the drill bit goes against the face of the User;
- Cut and bruised in the face of the user.

The manual recommended that the user should always use the drill in a perpendicular position to the action used, in this case is the wall, which should be 90 ° with respect to it. The angle adopted by the User may have resulted in the breaking of the drill bit. In the Fault Tree on the drill (Figure 13). These considerations can be checked detailed in accordance with the User-run.

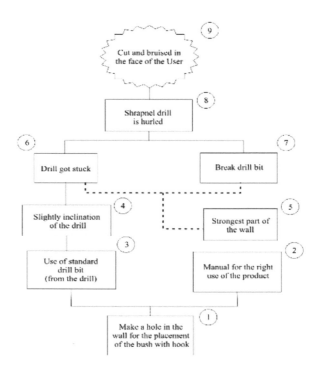

Figure 13 Fault Tree for the accident with the drill

If the User had some more experience with the activity performed (which he affirmed that he didn´t), he would know that a wall requires a stronger type of drill bit that, when it is thicker (using an bush to a thicker wall below) the drill bit should be stronger (other material).

3   QUESTIONNAIRE WITH THE USERS: APPLICATION OF SUS – SYSTEM USABILITY SCALE

SUS shows to the respondents some few questions to be answered by marking on a satisfaction scale according to the level of agreement or disagreement for each question. After answered and encoded the answers are calculated, through a coefficient given by the SUS, the degree of usability of a given product.

3.1 Implementation procedures of SUS

According to the recommendations made by Stanton and Young (2001) were followed all the procedures for the implementation of SUS, ranging from the explanation on that it was the questionnaire to their weighting and analysis. Through this procedure, it was possible to obtain acceptable results for the study with users. The calculations are presented in the tables 1 and 2 were made based on questionnaires answered by the SUS users followed the order presented by the authors. For a proper understanding of users' responses, it is the "0" (zero) as the lowest score for the level of usability and "100" a higher score.

3.2 SUS analysis for pressure cooker

As can be seen in Table 2, the pressure cooker had an index of usability of 40 out of 100. This level of usability means that the product in question leaves much to be desired in how performance and safety. According to the opinion of the user of the product, this product should be taken off the marker.

Table 2 Calculation of the score of usability for the pressure cooker

Score of odd-numbered items = Position in the scale − 1	Score of even-numbered items = 5 − Position in the scale
Item 1 $2 - 1 = 1$ Item 3 $4 - 1 = 3$ Item 5 $1 - 1 = 0$ Item 7 $4 - 1 = 3$ Item 9 $2 - 1 = 1$	Item 2 $5 - 3 = 2$ Item 4 $5 - 2 = 3$ Item 6 $5 - 5 = 0$ Item 8 $5 - 5 = 0$ Item 10 $5 - 2 = 3$
Sum of odd numbers (NI)	= 8
Sum of even numbers (NP)	= 8
Total sum of the items = NI + NP	= 16
Total score of usability for SUS Total of items x 2.5	= 16 x 2,5 = 40

3.3 SUS analysis for drill

For the second case, the drill, the calculations are presented as follows in table 3. In the case of the drill, it became a level of usability in the 50 value. Considering the scale from 0 to 100 showed, note that a level of 50 for usability in consumer products can now be considered below the moderate.

Table 3 Calculation for the level of usability of the drill

Score of odd-numbered items = Position in the scale – 1	Score of even-numbered items = 5 – Position in the scale
Item 1 2 – 1 = 1	Item 2 5 – 2 = 3
Item 3 4 – 1 = 3	Item 4 5 – 3 = 2
Item 5 2 – 1 = 1	Item 6 5 – 2 = 3
Item 7 5 – 1 = 4	Item 8 5 – 5 = 0
Item 9 3 – 1 = 2	Item 10 5 – 4 = 1
Sum of odd numbers (ON)	= 11
Sum of even numbers (EN)	= 9
Total sum of the items = NP + EN	= 20
Total score of usability for SUS Total of items x 2.5	= 20 x 2,5 = 50

It is evident that the dill in this study, presents a level of usability weak. What, similarly to the previous item, should be considered low to moderate. In general can be seen that the products are presented in a not acceptable scale of usability. This scale should be considered as very low, since the product must meet and satisfy their users (that means 100 points). SUS has through the points where users complained more, try to develop solutions that seek to address gaps in certain product area. As an example, if a User strongly disagree that a product or system is easy to learn to use, you should think about how a more friendly interface can be applied to such product or system, as quoted by Stanton and Young (2001) and Stanton and Barber (2002).

4   CONCLUSIONS AND LESSONS LEARNED

"No product can be considered as absolutely safe. To achieve an acceptable level of safety in the product is necessary to define a strategy for design and evaluation of appropriate products. "(SOARES AND BUCICH, 2000).

It may be noted that that despite the Users' reluctance to remake the incident by claiming difficulty of remembering what happened, or saying that they no longer had the products, the results were satisfactory. The simulations took place at the injured Users' home, bringing him closer to the real in relation to what had happened.

This was asked directly to each user that had any kind of accident. The methodology of Weegles (2001) was followed step by step, for the most part, for a correct analysis and evaluation of the study. Through the construction of the Fault Tree some lacks were observed, from both the product and the User. It is also clear that, despite the attempt at reconstruction (simulation) of the accident, all were aware that some detail or details might have been forgotten. SUS used in the interviews was the most important to certify that the products analyzed in this essay have relatively low level of usability, though one can contest the fact that all the opinions given in here came from the use of products that ended up in an accident.

The authors of the SUS does not define the contents as 40, 50 or 60 are low, but show that a scale from 0 to 100, those two would be the extreme negative and positive end respectively, which can ensure that levels below 70 to 80, attest, in general, products with low usability, or they could be better studied in their conceptions. It can be affirmed that the accidents analyzed occurred through the fault of the product, and when the lack of the User, attests to the insufficient attention given to some norms of security and usability.

According to Jordan (2003) and Vendrame (2000), is a mistake of designers and drafters develop a product for themselves thinking that users will act and think like them. Designing for children, the elderly, people with some type of disability or limitation, male or female, tall or short, requires a different degree insight in each case, and this is where is the difference between a good product and a bad product.

Consider the skills, needs and limitations of the user in the design development of consumer products, is not just a matter of survival in the competitive market of today, it is a matter of human respect and social responsibility (Correia, 2002)

References

ABBOTT, H. & TYLER, M. (1997) Safer by Design. A guide to the management and law of designing for product safety. England, Gower.

CHIOZZOTTI, A. (2002) Pesquisa em ciências humanas e sociais. 4 ed. São Paulo, Cortez.

CORREIA, W. F. M. (2002) Segurança do Produto: Uma Investigação na Usabilidade de Produtos de Consumo. Dissertação de Mestrado, PPGEP / UFPE.

DENARI, Z. (2000) Estudo Especial em Responsabilidade Civil por Danos Causados aos Consumidores. Revista de Direito do Consumidor. São Paulo: Revista dos Tribunais.

GUIMARÃES, L. B. M. (ed.) (2001). Ergonomia de Processo. Série Monografia e Ergonomia. Vol. 2. Porto Alegre - RS, PPGEP - UFRGS.

JORDAN, P. W. (2003) An Introduction to Usability. 2 ed. London, Taylor & Francis.

PERSENSKY, J. J.; GAGNON, J. L. (1998) Evaluation of a hand probe for use in a product regulation. In Proceedings of the Symposium, Human Factors and Industrial Design in Consumer Products. Santa Monica, CA, D. J. Human Factors Society

SOARES, M. M. & BUCICH, C.C. (2000). Segurança do produto: reduzindo acidentes através do design. Estudos em Design. v. 8, maio, p. 43-67.

STANTON, N A. & BARBER, C. (2002) Error by design: methods for predicting device usability. Design Studies. Vol. 23, N° 4. London, Elsevier Science Ltd, July. p. 363 - 384.

STANTON, N A. & YOUNG, M. S. (2001) A Guide to Methodology in Ergonomics. Designing for human use. London, Taylor & Francis.

VEIGA, A.R. (1999) Atitudes de consumidores frente a novas tecnologias. Dissertação de Mestrado, Campinas, SP, Brasil, PUC-Campinas.

VENDRAME, A. C. Acidentes Domésticos. São Paulo, LTr Editora. p. 9-13, 2000.

WEEGLES, M. F. (2001) Accidents Involving Consumer Products. Doctorate's Thesys. University of Delft.

CHAPTER 9

Drillis and Contini Revisited

John Brian Peacock[1,2], Manoharan Aravindakshan[3], Tong Xin[1],
Chui Yoon Ping[1], Low Wai Ping[1,], Fabian Ding[2], Tan Kay Chuan[2],
Markus Hartono[4,] Ng Yuwen Stella[2]

[1]SIM University, Singapore
[2]National University of Singapore
[3]Nanyang Technological University, Singapore
[4]University of Surabaya, Indonesia
isejbp@nus.edu.sg

ABSTRACT

Anthropometry has been the bedrock of ergonomics and human factors since the formalization of these fields of study some 70 years ago. There are various measurement conventions and many data bases (ISO, SAE, WHO, CDC). Perhaps the most comprehensive text on the subject is that by Pheasant and Haslegrave (2006). The methods used for workplace design involve convenient, usually boney, landmarks such as the acromion or patella. On occasion, soft tissues are included, especially for widths and girths. Perhaps the most highly developed application is for "occupant packaging" in automobile design (Roe, 1993). Contemporary methods include wand based pointers and whole body scanning. Contini and Drillis (1963) and Pheasant (1982) hypothesized that any of these measures could be predicted with sufficient accuracy from a single measure of stature. Their results have been widely published and applied. The purpose of this paper is to revisit these findings by analyzing the correlations among many measures from data obtained in various recent anthropometric surveys. A more detailed approach used multiple regression and structural equation modeling to improve the reliability of prediction. As expected the dimensions that had the highest associations were limb segment lengths. Width and girth measures were inter-correlated and more associated with weight than stature. Head, foot and hand measures were not highly correlated with other measures. Whereas the Drillis and Contini ratios were found to be sufficient for many practical purposes, the addition of weight as a predictor provided greater accuracy.

Keywords Anthropometry, Drillis and Contini, correlations, segment proportions

1   INTRODUCTION

Anthropometry has been the bedrock of ergonomics and human factors since the formalization of these fields of study some 70 years ago. Usually the term anthropometry is limited to the static measurement of human body segment dimensions, such as stature, popliteal height, reach or shoulder width. There are various measurement conventions and many data bases (ISO, SAE, WHO, CDC) . Perhaps the most comprehensive text on the subject is that by Pheasant and Haslegrave (2006). The methods used for workplace design involve convenient, usually boney, landmarks such as the acromion or patella. On occasion, soft tissues are included, especially for widths and girths. Perhaps the most highly developed application is for "occupant packaging" in automobile design (Roe, 1993). Also NASA scientists routinely collect data from astronaut candidates for use in space suit and other equipment designs (Rajulu, 2009). A slightly different set of landmarks is used by the physical education community. The anthropomorphic modeling community uses joint centers of rotation as reference points for the development of avatars.

A formal description of body shape was developed by Sheldon (1940), who used scales of endomorphy, ectomorphy and mesomorphy to describe the relative degrees of fatness, lengthiness and muscularity. Skin fold measurements, underwater weighing and electrical resistance methods were later added to describe body composition. Contemporary methods of wand based landmark identification tools and whole body scanning can provide complete descriptions of body or segmental size and shape – such as heads, hands and feet for specific equipment and clothing design purposes (ISO, 2010, Harrison and Robinette, 2002).

Contini and Drillis (1963) and Pheasant (1982) hypothesized that any of these measures could be predicted with sufficient accuracy from a single measure of stature (Figure 1). Their results have been widely published and applied. The Drillis and Contini tables refer to segment ratios based on joint centers. This approach begs the question of what is "sufficient accuracy"? Parents have their unshod children on their birthdays "stand up straight" with their back to the door jamb, then with the aid of a book placed on the child's head, make a pencil mark to record the annual growth. Similar methods may be used in the evening and the morning to measure overnight "growth" in adults. Astronauts play the same game, although here the stakes are higher if micro gravity exposure results in their not being able to fit into the reentry or emergency evacuation suit. A pragmatic strategy by the space agencies is to reject very tall and very small candidates for selection to the prestigious astronaut corps; corpulence is usually not an issue among these highly driven individuals. Another major source of inaccuracy in the collection of population data is due to the universal use of convenient samples; even the most elaborate, expensive national surveys may fail this basic scientific assumption and test of random sampling. But this is the lot of most if not all ergonomics research.

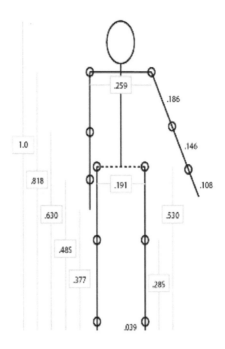

Figure 1 The Drillis and Contini (1966) model modified by Penn State University Open Lab

(http://www.openlab.psu.edu/tools/proportionality_constants.php)

Although the technology and methodology of anthropometry has advanced considerably, in practice product designers employ much less precise measures and use many other criteria in their designs. For example mass transit seat designers must address not only sitting comfort but also the ease of moving into and out of a seat during short journeys. Automobile designers (Roe, 1993) accommodate occupant size and shape variability by including adjustable seats, steering wheels and sometimes pedals into their designs. Fortunately our customers – the equipment designers - rarely complain about our lack of precision. A ball park number, based on tradition, will often suffice. Does it really matter whether we accommodate the 5th or 10th or even the 20th percentile lower leg length in transportation, auditorium or public restaurant seating? Around the world travelers may offer a more critical analysis of seat design, especially if they come from some Eastern countries where ethnic anthropometric differences are observed. Usually door head clearances are a little more generous – their dimensions are based more on tradition and industry norms than precise anthropometry. Clothing and shoe manufacturers base their initial dimensions on tradition and adjust their batch production based more on sales than sizing; contemporary "big and tall" shops make proportional adjustments; short people can often use the children's or teen age section. Aftermarket tailored adjustments sometimes resolve the design or purchase error. The clothing market and the forces of fashion are generally somewhat forgiving of small anthropometric errors in design.

One shortcoming of the "Drillis and Contini" segmental proportions method is that the correlations between stature and segment lengths are generally more reliable than those between stature and width or girth (Table 3); these latter measures are more highly correlated with body weight or a single measure of abdominal girth, albeit with problems of precise definition and implementation. Furthermore measures of heads, feet and hands appear to be less related to stature or width. It is also likely that there are gender and ethnic differences in body segment proportions.

The purpose of this paper is to revisit the findings of Drillis and Contini (1966), Roebuck et al (1975) and Pheasant and Hazelgrave (2006) by analyzing the correlations among many measures from data obtained in recent anthropometric surveys of Singaporean residents. The results of these analyses will be applied to the development of a minimal set of measures from which sufficiently reliable proportional predictions of other measures can be made.

2   METHODS

Three anthropometric surveys were conducted in Singapore. Hartono and Tan, (2010), using conventional anthropometry calipers, recorded 42 dimensions, using standard procedures (Table 1) from 194 male and 99 female Singaporeans and 156 male and 67 female Indonesians. Tong Xin et al, using a purpose made measurement rig, very precise skin markings and 3 repeated observations for each of 43 dimensions on each subject, surveyed 57 male Singaporeans and 56 females. These totals were after the elimination of subject records for missing and clearly spurious data, due to either measurement or recording error. The methods of rejection were by observation of the skewness and variance of the raw data. The third survey involved the use of a simple tape measure in a class exercise in which the class members surveyed 8 dimensions of family and friends with considerably less precision. This rapid exercise resulted in more than 600 Singaporean and Chinese subjects after, as expected, the rejection of at least 10% of the measures for missing data or clearly spurious observations.

Table 1 ISO sources for anthropometry procedures

ISO (7250-!:2006 Basic human body measurements for technological design – Part 1: Body measurement definitions and landmarks.
ISO/TR 7250-2 2010 Basic human body measurement for technological design – Part 2: Statistical summaries of body measurements from individual ISO populations.

Both the traditional and contemporary methods of anthropometry strive for accuracy by applying precise, calibrated equipment and procedures (Roebuck et al, 1975). But even these efforts sometimes fail in the hands of inexperienced investigators, and both within and between observers errors abound. What exactly is "shoulder width?" Other errors are introduced by the subject sampling procedures, especially with regard to gender, ethnicity, nationality and age.

3   DATA ANALYSIS

The data analysis first consisted of the construction of a correlation matrix relating all measures obtained from the three surveys with each other. These Pearson Product Moment Correlation Coefficients were tested using the Students T test and grouped according to high, medium and low significant correlations. A second analysis used the structural equation modeling to extract significant multiple predictor variables.

4   RESULTS

The results in Tables 2, 3 and 4 show a section of the raw data, correlations and correlation coefficient significance levels respectively.

Table 2 Typical raw data from the UniSIM data set

Subject No	Age	Neck Circumference	Shoulder Circumference	Waist Circumference	Hip Circumference	Weight (Kg)	Stature	Shoulder Height	Overhead Reach
1	39	37.7	111.3	82.5	95.2	70.0	161.9	132.1	201.8
2	31	37.1	105.1	70.3	87.0	53.3	165.5	132.9	206.2
3	45	37.6	109.9	90.7	98.1	70.4	175.0	140.6	220.0
4	26	34.4	99.7	66.4	85.4	52.4	169.0	132.9	207.9
5	33	38.9	114.0	94.3	105.0	80.9	172.3	141.2	215.3
6	27	34.6	103.1	85.5	93.4	58.7	169.5	137.7	212.2
7	38	37.9	106.6	90.0	97.5	71.4	176.3	141.7	222.0
8	37	42.8	124.9	108.1	108.2	97.6	176.0	143.9	225.0
9	48	41.5	119.9	98.1	101.5	80.8	171.4	136.2	212.1
10	30	36.7	111.5	88.1	91.8	65.7	171.1	136.5	217.1

Table 3 Typical correlation coefficients between segments

	Age	Neck Circumference	Shoulder Circumference	Waist Circumference	Hip Circumference	Weight (Kg)	Stature	Shoulder Height	Overhead Reach
Age	1.00	0.43	0.23	0.48	0.15	0.21	-0.23	-0.16	-0.17
Neck Circumference	0.43	1.00	0.79	0.83	0.68	0.82	0.00	0.09	0.06
Shoulder Circumference	0.23	0.79	1.00	0.76	0.78	0.88	0.12	0.20	0.16
Waist Circumference	0.48	0.83	0.76	1.00	0.83	0.87	0.00	0.10	0.06
Hip Circumference	0.15	0.68	0.78	0.83	1.00	0.90	0.16	0.25	0.20
Weight (Kg)	0.21	0.82	0.88	0.87	0.90	1.00	0.26	0.36	0.31
Stature	-0.23	0.00	0.12	0.00	0.16	0.26	1.00	0.96	0.95
Shoulder Height	-0.16	0.09	0.20	0.10	0.25	0.36	0.96	1.00	0.94
Overhead Reach	-0.17	0.06	0.16	0.06	0.20	0.31	0.95	0.94	1.00

Table 4 Typical Students T tests for the significance of the Correlation Coefficient

Dimension	Age	Neck Circumference	Shoulder Circumference	Waist Circumference	Hip Circumference	Weight (Kg)	Stature	Shoulder Height	Overhead Reach
Age		2.202	0.812	2.944	0.464	0.729	-1.635	-1.211	0.400
Neck Circumference	2.202		9.209	9.447	6.636	10.097	0.723	1.187	0.390
Shoulder Circumference	0.812	9.209		7.709	9.061	14.107	1.523	1.964	0.960
Waist Circumference	2.944	9.447	7.709		9.714	10.873	0.449	1.038	1.147
Hip Circumference	0.464	6.636	9.061	9.714		14.649	1.635	2.213	1.486
Weight (Kg)	0.729	10.097	14.107	10.873	14.649		2.381	2.976	1.526
Stature	-1.635	0.723	1.523	0.449	1.635	2.381		21.879	2.813
Shoulder Height	-1.211	1.187	1.964	1.038	2.213	2.976	21.879		2.930
Overhead Reach	0.400	0.390	0.960	1.147	1.486	1.526	2.813	2.930	

Table 5 and 6 show the means and standard deviations of anthropometric measures from the UniSIM (Tong Xin, 2011) data, the correlations with weight and stature, and the ratios with weight and stature for males and females respectively. The highlighted correlations are significant at the P< .05 level.

Table 5 Male Data showing segment means and standard deviations, and correlations and ratios with Stature and Weight; the highlighted values are statistically significant based on the Student's T test.

	Avg	SD	Weight Corr	Stature Corr	Weight Ratio	Stature Ratio
Age	30.89	6.24	0.21	-0.23	0.45	0.18
Neck Circumference	37.17	2.92	0.82	0.00	0.54	0.22
Shoulder Circ	112.60	7.76	0.88	0.12	1.65	0.66
Waist Circ	84.02	10.16	0.87	0.00	1.22	0.49
Hip Circumference	96.81	7.15	0.90	0.16	1.42	0.56
Weight (Kg)	69.28	14.96	1.00	0.26	1.00	0.40
Stature	171.73	7.27	0.26	1.00	2.53	1.00
Shoulder Height	139.11	6.89	0.36	0.96	2.05	0.81
Overhead Reach	215.05	12.99	0.31	0.95	3.17	1.25
Span	172.89	9.93	0.30	0.86	2.55	1.01
Shoulder to Elbow	34.56	2.08	0.37	0.79	0.51	0.20
Chest Breadth	31.35	3.34	0.81	0.11	0.46	0.18
Chest Depth	20.06	2.98	0.83	0.01	0.29	0.12
Sitting Waist Depth	21.60	2.75	0.80	-0.03	0.31	0.13
Shoulder Breadth	45.92	2.69	0.77	0.22	0.67	0.27
Biacromial Breadth	40.23	1.35	0.30	0.34	0.59	0.23
Forearm To Forearm	47.21	5.87	0.91	0.10	0.69	0.28
Head Length	19.05	0.94	0.64	0.06	0.28	0.11
Head Breadth	16.12	0.35	0.22	0.21	0.24	0.09
Head Circumference	57.79	1.66	0.59	0.24	0.85	0.34
Hand Length	18.79	1.04	0.33	0.53	0.28	0.11
Hand Breadth	8.08	0.42	0.68	0.34	0.12	0.05
Wrist Circumference	16.59	0.71	0.80	0.16	0.24	0.10
Buttock To Popliteal	45.75	3.25	0.43	0.63	0.67	0.27
Buttock To Knee	58.78	3.32	0.45	0.71	0.86	0.34
Elbow To Fingertip	46.13	2.67	0.30	0.77	0.68	0.27
Forward Reach	83.62	5.48	0.43	0.75	1.23	0.49

Sitting Eye Height	78.86	2.99	0.31	0.78	1.16	0.46
Shoulder Height	58.98	3.29	0.35	0.67	0.87	0.34
Elbow Rest Height	25.47	3.43	0.15	0.12	0.37	0.15
Thigh Clearance	14.99	1.34	0.61	-0.02	0.22	0.09
Sitting Knee Height	52.66	2.64	0.30	0.73	0.78	0.31
Popliteal Height	40.67	1.71	0.03	0.72	0.60	0.24
Sitting Height	91.24	2.95	0.26	0.83	1.34	0.53
Overhead Reach	134.81	6.97	0.23	0.84	1.99	0.78
Sitting Hip Breadth	36.46	3.66	0.87	0.13	0.53	0.21
Foot Length	25.44	1.35	0.35	0.67	0.37	0.15
Ankle Height	6.98	0.31	0.15	0.35	0.10	0.04
Heel Breadth	5.86	0.38	0.55	0.23	0.09	0.03
Ball Of Foot Width	9.75	0.41	0.61	0.06	0.14	0.06
Foot Circumference	24.28	1.06	0.47	0.01	0.36	0.14
BMI	23.52	4.56	0.88	-0.22	0.34	0.14

Tables 7 and 8 present summaries of the correlations and ratios with weight and stature respectively from the UniSIM data (Tong Xin (2011).

Table 7 Summary of rank ordered significant mean correlations and ratios with weight for males and females

Female	Mean Ratio with Weight	Correlation	Male	Mean Ratio with Weight	Correlation
Shoulder Circumference	1.81	0.86	Shoulder Circumference	1.65	0.88
Hip Circumference	1.73	0.87	Hip Circumference	1.42	0.90
Waist Circumference	1.35	0.80	Waist Circumference	1.22	0.87
Shoulder Breadth	0.75	0.84	Forearm To Forearm	0.69	0.91
Forearm To Forearm	0.75	0.83	Shoulder Breadth	0.67	0.77
Sitting Hip Breadth	0.67	0.85	Neck Circumference	0.54	0.82
Neck Circumference	0.59	0.76	Sitting Hip Breadth	0.53	0.87
Chest Breadth	0.50	0.75	Chest Breadth	0.46	0.81
BMI	0.39	0.83	BMI	0.34	0.88
Sitting Waist Depth	0.35	0.68	Sitting Waist Depth	0.31	0.80
Chest Depth	0.34	0.76	Chest Depth	0.29	0.83
Wrist Circumference	0.27	0.75	Wrist Circumference	0.24	0.80
Thigh Clearance	0.25	0.69	Thigh Clearance	0.22	0.61

Table 8 Summary of rank ordered significant mean correlations and ratios with stature for males and females

Female	Mean Ratio with Stature	Correlation	Male	Mean Ratio with Stature	Correlation
Overhead Reach	1.24	0.94	Overhead Reach	1.25	0.95
Span	0.99	0.75	Span	1.01	0.86
Shoulder Height	0.81	0.94	Shoulder Height	0.81	0.96
Sitting Overhead Reach	0.78	0.92	Sitting Overhead Reach	0.78	0.84
Sitting Height	0.54	0.82	Sitting Height	0.53	0.83
Forward Reach	0.48	0.69	Forward Reach	0.49	0.75
Sitting Eye Height	0.46	0.78	Sitting Eye Height	0.46	0.78
Sitting Buttock To Knee Length	0.35	0.66	Sitting Shoulder Height	0.34	0.67
Sitting Shoulder Height	0.35	0.66	Sitting Buttock To Knee Length	0.34	0.71
Sitting Knee Height	0.30	0.55	Sitting Knee Height	0.31	0.73
Sitting Buttock To Popliteal Length	0.27	0.66	Elbow To Fingertip	0.27	0.77
Elbow To Fingertip	0.26	0.70	Sitting Buttock To Popliteal Length	0.27	0.63
Sitting Popliteal Height	0.24	0.68	Sitting Popliteal Height	0.24	0.72
Shoulder to Elbow Length	0.20	0.75	Shoulder to Elbow Length	0.20	0.79

It is noted that the dimensions that had the highest associations with stature were limb segment lengths. Limb lengths were more highly correlated with stature and (standing) shoulder height than was sitting height. Width and girth measures were inter-correlated and more associated with weight than stature. Head measures were not highly correlated with other measures nor were hand and foot measures. These associations also showed differences between the sexes. There were not enough data to draw reliable conclusions regarding the differences or similarities among subjects of different ethnic backgrounds. Whereas the Drillis and Contini (1966) ratios were found to be sufficient for many practical purposes, the addition of a weight or girth measure plus more sophisticated statistical methods provides greater predictive capability. The results of the Tan et al (2010) study and the large, less rigorous class survey provided similar results regarding correlations and proportions.

5   DISCUSSION

Scientists and practitioners are often found to diverge in their pursuit of accuracy and practicality. The science and practice of human factors, ergonomics and anthropometry are no exception. Researchers strive for accuracy and spend money and time developing equipment and procedures that more accurately describe their subject of interest. Such processes are essential ingredients of science. Practitioners, on the other hand, depending on the situation, use rapid access to sufficiently accurate data and procedures, often through look up tables, then round the values to convenient whole numbers. Next they take into account other aspects and constraints of the applied situation and design accordingly. Designers and engineers who develop products for the mass market prefer round the numbers and rely on human variability and versatility to correct for residual "errors." What really is the practical difference between a fifth percentile reach and a tenth percentile reach, especially when the operator can move his or her feet? This issue was highlighted in a recent court case related to a collapsed store shelf after a shopper had used the bottom shelf as a step to reach a product. How high should the seat be in a mass transit train given relatively short journey times and the associated acts of sitting down and getting up easily? One also may recall the sticks provided for Soyuz cosmonauts to manipulate their controls, the aftermarket extensions for some automobile controls or the devices developed for wheel chair users to reach the elevator buttons. Also, even if we do get the anthropometry "right" for one sub task, what about compromises between reach and fit? What about frequency of use and importance?

It is usual for practicing ergonomists and designers to use anthropometric data developed by specialists. These data are likely to be based on reliable measurements using standard procedures. The ISO TC 159/SC3 committee (ISO, 2012) is the principal source for methods, data and applications of anthropometry. Despite the reliability of these data and ergonomics methods, engineers and designers, perhaps because of the lack of availability of trained ergonomists, are more likely to use simpler methods. Larger organizations make use of anthropometric modeling methods, but these again are more effectively used in the hands of skilled ergonomists. Contemporary integrated anthropometric and cognitive models such as MIDAS (NASA, 2008) take the sophistication of human modeling to another dimension. However the methods of Drillis and Contini, albeit amplified by adding in weight as a predictor of width and girth measures, are sufficient for many practical purposes.

6   CONCLUSIONS

The correlations and ratios described in this study advance the model offered by Drillis and Contini (1966) and Roebuck et al (1975) by adding weight as the main predictor of width and girth measures. For some measures, shoulder height was found to be a marginally better predictor than stature. In depth data analysis, using structural equation modeling looked at multiple predictor variables. Greater refinement of this study is possible by adding in more predictor variables but for practical purposes the simple measures of stature and weight should suffice. Some of the apparent discrepancies in this study may be due to differences between the segmental reference points used by Drillis and Contini (1966) and those currently used in anthropometry studies (Table 1). An ergonomics practitioner who comprehends the many challenges of human and situational variability in design may use this practical extension of the Drillis and Contini (1966) model to estimate most other dimensions of size and shape (Tables, 5, 6, 7, 8). The possibility of differences among populations of different ethnic origins needs to be addressed by controlled sampling of these cohorts.

REFERENCES

Contini, R., R. Drillis, and M. Bluestein, (1963) Determination of body segment parameters, Human Factors, 5 (5)

Drillis R and R Contini (1966) Body segment parameters. DHEW 1166-03. New York University, School of Engineering and Science, New York

Harrison C. R. and K. M. Robinette, (2002) CAESAR: Summary statistics for the adult population (ages 18-65) of the United States of America. AFRL-HE-WP-TR-2002-0170. United States Air Force Research Laboratory.

ISO (2012) http://www.iso.org/iso/iso_catalogue/catalogue_tc/catalogue_tc_browse.htm?commid=53362

ISO 20685:2010 "3-D scanning methodologies for internationally compatible anthropometric databases"

NASA (2008) http://human-factors.arc.nasa.gov/awards_pubs/news_view.php?news_id=54

Pheasant, S and C. M. Haslegrave (2006), Bodyspace, Taylor & Francis

Pheasant S (1982) A technique for estimating anthropometric data from the parameters of the distribution of posture, Ergonomics, 25, 981:982

Rajulu, S (2009) http://ntrs.nasa.gov/archive/nasa/casi.ntrs.nasa.gov/20090026488_2009026743.pdf

Roe R (1993) "Occupant packaging," in Automotive Ergonomics, Peacock and Karwowski, Taylor and Francis

Roebuck, J.A., Kroemer, K. H. E., and Thompson, W. G. (1975) Engineering Anthropometry Methods, New York, Wiley

Sheldon, William H. (1940), The Varieties of Human Physique (An Introduction to Constitutional Psychology), Harper & Brothers

Tan, K. C., Hartono, M and Kumar, N (2010), Anthropometry of the Singaporean and Indonesian populations, International Journal of Industrial Ergonomics 40, (757-766)

Tong Xin (2011) Singapore Anthropometry, Project Report, SIM University.

CHAPTER 10

Usability Evaluation of a Virtual Reality System for Motor Skill Training

Wooram Jeon, Michael Clamann, Biwen Zhu, Guk-Ho Gil & David Kaber

Edwards P. Fitts Department of Industrial and Systems Engineering,
North Carolina State University, Raleigh, NC, 27695-7906, USA
{wjeon, mpclaman, bzhu, ghgil, dbkaber} @ncsu.edu

ABSTRACT

The objective of this study was to conduct a preliminary usability evaluation of a virtual reality (VR) simulator designed for training or retraining fine motor skills. The simulator was to be used to investigate the effects of visual display and haptic technology features on motor control rehabilitation and learning. The simulator modeled a block design task from the Wechsler Abbreviated Scale of Intelligence with augmented haptic control and visual feedback. Participants subjectively evaluated the usefulness and effectiveness of the simulation and provided comments regarding enhancements to the original design. Usability problems were categorized as haptic or visual-spatial. Some users had difficulties grasping and rotating a pen-like stylus interface through more than 90 degrees at the wrist. Adjusting the system control-response (C/R) ratio and simplifying virtual block grabbing mechanisms were proposed as potential solutions. In terms of the visual-spatial challenges, field dependent users were expected to have difficulty in recognizing how the position and orientation of each virtual block uniquely contributed to overall designs in the simulation. Furthermore, since a user's visual perspective in the VR was fixed they were expected to have less awareness of color patterns on blocks. Superimposing a grid on the stimulus design and making blocks translucent to reveal patterns on each side were proposed as potential solutions. Through system testing, a preferred C/R ratio was determined and grasping blocks by contact appeared useful for reducing unnecessary hand reorientations with the stylus. Performance problems related to haptic input and feedback were reduced to a greater extent than those related to visual-spatial issues.

Keywords: haptic simulation, usability analysis, motor skill training, block design task

1   INTRODUCTION

The domain of motor rehabilitation may benefit from the growth of virtual reality (VR) technology. Practical advantages of using VR systems for motor skill training include increased safety, reduced time on task for therapists, consolidation of space and equipment, cost efficiency as a result of reduced personnel needs and automated help (Holden, 2002). A variety of interfaces can be used to present virtual environments (VE) to patients or learners with perfect repeatability. The interfaces can also be used to provide immediate feedback to patients on performance in novel environments, which may serve to motivate further active participation in training programs. In general, previous research has demonstrated the efficacy of VR-based haptic simulation as a therapeutic tool for rehabilitation to enhance motor function (Holden, 2005; 2007).

The current study is part of a broader research project sponsored by the National Science Foundation (NSF). The target context is motor skill rehabilitation and training for persons attempting to recover from minor Traumatic Brain injury (mTBI) or seeking to develop new motor skills for work and societal activities. As part of this broader effort, our research team designed and implemented a VR reproduction of the block design (BD) subtest from the Wechsler Abbreviated Scale of Intelligence (WASI; The Psychological Corporation, 1999). In the BD test, participants are given a set of nine identical red and white blocks printed with either solid or cross-sectional patterns on each side. The goal for each trial is to arrange the blocks to match a stimulus figure as quickly and accurately as possible. In the VR version, participants manipulate virtual representations of blocks arranged on a work surface using a stylus (SensAble Technologies PHANTOM® Omni® Haptic Device) to place them in a target grid and to reconstruct the design stimulus. Figure 1 shows the traditional WASI BD materials and the VR representation.

(a) Block Design task (native)   (b) VE setup

Figure 1. The native block design task and a VR workstation and VE model of the same

The BD test was selected, in part, because it includes fundamental motions and sequences that occur in many occupational activities. The overall design was previously validated through an experiment conducted with unimpaired participants (Kaber et al., 2011), which provided justification for the use of augmented haptic features for motor skill training. While the study served as a proof of concept exercise for the VR BD interface, it was determined that some aspects of the haptic and visual assistance modes could be improved. The present study focused on prototyping and testing of the improved features. Usability testing was identified as a method for objectively assessing and selecting among several preliminary designs as a basis for overall system improvement.

Related to this, usability testing has been shown to be an effective evaluation tool for rehabilitation systems. Lewis, Deutsch and Burdea (2006) developed a VR telerehabilitation application that enabled therapists to remotely communicate with patients while monitoring and controlling therapy procedures. A usability test was conducted to assess the system set-up and instructions from a therapist's perspective. Lange, Flynn and Rizzo (2006) assessed usability of off-the-shelf video game consoles for motor rehabilitation. The author focused on the interaction, game play and game mechanics instead of the interface technology. For example, the Nintendo Wii was tested to determine compatibility with therapeutic goals for different patient populations. The author concluded that use of off-the-shelf games offers potential for rehabilitation, and a new software development effort was initiated for the Wii. Related to this research, Feys (2008) conducted a usability test of a robot-assisted rehabilitation tool incorporating a PHANTOM haptic device. In this study, and the work by Lange et al., usability tests were performed in an early planning stage of development of a rehabilitation tool (Feys, 2008; Lange, Flynn and Rizzo, 2006). In the case of Lewis et al. (2006), tests were used to identify system technical issues and for fine-tuning of parameters.

1.2   Usability Problems

Usability problems with the VR simulator, identified by focus groups during the 2011 study, were associated with either the haptic interface or visual displays and user spatial processing. Among the observed problems, three were considered to be critical as a basis for system redesign.

(1) Block rotation. The main haptic challenge identified was that users had difficulty grasping and rotating the pen-like stylus more than 90 degrees at the wrist, which limited block rotation. In the BD task, when a block is being manipulated and the visible surface does not match a part of the stimulus figure, it needs to be rotated a minimum of 90 degrees to orient the block to a different side. It is important, therefore, that movements of greater than 90 degrees are easily attainable. Figure 2(a) illustrates that, when gripping the stylus, the maximum pronation angle is less than 90 degrees from a normal handwriting posture position. Furthermore, due to the limitation of gimbals the stylus head, 180-degree rotation is impossible without prepositioning the stylus. Reorientation of the stylus was not possible because participants were required to hold a control button on the stylus with the index

finger while manipulating each block. These issues resulted in excessive fatigue or participants reoriented blocks by repeatedly picking them up and putting them down, which increased task time.

(2) Block grasp. Another haptic usability problem related to block manipulation was the requirement to hold down a button on the stylus while manipulating blocks. The original concept was that pressing and releasing the button would represent a pinch grip, however, this introduced additional problems. The button was not large enough to be easily located and held by a finger, and the requirement for constant pressure at the button increased fatigue. Figure 2(b) presents the button and hand grip.

(3) Stimulus configuration. The first of the observed visual-spatial problems suggested that field dependent (FD) users might have difficulty recognizing how individual block position and orientation in a grid pattern contributed to the overall design (Figure 2(c)). This issue has been identified by prior research (Witkin, 1962). With respect to the present program of NSF research, there was a concern that learning the effect of motor skill training cannot be maximized by FD users because of the excess cognitive effort

(4) Block visibility. Due to limitations on perspective views in the VR, there were concerns that users might not be able to develop accurate internal representations of block configuration. In the native BD task, the side of each block could be viewed not only by rotating the block with the fingers, but also by moving the head. The VR did not support the latter mode of viewing and, therefore, task performance time was increased.

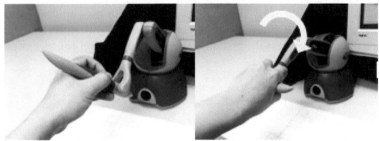

(a) Illustration of 90 degree rotation while holding a button on the stylus

(b) Pressing a button on the stylus

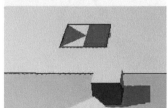

(c) BD task stimulus design (captured in VR)

Figure 2 Identification of usability problems

1.3 Design Proposal

The project team investigated the usability problems with the existing VR training system and proposed four enhancements to the original design. The enhancements were aimed at improving user experience with a new version of the simulator.

(1) Block rotation. The first design enhancement was to optimize the control-to-response (C/R) ratio (gain) of the haptic device to reduce the block rotation challenge. In the present VR system, this meant decreasing the C/R ratio. Two levels, 1:2 and 1:3 were implemented for usability testing. This meant that a 90 degree wrist rotation would translate to 180 or 270 degrees of block rotation (see Figure 2).

(2) Block grasp. The second enhancement was to simplify the block grasping mechanism. When contacting the cursor to a block, instead of holding a button on the stylus, the block could be grasped by merely pressing the cursor against a block. The system update required identical hand movements, as with the prior grasping technique, except the button press was no longer necessary. A block could be returned to the work surface by simply contacting it to the table.

(3) Stimulus configuration. To solve the problem faced by FD participants, a virtual line grid was superimposed on the stimulus figure to reveal the BD construction. Touching the stimulus figure with the cursor could activate this form of assistance. The grid would then continue to appear briefly once every 3 seconds until turned off by touching the stimulus with the cursor a second time.

(4) Block visibility. The last proposal was to make blocks translucent in order to reveal the patterns on every side. Blocks grasped with the simulation cursor would become translucent and then return to normal (opaque) when returned to the work surface. Figure 3 presents illustrations for each of the design enhancements.

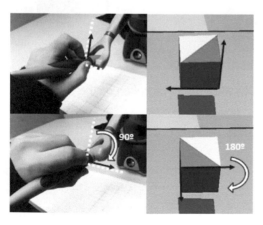

(a) Illustration of changing C/R ratio (1:2).   (b) Grasping a block without holding

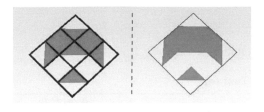
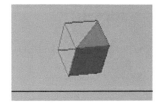

(c) Stimulus figure with and without grid   (d) Translucent block

Figure 3 Enhanced simulation design features

2   METHODOLOGY

2.1   Apparatus

The enhanced VR system simulated the BD subtest from the WASI with augmented haptic control and visual feedback based on the above design proposals. In general, haptic control was provided through an Omni haptic device, which featured a boom-mounted stylus providing 6 degrees of freedom (DOF) for movement and 3 DOF in force feedback. The VE was presented with a stereoscopic display and an NVIDIA® 3D Vision™ Kit, including 3D goggles and an emitter. All VE prototyping was done using the Virtual Environment Software Sandbox (VESS; University of Central Florida, Orlando, FL) and the haptic interface enhancements were programmed using OpenHaptics Toolkit (Sensable Technologies, Wilmington, MA).

2.2   Users, Dependent Variables and Procedure

Four expert VR users (2 males and 2 females) participated in the study. All had extensive prior knowledge of the original training system.

All participants evaluated the usefulness and effectiveness of the redesigned system by responding to a survey. The instrument included 12 questions (e.g., "The translucency of blocks helped me recognize patterns."), which were associated with 7-point rating scales of agreement/disagreement (ranging from 1=Strongly Disagree to 7=Strongly Agree). User comments were also requested on the four new forms of aiding as part of the system. Each usability test was videotaped following the consent of a user. The videos were reviewed and results were used to further inform the feature improvements.

The users were asked to complete a predefined scenario using the VR BD system. Each scenario was designed to incorporate all the new features. Before performing the tasks, a moderator explained the objective and provided instructions on the new features. Users completed between two and three BD trials. There was no time limit, and participants were encouraged to think aloud and ask questions during testing.

3   RESULT

The results of the usability test are summarized in Table 1.

Table 1 Test results

Issue	Feature	Survey Question	Average Agreement Score
Block rotation	Decreased C/R ratio to 1:2	Allowed blocks to be easily rotated into position.	6.25
Block grasp	Grab block by contact	Allowed for blocks to be easily picked-up.	4.5
Stimulus configuration	Flashing grid	Made clear individual block placement and orientation in stimulus figure.	3.25
Block visibility	Transparent block on contact	Partially transparent wireframe made block configuration clear.	3.25

The decreased C/R ratio for the haptic device received the highest rating among the proposed enhancements. Users unanimously agreed that the feature was helpful in terms of shortening task completion time and promoting convenience of hand grip. The average agreement score across the C/R ratio settings was 6.00 (SD = 0.96). Most users preferred the 1:2 C/R ratio, while one preferred the 1:3 ratio (M = 2.0, SD = 2.83). This user offered that the 1:2 C/R ratio might be more comfortable for less experienced users (although all other users were also experts).

Users were divided on the simplified block grabbing mechanisms. Two agreed that it was easier to pick-up a block without pressing a button than by holding a button; whereas, the other two disagreed (M = 4.5, SD = 2.89). However, three users reported inadvertently moving blocks due to accidental contact with the cursor, which they found frustrating.

The flashing grid was not considered to be useful for visual cues on stimulus figure construction. Users generally felt the grid was distracting (M = 3.25, SD = 2.06); however, they also believed that it could help FD users recognize block arrangement within the stimulus pattern (M = 5.0, SD = 1.91). Such comments were contradictory in nature. One user commented that the visual aids might be helpful for novice users but distracting for experts. Another user raised a concern about increased movement time to complete the task because of the additional time required to touch the cursor to the stimulus figure in order to reveal the grid aid.

The block transparency also produced negative user responses. The users did not feel that translucent blocks helped them recognize patterns (M = 3.25, SD = 1.89), and the aid was not considered to effectively communicate block configuration (M = 2.75, SD = 2.22). One participant observed that the transparent blocks were less visible and more distracting from the task as compared to opaque blocks. Another user did not understand the intent of the transparent block design.

4   DISCUSSION AND CONCLUSION

All the users in the present study understood the implementation requirements of the VR training simulator and foresaw several situations when subjects could benefit from the enhanced features presented above. However, there were some negative opinions on certain features that provide a basis for additional future system design improvements.

From the results of the usability survey, performance problems related to haptic input and feedback appeared to be reduced to a greater extent than those related to visual-spatial issues. No users inadvertently dropped virtual blocks during the usability test trials. There was substantial improvement in user hand comfort and reduced task completion time.

Regarding the revised block grabbing mechanism (by contact), most participants reported inadvertently grasping blocks, which was frustrating. The design revision was intended to reduce user fatigue by removing the requirement to hold a button with the index finger when moving a block. In order to address the remaining usability issue, the design team proposed clicking on a block (pressing and releasing the button) instead of automatic grasping on contact with a cursor. Since this form of grabbing a block requires user action, blocks will not be grasped unintentionally. Block movement and manipulation will also be simplified by removing the need to hold a stylus button.

Although most users found the superimposed flashing grid to be distracting to task performance, some agreed that novice or FD users might benefit from the aid. To maintain the potential benefits while reducing any visual distraction, the design team proposed to only present gridlines around a single block in the stimulus figure when selected by a user. The gridlines will disappear when any surface outside of the stimulus figure is contacted with the simulation cursor. Since gridlines are only shown following user input, field-independent users may not be distracted by this implementation.

To some extent, the increase in the haptic device control gain also reduced the limited perspective viewing problem. As discussed previously, users had difficulty comprehending block color patterns across sides, even with transparent blocks were used. However, promoting ease of rotation of blocks with the stylus allowed users to more quickly view opposite sides of a block and to form an accurate internal model of block configuration.

In general, this study provides an example of how a simplified usability test can be implemented to quickly provide effective feedback on VR system design and a basis for enhancing features to support user performance. The target system included a haptic interface for motor-control task training. The usability test allowed the design team to rapidly identify benefits and shortcomings of haptic and visual aids and to propose additional design changes. Such testing can be easily integrated in the development cycle of VR training systems with limited time and resources for execution, interpretation of results and proposal of design revisions. Results are also highly useful for informing modifications to better accommodate user needs.

ACKNOWLEDGMENTS

This study was sponsored by the NSF (Grant No. IIS-0905505). The technical monitor was Ephraim Glinert. The views and opinions expressed are those of the authors and do not necessarily reflect the views of the NSF.

REFERENCES

Chen, E., 1999. Six degree-of-freedom haptic system for desktop virtual prototyping applications, *Proceedings of the First International Workshop on Virtual Reality and Prototyping* :97-106.

Feys, P., Alders, G., Gijbels, D., De Boeck, J., De Weyer, T., Coninx, K., Raymaekers, C., Annegarn, J., Meijer, K. and Savelberg, H., 2008. Robotic rehabilitation of the upper limb in persons with multiple sclerosis: A usability and effectiveness study, Proceedings of the Robotic Helpers: User interaction, interfaces and companions in assistive an therapy robotics: 49-52.

Holden, M.K., 2005. Virtual environments for motor rehabilitation: review, *Cyberpsychology and behavior,* vol. 8, no. 3:187-211.

Holden, M.K. and Dyar, T., 2002. Virtual environment training: a new tool for neurorehabilitation, *Journal of Neurologic Physical Therapy,* vol. 26, no. 2: 62.

Holden, M.K., Dyar, T.A. and Dayan-Cimadoro, L., 2007. Telerehabilitation using a virtual environment improves upper extremity function in patients with stroke, *Neural Systems and Rehabilitation Engineering, IEEE Transactions on,* vol. 15, no. 1: 36-42.

Kaber, D. B., Lee, Y-S & Tupler, L., 2011. *HCC: Medium: Haptic Simulation Design for Motor Rehabilitation and Skill Training* (Annual Rep.: NSF Award #IIS-0905505). Arlington, VA: National Science Foundation.

Lange, B., Flynn, S. and Rizzo, A., 2009. Initial usability assessment of off-the-shelf video game consoles for clinical game-based motor rehabilitation, *Physical Therapy Reviews,* vol. 14, no. 5: 355-363.

Lee, T.D., Swinnen, S.P. and Serrien, D.J., 1994. Cognitive effort and motor learning, *Quest-Illinois-National Association for Physical Education in Higher Education,* vol. 46: 328-328.

Lewis, J.A., Deutsch, J.E. and Burdea, G., 2006. Usability of the remote console for virtual reality telerehabilitation: formative evaluation, *CyberPsychology and Behavior,* vol. 9, no. 2: 142-147.

Massie, T.H. and Salisbury, J.K., 1994. The phantom haptic interface: A device for probing virtual objects, *Proceedings of the ASME winter annual meeting, symposium on haptic interfaces for virtual environment and teleoperator systems*: 295.

Wechsler, D., 1997. WAIS-III, *Administration and scoring manual.San Antonio, TX: The Psychological Corporatio, .*

Witkin, H.A., Moore, C.A., Goodenough, D.R. and Cox, P.W., 1977. Field-dependent and field-independent cognitive styles and their educational implications, *Review of educational research,* vol. 47, no. 1: 1-64.

CHAPTER 11

Proposal of Human Interface Architecture Framework
-Application Human-centered Design Process to Software/ System Development Process-

Shin'ichi Fukuzumi and Yukiko Tanikawa

NEC Corporation
Tokyo, Japan
s-fukuzumi@aj.jp.nec.com

ABSTRACT

Software / system quality is the most important issues in their values. Recently, usability is considered as one of elements for quality. However, as previous usability evaluation was carried out right before shipment, usability evaluation results could not be difficult to reflect products. To solve this issue, applying Human-centered design (HCD) process to development process, especially upper phase, is necessary to improve usability. This paper proposes a framework for HCD and related technology, they are "HCD process application support technology", "Usability quantification technology" and "Human model development and usage technology". System / software could be developed for the view point of HCD efficiently to applying this framework to system /software development process.

Keywords: Human centered Design Process, usability, development process

1   INTRODUCTION

Software / system quality is the most important issues in their values (ISO/IEC 25000, 2005). Recently, usability is considered as one of elements for quality (ISO/IEC 25030, 2007). Usability is defined as "Extent to which a product can be used by specified users to achieve specified goals with effectiveness, efficiency and satisfaction in a specified context of use" (ISO 9241-11, 1998). This definition means that usability shall be measured as quantitatively. They are not only operation time, achievement level and subjective rating but also reduce product introduce periods and low human error rate (Smith and Mosier, 1984), (Ravden and Johnson, 1989). However, even though these quantitative data could be collected, as previous usability evaluation was carried out right before shipment, usability evaluation results could not be difficult to reflect products, even more, system integrations or service developments. To solve this issue, applying Human-centered design (HCD) process to development process, especially upper phase, is necessary to improve usability.

For interactive system, HCD process has been prescript as International standard since 1999 (ISO 13407, 1999) and revised in 2010 as a new version (ISO 9241-210, 2010). In this HCD process standard, four activities are defined, they are 1) "Understand and specify the context of use", 2) "Specify the user requirements", 3) "Produce design solutions to meet user requirements" and 4) "Evaluate the designs against requirements". Figure 1 shows an interdependence of HCD activities described in ISO 9241-210 (ISO 9241-210, 2010).

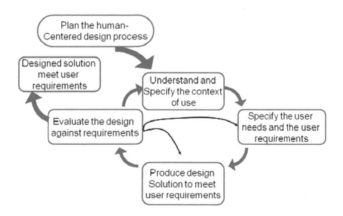

Figure 1 Interdependence of HCD activities (ISO 9241-210, 2010)

Moreover, related standards (Common Industry Format for Usability: CIF) about these four HCD activities are discussed internationally (ISO/IEC TR25060, 2010), (ISO/IEC 15062, 2005). These standard series have been discussed as software / system quality area called SQUARE (ISO/IEC 25060, 2010). These CIF standards are constituted correspond to HCD activities (Theofanos, Stanton and Bevan, 2006). Figure 2 shows the relationship of CIF documents and HCD

(Theofanos and Stanton, 2011). Currently, four documents correspond to above HCD activities are deliberated as standards. They are "Context of use description", "User needs reports", "User requirements specification" and "Evaluation report"

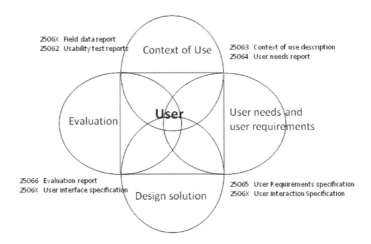

Figure 2   Relationship of CIF documents and HCD (Theofanos and Stanton, 2011)

ISO9241-210 is more concrete and more understandable than the first version. However, to apply these standards to actual software development or system integration process, each item in these standards has to be more concrete.

2   APPLICATION OF HCD PROCESS TO SYSTEM DEVELOPMENT PROCESS

These standards are kinds of guideline documents for applying HCD process to development of software or system. Though they are much important for HCD process, it is difficult without HCD expertise (Bellotti, Fukuzumi, Asahi and et. al., 2009). Thus, not only these documents but also some new technology is necessary for applying HCD process. The authors assumes three technology, they are, "HCD process application support technology", "Usability quantification technology" and "Human model development and usage technology". First technology has three themes, "determination of purpose of usability", "agreement of usability" and "specification of usability" (Lutsh, 2011), (Constantine, 2004), (Vredenburg, 2003), second technology also has three theme, "determination of quantitative usability axis", "usability evaluation methods" and "user interface elements correspond to systematization of usability". These two technology are supported by third technology. Main theme of this technology is how to use any human model like cognitive model or behavior model.

3   PROPOSAL OF HUMAN INTERFACE ARCHITECTURE FRAMEWORK

Relationship between these assumed technology and software / system development process is as follows:

HCD process application support technology is mainly useful for upper phase in development process, e.g. planning, definition of user requirements and external design. Usability quantification and evaluation technology is also mainly useful for same phase, for about usability evaluation, it is also useful for "test" phase. The authors propose Human Interface architecture framework which applied these technology to HCD Process. The purpose of this framework is to enrich customers' value by developing products and/or system which meets customers' requirements for usability. Figure 3 shows a Human Interface architecture framework.

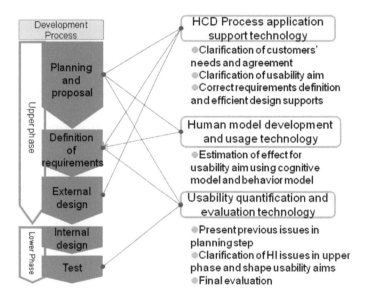

Figure 3 Human Interface architecture framework

At first, software development teams or system vendor have to be extracted customers' needs for usability accurately. For these needs, they have to clarify usability purpose, specify usability requirements obtained from user needs. Secondly, development division designs these requirements concretely. In this phase, usability quantification technology is required. Before and after development, some kinds of evaluation are carried out. To evaluate usability more upper phase, usability evaluation method applied in upper phase of development process is also required.

3.1 HCD process application support technology

Goal of this technology is to be able to apply HCD process to software /system development process by developers or system engineers themselves. Concretely, to clarify customers' needs, to agree with human interface and to clarify usability aim, we use design and evaluation support documents.

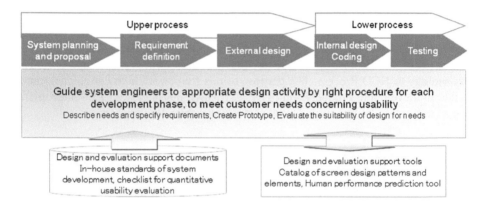

Figure 4 Outline of HCD process application support technology

3.2 Usability quantification and evaluation technology

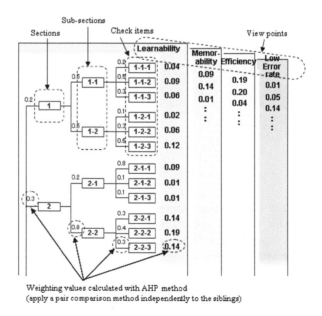

Figure 5 Outline of checklist weighting method

Goal of this technology is to show usability level quantitatively. Nielsen proposed that usability has five elements, they are, efficiency, low error, memorability, learnability and satisfaction (Nielsen, 1993). We try to quantify usability by using these four elements except satisfaction. Figure 5 shows the outline of our evaluation method for usability quantification (Asahi, Ikegami and Fukuzumi, 2009), (Johnson and Shneiderman, 1991) and example of evaluation result is shown in figure 6 (Fukuzumi, Ikegami and Okada, 2009).

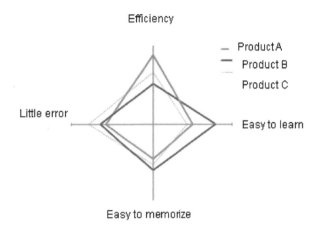

Figure 6 Example of evaluation result

3.3   Human model development and usage technology

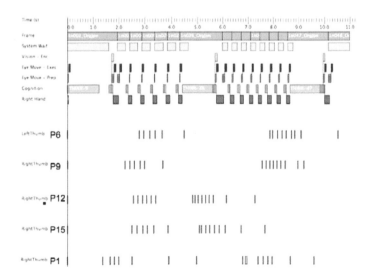

Figure 7 An example of usability estimation result by using cognitive human model

Goal of this technology is to develop usable human interface efficiently and reliably (Ham, 2007, Card, 1983, Pirolli, 1995). To use human model, usability level according to human characteristics can be estimate before developing prototype or products. Figure 7 shows an example of operation time estimation results for mobile phone operation by using cognitive model (John, 2004).

4   SUMMARY

This time, we proposed a human interface architecture framework to develop usable system / software products by human centered design approach. This framework is in correspondence between development process with related technology. They are "HCD process application support technology", "Usability quantitative technology" and "Human model development and usage technology". It is much important for usability engineering division to apply this framework to development process by development division members themselves. To realize this, these technology have to be completed that development division member could use these technology directly..

REFERENCES

Asahi, T., Ikegami, T. and Fukuzumi, S., 2009, A Usability Evaluation Method Applying AHP and Treemap Techniques, *J. A. Jacko (Ed.): Human-Computer Interaction, Part 1, HCII 2009, LNCS 5610*, pp195-203.

Bellotti, V., Fukuzumi, S., Asahi, T. and Suzuki, S., 2009, User-Centered Design and Evaluation –The Big Picture, *J. A. Jacko (Ed.): Human-Computer Interaction, Part 1, HCII 2009, LNCS 5610*, pp214-223.

Card, S. K., Moran, T. P. and Newell, A., 1983, *The Psychology of Human-Computer Interaction*, Lawrence Erlbaum Associates, Hillsdale.

Constantine, L. L., 2004, Beyond User-Centered Design and User Experience: Designing for User Performance, *Cutter IT Journal*, 17(2), pp16-25.

Fukuzumi, S. , Ikegami, T. and Okada, H., 2009, Development of Quantitative Usability Evaluation Method, *J. A. Jacko (Ed.): Human-Computer Interaction, Part 1, HCII 2009, LNCS 5610*, pp252-258.

Ham, D. H, Heo, J., Fossick, P., Wong, W., Park, S. H., Song, C. and Bradley, M., 2007, Model-based approaches to quantifying the usability of mobile phones, Jack, *J. A. (Ed.) HCII 2007, LNCS, Vol. 4551*, pp288-297.

ISO/IEC TR25060: Software engineering – Software product Quality Requirements and Evaluation (SQuaRE) –Common Industry Format (CIF) for Usability –General Framework for Usability –related Information, 2010, *International Organization for Standardization*.

ISO/IEC 25062: Software engineering – Software product Quality Requirements and Evaluation (SQuaRE) –Common Industry Format (CIF) for usability test report, 2005, *International Organization for Standardization*.

ISO 9241-210, Ergonomics of human-system interaction –Part 210: Human-centred design for interactive systems,2010, *International Organization for Standardization.*

ISO 13407 (Withdraw), Human-centred design processes for interactive systems, 1999, *International Organization for Standardization.*

ISO 9241-11, Ergonomic Requirements for Office Work with Visual Display Terminals (VDTs) - Part 11: Guidance on Usability, 1998, *International Organization for Standardization*

ISO/IEC 25000, Software engineering – Software product Quality Requirements and Evaluation (SQuaRE) -Guide to SQuaRE, 2005, *International Organization for Standardization.*

ISO/IEC 25030, Software engineering – Software product Quality Requirements and Evaluation (SQuaRE) -Quality requirements, 2007, *International Organization for Standardization.*

John, B. E., Prevas, K., Salvucci, D. D. and Koedinger, K., 2004, Predictive Human Performance Modeling Made Easy, *Proceedings of the SIGCHI Conference on Human Factors in Computing Systems, CHI 2004*, pp455-462, ACM, New York.

Johnson, B. and Shneiderman, B., 1991, Tree-Maps: A Space-Filling Approach to the Visualization of Hierarchical Information Structures, *Proceedings of the 2nd Conference on Visualization*, pp284-291.

Lutsch, C., 2011, ISO Usability Standards and Enterprise Software: A Management Perspective, *A. Marcus (Ed.): Design, User Experience, and Usability, Pt 1, HCII 2011, LNCS 6769*, pp154-161.

Nielsen, J., 1993, Usability Engineering, *Academic Press*

Pirolli, P. and Card, S., 1995, Information Foraging in Information Access Environments, *Proceedings of the SIGCHI Conference on Human Factors in Computing Systems (CHI 1995)*, pp51-58, ACM Press/ Addison – Wesley Publishing Co., New York.

Ravden, S. and Johnson, G., 1989, Evaluating Usability of Human-Computer Interfaces: *A Practical Method*, Prentice Hall

Smith, S. L. and Mosier, J. N., 1984, A Design Evaluation Checklist for User-System Interface Software, *Technical Repoert ESD-TR-84-358*, MITRE, MA.

Theofanos, M. F. and Stanton, B. C, 2011, Usability Standards across the Development Lifecycle, *M. Kurosu (Ed.): Human Centered Design, HCII 2011, LNCS 6776*, pp.130-137.

Theofanos, M.F., Stanton, B. C., Bevan, N., 2006, A practical guide to the CIF: usability measurements, *Interactions*, pp34-37.

Vredenburg, K., 2003, Building ease of use into the IBM user experience, *IBM Systems Journal*, Vol. 42, No. 4, pp517-531.

CHAPTER 12

A Study on the Interaction Styles of an Augmented Reality Game for Active Learning with a Folk Festival Book

Cheng-Wei Chiang
Tatung University
Taipei College of Maritime Technology
Taipei, Taiwan, ROC
mailtowaylan@gmail.com

Li-Chieh Chen
Tatung University
Taipei, Taiwan, ROC
lcchen@ttu.edu.tw

ABSTRACT

In order to increase the learning interest and assistance in active learning with a folk festival book, the objective of this research was to develop a tablet game based on the technology of Augmented Reality (AR). The authors first conducted a field study through observing the college students majoring in design. In addition, after interviewing some students and department instructors, the requirements for designing the active learning game were obtained. Two prototype games with different interaction styles were developed for comparative studied. The purpose of these games was to learn festival stories and important elements in the book about Lantern Festival. Based on the result of experiments, the participants appreciated the AR Interactions approach. The learning of the curiosity and the folk festival knowledge did improved dramatically, and users expected to have reviewing and sharing functions on the tablet in order to learn it extensively.

Keywords: E-Learning, Augmented Reality, Active Learning

1   INTRODUCTION

Reading is an important part of human civilization both in education and heritage. Traditional reading by books enables multi-sensory experiences including expression (text narrative) and vision (seeing the picture). However, comprehensive reading may be difficult while knowing limited vocabulary and organizing content is definitely challenging to most people. This situation leads readers to misunderstanding the content and also makes people frustrate. Frustrating thinking reduces the learning ability and enjoyment of reading. One solution is to improve the reading skill by better understanding the whole content through storytelling, such as the folk festival book. Another solution is to apply information technologies that allow people to actively get the transmission of the intended content in the digital world. In order to enhance and increase culture reading and learning, the objective of this research is to develop a tablet game based on the technology of Augmented Reality (AR). With such a system, the content can be offered by animation and games. Readers can read less text and experience the rebuilt festival stories when they have difficult time to read and need more information update.

2   LITERATURE REVIEW

Recently, the applications of using AR technology for education were increasing, such as teaching children the knowledge of animals (Juan et al., 2008; Hornecker and Dunser, 2009), or demonstrating the interface for storytelling in AR system (Zhou et al., 2004). Moreover, AR technology had been applied in many devices successfully. Diinser and Hornecker (2007) created and manipulated virtual 3D objects in a real-world environment that enhanced books with interactive visualizations, animations, 3D graphics and simulations. Zhou et al. (2004) applied AR into storytelling in 3D with multimedia supports (human voice, sound, and music). Kaufmann and Meyer (2008) presented an AR application for mechanics education. It utilized a recent physics engine developed for the PC gaming market to simulate physical experiments in the domain of mechanics in real time. Woods et al. (2004) had increased the viability of applying AR to educational exhibits for use in science centers, museums, libraries, and other education centers. Henrysson et al. (2005) constructed a system in which 3D contents could be manipulated using both the movement of a camera tracked mobile phone and a traditional button interface as input for transformation. Although many AR edutainment applications had been proposed in the literature, the studies of interaction styles with AR contents on tablet computers were lacking. In order to increase the learning interest and assistance in tablet learning through AR technologies, the issues relevant to active learning deserved in-depth studies.

3 FIELD STUDY AND OBSERVATION

The authors first conducted a field study through observing the students' behaviors in a reading situation, i.e., reading a Lantern Festival book. Moreover, the authors interviewed some students and instructors to elicit the requirements and ideas of designing the active learning game. Lantern Festival holiday is one of the famous holidays in Chinese tradition. It is famous for light sky lanterns, a huge display of decorative lanterns, lantern riddle party, glutinous rice ball, and families united in the joyful atmosphere. Lantern Festival activities contain the origin of the historical allusions, religious worship, diet habit, stories, and sages, and these make designers understand more culture elements and have more abilities to do cultural and creative design.

4 SYSTEM DESIGN CONCEPT

Based on the results of observation and interview, the user interfaces were built with paper prototypes that simulated a tablet computer with a touch-screen. Students could use the tablet to capture the image of tags located near the pictures inside the Lantern Festival book. The media introducing the important element of Lantern Festival would show up to provide augmented information and 3D models on the tablet screen. In the media, animated characters and culture elements in the folktales were re-built to deliver the story of the ancient. These contents help students to understand the history and traditional culture of Chinese Folk Festival holidays. The user interface of the experimental AR game was constructed based on paper prototypes. The models of characters and culture elements were first created using Maya and Illustrator, and then converted to the format appropriate for demonstration.

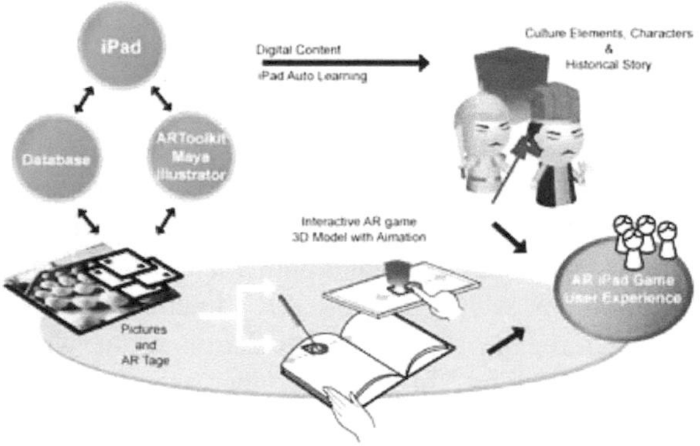

Figure 1 System Diagram

Furthermore, two prototype games with different interaction styles were developed for comparative studies. The purpose of these games was to learn the festival stories and important elements in the book about Lantern Festival. In the first prototype, named Simple Interaction Styles, multiple tags were placed on the pictures inside the Lantern Festival book. When the tags and the pictures were within the range of the tablet camera, the users had to touch the digital contents to trigger the animation of a story. The animation would stop at some points waiting for user inputs. Then users needed to drag the important elements of Lantern Festival on the touch screen, such as picking up a representative elements among multiple choices related to the main character. When the choice was correct, the system delivered the rest of the animation.

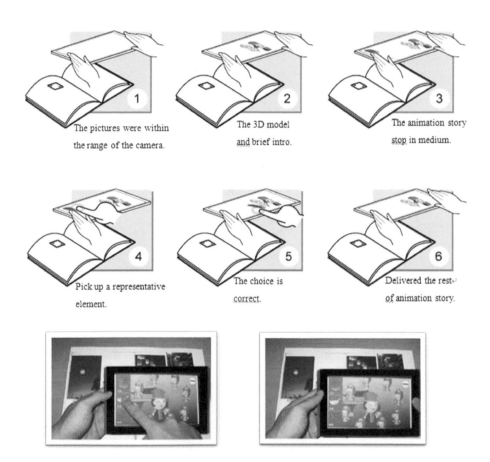

Figure 2 Simple Interaction Styles

In the second prototype, named Intensive Interaction Styles, the animation of a story would automatically start to play while the pictures were within the range of the tablet camera. Similarly, the animation would stop at some points waiting for user inputs. In order to make animation continued, the user needed to search inside the book and detect the picture, which is a representative element related to the main character. When the screen showed the 3D model representing important elements of Lantern Festival, the user needed to move the book or the tablet, so that the 3D model could get close to the main character. If the choice was correct, the system delivered the rest of animation.

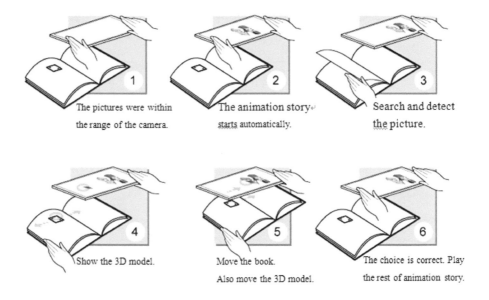

Figure 3 Intensive Interaction Styles

5 RESULTS AND DISCUSSION

During the period of experiment, ten students participated in the game. They were asked to complete two sessions with different prototypes to experience different interaction styles. In order to avoid the learning effect, half of them started with the first prototype, and the others started with the second one. During the experiment, the participants were asked to search for pictures within the Lantern Festival book and watch as many stories as possible. Their scores were counted based on the number of correct match between characters and representative elements as well as the key events they could remember after they watched the stories and play the games.

Through the analysis of post-session questionnaire, 60% of the participants preferred the first prototype. Although searching and detecting the picture were more like treasure hunting, most participants preferred the experience of focusing on watching the animation stories and learning Lantern Festival important elements. In addition, keeping the relative picture and animation story at the same pages that will help the users to enhance their organization and construction of the contents. The detailed comments and evaluations of the prototypes were shown in Tables 1 and 2.

Table 1 The Advantages and Disadvantages of Two Interaction Styles

Styles	Advantages	Disadvantages
Simple Interaction Styles	All digital contents display at the same time. Full cover screen, it makes the characters and the important elements easy to control. Users could really pay attentions on the interactive game and learning the knowledge. It's easy to find out the Lantern Festival elements in order to play the game.	The process is lack of challenging from the view point of playing game. The process makes users feel that the game separates from the book.
Intensive Interaction Styles	The process of searching and detecting the pictures and tags is interesting. The process allows users to view and learn Lantern Festival important elements as many as possible while playing the game.	It is difficult to move the book and hold the tablet at the same time by one person. The user needs adjust the angle of the tablet carefully in order to show up the 3D Lantern Festival elements. The background is contained by too many graphics. It's not

easy to search and detect the
Lantern Festival elements in
order to play the game.

Table 2 Evaluations of Two Interaction Styles

Criteria	Simple	Intensive	Significant differences
Easy Understand The Stories	5.60(0.89)	6.20(0.44)	
Easy to Move The Characters and Elements	6.20(0.83)	6.20(0.83)	
Easy Find Out The Lantern Festival Elements	6.40(0.89)	5.60(1.51)	*
Pay Attention on Learning	6.00(1.00)	6.20(0.83)	
Willing to Explore Other Stories	5.80(1.30)	6.40(0.54)	*
Save Time on Reading	5.80(1.30)	6.20(0.83)	*
Operate The System without Thinking	6.40(0.54)	6.60(0.54)	
The Process and Operation Is Very Smooth	6.40(0.54)	6.00(0.54)	
Like to Play More Games	5.80(1.09)	4.60(0.89)	*
Encourage to Learn	5.80(1.09)	5.00(1.41)	*
Know The Knowledge profoundly	6.20(0.83)	6.00(0.00)	*

* Significant differences

6   CONCLUSION AND RECOMMENDATION FOR FUTURE RESEARCH

In this study, two prototype games with different interactive styles were constructed. Overall, the participants appreciated the AR approach. The knowledge to the scenes, learning ability, festival stories and important elements about Lantern Festival did improved dramatically. These results encouraged the research team to develop more interactive games and systems of active learning about Chinese traditional festivals around our environments.

Nowadays, although e-books, consisting of text, images, video and animation, are a current tendency, unfortunately it still needs good reading skills and patience to read. Applying AR games with books is interesting, it provides with updated contents for books, attracts reader attention by multimedia, helps people to read in different options they want, and reduces the content to read. Moreover, this can apply in many different ways in our daily life. Advertisements, restaurant menus, or museum brief introduction, for examples, can also use AR interaction in order to obtain more interactive information. Therefore, in this research, there are some practical or research issues deserve further studies. First, reading depends on short term memory. Therefore, reviewing knowledge and notes about festivals are very important. Users should not only get multiple festivals info from AR, but also be able to take notes inside the contents in order to review the notes and enhance the knowledge they have learned around their environment at anytime.

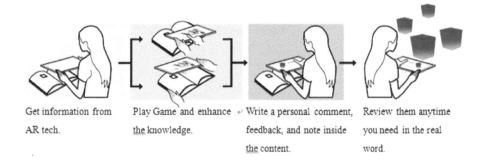

Get information from AR tech.

Play Game and enhance the knowledge.

Write a personal comment, feedback, and note inside the content.

Review them anytime you need in the real word.

Figure 4 Auto Learning process

Second, sharing notes and information can encourage people to accumulate festival knowledge and get different aspects. Therefore, collecting the notes and connecting to the network through Cloud Technology will help people to learn it more efficiently. When people read the festival book by AR, the system will deliver the notes and new information from different users, who are actively reading the same book and offering the notes in some subjects but at different places. This future system should be able to share festival knowledge among different users and to learn it extensively.

112

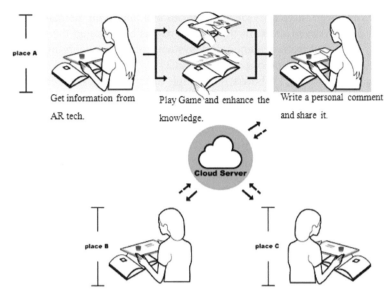

Get information from
AR tech.

Play Game and enhance the
knowledge.

Write a personal comment
and share it.

Read the different notes and share them with another people.

It makes learning extensively.

Figure 5 learning Extensively

REFERENCES

Diinser, A. and Hornecker, E. 2007. *Lessons from an AR Book study TEI'07*, 15-17 Fed 2007, Baton Rouge, LA, USA.

Hornecker, E. and Dunser, A. 2009. Of pages and paddles: children's expectations and mistaken interactions with physical-digital tools. *Interacting with Computers*, 21, 95-107.

Henrysson, A., Ollila, M., and Billinghurst, M. 2005. Mobile phone based AR Scene assembly. *Proceeding of the 4th international conference on mobile and Ubiquitous multimedia, Christchurch*, New Zealand, 95-102.

Juan, C., Canu, R., and Gimenez, M. 2008. Augmented reality interactive storytelling systems using tangible cubes for edutainment. *Eighth IEEE International Conference on Advanced Learning Technology,* ICALT 08, 1-5 July 2008, 233 – 235.

Kaufmann, H. and Meyer, B. 2008. Simulating Educational Physical Experiment in Augmented Reality. *Siggraph Asia 2008,* Singapore, December 10-13, 2008.

Vatavu, R. 2010. Presence bubbles: supporting and enhancing humam-humam interaction with ambient media. *Springer Science + business Media, IIC 2010.*

Wang, K., Chen, L., and Chu, P. 2009. A study on the Interaction styles of an augmented Reality Game for mobile learning in a Heritage temple.

VMR '09 Proceedings of the 3rd International Conference on Virtual and Mixed Reality: Held as Part of HCI International 2009.

Woods, E., Billinghurst M., and Aldridge, G. 2004. *Augmenting the Science Centre and Museum Experience*. by the Association for Computing Machinery, inc.

Zhou, Z., Cheok, A., and Pan, J. 2004. Magic Story Cube: an Interactive Tangible Interface or Storytelling. *TEI'07*, February 15-17, 2007, Baton Rouge, Louisiana, USA.

CHAPTER 13

UX Embodying and Systematizing Method to Improve User Experience in System Development
-Applying to Planning and Proposal Phase

Ryosuke Okubo, Yukiko Tanikawa, Shinichi Fukuzumi

NEC Corporation
Tokyo, Japan
r-ookubo@cq.jp.nec.com

ABSTRACT

We developed a user experience (UX) embodying and systematizing method to improve customer satisfaction and efficiency in system developments. This method embodies customers' demands concerning UX (UX demands) as concrete operable human interfaces (HI) and systematizes the HI. This time, targeting the first phase (planning and proposal), we systematized HI on the basis of look-and-feel and operability, enabling customers to experience the image of the completed system in advance. We evaluated the practical validity of this method by interviewing engineers.

Keywords: user interface, human interface, screen layouts, prototyping, user experience, usability

1   INTRODUCTION

Recently, in system developments, customers have become able to emphasize usability and comfort more than ever (Hassenzahla, and Tractinskyb. 2006). In terms of technical aspects, web applications called Rich Internet Applications (RIAs) have become common. To use these technologies, we can realize advanced operability and expression for any application. (Farrell, and Nezlek. 2007). Accordingly, although before HI requirements were so simple that they could be specified as functions, they now involve demands about abstract "user experience" (UX) such as users' feelings and thoughts toward systems. This makes it hard for customers to communicate their demands to developers and for them to agree on the customers' exact demands and requirements (Hellman, and Rönkkö. 2008). This risks a tricky-to-use or uncomfortable system being built and back tracking occurring in the later phases of developments. On the other hand, customers' demands (for example, to consider usability more, to realize advanced operability and expression) increase the development cost.

To solve these problems, we take an approach to provide samples of screen layouts and their components to meet the customers' demands and to decrease the development cost. The samples are made to be actually operable so that customers can experience an operational feeling of systems in advance. This enables customers to depict their UX demands beyond descriptions and illustrations and to share their demands with developers. Moreover, because the samples are made to be usable for design and implementation of system development, developers can develop systems more efficiently by using them.

On this approach to provide samples of HI, we began to develop a UX embodying and systematizing method. First, this method embodies customers' abstract UX demands as concrete HI, such as the screen and its components. Next, it organizes the HI according to development proceeding, for example, to share customers' demands with developers and to make developments efficient.

This time, In order to build this method targeting the first phase of system development, we systematized HI on the basis of look-and-feel and operability to provide these samples of HI. This paper purposes to show effectiveness this method using look-and-feel and operability.

2 UX EMBODYING AND SYSTEMATIZING METHOD

The UX embodying and systematizing method embodies customers' UX demands as concrete operable HI and systematizes the HI according to development proceeding. Its intended effects are as follows.

-To share true customers' UX demands with developers certainly by enabling customers to experience an operational feeling and look-and-feel of systems in advance.

-To realize the customers' demanding UX efficiently by providing developers bases of design and implementation

We built screen design patterns and sent them to development sections in 2009, prior to the development of the UX embodying and systematizing method. These patterns are HI catalogs organizing screen layouts and their components by their functions and forms. Then, interviews with the developers who used the screen design patterns confirmed that the method makes developments to realize customers' demands more efficient, which was the second of our two aims. On the other hand, they found that the first, to share the customers' demands, was strongly needed but insufficiently achieved. Moreover, the interviewees said that the contents that customers could experience were needed in addition to the contents for design and implementation, so as to share the customers' demands with developers. They also said that a customers' UX demand has different concreteness according to development phases, such as proposal and designing.

To achieve both aims, we suppose that we need to provide not only HI parts for implementation but also visible and operable HI embodying customers' UX demands. Consequently, we propose the idea of the UX embodying and systematizing method shown in Figure 1.

The proposed method embodies demands and systematizes HI. Demands are embodied by what customers (including end-users) need when using systems. We define HI to fulfill what UX customers demand, such as operation and expression on systems. The defined HI are, for example, screen transitions, screen layouts, screen's components, and look-and-feel. HI are systematized by what system developers need when designing and implementing. We define the axes to classify HI, for example, platforms, targeted users, data scale, and devices. These axes are defined for each development phase, such as proposal and designing. On the basis of these axes, we build the defined HI as visible and operable contents to systematize HI.

As above, the proposed method has contents according to development phases, because abstract customers' UX demands become concrete as development progressed. Figure 2 shows support of the proposed method organized by these system development phases.

In proposal and planning, customers want to communicate with developers their abstract demands, that greatly affect customers' impressions about UX and costs, which correspond to the most important HI requirements to agree for both customers and developers. Thus, it is important to share these requirements as visible and operable HI between customers and developers (Moscove. 2001). To support this, the UX embodying and systematizing method provides screens that customers can actually operate. Customers can pass on demands beyond descriptions by trying these screens and saying "I want this interaction" or "I want this look-and-feel".

In requirement definition and external design, customers want to confirm the image of the completed system reflected in customers' UX demands. This image becomes specific HI requirements for developers to begin concrete design. Thus, in these phases, it is important for developers to make prototypes that are more specific than those in the previous phase by including the features of customers' work. To support this, the UX embodying and systematizing method provides easy-to-customize screens and their components to make prototypes.

Moreover, in external design and internal design, customers want to obtain the completed system in which their demanding UX are realized early. Thus, it is important for developers to design and implements consistent screens on the basis of requirements on which they agreed in previous phases. To support this, the UX embodying and systematizing method provides source codes to be utilized on the basis of design and implementation. To do this, we put the previously built screen

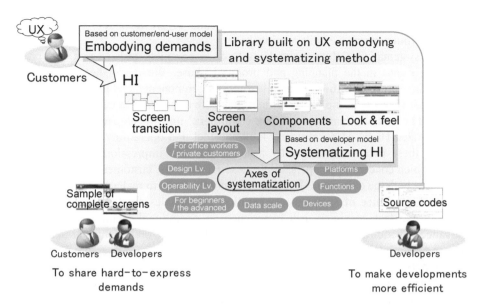

Figure 1 Concepts of UX Embodying and Systematizing Method

	1. Proposing and planning	2. Requirement definition	3. External design	4 .Internal design	Coding, test, maintenance
Purposes of a method	To transmit cutomers' important UX demands influencing in costs..	To confirm the image of the completed system reflected in demands.	To obtain the completed system realizing demands early.		
Support by a method	To share customers' outline images with developers using operable screens.	To make prototypes efficiently and to derive specific requirements to achieve the prototypes.	To make usable and consistent screens efficiently using software templates and guides corresponding the requirements .		

Figure 2 Purposes of UX Embodying and Systematizing Method sorted by system development phases.

design patterns on the systematization and contents in this phase of the proposed method (Okubo, et al. 2010).

As above, we are developing the UX embodying and systematizing method so as to make customers' UX demands more concrete as development progressed.

3   FEATURES OF SYSTEMATIZATION AND CONTENTS IN PLANNING AND PROPOSAL PHASE

We systematized HI and built contents of the systematization targeting the first phase (planning and proposal phase) of the systems development phases we stated in the previous section. The reason is that discrepancies between the UX the customers want and developers give them in this phase can be most risky. Moreover, we found this phase needed the most support for activities of proposal and estimating costs. We learnt this from the results of the interview with users of screen layout patterns we built previously.

We systematized HI to make HI sets do a complete task (e.g. to reserve a meeting room) in the planning and proposal phase. We made these HI sets by integrating screen transitions, screen layouts, its components, and look-and-feel. We call these HI sets HI samples. HI samples have the three features below.

(1) **The systematization based on customers' and developers' interests**: We systematized HI based on systems' look-and-feel and operability (Section 4). We assume these two elements greatly affect customers' impressions about UX and costs, which concern developers. With these HI samples, customers and developers can share both the customers' UX demands and the costs to realize them.

(2) **Experience of systems' operability:** HI samples are made to be actually operable so that customers can experience an operational feeling of systems in advance (Section 5). This enables customers to depict their demands about UX beyond description and illustration and share their demands as operable forms with developers.

(3) **Derivation of requirements:** We link each HI sample to requirements of HI. This enables developers to share customers' demands as clearly stated requirements among the members in the development.

With these three features, both customers and developers can together figure out embodied look-and-feel and interactions on the HI samples. Consequently, this makes it possible for customers' demands to be derived properly and efficiently.

4   SYSTEMATIZATION IN PLANNING AND PROPOSAL PHASE

Figure 3 shows systematization of HI in the planning and proposal phase. First, we classify HI sets into the axes "Proficiency Level in IT (Information Technology) and Tasks". Next, we classify them more finely into the axes "Richness Level in Look-and-feel and Operability". The reason, we suppose, is that proficiency levels in IT and tasks are basic user characteristics deciding requirements, and richness levels in look-and-feel and operability greatly affect customers' impressions and costs. We explain each axis in detail below.

4.1   Proficiency Levels in IT and Tasks

We define proficiency levels in IT as HI sets according to users' IT skills and proficiency levels in works as HI sets according to users' knowledge and experience of targeted tasks. Moreover, we divide each axis in the two levels into stages: "For Beginners" and "For the Advanced". We explain the levels and the stages in detail bellow.

Proficiency Levels in IT: HI sets according to users' IT skills
-For Beginners: Targeted users sometimes browse the web but rarely use e-mail

1. Proficiency Levels in IT and Tasks

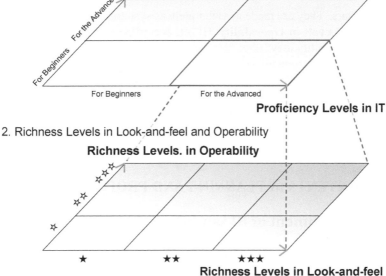

Figure 3 Systematization in planning and proposal phases

or office applications. Provided HI are the simple ones not supposing general usages of GUI (e.g. not using scroll bar).

-For the Advanced: Targeted users usually use e-mail or office applications. Providing HI are the ones supposing general usages of GUI (e.g. using scroll bar, using radio button as alternative selection, etc.).

Proficiency Levels in Tasks: HI sets according to users' knowledge and experience of targeted tasks.

-For Beginners: Targeted users have little knowledge of phases of the object tasks (tasks themselves, task procedures) or experience. Provided HI are the easily understood by looking.

-For the Advanced: Targeted users have sufficient knowledge to process the object tasks (tasks themselves, task procedures) and experience doing the tasks. Provided HI are efficient.

4.2   Richness Levels in Look-and-feel and Operability

We define richness levels in look-and-feel as HI sets according to how elaborate they are and richness levels in operability as HI sets according to how substantial functions to improve operability they have, such as input supports and intuitive interactions. Moreover, we divide each axis in the three levels into stages (One Star, Two Stars, and Three Stars). We explain the levels and the stages in detail bellow.

Richness Levels in Look-and-feel: HI sets according to how elaborate they are.

-One Star: They are made very simply with HTML and a few images.

-Two Stars: They are made somewhat elaborately with images on buttons or shades.

-Three Stars: They are made as elaborately as systems for consumers.

Richness Levels in Operability: HI sets according to how substantial functions to improve operability they have.

-One Star: They are made very simply with HTML and a few scripts.

-Two Stars: Input supports and intuitive interactions are made with standard components provided by RIA.

-Three Stars: Specialized supports for particular purposes are made with handmade developments.

In the systematization, the more stars, the higher the cost.

5   CONTENTS IN PLANNING AND PROPOSAL PHASE

5.1   Development of HI Samples

We developed HI samples as contents of the proposed method in the planning and proposal phase. They are made on the basis of the systematization we explained in Section 4. We made one HI sample in each stage of each axis. Note, however, that there are two exceptions.

-"Richness Levels in Operability – Three Stars" is excluded considering practical frequency, because it has only limited application specialized for particular purposes (e.g. mission critical works, large scale data processing, etc.).

- The combination of "Proficiency Levels in IT – For Beginners" and "Proficiency Levels in Tasks – For the Advanced" is excluded. The reason is that the former targets average citizens or consumers, who are not expected to have knowledge about the object tasks.

-The combination of "Proficiency Levels in IT – For Beginners" and "Richness Levels in Look-and-feel" is excluded, because the systems for citizens or consumers need a somewhat elaborate look-and-feel. The reason is that we must enhance the customers' subjective satisfaction with the systems and encourage customers to use them from the viewpoint of business.

Consequently, we made 16 HI samples.

5.2   Preconditions of HI Samples

We define object systems, object tasks, and screen transitions as preconditions of HI samples as shown in Figures 4. We have two different preconditions for each proficiency level in IT. The reason is that these two types of HI samples have greatly different features. For example, concerning screen transitions, HI samples for beginners have one function in one screen emphasizing understandability. In

Proficiency Levels in IT: For Beginners
Targeted system: institution reservation system
Targeted task: to reserve institutions on kiosk terminals
Screen transitions: search, list, detail (editable)

Proficiency Levels in IT: For the Advanced
Targeted system: order management system
Targeted task: to confirm / renew orders
Screen transitions: search, list, detail (editable), graph

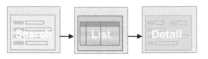

Figure 4 The preconditions for beginners in IT

contrast, HI samples for the advanced have multiple functions in one screen emphasizing efficiency.

5.3   Use of HI Samples

HI samples are made actually operable so that customers can experience operability. Moreover, each HI sample has a description page to explain its features. Each description page has a video that explains these features in more detail. Due to this, HI sample users can actually experience operability of HI samples and developers can easily explain to their customers about features of HI samples with the description pages.

5.4   Derivation of Requirements

Both customers and developers will be able to share customers' demands about UX with embodied look-and-feel and interactions on HI samples. What is more, developers need to obtain demands contained in selected HI samples as clearly stated requirements so that the demands can be properly shared among all the members in the development. we, therefore, established rules to derive the requirements from each HI sample. These rules dictate which HI sample embodies requirements established as items agreed to in the planning and proposal phase.

6   EVALUATION

6.1   Purpose

The aim in the planning and proposal phase is to reach agreement about the HI requirements that greatly affect customers' impressions about UX and costs. We conducted interviews to evaluate the practical effectiveness of the proposed method which systematized HI on the basis of look-and-feel and operability for achieving this aim.

Table 1 The result of the interviews

Good Points
[Problems] **G1** The problems (back tracking occurring in the later phases of developments because of not sharing customers demands properly) are important, and the method (To express customers' demands and to create evidence for the decisions using operable HI samples) is effective (**5 people**)
[Effects] **G2** HI samples enable customers and developer to share demands and make their decision-making easier (it enable decision about desired operability or the quality level of look and feel) (**4 people**) **G3** HI samples improve efficiency of developments (it enables proposals of combinations of standard UI components or creation of screens for proposal documents.) (**3 people**)
[Functions] **G4** Videos on description pages enable comprehension of the features of HI samples (**1 person**)
Shortage Points
[Application and Operation Ways] **S1** To show all the HI samples to customers risks customer beating down prices and imposing impossible demands, because they think that HI samples are as good as a completed system. (**2 people**)
[Untargeted Supports] **S2** HI samples need to be expended (e.g. it is hard to consider HI samples as different types of work system). Present lineup of HI samples is not sufficient to apply to own systems (systems of operational monitoring, information analysis, information reference, and so on)). (**6 people**) **S3** HI samples to decide a color scheme are demanded. (**1 person**)
[Functions] **S4** A function is demanded that shows the grounds on which to select the HI sample. (**1 person**)

6.2   Method

We then interviewed a total of 20 people five times about the targeted problems, effective/deficient points, scenes needed to help, and so on. These interviews were conducted as conversations without preplanned questions in order to derive potential needs and demands.

6.3   Result

First, we broadly classified the comments obtained from the interviews into good points and shortage points. Next, when possible, we classified them more finely and labeled them. The results of the interviews are in Table 1.

6.4   Discussion

The comments about the good point are classified into three groups: "the problems" of the study (G1), "the effects" of the method (G2, G3), and "the functions" of the method (G4). The results concerning the problems show the importance of targeted problems and the support for the fundamental concept operable HI samples enable decision about customers' UX demands (five in twenty comment) Moreover, the results concerning the effects show that this method has the intended effects (e.g. to share customers' demands and to realize the UX efficiently) (seven in twenty comment). These results show that HI samples this method systematizes on the basis of look-and-feel and operability have efficacy for the important problems to reach agreement about the HI requirements.

The comments about shortage points are also classified into three groups: "the application and operation ways" (S1), "the untargeted support" for the phases excluding planning and proposal phase (S2, S3), and "the functions" of the method (S4). Especially, the results concerning untargeted support show that HI samples strongly need to be customizable (e.g. to accord to customers' works and to have color and look-and-feel variation) (seven in twenty comment). For this reason, we consider that customers' demanding UX become more concrete by provided these HI samples to customers.

Considering these needs, we will develop the UX embodying and systematizing method in the next planning and proposal phase from now on in order to support making prototypes embodying more concrete customers' demands and to support realizing the demands efficiently in designs and implementations.

7 CONCLUSION

We developed a UX embodying and systematizing method targeting the first phase (planning and proposing). This method embodies customers' UX demands as concrete operable HI and systematizes the HI according to development proceeding.

In this paper, we showed that this method which systematized HI on the basis of look-and-feel and operability enabled to share customers' UX demands with developers and improve efficiency of developments.

After targeting the planning and proposing phase this time, we will extend this method to the requirement definition phase and external design phase. Therefore, we aim to enable customers to experience systems that are more specific in advance and developers to realize the experiences more efficiently.

REFERENCES

Marc Hassenzahla and Noam Tractinskyb. 2006. User experience - a research agenda. *Behaviour & Information Technology Volume 25 Issue 2*: 91–97.

Farrell, J.and Nezlek, G.S.. 2007. Rich Internet Applications The Next Stage of Application Development. *Information Technology Interfaces 29*: 413–418.

Mats Hellman and Kari Rönkkö. 2008. Is User Experience supported effectively in existing software development processes?. *Valid Useful User Experience Measurement*: 32-37.

Stephen A. Moscove. 2001. Prototyping: An Alternative Approach To Systems Devleopment Work. *Review Of Business Information Systems Vol. 5, No. 3*: 65-72.

Ryosuke Okubo, et al. 2010. Systematization of screen layouts and elements ~Building of Screen-Design-Pattern. *Human Interface Symposium*: 1333. (in Japanese)

CHAPTER 14

Ergonomics and Kaizen as Strategies for Competitiveness: Theoretical and Practical Research in an Automotive Industry

VIEIRA Leandro, BALBINOTTI Giles, GONTIJO Leila.

Universidade Tecnológica Federal do Paraná - UTFPR
Curitiba, BRAZIL
l_vaz_vieira@hotmail.com

ABSTRACT

With increased international competitiveness in the automotive industry, came the concern of the companies save costs and lower production costs. For this purpose many ways are designed to reduce costs and waste of raw materials and reduce activities that do not aggregate value to manufacturing processes. In the early XVII appears the manufacturing system, which processes were hard with little concern for the health and safety of employees and conditions of the workplace. After the advent of the production system called lean manufacturing, a new paradigm in terms of production system capable of providing high levels of productivity and quality. It is based on waste elimination that occur during the production process. After began a new way of thinking, creating a culture of continuous improvement and lean process with no waste and reducing costs, without neglecting the welfare worker and improving the conditions of their work environment. This paper presents a reflection on the application of ergonomics in a lean production system of an automotive industry, using methodology based on the Kaizen (Continuous Improvement) to gain performance and improving the conditions of the workplace, also will be presented with positive and negative points in using this methodology in relation to ergonomics. The research will be conducted by collecting data 'in loco' and interviews with workers. Some studies show that in companies that are lean system and using the methodology of Kaizen, the results of product quality, levels of absenteeism and accidents are better than those obtained in companies that do not apply the same concept.

Keywords: Working condition, Lean Manufacturing, Performance, Kaizen Methodology

1. INTRODUCTION

Lean production is the third revolution of the automobile in order to produce vehicles. As said Womack, Jones and Roos (1990), lean production represents a new paradigm in terms of production system capable of providing high levels of productivity and quality. It is based on waste elimination that occur during the production process. After emergence of the system of Henry Ford , the volume per vehicle has risen sharply to 2 million units a year the Model T, but the departure of virtually all producers craft market did drop the variety of products from thousands to tens of offers.

Lean production began in Japan, as he comments Womack et ALL (1990), it originated with the Japanese engineer Eiji Toyoda, he left for a three-month study by the Ford Rouge plant in Detroit, after studying carefully the system of factory production, the largest and most efficient manufacturing complex in the world, after much analysis and studies he came to a conclusion that mass production would never work in Japan "In this early experiment was born what Toyota came to call Toyota Production System, and finally lean production". Came with the system of analysis methodologies and improvement of works, among which we highlight the Kaizen, a tool for continuous improvement system that covers all the needs of those involved in a production process.

2. ERGONOMICS AND THE KAIZEN METHODOLOGY

Ergonomics was a great evolution in the systems of mass production and lean because of the race for quality and productivity. According to IEA (2007), ergonomics is a scientific discipline that studies the interactions of men with other elements of the system, making application of theory, principles and design methods with the aim of improving human well-being and overall system performance.

Another important aspect is ergonomics as Balbinotti (2003) is that it seeks not only to prevent workers in jobs stressful and/or dangerous, but seeks to put them in the best possible working conditions to avoid accidental injury or fatigue excessive and improve performance.

The relationship of ergonomics in lean production can be observed in Figure 1 that the rate of absenteeism of Japanese companies is lower than the European and

North American, it is arguable that there was action for this reduction. Within this line as Womack et al (1990) summarizes several indicators as well as yield and quality of the current performance, the assembly activity of the large producers. It's amazing the difference between the average performance of Japanese and Americans and Europeans, the size of areas needed repair, the percentage of workers in teams, suggestions, and the amount of training given to new workers in the assembly.

	JAPONESAS NO JAPÃO	JAPONESAS NA A. NORTE	NORTE-AMERICANAS NA A. NORTE	TODA EUROPA
Desempenho:				
Produtividade (horas/veíc.)	16,8	21,2	25,1	36,2
Qualidade (defeitos de montagem/100 v.)	60,0	65,0	82,3	97,0
Layout				
Espaço (m²/v./ano)	0,53	0,85	0,72	0,72
Área de Reparos (% do espaço de montagem)	4,1	4,9	12,9	14,4
Estoques (dias para amostragem de 8 peças)	0,2	1,6	2,9	2,0
Força de Trabalho:				
% da F.T. em Equipes	69,3	71,3	17,3	0,6
Rotação de Tarefas (0 = nenhuma, 4 = freq.)	3,0	2,7	0,9	1,9
Sugestões por Empregado	61,6	1,4	0,4	0,4
Nº de Classificações no Trabalho	11,9	8,7	67,1	14,6
Treinamento de Novos Trabalhadores (horas)	380,3	370,0	46,4	173,3
Absentismo	5,0	4,8	11,7	12,1
Automação:				
Soldagem (% passos diretos)	86,2	85,0	76,2	76,6
Pintura (% passos diretos)	54,6	40,7	33,6	38,2
Montagem (% passos diretos)	1,7	1,1	1,2	3,1

Fonte: Pesquisa Mundial das Montadoras do IMVP, 1989, e J. D. Power Pesquisa Inicial de Qualidade, 1989.

Figure 1: Characteristics of Japanese automakers, North American and Europe – 1989 (From: The Machine That Changed the World.)

Two important comparisons between the systems and lean mass is what the authors said Womack, et al (1990), in the old mass-production factories, managers were hiding information about the condition of the factory, because they have such knowledge to the key its power. In a lean factory as Takaoka, all information - daily production targets, cars built to date, equipment breakdowns, personnel shortages, overtime requirements that are displayed in frames andon (electronic boards bright) visible in all seasons the factory.

No doubt it is important to analyze the lean system has resulted in a great company it is important to adopt an ergonomic program, they complement each other. According Balbinotti, (2003), the dissatisfaction of people at work, often neglected or unknown, arising from a mismatch between the content of an ergonomic work to men.

The methodology and the bases of a lean production systems, according to Martins et al (2006), the term kaizen is formed from KAI, which means changing, and ZEN, which stands for the better. Kaizen has expanded to an organizational philosophy and behavior, a culture focused on continuous improvement focusing on eliminating waste in all systems in an organization and involves application of two elements in the improvement, understood as a change for the better and continuity understood as acts as a permanent change. Thus, there should be a single day without some improvement in the company.

The Kaizen philosophy is the key to success of organizations to ensure competitiveness, as defined Masaki Imai (1994), "Kaizen, the Key to Success", ie continuous improvement in their personal, domestic, social and professional. When applied to work or say, the improvement that involves everyone. KAIZEN, business strategy involves everyone in an organization working to make improvements with low or no investment. with KAIZEN, an involved leadership guides people to improve the ability to meet expectations continuously high quality and delivery time.

Another important aspect that says Martins et al (2006), kaizen management philosophy can be applied in specific parts of the targeted organization, such as Kaizen project: to develop new concepts for new products, Kaizen planning: developing a planning system for both production to finance or marketing, manufacturing and Kaizen: developing actions that aim to eliminate waste in the factory-floor and improve the comfort and safety.

The organization should be to create a culture of continuous improvement, but without neglecting the welfare and quality of life of the employee, as defined Martins et al (2006), Kaizen is a management philosophy as it covers the continuing need for managers, workers in all aspects of life.

According to Matthew (2007) Kaizen aims to develop curiosity and creativity of people and direct them to the process of adding value to customers. Kaizen is not an attempt to light a fire under people, Kaizen turns the light on inside people. Know that the bottom Kaizen is about people. People who are not businesses innovate. You must change attitudes for Kaizen to work, which requires a great commitment and a long time, and much study.

Another important aspect that says Matthew (2007) when we improve a little each day, with time great things happen. When you improve conditioning a little each day, with time we will have great results in terms of conditioning. Not tomorrow or after tomorrow, but over time, we get a huge profit. Do not try to improve the lot overnight. Stick to the small daily development. It's the only way it works and when it happens, it's durable, it is necessary to Kaizen become imperative.

In the methodology of kaizen can not forget the concept value-added productivity, and informs Balbinotti (2003), in a company that seeks to produce more and better with less, always increase the effectiveness (purpose) and efficiency (means) should be concerned as quality planning (setting new standards) and the maintenance of quality (ensuring compliance with the standards) with the quality improvement (continuous improvement). This means that increasing quality and reducing costs increases the value, through the concept of total quality, which means satisfaction for all.

The organizations work with people that influence productivity and can increase the value of the organization, according Balbinotti (2003), people influencing productivity, productivity change, productivity depends on the performance of people. The performance by changing the productivity of people, puts us in direct contact with the ergonomic issue is evident and the contribution of ergonomics in this context, since the ergonomics seeking better working conditions, so that work can be developed without the reduction of health of workers and therefore with lower rates of absenteeism and turnover, and this contributes to reduced productivity.

3. METHODOLOGY

The research presented was applied in an auto factory in Paraná, Brazil, the company works with the lean production system based on Toyota production system, the system is being used for almost 10 years. Ergonomics is part of this system to obtain results, and since the implementation of the company achieved many improvements in working conditions, as will be presented in the discussion of results. The foundations of this system are:

5'S: the application of the 5's of Japanese origin (Seiri, Seiton, Seiso, Seiketsu and Shitsuke) Apply the 5's will reduce waste, and jobs organized will reduce the offsets, improving safety, improving motivation of teams with a pleasant working environment, with the 5'S is possible to improve equipment performance.

Dexterity: is learning the operation of the workplace through training. The field of Dexterity allows the repeatability of gestures, which reduces the dispersion of implementation and the risks of non-quality, skill favors optimizing operations, improving the fluidity of movement. The relationship with the dexterity of ergonomics is the teaching of correct postures through training applied to employees.

Standardization: Standardization is the default operation being the best method of producing at the moment but there is no reason why there is no pattern change. Implementing the standard in the workplace there is the Standard Operating Sheet. The ergonomics is linked to standardization through the development of operational procedures that take into account the know-how and experience.

Ergonomics: the production system, ergonomics is the basis, along with standardization, dexterity and 5's. They are all interlinked so that you have a good working condition for the developer. The goal of ergonomics in this production system is to ensure the adequacy of the operator (human capabilities) and the jobs or job offers. Improve performance while preserving the Health Delete musculoskeletal disorders (TMS) related to work and improve the conditions under Labor.

Kaizen: This is the subsystem that has a greater connection with the ergonomics, as with continuous improvement or Kaizen ergonomics tends to evolve in the enterprise, making it an improvement in jobs and in most cases to improve the working conditions of developer, who on the increase employee satisfaction and company productivity, obtaining higher results. The subsystem Kaizen is a method based on the cycle SDCA (Standardize, Do, Check and Act) detailing for better understanding, S (Standard) means establishing the best standard for the operation time, D (Do) application form operations effective, C (Check) to observe the operations, find problems, improve posture, improve processes, A (Action) found after the improvements should act.

This is the subsystem that has a greater connection with the ergonomics, as with continuous improvement or Kaizen ergonomics tends to evolve the company, making it an improvement in jobs and in most cases to improve the working conditions of the employee, who forth to increase employee satisfaction and company productivity, achieving better results. The subsystem Kaizen is a method based on the cycle SDCA (Standardize, Do, Check and Act) detailing for the best understanding, the S (Standard) means establishing the best standard for the time of operation, D (Do) application form operations effective, C (Check) to observe the operations, find problems, improve posture, improve processes, A (Action) found the following improvements should act. There is an important relationship between the PDCA cycle (Plan, Do, Check and Act) and SDCA cycle, as will be shown in Figure 2, SDCA cycle, the method is applied in a stable process for small changes and is PDCA cycle is applied during a improve and / or change. According Balbinotti (2003), Kaizen means of production, reduces the physical effort at the time, through the installation of mechanical assistance, for example, ensuring the proper gesture and poise, as well as the correct use of tools, through plans and also skills development for managers, aimed at leadership in the animation teams.

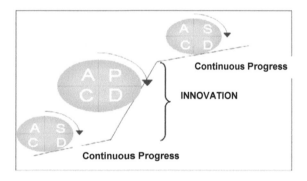

Figure 2 Characteristics of Japanese Car Manufacturers, North American and Europe – 1989 (From: Automotive industry documentation.)

Measure of Time: Everything is time-based manufacturing, the time of the job you should take the cycle time of the production line, and all synchronized so that the customer receives the product on time. Ergonomics is also concerned with time, for improved ergonomics reduces losses of income and therefore of time.

Quality Control: For quality control, or better, quality management, relies on some tools to deal with quality or ergonomic problems, a method is the Qc-story (method of problem solving), the 7 quality tools (Pareto Diagram, Cause and effect diagrams, histograms, check sheets, scatter charts and control charts). Ergonomics has a relationship that through these tools is necessary to address problems with ergonomics.

Performance Management of Resources: serves to avoid problems with equipment, being necessary to perform preventive maintenance. The TPM is aimed at reducing and avoiding any loss of production-related equipment that could break. Ergonomics can follow this performance with the means to identify critical points for a man to avoid accidents.

Just in Time: Is the customer to manufacture the required products in the required time as required. The importance of JIT in a lean production system is to regularly identify and eliminate waste, low inventory.

Guidelines for the Management: is a management system that allows you to Target all efforts and resources on one goal to success for the company. Based on strategic planning, identifying the organization's targets, according to Falconi (1997), is a subsystem of TQM (Total Quality Management) and facing competition not only includes the improvement of existing products and processes, but mainly the innovation represented by new technologies.

The lean production system is based on Japanese tools to reach excellence in everyday life, aims to ensure the quality demanded by customers, reduce costs, produce the required products and be responsible and respect the man.

4. RESULTS

This survey was conducted in an automobile, which has about 6.000 employees and 3.000 employees of partners. To analyze the company's results, we applied a form to identify the knowledge level of the lean production system and what the relationship with the ergonomics program implemented in the company. The form with 12 questions, 3 of 9 closed and opened to identify the views of officials in the relationship between ergonomic factors and work conditions and production system of the organization. This form was administered to 10 employees, that is Manufacture 4, Human Resources 2, Quality Control 1, Logistics 1, Performance 1 and 1 Communications Department.

Most of the survey was answered by the area of production, and even be the largest public company and where the lean system is more powerful. It was observed that there is a difference between the concept of lean production system and system of mass production among company employees, of the 10 employees of eight research think the company is using lean production and two mass production. The Graphic 1 the results in graphic form (closed questions) and feedback sequence divided by area.

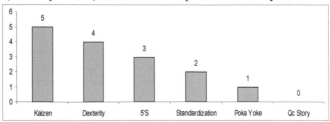

Graphic 1 Question about the degree of importance of the tools that support ergonomics.

We can see through the Graphic 1 that the kaizen, the Knack, and 5's are really the most striking evidence for the relationship with the ergonomics. The following are the main comments from employees.

Viewpoint of the Logistics area: The participation of the supervisor, ergonomist and workplace safety is to identify which points / stations with ergonomic problems and act on improvements to eliminate them (or at least reduce). This, and make technical improvements to the station will also bring the benefit of "trust and credibility" among all parties, also reflecting on productivity and work quality.

Viewpoint of Human Resources: In the better ergonomic design of the workplace (ergonomics in design), or better, are suited to the employee (ergonomics series, correction of problems), the likelihood of having products with best quality is higher, in order to decrease the physical and psychological operator allowing better perform the activity of the job. With regard to what has been predictable as, for example, occupational diseases, more specifically on musculoskeletal disorders, guidelines and actions to ergonomics are essential for improving QWL (quality of work life) and also for reducing the impact on cost for companies facing legal problems.

Viewpoint of the area of Manufacture: The system currently used in all car companies to achieve a high degree of competitiveness and performance, seeking a greater return of profits to the detriment of work in production jobs, to reduce losses and increase productivity with lower labor and possible mainly targeting the most important factor is that the proper treatment of people. Absenteeism has a direct impact with ergonomics, if the operator begins to miss work that can be put in where it performs the operations are no conditions that compromise their posture, causing muscle fatigue and subsequent absence from work.

Viewpoint of the area of Performance: I understand that the lean production system recommends a suitable job for execution of its activity, in addition to 5'S must be ergonomically aligned. When deploying the ergonomics the company will be providing a better quality of life to the contributor (fatigue) will avoid people to depart / problems related to lack of ergonomics at work, avoid future labor actions, with a more favorable job may occur improvement or even elimination of a quality problem, among others.

Viewpoint of the Communications Department: All the work to improve the ergonomics of a workstation must occur in conjunction with all the operators who work in that post. You must create a culture in the companies to conduct regular reviews on all jobs, with the goal of establishing a "preventive" of future problems.

It can be seen from the viewpoints of the employees that the company's ergonomic program, has affected several areas of the company, raising awareness about the health of employees. Another important point that all areas have a preventive target to work on, avoiding the risk of accidents in the workplace. A negative point is the research division of knowledge production system that the company currently works.

5. DISCUSSION AND CONCLUSION

Approach of the case it was found that the company is concerned about the continuous improvement of working conditions for operators, as presented indicators of Human Resources and Quality, listed below:

The company's ergonomic program, the jobs are divided into two types of critical posts and the posts are less critical than with the low level of risk for accidents and muscle problems. The most critical positions where there are medium and/or high likelihood of the employee having trouble muscle. According Balbinotti (2003), ergonomics seeks not only to prevent workers in jobs stressful and/or dangerous, but seeks to put them in the best possible working conditions to avoid accidental injury or excessive fatigue and improve income.

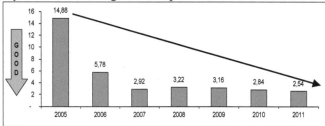

Graphic 2 Absenteeism Index.

Absenteeism is the absence of employees in the workplace. It is noticeable that from 2011 the level dropped drastically, with reduction - 83%, many activities were organized to reach this number low. It is important to note in Graphic 2 that the level of absenteeism of Japanese firms that had a lean production system is related to this indicator of the company studied.

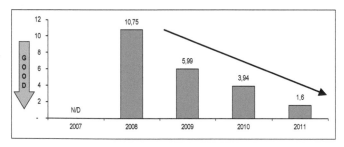

Graphic 3 Index Accident

The company's accident rate is calculated as the proportion of accidents for the hours worked by employees. The company has reduction of - 85% the improvement. This indicator includes all types of accidents with and without leave the company. All efforts to improve the ergonomics and conditions of employment are to prevent and avoid the level of accidents, and the company's goal is equal to zero.

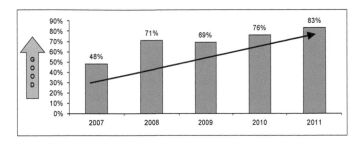

Graphic 4 Vehicle without rework (%)

It can be seen in Graphic 4, the 35% increase in the level of quality vehicles, it is clear focus on quality to stay competitive with the competition. Performing a comparison with Figure 1 we can observe that the Japanese companies that implement the Kaizen methodology, has a higher level of quality companies considered mass system.

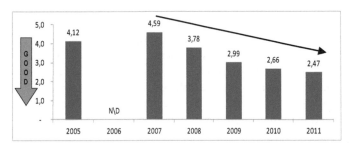

Graphic 5 Performance Production

The Graphic 5 shows the level of performance of the company to produce more at lower cost and less resource as possible. The company have reduction – 47%. This indicator is important for productivity in relation to competitiveness, productivity is mainly defined as the ratio between output and factors of production used, competitiveness involves obtaining greater competitive advantage, or be the best at what it produces. When working on Ergonomics and working conditions of employees to search for productivity is achieved, it is noticeable that in recent years the company aims to improve working conditions and consequently increase the productivity of employees.

It is the indicator that measures the percentage of cars that can carry all the way from the assembly line without the need to be taken off line to perform rework. As said Womack, et all (1990). Lean producers, in turn, openly aspire to perfection, always declining costs, absence of defective items, no stock and a myriad of new products. Another aspect seen in Figure 1 of the authors' analysis is the quality level of the industries that lean production system has, always striving for perfection and when comparing with the result is visible to the search company in terms of perfection in product quality.

According to research presented, it is possible to observe that today there is a high level of concern about the quality of working life in business, because people are not worried about their own health, only when problems arise and this will be bad for both company and the employee because the employee will have to move away due to health problem, and the company will lose one of its employees, which in turn will have to hire another employee and training him, thereby generating cost more to the company.

You can see that working in the lean system, the company may have better results are reconciled with the ergonomics. The company's employees have a need to use and the ability to solve problems in the system "lean". The subsystem used for improvement of working conditions and also for process improvement is Kaizen, workers use this tool to improve their own job and their colleagues.

Based on research in the field was possible to identify that there are developers with different ideas about the system of the company, which were targeted by this research on the main characteristics of a lean production system. And most of the employees interviewed agree with the reconciliation of the lean system ergonomics, relating mainly to complement each other.

Employee productivity is directly related to the company's ergonomic program, you can see in the results discussion and point of view of employees, which officials acknowledge that the company does for them.

Through this research is to identify possible that by creating a culture of continuous improvement (Kaizen) employees, effective improvements with simple and to the process of organization, are use for their own benefit that is improving ergonomics and working conditions of the job.

Just as some studies have already proposed the same argument in this article, other issues could be improved and enhanced with the topics discussed, aiming at the real situations of the working environment.

REFERENCES

Balbinotti Giles. 2003, A Ergonomia como Princípio e Prática nas Empresas, 1° Edição. Curitiba: Editor Gênesis.

Campos, V. Falconi 1997, Gerenciamento pelas Diretrizes, Belo Horizonte: Fundação Christiano Ottoni.

Imai, Masaaki. Kaizen, A estratégia para o sucesso competitivo. São Paulo: Editora Imam, 1994.

Womack, J. P. e Jones, D. T. A. 1990, A Maquina que Mudou o Mundo, Rio de Janeiro: Editora Campus.

Womack, J. P. e Jones, D. T. A. 2004, Mentalidade Enxuta nas Empresas. Rio de Janeiro: Editora Campus.

Martins, Petrônio G. E Laugeni, Fernando Piero. 2006, Administração da Produção. São Paulo: Editora Saraiva.

Matthew E. May. 2007, Toyota A formula da Inovação. Rio de Janeiro: Editora Campus.

Automotive industry documentation, 2010. Curitiba.

IEA – International Ergonomics Association, Retrieved May 02, 2007, from http://www.iea.cc.

CHAPTER 15

Thinking-Action-Information Model-based Research on Knowledge Work Processes

WEN Peihan, ZHANG Yanhong, YI Shuping, XIONG Shiquan

The State Key Laboratory of Mechanical Transmission
College of Mechanical Engineering
Chongqing University, Chongqing 400044, P.R. China

ABSTRACT

Taking knowledge workers as the research object, three key elements of knowledge work processes, thinking, action and information, were proposed to analyze knowledge work processes in-depth and study individual differences. Then a relational "Thinking-Action-Information" (TAI) model was established. By observing and analyzing knowledge work processes, the thinking activity set and the action set were summarized and extracted respectively. To study the matching feasibility of the TAI model with knowledge work processes, a knowledge work experiment about the dimension design was performed to analyze the whole design processes of participants with method of the TAI model, so as to obtain the analysis graphics of knowledge work processes. Finally, the statistical information about knowledge work processes was discussed, and four kinds of differences between "novices" and "experts" were revealed.

Keywords: knowledge work processes; Thinking-Action-Information model; knowledge management; thinking activity set; action set

1   INTRODUCTION

Knowledge work processes are becoming a research focus of the knowledge management area. Typical knowledge works consist of the design, management and productive plan, the computer programming, the medicine diagnosis, the process control, the air traffic control as well as the equipment failure diagnosis. Maier, R indicated the implementation required adequate modeling techniques that consider the specifics of modeling context in knowledge work. Zheng, LY presented a systematic knowledge model for manufacturing process cases. The model represented process knowledge at different levels of granularity to facilitate process configuration. With this knowledge Model the process knowledge was categorized into 6 levels:(i) core process skeletons, (ii) process networks, (iii) process routes, (iv) process segments, (v) processes/work plan, and (Vi) operations/working step. Nielsen, J et al. built a resource capability model to support product family analysis. Considering the knowledge work process as the production process, inputs are the knowledge skill, time, software and hardware resources, etc., and outputs are the solution, knowledge growth, expertise level, etc. The outputs are not predicted from the inputs directly, since the knowledge work process is also the knowledge flowing process, simultaneously may be regard as the cognition process, namely information processing course.

There are three main angles regarding to the research of knowledge work processes, corresponding to three different studying objects separately. (1) Take knowledge work itself as the research object. Hart-Davidson et al. have conducted visualization research for the knowledge work, whose methodological combined process tracing methods (preliminary interviews & training sessions, participant diaries, computer event logs) with stimulated recall interviews to produce rich accounts of proposing and grant-seeking activity. Pan et al. thought that various different process activities included structured process activities and knowledge-intensive activities. A unit of work is made up of Metadata, roles, services, contents, methods, assessments, work plans, which is described by work process specification language (WPSL). (2) Take knowledge as research object. Zhuge et al. proposed a pattern-based approach to knowledge flow design for more effective and efficient planning. The basic idea is to adapt and control logistical processes for knowledge flow within teams. Kamhawi et al. proposed a knowledge flow which formed a cycle of knowledge sharing, conversion and innovation to promote the accomplish of knowledge work. Chen et al. combined work processes and knowledge to establish a knowledge structure, so that knowledge workers develop and accumulate their own process knowledge base effectively to improve the efficiency. (3) Take brain as research object. Mladkova et al. indicated that knowledge workers used the brain more than body. Jo et al. established a shared mental model and did some research on 51 teams to confirm the hypothesis that the knowledge sharing mental pattern is very important for a team.

Because of the knowledge workers' personal characteristics, it's inevitable that different people have different work processes even facing the same knowledge work. This paper takes knowledge workers as the research object, and proposes the three essential factors of knowledge work processes: thinking, action and information, where thinking plays the leading role, the information supports and guides the thinking, and action is the effective method for gaining information. Then, a relational model of the three essential factors, the Thinking-Action-

Information model, is established, and sets of thinking activities and action are concluded, extracted and collected respectively. Also, an experiment of dimension design is performed to further analyze the knowledge work process and reveal the differences of working processes between the "novices" and the "experts".

2 THINKING-ACTION-INFORMATION (TAI) MODEL

(1) Key elements of knowledge work processes

The knowledge work process is composed of two states: an initial state and a corresponding target state. Knowledge workers characterize the initial state including uses of concepts, theory and thought, and then reach the target state through a series of knowledge activities. There are only a few determined rules on the knowledge work, workers should decide how to accomplish it specifically, which will influence the working quality and completion time. Hence, it is especially meaningful to study knowledge work processes in the view of workers.

By observing and analyzing the knowledge workers' behaviors, we realize that knowledge work processes contain three key elements: thinking, action and information.

① Thinking. Compared with the operation tasks, the knowledge work does not have specific and fixed procedures, and depends on workers' thinking more than operation tasks. Different thinking processes result in different results. Thinking plays a leading role in the working process, and a series of thinking activities is called a thinking chain.

② Action. Similar to operation tasks, the knowledge work also requires the involvement of human body. Under the close cooperation and natural auxiliary of senses (ears, mouth, nose, eyes, etc.) and limbs (hands, feet, etc.), the knowledge work is executed purposefully. Action, the general designation of this kind of activities, is the only approach to communicate with the external. It is determined by thinking, varying with demands of thinking and having no obvious regularity. During knowledge working processes, action plays a supporting role.

③ Information. Include knowledge and messages. In operation tasks, work procedures and job specifications are shared information, which is researched long-term and proved to improve the efficiency by working on the established procedure. However, in knowledge works, workers search, apply, or (and) generate information from himself or (and) the external according to the real situation. Along with the whole process of thinking evolvement, the information plays the important guidance function. For the operators, the information used in working process is a kind of passive acceptance; where for the knowledge workers, the information produced and managed in working process is positive thinking. In summary, the knowledge work process is a series of purposed activities dominated by the thinking and assisted by the action. Knowledge workers exercise information to solve problems and achieve the objective eventually.

Knowledge work processes are complex and difficult to detect, especially

138

for thinking and information. Understanding knowledge workers' thinking without disturbing their normal working is still a great challenge. In this research, the think-aloud protocols is adopted, which takes advantages of the language to express thinking during knowledge working, and is generally used for process research to mine information and reasoning method about activities and behaviors at a higher level. Observers could comprehend the instant changes of knowledge workers by video analysis and face-to-face communication, obtain the track of working processes, and record the thinking and information one by one.

(2) Thinking-action-information model

The knowledge work is completed through the natural coordination and close cooperation of the above three elements. One or more actions associated with the thinking activity carried out, assisting thinking to obtain, use or (and) generate information for reducing the distance between the current and target states. For example, in the progress of "searching" activity, looking, moving, waiting, etc. will accompany with thinking to acquire the needed information, which promotes activities of the thinking chain till the end of the work. During the process, information and action couple with the whole thinking chain.

Denote the limited set of thinking activities as $\{T_1, T_2, ..., T_n\}$, the limited set of body action as $\{A_1, A_2, ..., A_n\}$, the infinite set of information as $\{I_1, I_2, ..., I_n\}$, respectively. The Thinking-Action-Information (TAI) model can be illustrated as Fig. 1, where $\{t_1, t_2, t_3, ..., t_x\} \subset \{T_1, T_2, ..., T_n\}$ represents the thinking chain, and $\{i_1, i_2, ..., i_y\} \subset \{I_1, I_2, ..., I_\infty\}$ represents the information set. The thinking activity and information could come forth repeatedly in the thinking and information chain. For the action set, one or more actions may turn up for each thinking activity.

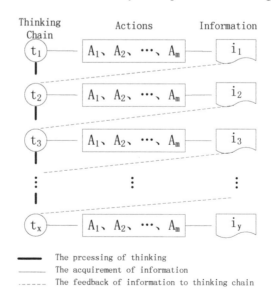

Fig.1 Thinking-Action-Information model of knowledge work processes

During the knowledge work process, the thinking activity t_1 starts firstly, obtaining the information from the internal or (and) external with the support of action, and sending feedback to the thinking chain; then the thinking activity t_2 executes, receiving information and feedback; ...; when the thinking activity t_x completes after getting the information i_y, the whole knowledge work is accomplished. t_1, t_2, ..., t_x is the thinking chain, which takes the task identification as the origin and the solution as the destination.

The above TAI model describes the individual knowledge work, the thinking chain belongs to individual thinking process. For a team, thinking chains of members will influence each other, which will be researched in future work.

3 THINKING ACTIVITY SET & ACTION SET

The following takes the design work under the dynamic network environment as an example of knowledge works to introduce the thinking and action elements, as well as further study the TAI model.

In domains of psychology and cognition, thinking activities are deeply studied and represented, such as identification, summarization, imagination, plan, decomposition, memory, abstraction, inquiry, calculation, analysis, inference, comparison, modification, contraposition, synthesis, evaluation and information expression. Based on the above, we tracked the knowledge work process and interviewed with knowledge workers in the research and design department of some manufacturing enterprises. Then, 10 kinds of thinking activities of designed knowledge work are put forward as shown in Table 1.

Table 1. Thinking activity set

	Thinking activity	Explanation
1	Identification	Identify and confirm the design task, including identify activity, such as concept understanding and problem determining.
2	Decompose	Analyze designed targets, and break it into subtasks.
3	Memory	Extract knowledge or experience from memory actively and consciously.
4	Association	Think of one object from another.
5	Search	Seek for the need by looking for the information.
6	Calculate	Model and formulate by using the related theory and algorithm.
7	Reason	Derive the conclusion from one or more premise.
8	Comprehensive	Form the overall conclusion by combining the analyzing result of all subtasks. It is contrary to decomposition.
9	Compare	Contrast all results, and adopt the optimized process and solution.
10	Result expression	Present the result in some way

Although the operation task could not avoid the conductor and command of the brain, its work efficiency is determined mainly by the speed and range of action, so

Gilbreath promotes the concept of therblig, and improves work efficiency using the motion economy principle. Different from operation tasks, the knowledge work is led by thinking where action is just the assistance of communicating with the external, it is not essential to divide the knowledge into therblig. So we generalize 6 kinds of common action by analyzing the senses (eye, ear, mouth and nose, etc.) and body (hand, foot, etc.), as shown in Table 2.

Table 2. Body action set

	Action	Explanation
1	Hand-action	Action of hands, such as key pressing and page turning.
2	Look	Action of eyes, such as information search, read and observation
3	Say and listen	Action of ears and mouth, such as speak and hear.
4	Smell	Action of nose, such as odor perception, etc.
5	Foot-action	Action of feet, such as trample and walk etc.
6	Wait	Without body action, including thinking pause etc.

4 EXPERIMENT ANALYSIS

4.1 Experiment description

In order to inspect and explore the matching ability of the TAI model with knowledge work processes, an experiment of dimension design is performed and the volunteers' design processes are analyzed by exerting the above model.

Subject: 20 graduate students with the same scientific background (mechanical engineering), who are familiar with the ANSYS 9.0 software. 8 persons are experts who experience two or more dimension design projects while others are novices having no similar experience.

Task: In view of the ideology of reducing cost and reasonable resource configuration, the experiment is about dimension design of desks based on the secondary development of ANSYS. All participants must select one from the four given models (1: square desktop and square legs; 2: square desktop and round legs; 3: round desktop and square legs; 4: round desktop and round legs). The length of desk legs is $60cm$, and the desktop square is not less than $0.64cm^2$. The desk should carry a $100\ kN$ uniform load with allowable stress of $200MPa$. The objective is to minimize the volume.

Equipment: A Lenovo notebook computer (Intel binuclear CPU1.66G; 1G memory; NVIDIA GT400M independent display card), 4 RED-IV cameras, a quick fish monitor device, a WV-CZ352/302 Panasonic Camera of Color and variable magnification integration, some A4 paper, a stopwatch, a gel-ink pen, the finite element analysis software ANSYS 9.0.

Requirement: All participants are required to adopt thinking-aloud protocols, which is to speak out their thinking and ideas during the whole design process.

Process: Two experiment observers could always ask or remind the subjects, answer the relational issues, and keep the necessary records during design processes.

During the experiments, there are two position-fixed and point adjustable cameras and a close distance video camera recording the whole process from various view points. The video, draft, experiment result of each participant and note-taking of every observer are analyzed to shape the chart of knowledge work processes. Experiments show that matching the TAI model with the knowledge work processes could not only obtain the thinking process of knowledge work, but also mine the information obtained from the working process.

4.2 Discussion

Based on the analysis of experiment data, four knowledge work differences between experts and novices are identified: the general difference, the thinking difference, the action difference (shown in Table 3) and the information difference.

Table 3 The statistical information of differences between novices and experts

(Thinking chain length: step, Time: min, Volume: cm3)

Difference Category		Expert	Novice	D-Value	Percentage
General	Model Volume	7130.7	10314.72	-3184.02	-30.87%
	Design Period	52.88	55.00	-2.12	-3.87%
	Thinking Chain Length	36.63	21.58	15.04	69.69%
	Identification Num	2.25	3.42	-1.17	-34.21%
Thinking	Identification avg time	4.11	2.51	1.60	63.75%
	Memory Num	1.63	0.25	1.38	552.00%
	Operation Error Num	0.88	2.50	-1.63	-65.20%
Action	Waiting Time	4.31	6.01	-1.69	-28.17%
	Proportion of Wait Time	87.50%	25.00%		62.50%

(1) General difference. The average volume of novices is 30.87%, while the average designed period is 3.87% larger than that of experts. Novices have no concepts about reducing the volume and no optimization experience, so there are some differences between the average volumes. There are not so many differences among the designed periods of novices as the experience has little constraint and pressure. The thinking chain of experts is 69.69% higher than that of novices, which shows that the average thinking activity time of novices is longer and the thinking load is relatively higher.

(2) Thinking difference. Based on the statistics of two thinking activities (identification and memory), which have great effect on knowledge works, we find that the average identification number of novices is 34.21% more than that of experts, while each identification time is 63.75% less than that of experts. Novices have relative crude representation of problems and require repeated identification because they have less knowledge module about the problem than experts.

Memory, which is to research information from internal knowledge base, is a low load of thinking activity and also the biggest thinking difference between experts and novices. The average memory number of experts is 1.63, the most content is experience related, and has great promotion to task progress. However, novices have almost no memory activities.

(3) Action difference. The action has something to do with design results. (i) Exploration of software features, such as three-dimensional rotation function, experts do not need to explore it. (ii) Mistakes, such as the missing modification of some parameters and the records of results. We find that experimental mistakes of novices in the average are 2.50 times, 65.20% higher than that of experts. (iii) The use of waiting time, such as the waiting of software running or other emergencies, 87.50% of the experts recorded data during the waiting time, and 75% of the novices didn't take full use of the waiting time, which makes the experts' actual average waiting time decreased 28.17% than that of novices. Above all, (ii) and (iii) played a greater impact on the design results and can be optimized, such as reducing the mistakes times, recording parameters during the software, only recording the change parameters of each experiment, etc., then the time can be reduced and the change of results can be found easily.

(4) Information difference. There are three information differences between novices and experts. (i) Since they are unfamiliar with ANSYS, novices observe the change of volume and stress by adjusting the size in two directions, rather than analyzing the stress distribution chart. However, experts judge the optimization direction through the color distribution and variation. (ii) Experts have the experience to analyze observation information and data, which is versatile in this experiment. (iii) Experts can mine some regular information with their high learning and acceptance ability and sensitivity to the experiment variation.

The above differences concluded from the analysis of knowledge work processes can help novices to improve disadvantages. Some valuable information can be obtained from the think-aloud protocol and questioning to experts. The observer summed up some key information for completing this task, for example, (i) optimize the leg location to find the minimum stress point under the same circumstances firstly; (ii) the desktop adjustment range ought to be reduced to 1 or less since it has great influence on stress and volume; (iii) the thick leg is very beneficial for reducing stress while the thin one is good for reducing volume; (iv) the model color of ANSYS should be as uniform as possible; etc. Promoting the above information is equivalent to reduce the difficulty of knowledge work, optimize the novice path of thinking, simplify thinking steps, decrease the explore time and improve the efficiency of knowledge work processes.

5 CONCLUSION

The process analysis can help novices find the enhancement way of working efficiency while mining the valuable process information of experts to form standard material gradually which provides novices to reduce the thought load and raise the working efficiency. Hereby, this paper proposes the TAI model to draw up the chart for analyzing knowledge work processes, reappearing the whole working processes from the interactive view and promoting the thinking chain evolution from identification to final result expressing under the assistant of information and action.

In order to confirm the matching ability of the TAI model with knowledge work processes, an experiment about the dimension design is performed and the entire design processes of 20 participants are record using the thinking-aloud protocol, field observation and cameras. The general, thinking, action and information differences are obtained from analyzing the process diagram and statistical data. Based on the above achievement, future research will mainly focus on process factors influencing knowledge work efficiency and team knowledge work processes.

REFERENCES

Borsci S,FedericiS (2009).The Partial Concurrent Thinking Aloud: A New Usability Evaluation Technique for Blind Users. Emiliani PL, Burzagli L, Como A, et al.Assistive Technology From Adapted Equipment to Inclusive Environments, 421-425

Chen WL, Xie SQ, Zeng FF, et al (2011). A New Process Knowledge Representation Approach using Parameter Flow Chart.Computers in Industry, 62(1):9-22

Duncker K (1945). On Problem Solving.Psychological Monographs, 5:58-62

Han SH, Fang F (2008).Linking Neural Activity to Mental Processes.Brain Imaging and Behavior,2(4):242-248

Hart-Davidson W, Spinuzzi C, Zachry M (2007). Capturing&Visualizing Knowledge Work:Results&Implications ofa Pilot Study of Proposal Writing Activity. ACM.Proceedings of the 25th ACM International Conference on Design of Communication, 113-119

IshinoY,Jin Y (2006).An Information Value Based Approachto Design Procedure Capture.Advanced Engineering Informatics, 20(1):89-107

Jo IH (2011). Effects of Role Division, Interaction, and Shared Mental Model on Team Performance in Project-Based Learning Environment. Asia Pacific Education Review, 12(2):301-310

Kamhawi EM (2010).The Three Tiers Architecture of Knowledge Flow and Management Activities. Information and Organization ,20(3-4):169-186

Lazarev VV (2006).The Relationship of Theory and Methodology in EEG Studies of Mental Activity. International Journal of Psychophysiology,62(3):384-393

Maier R (2005). Modeling knowledge work for the design of knowledge infrastructures.JOURNAL OF UNIVERSAL COMPUTER SCIENCE, 11 (4) : 429-451

Mladkova L(2011). Knowledge Management for Knowledge Workers. TurnerG,MinnoneC.Proceedings of the 3rd European Conference on Intellectual Capital, 260-267

Nielsen, J ,Kimura, F (2006). A resource capability model to support product family analysis.JSME INTERNATIONAL JOURNAL SERIES C-MECHANICAL SYSTEMS MACHINE ELEMENTS AND MANUFACTURING ,49 (2): 568-575

Pan WD, Liu JX, HawryszkiewyczI (2008). A Method for Describing Knowledge Work Processes. International Workshop on Advanced Information Systems for Enterprises,Proceedings, 46-52

Pottier P,Hardouin JB,Hodges BD, et al (2010). Exploring How Students Think:a New Method Combining Think-Aloud and Concept Mapping Protocols. Medical Education, 44(9):926 -935

Robert DA (2005). Knowledge Work. Encyclopedia of Social Measurement, 2:417-422

SalvendyG (2001). Handbook of Industrial Engineering: Technology and Operations Management. 3rd Edition. NewYork: JohnWiley& Sons Inc,2001

Schellings G, Broekkamp H (2011). Signaling Task Awareness in Think-Aloud Protocols From Students Selecting Relevant Information From Text.Metacognition and Learning, 6(1):65-82

Sudakov KV (2010).Systemic Mechanisms of Mental Activity.ZhurnalNevrologII I PsikhiatrIIImeni S S Korsakova, 110(2):4-14

Wagner HN (2008). The Chemistry of Mental Activity. Neuroimage, 208

Yagolkovsky SR (2009). A Systemic and Structural Approach to Analyzing a Subject's Thinking Activity. Voprosy PsikhologII, 6:43

Zheng L. Y, Dong H. F, VichareP (2008).Systematic modeling and reusing of process knowledge for rapid process configuration.ROBOTICS AND COMPUTER-INTEGRATED MANUFACTURING, 24 (6): 763-772

Zhuge H (2006).Knowledge Flow Network Planning and Simulation.Decision Support Systems, 42:571-592

CHAPTER 16

A Study on R&D Team Synergy-oriented Knowledge Integration

YI Shuping, SU Li, WEN Peihan

The State Key Laboratory of Mechanical Transmission
Mechanical Engineering College
Chongqing University, Chongqing 400030, China

ABSTRACT

Team synergy is the main mode of R&D team knowledge work. How to effectively share ideas and knowledge among team members is the key problem of knowledge integration. We analyze the general working process of R&D team and find the problems, such as strong independence among subtasks, the lack of effective communication among members and product design process that is difficult to control. According to the problems, we put forward the knowledge working processes in team synergy modes. Then, the knowledge integration mechanisms, such as 3D interaction, multi-dimensional resources synergy, knowledge increment and conflict resolution, are discussed from the cognitive perspective. Finally, with the system theory, the model of knowledge integration oriented to team synergy is put forward from 5W1H (Who, What, Why, Where, When, How) aspects.

Keywords: R&D Team, Team Synergy, Knowledge Integration, Knowledge Activity

1   INTRODUCTION

New product development is a complex decision problem with multi-disciplinary, multi-member participation and multi-objective optimization. Team synergy is the main mode of R&D knowledge work. However, team synergy was composed of different design workers. They form different thinking set and cognitive ability basing on different knowledge background, cultural background and work experience. Thus barriers of knowledge exchanging, knowledge conversion and knowledge sharing will be formed among team members. Also team members, who are usually from different departments and are assembled temporarily, are not familiar. Barriers from different professional background and unfamiliar relationship result in lack of communication, which means the product design process is difficult to control and greatly reduce the efficiency of the team synergy work. Therefore, it is important for new product development to effectively and efficiently synergy ideas and knowledge among team members, which is the problem of knowledge integration too.

At present, knowledge integration in new product development have accomplished many achievements home and abroad. Sivadas and Dwyer (2000) believed that why the enterprise knowledge can not be effectively integrated is that the product design can not achieve the goal of product innovation. Yang (2005) discussed the impact of knowledge integration on new product development, and proposed that knowledge integration had a very active role on the high-tech industry performance. Becker and Zirpoli (2003) analyzed the knowledge integration of the supply chain in new product development from the resource-based perspective. Lee, et al. (2005) proposed object-based knowledge integration system to assist the initial stage of the product development using XML as a data exchange technology. Wang et al. (2007) proposed a knowledge active push framework based on knowledge management and driven by workflow. Chen, et al. (2009) put forward a mechanism for ontology-based product lifecycle knowledge integration and developed integration technology. Shi, et al. (2010) proposed a reengineering and estimation method of the process activities presented under the environment of knowledge integration based on traditional DSM optimization methods. Chen (2010) presented a systematic approach to developing knowledge integration and sharing mechanism for collaborative molding product design and process development. Wu and Ragatz (2010) viewed ESI as a knowledge integration process and examined the role of the purchasing firm's integrative capabilities in the process. Kleinsmann, et al. (2010) investigated the factors influence the creation of a shared understanding in Co-NPD, such as the actor, project and company level. Liu, et al. (2012) put forward a general framework for the analysis of knowledge integration and diffusion using bibliometric data to assess knowledge integration and diffusion. However, the above researches on knowledge integration have focused mainly on resources, objects and methods. Little work has been done on the knowledge integration process in team synergy mode, which is the active behavior of the team members. The above fact motivates us to study the R&D team synergy- oriented knowledge integration mechanism.

The rest of this paper is organized as follows. First, working process of new product design team is analyzed. Second, the knowledge integration mechanisms of team synergy from the cognitive perspective are followed to discuss. Finally, team synergy-oriented knowledge integration model is built.

2   TEAM SYNERGY WORKING PROCESS

(1)General Working Process of R&D Team

New product design team is a distribution center and reservoir of knowledge, and is also the birthplace of new technologies and the market competition force in modern manufacturing enterprises. Team task will be decomposed into N subtasks in accordance with the professional fields. And subtasks will be assigned to domain experts. Domain experts in various fields will communicate regularly job scheduling and task matching. And then subtasks will be combined. Finally, the team task will be completed and team will do data filing. This is the general working process of a new product development team, as shown in figure 1.

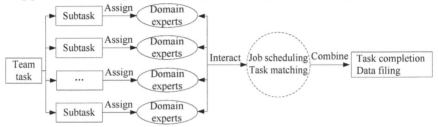

Figure 1 General working process of R&D team

The above figure shows that, concurrent engineering and other advanced management mode were applied to new product development process. However, due to task complexity of new products and strong independence among subtasks in various fields, the communication contents are usually only about job scheduling and task matching. Whereas there is no enough effective communication on professional knowledge among domain experts, which results in a series of isolated knowledge islands. Isolated knowledge islands are not conducive to knowledge integration and the formation of the overall concept of the new product development. The design task of new product is innovative, such as new principles, design, materials. And the innovation degree depends on the amount of knowledge of team member contributions. The new product design team members have different educational backgrounds, experience backgrounds and cultural backgrounds, or even belong to different functional departments. It increases the barriers to communication, which result in that the product design process is difficult to control. The above problems, such as strong independence among subtasks, the lack of effective communication among members and product design process that is difficult to control, will lead to ineffectively synergy working of the team members, the low efficiency and long product development cycle.

(2) R&D Team Synergy-oriented Knowledge Integration Process

Based on the characteristics of new product development process, we can see that the main reasons for the inefficient synergy working are that team members can not effectively communicate and team resources can not be synchronized with. The research contents of team synergy are to solve these problems. Team synergy knowledge working processes emphasize on knowledge innovation and exchanging as shown in figure 2. It is essentially team synergy-oriented knowledge integration process.

148

Knowledge integration is that ideas, schemes and knowledge among team members are merged, to reach interactive growth of knowledge in the R&D team working process. The differences between team synergy-oriented knowledge integration process and general knowledge integration lie in the team three-dimensional (3D) communication and resources synchronization. 3D interaction is that team members achieve sound, images and data transmission and knowledge transferring to share data, information and knowledge through software or hardware. That is "hear his voice, and saw its shadow", such as face-to-face communication and video conference. Compared with the general team process, the main interaction mode of team synergy is 3D interaction. The contents of 3D interaction is not only job scheduling and task matching, but also the communication aiming at specific expertise and team resources. Team members participate in the overall product development process. A variety of expertise and capabilities complement and promote each other. And then realize knowledge exchanging and sharing, so as to enhance the individual effects and result in the development of the overall team. 3D interaction of the team can grasp the use of various resources, ensure the coordination of various resources is synchronized in time and space, and form the consistent pace. Through 3D interaction, team members can overcome differences in knowledge and ability, achieve effective synergy of ideas and knowledge, and realize knowledge increment to improve the working efficiency.

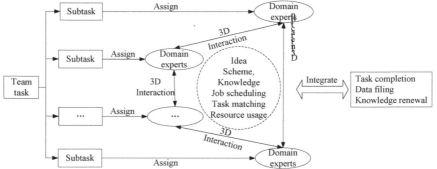

Figure 2 R&D team synergy-oriented knowledge working process

3   KNOWLEDGE INTEGRATION MECHANISMS

(1) 3D Interaction

3D interaction is the main interaction mode of R&D team synergy-oriented knowledge integration. In the team synergy environment, team members can be in the same or different places. However, Information transmission must be considered in space and time consistency. 3D interaction is a synchronous interaction and is divided into local synchronized interaction and remote synchronized interaction, as shown in figure 3. In synchronous manner, reliability and timeliness are equally important. So it is required to provide a synergy mechanism to ensure the consistency of spatial and temporal information. Local synchronized interaction, as a commonly used mode, it refers to team members located at the same place are assembled to interact in the physical environment and has high reliability and real-time, such as face-to-face meeting. Remote synchronized interaction, it refers to the team members located in different places are assembled to interact through internet in the virtual environment, such as video conference. Compared with other interactions, 3D interaction can quickly acquire all resources, and improve the efficiency of synergy working.

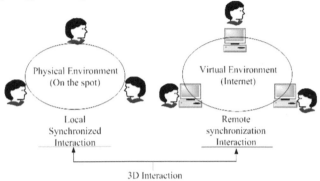

Figure 3 3D interaction

(2) Multi-dimensional Resources Synergy

Multi-dimensional resources synergy is the environment of R&D team synergy-oriented knowledge integration. New product development team synergy-oriented requires the members to share data, information, knowledge and resources in problem-solving process in real time. In the team synergy environment, distributed multi-dimensional resources work in the synergy mode. It includes information synergy, knowledge synergy, physical synergy, and other resources synergy.

Information synergy is that information related to product development among subtasks and subsystems must be coordinated and cooperated well to achieve unison. Not because of any information errors of the task and the system, make the whole product development lags behind, and even failures. In addition, because of the differences in business and personnel quality, it will increase the difficulty of coordination among the various subtasks and subsystems. And it will cause the information flow poor. However, the research in this area is more and there has been many mature network technology and database technology to solve the problems of sharing. Knowledge synergy refers to the dynamic process that knowledge subjects transfer the right information and knowledge to the appropriate object though the 3D interaction and achieve knowledge innovation. Team members generate knowledge conflict in the interactive process because of knowledge activities and achieve knowledge increment in the conflict resolution process. With different cultures, education, and experience backgrounds, team members have individual cognitive differences, making the synergy of team knowledge more difficult. Tool synergy refers to the sharing of problem-solving tools, software and other tool resources in the team synergy knowledge working. It needs sharing mechanism to come true, such as the business office sharing platform, whiteboard,

150

application sharing, and even modern network sharing software. Sharing mechanisms provide a mutual workspace for team members. So we can grasp the shared resources in real time and maximize the use of the vast resources provided by the synergy environment. Team synergy, emphasizes team collaboration as a whole. In addition to information, knowledge, tools, the synergy of other resources are also essential.

(3) Knowledge Increment

Knowledge is held by individuals, but is also expressed in regularities by which members cooperate in a social community (Kogut and Zander, 1992). The overall trend of knowledge in the integration process is the rise of the "spiral", as shown in figure 4.

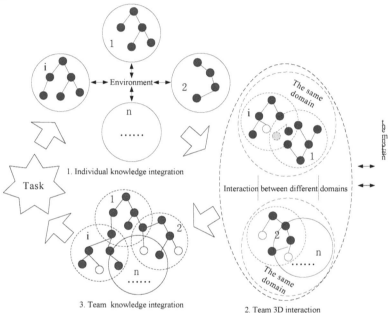

Figure 4 Knowledge change in team synergy working mode

●is existed knowledge of team members.○is newly added knowledge. ⬭ is the lost knowledge. A circle of solid line is a team member. A circle of dotted line is the lost knowledge boundary.

First of all, after receiving the assigned task, team members integrate related knowledge according to their own abilities. It includes the existing knowledge, knowledge in external environment, and new knowledge created by them. It is the first stage, individual knowledge integration. Second stage is 3D interaction of team. Due to the interaction among team members, the individual knowledge breaks its boundaries into the team knowledge. It includes knowledge integration with the same domain experts and knowledge collision among different domains. New knowledge conflict is produced and knowledge synergy is achieved through the interaction in the environment. 3D interaction process is along with the knowledge added, discarded and forgotten. The third stage is team knowledge integration. All kinds of the team knowledge are integrated into systematic and structural knowledge system. The second and the third stages belong to team knowledge integration. Knowledge is increased in quality and quantity through individual and team knowledge integration processes.

(4) Conflict Resolution

New product development is the process of conflict production and conflict resolution. On the one hand, the conflict will hinder the smooth progress of the synergy design. On another hand, the conflict will have a beneficial effect of product development. By different causes, the conflict can be divided into conflicts of resources use, conflicts of division of labor and timing, inconsistent information conflicts, the conflicts caused by the knowledge and ability and so on. The first three conflicts will affect the synergy development efficiency, and are negative conflicts. It is not difficult to solve them applying existing management technologies and tools. Conflicts caused by the knowledge and ability will directly affect the knowledge level and the innovation degree, are the focus of conflict resolution. However, the conflicts caused by the knowledge and ability will force team members to strengthen exchanges, thereby contribute to the knowledge integration within the team, which is conducive to the development of new products. So they should belong to positive conflicts. Conflict resolution strategies are made based on the causes of the conflicts. The quality of conflict resolution is related to team members' knowledge activities. Knowledge activities can be divided into specific knowledge units, such as listen, question, answer and so on. The size of the conflict is different, so is the frequency of knowledge activities. Different conflict resolution strategies will be made according to the different conflicts.

Conflicts caused by the knowledge differences. New product development team focused on experts in different fields. Each expert considers the issue from their own domains and has different views on the aspects of the products. Thus conflicts are produced. Generally speaking, moderate conflicts play an active role in the product development process. These conflicts, too small or too more, will hinder the synergy design process. To deal with such conflicts, conflict resolution strategies are the establishment of the knowledge base of product knowledge, product knowledge training on a regular time, and optimal performance of the product as the rules of conflict resolution.

Conflicts caused by cognitive ability. Team members have the different cognitive abilities. On the one hand, with differences in cognitive abilities, team members will gain the different amounts of knowledge in individual knowledge integration. On another hand, it will bring conflicts of knowledge transferring and access in 3D interaction. Such conflicts play a negative impact on the product development process. Within the team working process, the cognitive abilities of the members improved, such conflicts will diminish. However, the fundamental conflict resolution strategy is team building. Select the members of the corresponding ability according to the task characteristics.

4   KNOWLEDGE INTEGRATION MODEL

Based on the analysis of the knowledge working process and the knowledge integration mechanisms, with the system theory, R&D team synergy-oriented knowledge integration model from 5W1H (Who, What, Why, Where, When, How) aspects was built, as shown in figure 5.

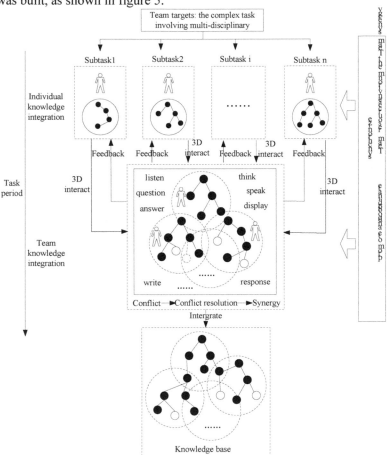

Figure 5 Knowledge integration model in team synergy working mode

In the model, "Why" is the goal of team synergy, new product development task involving multi-disciplinary fields. The new product design is the most knowledge-intensive business based on large amounts of data, information and knowledge. In this case, the team task must be assigned to various experts that are "Who" in the model. Due to the different cultures, experiences, educational backgrounds, team members have different expertise, cognitive abilities and behaviors in the individual knowledge integration and team interaction processes. There are two stages in the entire team task cycle that is "When" in the model. The one is individual knowledge integration. Team members gain individual knowledge through the different knowledge activities. With different professions and cognitive abilities, the amount and the content of knowledge are different. Another is team knowledge integration. In 3D interaction mode, team members form team knowledge through the knowledge activities, such as speak, think and listen and so on. With different professions and cognitive abilities, knowledge and activities vary in interaction process. After a number of feedbacks and interactions between individual knowledge integration and team knowledge integration, new knowledge system is integrated. The entire process is completed in the team synergy environment that is "Where" in the model. The team synergy environment, where team resources can cooperate and synchronize with each other, has different features from that of the general knowledge integration environment. Knowledge integration object is knowledge that is "What" in the model. Knowledge grows in quality and quantity. Relative to the knowledge increment, knowledge integration activities make conflicts reduced to achieve synergy, which is "How" in the model. Make the knowledge activities maintain a certain active degree using application technologies and management strategies. It is benefit to reduce conflicts and improve the efficiency of team synergy.

5   CONCLUSIONS

Team synergy is the main mode of R&D team knowledge work. The research on team synergy-oriented knowledge integration can effectively share ideas and knowledge among team members. We analyzed the general working process of R&D team and found the problems, such as strong independence among subtasks, the lack of effective communication among members and product design process that is difficult to control. To deal with the above problems, we bring forward knowledge working processes in team synergy modes, with the characteristics of 3D interaction and resource synergy. Then, the knowledge integration mechanisms of R&D team were analyzed, such as 3D interaction, multi-dimensional resource synergy, and knowledge increment and conflict resolution. 3D interaction is a synchronous interaction and is divided into local synchronized interaction and remote synchronized interaction. Multi-dimensional resources synergy is the environment where distributed multi-dimensional resources cooperate and synchronize with each other. Knowledge is added, discarded or forgotten, and combined in the process, which extend the team knowledge in quantity and quality. The degree of conflict resolution is related to the frequency of knowledge activities. Finally, with the system theory, from 5W1H (Who, What, Why, Where, When, How) aspects, the model of knowledge integration oriented to team synergy was put forward. Within the model, "Who" are team members with professional differences and ability differences. They contribute different knowledge and show different knowledge activities. "Why" is the goal of team synergy, complex tasks involving multidisciplinary knowledge for the team to accomplish with knowledge activities. "Where" is the environment of team synergy. Team members cooperate, coordinate and synchronize resources under this environment. "When" is the time to complete the task in team synergy working mode that will be shorter than in other modes.

"What" and "How" are the above-mentioned knowledge and knowledge activities. This model has an important role to analyze the knowledge increment and conflict resolution and can be guiding principles to improve work efficiency.

This paper has drawn attention to exploratory research on R&D team synergy-oriented knowledge integration. Yet, more details remain to be explored (e.g. influence of individual cognitive difference on R&D team synergy-oriented knowledge integration, application of R&D team synergy-oriented knowledge integration model, knowledge integration strategies, etc.).

ACKNOWLEDGMENTS

This article is supported by the National Nature Science Foundation of China (Grant No. 70871127). The authors would like to express their appreciation to the agencies.

REFERENCES

Becker M C, Zirpoli F (2003). Organizing new product development: knowledge hollowing-out and knowledge integration-the FIAT auto case. International Journal of Operations& Production Management, 23 (9):1033-1061

Chen Y J (2010). Knowledge integration and sharing for collaborative molding product design and process development. Computers in Industry, 61(7): 659-675

Chen Y J, Chen Y M, Chu H C (2009). Development of a mechanism for ontology-based product lifecycle knowledge integration. Expert Systems with Applications, 36 (2):2759-2779

Kleinsmann M, Buijs J, Valkenburg R (2010). Understanding the complexity of knowledge integration in collaborative new product development teams: A case study. Journal of Engineering and Technology Management, 27(1-2):20-32

Kogut B, Zander U (1992). Knowledge of the firm, combinative capabilities, and the replication of technology[J]. Organization Science, 3(3):383-397

Lee C K M, Lau H C W, Yu K M (2005). An object-based knowledge integration system for product development: a case study. Journal of Manufacturing Technology Management, 16 (2):156-177

Liu Y, Rafols I, Rousseau R (2012). A framework for knowledge integration and diffusion. Journal of Documentation, 68(1):31-44

Shi H, Wu J, Li J, Yu H (2010). Process management approach with design knowledge integration. Computer Integrated Manufacturing Systems, 16(3):527-535

Sivadas E, Dwyer F R(2000). An examination of organizational factors influencing new product success in internal and alliance—based processes. Journal of Marketing, 64(1): 31-49

Wang S, Gu X, Guo J, Ma J, Zhan H(2007). Knowledge active push for product design. Computer Integrated Manufacturing Systems, 13(2):234-239

Wu S J, Ragatz G L (2010). The role of integrative capabilities in involving suppliers in New Product Development: a knowledge integration perspective. International Journal. Manufacturing Technology and Management, 19(1/2):82-101

Yang J (2005). Knowledge integration and innovation: securing new product advantage in high technology industry. The Journal of High Technology Management Research, 16(1):121-135

CHAPTER 17

Assessment of Parameters of Accessibility Based on the Intuitive Use

Raphael Aguiar Barauce Bento[1], Caio Márcio Almeida e Silva[2], Eduardo Candido da Silva[3], Maria Lúcia Leite Ribeiro Okimoto[4]

Federal University of Paraná
Curitiba, BR
[1]raphaeel@gmail.com
[2]caiomarcio1001@yahoo.com.br
[3]eduhcandido@gmail.com
[4]lucia.demec@ufpr.br

ABSTRACT

The accessibility emerged in the early sixties, with a concept of "barrier-free design" and in the nineties, due to advances in technology, internet usage, among other factors, has been expanded to "universal design" (Mazzoni et al., 2001). In the city of Curitiba-PR Brazil, for example, a platform lift is used, allowing the entry and exit for wheelchair users at certain bus stations (known as "tube-stations"), as these bus stations are located above ground level. Evaluations of environments/objects for their accessibility are being developed utilizing different approaches, such as questionnaires (Mont'alvão, 2006; Bacil and Watzlawick, 2007; Carvalho e Silva, 2007), checklists (Aragão, 2004; Ely et al., 2006), post-occupancy evaluation (Marghani et al., 2010), etc. Therefore, this article aims to evaluate the accessibility of tube-stations' platform lifts in the city of Curitiba-PR, through four parameters developed by Dischinger et al., 2009 (apud Ely and Silva, 2009): displacement, spatial orientation, use and communication. For this purpose, each parameter was related to design principles proposed by Lidwell, Holden and Butler (2010), for the development of a questionnaire. Ten (10) individuals were analyzed (mostly university students), that fit the type of public appropriate to the study. The ethical aspects of this research were answered by "standard ERG BR-1002 - Code of Ethics of the Certified Ergonomist (Abergo, 2003)". The participants were informed about the objectives and procedures of the study and then signed the Term of Consent. After the instructions of the task, a questionnaire was applied, aiming to evaluate the various elements of the platform based on their post-use evaluation. Each element was related to a principle of design that, in turn, was related to one of the four parameters of accessibility. This questionnaire was applied after the first use of the lifting platform for individual "non-wheelchair" users, and who had no familiarity with its technology, in a simulated procedure. The participants went through the process of entering and exiting of the tube-station, operating the lifting platform, using a wheelchair. The method, therefore, was the post-task evaluation, with the use of questionnaires, since this allowed to obtain the users' opinion about the product. Were, then, analyzed metrics based on specific issues of usability of the product, guided by the evaluated parameters. As there are several different kinds of platform in use within Curitiba, the tube station located at "Polytechnic Center", at Federal University of Paraná, was chosen for the evaluation process. This platform was chosen because it can be autonomously activated by the user, unlike other types of platform that can only be activated by the agent of the public transportation system that works in the tube-station. The obtained data were statistically analyzed based on their means and standard deviations. Finally, the results were discussed.

Keywords: accessibility, platform lift

1   INTRODUCTION

In the late twentieth century, society has come to understand and be concerned with forms of life on the planet. This led to a better care about the human being, as well (Lima and Silva, 2004). In the early 60's, specifically, emerged a concept of "barrier-free design" (Mazzoni et al., 2001).

This concept is applied to circulation problems that affected, and still affecting, wheelchair users in buildings, urban spaces, etc. In the 90's, the use of the internet to brought to disabled persons new possibilities for social inclusion with regard to study, work, etc. This technological breakthrough has also allowed the development of speech synthesizers, electronic magnifiers, Braille code, mouse and keyboards simulators with various forms of control, etc. (Mazzoni et al., 2001).

This has broadened the concept of accessibility, by passing from a "barrier-free design", in physical space, to the current concept that also involves aspects of the digital world, known as "designing for all" or "universal design". It is noteworthy that the creation of unique environments for people with disabilities can be considered a form of discrimination, that's why should be a concern from the project, developing environments of joint use to disabled and non-disabled persons (Mazzoni et al. 2001).

This article aims to evaluate the accessibility of the lift platform used in tube-stations of the city of Curitiba - PR, and a bibliographic study, as well.

2   FUNDAMENTALS

An evaluation of accessibility can be conducted in various ways. To carry out this research was done a bibliographic study aimed to analyze the way in which accessibility assessments have been conducted and how they are being addressed. Examples of studies on accessibility in buildings, on public transportation and street furniture will be cited.

The first example, in buildings, is a study by Bacil and Watzlawick (2007) that evaluated the urban and environmental accessibility of the leisure area of the Aquatic Park in the city of Irati - PR, identifying existing barriers to propose a accessible location to people with reduced mobility. For this purpose was used questionnaires that were intended to participants of the Pro-Love Association of Physically Handicapped of Irati ("Associação Pró-Amor de Deficientes Físicos de Irati" - APADEFI), people with disabilities and reduced mobility. It was found, with this research, that many recreation areas, such as the Water Park Irati / PR, do not meet the requirements for accessibility to persons with reduced mobility.

Similarly to the study of Bacil and Watzlawick (2007), Aragão (2004) aimed to review the architectural conditions of access for disabled persons to hospital services in the city of Sobral-CE and identify architectural barriers inside the hospitals. However, in this case, was used checklist, verifying compliance or non-compliance in accordance with Law n° 7853, of 24/10/1989, that establishes standards that ensure the exercise of individual and social rights of people with disabilities and their effective social integration. It was concluded that there are architectural barriers on the way home/hospital, direct access to the hospital and in their internal environments, disregarding the disabled people and the law.

Unlike the methodology used by Bacil and Watzlawick (2007) and Aragão (2004), Carlin and Ely (2005) used observations, interviews and accompanied tours, to assess the conditions of spatial accessibility (orientation, displacement, use and activities participation) of the Itaguaçu Shopping - São José - SC. Deficiencies were found in the environment with respect to spatial accessibility, particularly with respect to the spatial orientation of the user.

Unlike the previous, Mont'alvão (2006) used a more subjective form of evaluation. The author developed a survey, destined to visually impaired persons, aiming to verify the perception of these citizens regarding the accessibility of the built environment in the city of Rio de Janeiro - RJ. The location chosen to the accomplishment of this research was the Benjamin Constant Institute ("Instituto Benjamin Constant" - IBC), which meets and hires people with visual impairments. Forms were used to obtain information about the profile of the respondent and about any problems he finds in your routine. This research showed that the environment creates false expectations and that the auditory stimuli are incomprehensible or zero.

With a more comprehensive proposal than only evaluate some environment, Ferreira and Sanches (2010) described a procedure to quantify the costs of suitability of the sidewalks and crossings to the needs of all users, with the aid of a software. For this, was measured the level of accessibility of sidewalks in a urban area of São Carlos - SP and developed a program that calculates the cost of the interventions. The evaluation of the level of accessibility occurred in three steps: technical assessment (qualitative analysis of the attributes of physical characterization of the sidewalk); weighting of these attributes according to the viewpoint of wheelchair (obtained by an opinion survey); and the definition of a quality indicator (called the Accessibility Index). This procedure was used in two different situations in a midsize Brazilian city, proving its easy application, and the results are adequate, compared with the observed reality of the place.

The first example that comes to transportation, but deals specifically with street furniture, is the study of Carvalho and Silva (2011). In this study, the authors sought to identify the conditions for accessibility of bus stops in the Distrito Federal - DF. The place chosen was the Pilot Plan of Brasilia, as it concentrates more than 70% of all destinations of the buses in the city. For this, questionnaires were applied to both agents and users of public transportation system. Interviews were also conducted in some bus stops. With these results, was observed divergences between the views of the user and the agents of the public transportation system and the low accessibility for the user.

Methodologies such as those mentioned above are not sufficient to assess the accessibility of street furniture, according to Marghani et al. (2010). In this study, the authors present concepts and procedures for evaluation of street furniture according to their accessibility and, for them, the proposed methodologies, which consider the questions of Post-Occupancy Evaluation ("Avaliação Pós-Ocupação" – APO), allow a better assessment street furniture, facilitating the development of accessible objects projects, answering a bigger portion of the population.

Now, specifically in transportation, Lobo et al. (2010) propose to evaluate the efficiency of accessibility provided by bus transportation in the city of Belo Horizonte - MG. This efficiency was analyzed based on variables such as distance, time and speed of movements by buses between spatial units. The result was represented by Efficiency Ratio Accessibility ("Índice de Eficiência de Acessibilidade" - IEA) and showed differences between the regions of the city, and the best indicators refer to the North and South regions.

Unlike Lobo et al. (2010), Miranda, Okimoto and Silva (2011) conducted a literature review on accessibility, for people with reduced mobility, to the public transport system in Curitiba - PR. This study was focused on the platform lifts used in the tube-stations (platform that assist in the embarkation and disembarkation of passengers). The research was based on National Technical Standards, Guidelines and Recommendations for International and theoretical literature of Inclusive Design, as a tool for social inclusion and accessibility. This study concluded that the Brazilian Technical Standards analyzed have design and operating (electro-mechanical) criteria of the lift platform, giving the minimum elements to user safety, but there are no recommendations to adopt the principles of universal design.

3   METHODOLOGY

The study was conducted by the Ergonomics Laboratory (LabErgo) of the Mechanical Engineering Department, of the Federal University of Paraná, Curitiba - PR - BR. As there are several different kinds of platforms in use within Curitiba, the tube-station located at "Polytechnic Center", at Federal University of Paraná, was chosen for the evaluation process. This platform was chosen because it can be autonomously activated by the user, unlike other types of platform that can only be activated by the agent of the public transportation system that works in the tube-station (Figures 1, 2 and 3).

Figure 1 Image of the platform lift chosen for the study.

Figure 2 Highlight of the instructions for use of the platform lift chosen for the study.

Figure 3 Highlight of the controls of the platform lift chosen for the study.

To evaluate the accessibility of the platform lift was applied questionnaires. To develop this questionnaire, was performed an analysis of the 125 design principles described by Lidwell, Holden and Butler (2010). Of these 125 principles described, 22 were selected, which may have some application in the platforms. These 22 design elements were separated according to the four parameters described by Dischinger et al., 2009 (apud Ely and Silva, 2009) to classify an environment according to their accessibility. These parameters mentioned above are: spatial orientation, displacement, use and communication.

The questionnaire presents thirteen criteria (separated according to the four parameters mentioned above, as shown in Table 1), related to some of these design elements, the participant should review with a note on a scale of zero to ten.

Table 1: Distribution of the criteria used in the questionnaires according to the parameters described by Dischinger et al., 2009 (cited in Ely and Silva, 2009)

Parameter	Criteria
Communication	1, 7, 9, 10, 12, 13
Displacement	4, 5
Spatial Orientation	2, 9, 11
Use	3, 6, 8

Then, was realized a simulation procedure of first use of the platform lift to individuals non-wheelchair users and without familiarity with the platform's technology. The individuals would make the route of entry and exit of the tube-station, through the platform lift, using a wheelchair. The post-task evaluation was assured by applying the questionnaire after the procedure.

Were analyzed eight individuals (mostly college students) who fit in that we sought. The ethical aspects of this research were met by "standard ERG BR-1002 - Code of Ethics of the Certified Ergonomist (ABERGO, 2003)" by completing a Letter of Consent signed by all participants prior to the procedure.

The analysis was made from the means of the ratings given by participants for each criterion established. Also, it was made an analysis of the criteria mean of each parameter.

4 RESULTS

The answers given by participants are shown in Table 2, with the means for each criterion.

Table 2 Results of questionnaires with the means by criteria

		Question												
		1	2	3	4	5	6	7	8	9	10	11	12	13
Individual	1	5	10	5	6	5	2	7	8	7	8	10	8	7
	2	8	10	8	5	9	4	6	10	8	7	9	6	9
	3	6	7	7	4	4	4	7	7	7	8	8	6	8
	4	2	8	6	7	7	0	3	7	7	3	8	4	7
	5	4	8	3	1	3	5	8	6	7	5	8	2	6
	6	5	7	3	9	3	6	6	7	8	5	8	5	7
	7	6	4	6	4	4	1	5	6	5	7	3	6	8
	8	10	10	8	7	8	2	2	10	10	4	10	6	7
Mean		5.75	8	5.75	5.375	5.375	3	5.5	7.625	7.375	5.875	8	5.375	7.375

The results of the questionnaires show that the platform lift fails with respect to ease of use. It's visible that the worst grade assigned by the user was the criterion of access to commands, resulting in a user evaluation mean equal to three. This occurred because the commands are covered by a visual element that complicates their identification.

Following this criterion, seven other criteria appear with a mean of about five. These criteria are: the symbol that was used in the instructions; the highlight of the emergency button; the sidebar of the platform, as an element of restriction; the sidebar of the platform, as an element of defense; operating instructions, present in the side panel platform; clarity of the use instructions; and ease of reading the instructions. These criteria indicate that the lift platform is flawed in the way the commands, the use instructions and the mechanisms that transmit user safety were developed.

The other criteria showed better evaluation by the user, obtaining a mean better than seven. This indicates that the platform does not have problems in these aspects.

To make an analysis by parameters, was calculated the mean of the criteria for each parameter, as shown in Table 3. Can be seen that the parameters "use" and "displacement" were the worst among the four (with a mean about five) and the orientation parameter achieved the highest rating (with a mean greater than seven).

Table 3 Means, by parameter, of the results obtained by questionnaire

Parameter (Criteria)	Means by parameter
Communication (1, 7, 9, 10, 12 e 13)	6,208
Displacement (4 e 5)	5,375
Spatial Orientation(2, 9 e 11)	7,792
Use (3, 6 e 8)	5,458

6   CONCLUSIONS

With this study, was possible to realize the broad scope of the issue of accessibility with regard to research, and demonstrated the wide variety of methods existents for both the bibliographic studies and for assessments. Also, is visible the concern in trying to make environments more accessible or at least provide the tools necessary to do it.

With respect to the platform lift, it was realized that, despite having most of the reasonable elements of design, the platform lift hasn't been well evaluated by the user. This was justified by flaws in the design/development of the platform lift.

Thus, for future research, is the possibility to find suggestions that can improve the accessibility of the tube-stations' platform lift, mainly on issues of use and communication, which, as seen, had a bad rating by the user.

ACKNOWLEDGMENTS

The authors would like to acknowledge the "Araucaria Foundation for the Support of Scientific and Technological Development of Paraná", the "Coordination of Improvement of Higher Education Personnel" (CAPES) and the "National Counsel of Technological and Scientific Development" (CNPq) for the financial support to the project and the research participants.

REFERENCES

Aragão, A. E. A. A. 2004. Acessibilidade da pessoa portadora de deficiência física aos serviços hospitalares: avaliação das barreiras arquitetônicas. 104 p. Dissertação (Mestrado em Enfermagem) – Universidade Federal do Ceará, Fortaleza.

Bacil, M. K. and L. F., Watzlawick. 2007. Análise da acessibilidade de pessoas com mobilidade reduzida no parque aquático, Irati-PR. Revista Eletrônica Lato Sensu – Ano 2, n°1, julho de 2007. ISSN 1980-6116 http://www.unicentro.br – Engenharia

Carlin, F. and V. H. M. B., Ely. 2005. A acessibilidade espacial como um dos condicionantes ao conforto de usuários em shopping centers – um estudo de caso. Alagoas: Anais do ENCAC- ENLAC 2005.

Carvalho. E. B. and P. C. M., Silva. Indicadores de acessibilidade no sistema de transporte coletivo: proposta de classificação em níveis de serviço. 15p. Dissertação (Mestrado em Transportes) – Universidade de Brasília, Brasília. Acessed August 7, 2011, http://www.turismoadaptado.com.br/pdf/trabalhos_e_pesquisas/acessibilidade_no_trans porte_coletivo.pdf

Ferreira, M. A. G. and S. P., Sanches. 2010. Melhoria da acessibilidade das calçadas – procedimento para estimativa de custos. PLURIS, 2010

Lobo, C.; L., Cardozo and R., Matos. 2010. Transporte coletivo em belo horizonte: a eficiência de acessibilidade com base na pesquisa domiciliar origem e destino de 2002. PLURIS, 2010

Marghani, V. G. R.; L. Z. T., Raffaela and C. F. M., Fernanda. 2011. Avaliação do mobiliário urbano com ênfase na acessibilidade. Revista Eletrônica Brasileira de Ergonomia – Volume 5, n° 1.

Miranda, C.; M. L. L. R., Okimoto and C. M. A., Silva. 2011. Acessibilidade de pessoas com mobilidade reduzida ao sistema de transporte público de Curitiba - PR: a plataforma elevatória da estação-tubo

Mont'alvão, C. 2006. Acessibilidade no ambiente construído carioca. Paraná: Anais do 7° Congresso de Pesquisa & Desenvolvimento em Design, 2006.

CHAPTER 18

Sustainable Communication Design: With Human Centered Research

Maria Cadarso, Fernando Moreira da Silva

CIAUD – Research Centre for Architecture, Urban Planning and Design
From FAUTL - Faculty of Architecture of Technical University of Lisbon
Lisbon, Portugal
mc@sustentadesign.pt

ABSTRACT

"The power to make fundamental improvements to our world."
Kyoto Design Declaration 2008

The Kyoto Design Declaration, signed in 2008, was the contribution from CUMULUS members to a shared responsibility in achieving a sustainable development. For Communication Designers this commitment means designing while seeking for a balance between social, environmental, and economical aspects. Moreover, to find a new paradigm which is more, respectful, transparent, and culturally meaningful, to the community

Sustainable communication designer is a recent discipline still trying to find a practice, a methodology, and a process fully in favor of sustainability. As we currently investigate it: the proposed path uses research to speak to communities, using their background, their context, to transparently address their needs, to help then finding the required solutions, while causing no waste, or at least the less environmental impact possible.

The design process. To find sustainable concepts in Communication Design, we need to go back to the project phase, and re-think the whole process. Resources are critical, energy, water, paper, etc. But not just that, the whole process, needs to be reworked. How designers communicate, is also major challenge that needs to be more immaterial and avoid, for instance, massive paper waste.

The right values and motivations. How Communication Designers, express messages, is also extremely relevant. A message is always embedded with culture. Our receptors should not be the "target", but the community; a group of interacting people that have a geographical relation and share a common understanding of culture, values and beliefs.

In summary, with this paper we hope to bring relevance in showing how designers and ergonomics may address sustainability. The new "measures" are the immeasurable ones, such as values, identity, engagement, or motivations. And like any person as his own measures, also human needs have different level of being fulfilled. What is more, addressing those correctly, and we may expect to liver better with less.

Keywords: communication design; process; research; sustainability

1 INTRODUCTION

"From education to global responsibility."
Kyoto Design Declaration 2008

The Kyoto Design Declaration (CUMULUS, 2011), was signed in March 2008, by 124 CUMULUS members, who declared committed to share global responsibility for building sustainability, human centered and creative societies. The Kyoto Design Declaration, is among others, an open invitation for designers to assume new rolls, re-think their paradigm, and propose new values, in order to make a sustainable difference to the word. For this paper we quote the seven guidelines, as a way to demonstrate how the Kyoto Design Declaration, can be applied to the emerging discipline of Sustainable Communication Design.

The expression "visual communication" was coined in the United Sates in 1966 to cover various areas: typography, posters, layout, illustration, logo creation and so on (Rouard-Snowman 1992). Jorge Frascara, explains that although "graphic designer" is a widely accepted term, has however, contributed to an obscure profile, of the profession. In alternative, he suggests, the use of Visual Communication Design, as a way to reflect, the three main areas of work of these professionals.

For this research, Communication Design is used in its extended definition, including: graphic design, advertising, photography, and illustration; but also other forms, such as media, sensorial, events, or social design, among others. These other areas have their own uniqueness and methodologies, but they can be used to convey a message, a concept, or an idea, with a specific intention, to a specific public.

Communication Designers play a central role in the sustainable quest, but also ergonomists, which have measured the human body, to draw from, and for it. However nowadays, we find ourselves with other scales, immeasurable ones. And far more important, such as, community engagement, fair society, equal opportunities, values, culture, or fulfillment of human needs. Sustainability is about those immeasurable scales, which need to be researched, and that bring human again to the centre of the equation, in balance and closed loop with nature.

How Sustainable Communication Designers convey their messages? To whom? In what context? What arguments are reasonable? Which are truly achievable?

What cost? These are questions a professional should pose himself before starting any project. The paper here presented hopes to clarify some of these questions, in order to enable designers to find more sustainable solutions in their practice.

2 SUSTAINABLE DEVELPMENT

"Seeking collaboration in forwarding the ideals of sustainable development."
Kyoto Design Declaration 2008

The first attempt to analyze, what were the consequences of the economical development to the world, was done by The Ecologist magazine, in 1972, in it's pioneering issue "Blueprint for Survival", where they stated the need to reject development (O'Riordan 2000). This concept was highly criticized, because it was radical and problematic; "it is based on an assumption that frugality, self-restrain and the joy of minimalism is somehow an ideal human condition" (O'Riordan 2000).

Later in 1983, The World Commission on Environmental and Development (WCED), was created under the chairmanship of Mrs Brundtland. This commission task was about "identifying and promoting the cause of sustainable development" (O'Riordan 2000) and the fist definition of sustainable development was:

> *"For development to be sustainable, it must take account of social and ecological factors, as well as economic ones; of the living and non-living resource base; and of the long-term as well as the short term advantages and disadvantages of alternative action." (WCED 1983 cited in O'Riordan 2000)*

Many important conferences and documents have come after, but it is important to understand the begging. Nowadays, it is estimated that there are 64 sustainable development definitions and many interpretations of the term "sustainability", and the number will continue to grow as the global debate on the topic widens. For some, it means maintaining the status quo. For others it is equated with notions of responsibility, conservation and stewardship. However for a growing number of people, sustainable development is a "triple bottom line" activity, based in economic, social and environmental impacts.

As O'Riordan (2000) concludes, in his book "Environmental Science for Environmental Management", "the transition to sustainability has begun. We may not be able to verify it independently of other modernizing drives, but that is an advantage". But he also argues that there are two different scenarios in this matter. In one side, there are significant changes, pointing out positive paths towards a sustainable development. But on the other one, if we fail to take action, that may result in an unbalanced society stressed with growing fringes living with "inequity, debilitating health and seemingly permanent exclusion from the human dignity of a reliable income, an education, a weatherproof shelter and a regular meal, and the respect of race, religion or gender. That growing disparity is occurring both globally and locally in all societies of the globe" (O'Riordan 2000).

In short, if for one side sustainable development has begun, for another one, we are, still quiet far, from a clear solution. However, some of the most import aspects are: first, acknowledge global interdependency, in all scales. Secondly, engaging with others to find common paths and solutions. And finally, sharing information: academics with general society, governance with non-governmental organizations (NGO), corporate with academic, and so on.

2   STEPS TOWARDS A DESIGN FOR SUSTAINABLITY

"The imperative for designers to assume new roles."
Kyoto Design Declaration 2008

In 1971 Papanek, in his book "Design for the real world", challenged designers, for the first time, to act upon a social responsibility. He wrote that designers could propose from simple solutions, to products, or services to be used by the community and the society (Papanek 1985). In 1971 the world faced the first energetic crisis, and in 1974 the petroleum barrel was costing more then ever before. Since then, there has been a rising environmental awareness. It was facing the need to produce eco-solutions that eco design first emerged. In the early 90s TuDelf University and Philips designers' created a method for Life-Cycle Assessment (LCA), which is the evaluation of the environmental impact cost of products. They are analyzed during production, duration / usage and disposal, in what concerns resources, and energy, (Faud-luke 2002).

Taking into account that we need to consume fewer resources the model "Factor 4" was first proposed in 1995 by Weizsacker, Lovins, H, Lovins, A (Faud-luke 2002), as a model that would hold the key to sustainability. However "Factor 4", was soon proved not to be sufficient and the Factor 10 was recommended, meaning we need to reduce by 90% the use of natural resources, in a global scale by 2050 (Faud-luke 2002).

Ezio Manzini and Carlo Vezzoli (2002) took Design for Sustainability a step further. They proposed an interconnected system of services instead of physical objects for a sustainable quotidian. Also, Ezio Manzini with François Jégou (2003), have been researching in sustainable lifestyle options. The concept is based on communities that engage themselves in finding sustainable solutions for their daily problems, like taking care of children, older people, cooking for the community, or lift sharing. By proposing this, Ezio Manzini, wishes to empower people to find the necessary solutions by using the available resources at hand (Manzini, Collina, Evans 2004). When Manzini proposes shared products or services as an alternative, he is looking for less production and less consumption of resources, but he also brings design to a new levels: service design, social, community design, just to mention two of then; and those are very interesting new areas for designers to research and work.

Parallel actions were emerging among graphic and communication designers. In 1964, twenty-two Graphic and Communication designers signed the "First Things First Manifesto", led by Ken Garland (1964), that in short claimed that Graphic Designers should be able to work independently and regardless marketing, and advertising objectives, to pursuit more valuable causes. Later in 2000, Max Bruinsma, re-proposed the manifest and the debate to designers. In a short perspective, the manifesto can be perceived as naive, but in a wider sense it is an alert, and a call for action, in times of growing challenging conditions. From the other side of the argument, we have the majority of designers who claim no responsibility, as they work for a client and under their briefing. Although the "First Things First Manifesto" was a call for action, involvement and engagement to graphic and communication designers, it had less impact then the desired one.

However, over time, there has been a rising consciousnesses. Sets of principles and guidelines have emerged from declarations such as: Ceres Principles, in 1990, the Hanover principles, in 1992, or the EIDD Stockholm Declaration in 2004, od the one here mention, the Kyoto Design Declaration. Specifically committed to Sustainable Communication Design, is the Society of Graphic Design of Canada (GDC 2009), which in April 2009, during the annual general meeting, proposed the first definition for what the concept could be.

> *"Sustainable communication design is the application of sustainable principles to communication design practice. Practitioners consider the full life cycle of products and services, and commit to strategies, processes and materials that value environmental, cultural, social and economic responsibility".*

This definition came with a statement of values and principles to guide the GDC's members during their design practice (GDC 2009). The statement has three parts; the first one is about designers assuming their responsibility in this interconnect world. The second is about the in-house changes that can be done. And the third part is a set of guidelines for the design practice and client advising.

To sum up, these are times with unprecedented challenges. Designers had a role, which has changed dramatically over time. During the industrial times, forms and colors could follow function, as many times as desire could imagine, but not anymore. Assuming we all need to consume less, designers need to move gradually from eco to sustainable, exploring new roles, achieving more with less.

3   THE DESIGN PROCESS

"An era of human centered development."
Kyoto Design Declaration 2008

Most authors of design process have been, consistent, in dividing the communication design process, in three main areas: analyze, creativity and the execution. According to Frascara (2004) the sequence is: 1 commission of the project; 2 collection of information; 3 second definition of the problem; 4 definition of objectives; 5 third definition of the problems; 6 development of the design proposal; 7 presentation to the client; 8 organization of production: 9 supervision of implementation; and 10 evaluation of performance.

However, the design process for communication designers must be reviewed, to give place to a process where research plays a much central role. It becomes important because, when starting a process, everything needs to be re-questioned. Solutions are no longer obvious. Corporate have changed, more concerned about their image, and the potential impact from negative news (O'Rourke 2005), Consumer, is no longer a massive target, but a community of informed people (Nordstrom & Ridderstrale 2005). Functions and service should prevail over objects so we can have dematerialization (Manzini &Vezzoli 2002).

Using data from the United States, we can understand the need of Communication Designer working towards a Sustainable Development: "Americans receive over 65 billion pieces of unsolicited mail each year, equal to 230 appeals, catalogues and advertisements for every person in the country. According to the not-for-profit organization Environmental Defense, 17 billion catalogues were produced in 2001 using mostly 100 percent virgin fiber paper. That is 64 catalogues for every person in America" (Dougherty 2008). For Communication Designers paper is one support, but they use many other that also need to be researched.

In a word? Research! Research for new design processes which can guide designer better. Research to find a way to communicate, with no waste or without environmental and social impact. Research, to meet functions or services instead of products. Research about consumers so, we can find out, how to dematerialize their needs. Finally research the problem, so we can find truly new solutions.

4   HUMAN CENTERED RESEARCH

"Proposing new values and new ways of thinking."
Kyoto Design Declaration 2008

If designers play a central role in the artifacts that are massively produced everyday, communication designers, have it also when advertising them. Recent research has brought new insights about adverting in most industrialized countries, and its effects on consumers (Thorpe, 2007). To start, values seem to be subverted, and, owning appears to be more important than being (Thorpe, 2007). Commerce takes advantage from that and keep presenting consumers "new" alternatives of "being". Which brings us to another problem: identity, and how fast and superficially, is replaced (Robert, et al. 2008). Third, knowing people like to invest in their "façade", commerce and brands, invest a lot of money, creating perfect scenarios, with perfect people, and perfect lives, that remotely relate to our common ones. Although it is known that marketing "promises", are not real, consumers prefers, to indulge in believing them (Kapferer 2008). Fourth, is about community.

Usually, community is about, a common understanding of values, beliefs, and culture, that used to have a geographical limitation. Now with internet, they bound in global scattered cultural tribes; that may share common interest, but hardly ever engage as a real community (Nordstrom & Ridderstrale 2005). Fifth, and final, social networks, as net communities, invite people to share their lives, and oddly they accept. Sharing personal information instantly, indiscriminately, and quiet often shamelessly (Kenny, 2009).

Sustainable communication designers, when designing, should acknowledge these issues and change the paradigm. First, being should be more import than having. Secondly, identity cannot be replaced as quickly as a pair of Nike's shoes. And for that advertising needs to address the right motivations. Third, is about expectations, which need to be addressed with realism, as real life, with real people. Fourth, is about engaging, finding true connections, and not just shared interests. Meaningful relations are very important, for our sense of happiness. We enrich our lives, when we love, we care, we dedicate time to others, and them to us. Fifth, and final, is about fulfillment. Sharing our lives and reporting it by the minute, is like chasing our own shadow. Trying to prove that our lives are not empty has just the opposite effect. Publishing thousands of details, and instant pictures of no information, just puts in evidence, the absence of personal realization. It should be quality not quantity. But above all, growing and maturity should be an internal process, not external.

Concisely, we know designers, are meant to address human needs, it is just those "needs" that require research in a deeper level. In a sustainable role, communication Designers, should deal with the right expectations, and restrain to inspire frail identities. Motivations should seek a deeper fulfillment, and engaging with the community, should be local, and meaningful.

CONCLUSION

"Implementation."
Kyoto Design Declaration 2008

Looking in perspective the growing need to have a sustainable pattern is evident. For Communication Designers it means, breaking down the intervenient parts, define areas of interaction, but above all looking for a paradigm shift that would find balance between social, nature and economic aspects. Designers need to acknowledge their new roles, based on the growing and fast changes of the world. Environmental and social, limitations and frailties, should inspire groundbreaking outcomes. If research should be reason in heart, in order to re-think the sustainable communications design process. Then, spirit should be in motivations.

In conclusion, Sustainable Communication Designers show encourage that about being is more important than having. Individuals can still expect comfort in their lives, and will have ways to express their identity, but it should be with less, or more importantly, it should be differently. By achieving so, Sustainable

Communication Designers, perhaps, might develop in individuals areas of creativity, in which they find a deeper fulfillment.

ACKNOWLEDGMENTS

The authors would like to acknowledge the Foundation for Science and Technology, for their financial support, and CIAUD – Research Center in Architecture, Urban Planning and Design, from Faculty of Architecture, TU Lisbon.

REFERENCES

Dougherty, B 2008, *Green Graphic Design*, Allworth Press, New York
Frascara, J 2004, *Communication Design: principles, methods and practice*, Allworth Press
Fuad-luke, A 2002, *The eco-design handbook*, Thames & Hudson, London,
Kapferer, J-N 2008, *The new strategic brand management*, Kogan Page, London
Kenny, K 2009, *Visual Communication Research Designs*, Routledge, New York
Manzini, E and Vezzoli C 2002, *Product-Service Systems and Sustainability: opportunities for Sustainable Solutions*, United Nations Environmental Programme, Milano,
Manzini, E and Jégou, F 2003, *Quotidiano Sostenibile - Scenari di Vita Urbana*, Edizione Ambiente, Milan,
Manzini, E, Collina and Evans, S 2004, *Solution Oriented Partnership - How to design industrialised sustainable solutions*, Cranfield University, Bedfordshire,
Mcdonough, W & Braungart, M 2002, *Cradle to cradle*, North Point Press, New York
Nordstrom, K. & Ridderstrale, J. 2005, *Funky Business - O capital Dança ao Som do Talento*, Portuguese edition, Fubu Editores, S.A, Porto
O'Riordan, T 2000, *Environmental Science for Environmental Management*, Pearson Education Limited, Essex
O'Rourke, D 2005, Market Movements: nongovernmental organization strategies to Influence global production and consumption. *Journal of Industrial Ecology*, **9**(1-2), pp. 115-128
Papanek, V 1985, *Design for the real world*, Thames & Hudson Ltd, London
Robert, E, Wright, M & Vanhuele, M 2008, *Consumer behavior*, London
Rouard-Snowman, M 1992, *Museum Graphics*, Thames and Hudson Lta
Thorpe, A 2007, *The designer's atlas of sustainability*, Island Press, Washington

WEBSITES
CUMULUS, 2011, http://www.cumulusassociation.org
Ken Garland, http://www.kengarland.co.uk
SOCIETY FOR ENVIRONMENTAL GRAPHIC DESIGN 2009, http://www.segd.org/

CHAPTER **19**

Using Virtual Environment to Investigate Way-finding Behavior in Fire Emergency

Fanxing Meng, Wei Zhang

State Key Laboratory of Automobile Safety and Energy, Department of Industrial Engineering, Tsinghua University, Beijing, China

ABSTRACT

Way-finding is an important activity in people's daily life. Currently, many studies have been conducted on human way-finding behaviors under normal conditions. However, few studies are reported on human way-finding behaviors under emergent conditions, such as terror attack, fire incidents and other natural incidents. The objective of this study is to compare the difference of human way-finding behaviors under normal and fire emergency in virtual environment. Twenty participants were divided into two groups, and they were instructed to find an exit as soon as possible under normal and fire emergency in the same virtual hotel environment, respectively. Fire emergency was created by providing fire alarm, virtual fire and smoke, as well as smoke in real world. Results showed that participants had higher skin conductivity (SC) and heart rate (HR) in fire emergency than under normal condition during the process of way-finding. Participants need more escape time and travel longer distance in fire emergency than under normal condition. However, no significant difference was found in fixation duration between the two conditions. The present study provides more knowledge about the characteristic of human way-finding behaviors under emergent conditions, which is important for emergency management in public place and emergency evacuation in multi-agent simulations.

Keywords: Emergency; Way-finding; Eye tracker; skin conductivity; heart rate

1   INTRODUCTION

From the time when the word "way-finding" was first proposed by Lynch, 1960, way-finding studies have been given extensive attention by many researchers in different fields. Way-finding is the process of determining the route from the starting point to the destination and then successfully arriving at the destination (Gluck, 1991; Golledge, 1999). Indoor and outdoor way-finding behavior under normal conditions has been studied extensively (Evans, Smith et al., 1982; O'Neill, 1991; O'Neill, 1992; Lawton and Kallai, 2002). However, results reported by studies conducted under normal conditions may not be transferred to that in fire emergency directly. The study of human behavior in fires is still in its infancy, although beginning with the Wood's study completed in 1972 (Paulsen, 1984), and our knowledge of occupants' performances when confronted with fire is still very limited (Kobes, Helsloot et al., 2010). Hence, it is extremely important to study the way-finding behavior in fire emergency, where small variation in behavior can mean the difference between survival and extinction (Hancock and Weaver, 2005).

According to theories of information processing, given the same set of information, people may attend to information differently depending on the degree of stress and the amount of time pressure they experience (Janis and Mann, 1977; Miller, 1960). Stress and time pressure will affect people's perception of various environmental factors and thereby influence their action in a certain situation (Ozel, 2001; Nilsson, Johansson et al., 2009). Fires are perceived as very stressful and a person, who has to decide how to get out of a building, and away from an uncontrolled fire, is under extreme psychological stress (Benthorn and Frantzich, 1999). In such a stressful situation, people's emotional state is likely to influence how the individual interprets the situation and acts upon emergency information. A person's perception of the situation, such as interpretation of emergency information and other people's actions, as well as his or her subsequent behavior, i.e., decision to evacuate, choice of exit and pre-movement time, partly is a result of how he or she feels at the time (Nilsson, Johansson et al., 2009). Therefore, Fire emergency situation can cause psychological effects, and way-finding behavior may display different patterns with what we expect. Studies of evacuation from indoor buildings have shown that people do not always use the closest emergency exit (Nilsson, Johansson et al., 2009), but to move towards familiar places and use familiar exits in the event of a fire emergency (Sime, 1985). Thus, when a fire breaks suddenly, people usually choose to leave a building by the same way they came in, even if this is a poor alternative than others available (Benthorn and Frantzich, 1999). The study of Tang, Wu et al., 2009 conducted an experiment to determine the effectiveness of emergency signs to way-finding behavior. Results showed that most participants did not pay attention to the emergency exit signs during fire evacuation. All these results showed that people's way-finding behavior were not what they behaved under normal conditions and emergency evacuation system should be tested under stressful conditions before they could be relied upon in a real fire (Nilsson, Johansson et al., 2009).

In conclusion, human way-finding behavior in fire emergency takes place under considerable physical and psychological stress and time pressure, which can't be predicted from those in normal conditions simply. Currently, what we know about people's way-finding behavior in fire is still far from enough, and it is therefore important to know how people behave during fire evacuation by further experiments.

Fire emergency study is difficult. For moral and legal reasons, we are not permitted to deliberately expose normal experimental participants to real fire condition, which will pose a life-threatening degree of risk (Hancock and Weaver, 2005).Virtual reality (VR) technique is an effective tool offering the potential of addressing the requirement of fire emergency study and a number of formal studies have shown its application (Smith and Ericson, 2009; Smith and Trenholme, 2009; Ren, Chen et al., 2006; Cole, Vaught et al., 1998; Shih, Lin et al., 2000). With VR, participants could immerse themselves in the virtual building environment with virtual fire scenes (Ren, Chen et al., 2006), which could provide a stressful evacuation context, without exposing participants to life-threatening risk. Furthermore, the efficacy of VR in way-finding related experiments was justified by empirical studies carried out within the areas of cognitive psychology, neurosciences and psychophysiology as well (Bosco, Picucci et al., 2008). Also, numerous studies focus on the ability of people to transfer information about the environment acquired from either real or virtual navigation (Bosco, Picucci et al., 2008). They confirm that mental representations of space and move pattern in virtual environment are quite similar to those implicated in navigation of real environment (Skorupka,2009). So, measures of way-finding behavior in virtual fire emergency are predictive of subsequent performance in a similar real fire scene.

In order to know more about people's way-finding behavior in fire emergency, this study conducted a controlled experiment, where VR technology was used to build the virtual navigation environment. The present study has two aims: first, to evaluate the efficacy of VR technology in fire emergency studies, i.e., whether virtual fire scene can induce the feeling of stress for participants; second, to compare human way-finding behavior under normal condition with that in fire emergency.

2   METHOD

2.1   Participants

Twenty participants (10 males and 10 females) between the age of 20 to 25 (mean=22.8, s.d.=1.6) were recruited from Tsinghua University by online posters. They all came from engineering departments, who were divided into two groups, counterbalanced by gender. The two groups were required to complete a way-finding task in the same virtual hotel environment. Group 1, named "Normal Group", completed the task under a normal environment, and group 2, named "Fire Group", completed the task in a fire emergency.

2.2 Apparatus

The VR system used in present study was called Panorama Manifestation System (PM system), and was developed in 2011 (Meng, Zhang et al., 2011). It consists of six 47-inch LCD monitors, a computer with data collection system and a revolving chair with controlling appliances (see Figure 1). The six monitors are connected end by end, forming a ring configuration, which can display the virtual environment surrounding users. The display card of PC is ATI Radeon™ HD 5870 Eyefinity 6 Edition and it connects six LCD monitors with Mini-DisplayPort™ connector. Each monitor has a resolution of 1024×768 and the combined resolution can reach 6144×768. The revolving chair is placed in the middle. A photoelectric decoder is fixed in the revolving axis, which can read the real-time rotation angel of the chair. Users sit in the chair, controlling direction with body rotation. A small keyboard was attached on the right side of the chair, on which an acceleration button and a deceleration button were used to control the translation in virtual environment. When users travel around, the virtual environment keeps stationary. Users interact with environment by body rotation and hand controllers, which provides the same frame of reference with real environment. PM system can display surrounding environments of the user, which provides a 360° field of view.

Figure 1 Panorama Manifestation system in this study

During the experiment, participants' eye movements were recorded using a video-based eye tracker (SMI iView X, Senso Motoric Instruments, Teltow, Germany), sampling at 50 Hz. Participants' skin conductivity (SC) and heart rate (HR) were also recorded with a ProComp⁺ (Thought Technology, Ltd., Montreal, Canada).

2.3 Procedure

Upon arrival, participants were required to read a consent form and complete a background questionnaire. In the warm-up stage, participants were instructed to control their movement in the PM system, and some explanation about this system was given. Before the formal experiment, the eye-tracker and ProComp+ unit were set up to record participants' eye movement of physiological response.

The formal experiment contained a way-finding task in a virtual hotel environment. First, some text appeared in the front of participants and they were required to read it. A few minutes later, the text disappeared and a voice was given: "The hotel was on fire, please stop all your tasks and find the exit as quickly as you can". For the Normal Group, all the conditions in the hotel were normal, and there was no other person except for the participant. For the Fire Group, after the voice, a fire alarm would ring and virtual fire and smoke (Figure 2) appeared in the virtual scene. During the way-finding stage of Fire Group participants, virtual explosions would be activated at some locations and smoke in real environment was generated to provide olfactory and visual stimuli. The environment design in Fire Group was to create a virtual fire emergency, and provide a stressful evacuation environment. Along the corridor, some evacuation signs were set up to direct participant to the emergency exits. However, both the two groups were not told that they could rely on the evacuation signs to find the exit. When participants found the emergency exit and escaped from the hotel, the experiment was completed. During the reading stage and way-finding stage, several dependent variables were recorded, which included escape time, travel distance, fixation duration of eyes, heart rate (HR) and skin conductivity (SC).

Figure 2 The appearance of fire and smoke (left), explosion (middle) in virtual scene and smoke in real environment (right).

3   RESULTS

This study adopted a between-subject experimental design, and the sample size of each group was relatively small. Taking the sample variance into consideration, 10 percent was set as the significance level.

3.1   Physiological Response

First, we want to evaluate whether the virtual fire emergency could provide a stressful evacuation environment and increase the degree of stress for participants. Some previous studies have revealed that SC and HR will increase with the feeling of stress, thus these two indicators were tested in this study.

Result of SC was shown in Figure 3, and SC in reading stage and way-finding stage was compared in both Normal Group and Fire Group. For each participant, either in normal condition or in fire emergency, his/her SC was increased significantly in way-finding stage than that in reading stage (Normal Group: p = 0.01; Fire Group: p < 0.001). However, the SC increment of participants in Fire Group was obviously higher than that in Normal Group. SC ratio was calculated for each participant, by dividing SC in way-finding stage by SC in reading stage. Results showed that average SC ratio in Fire Group (mean = 3.19, s.d.= 2.50) was significantly higher than that in Normal Group(mean = 1.36, s.d.= 0.34), with a p-value of 0.047.

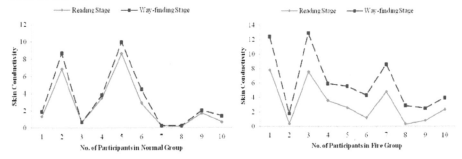

Figure 3 Skin conductivity (SC) in reading and way-finding stage: Normal Group (left) and Fire Group (right)

Similarly, HR was also analysed and compared, and results were shown in Figure 4. For each participant, no matter in Normal Group or in Fire Group, his/her HR was higher in way-finding stage. Paired t-test results showed that the HR increase was significant both in Normal Group (p = 0.001) and Fire Group (p < 0.001). HR ratio was also calculated by dividing HR in way-finding stage by that in reading stage. HR ratio in Normal Group was compared with that in Fire Group by a t-test. Results indicated that the HR increment in Fire Group (mean = 1.23, s.d.= 0.13) was significantly higher than that in Normal Group (mean = 1.12, s.d.= 0.07), with a p-value of 0.037.

Results of SC and HR revealed that participants had higher stress in way-finding stage than in readding stage, either Normal Group or Fire Group. Comparison between the two way-finding environment indicated that participants in Fire Group had significantly higher degree of stress and virtual fire envrionment was significantly more stressful than the environment under normal condition. Hence, it was valid to study way-finding behavior in virtual fire envrionment.

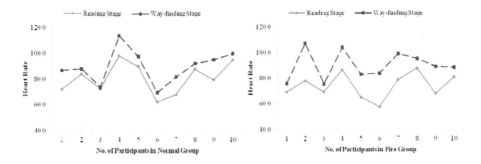

Figure 4   Heart Rate (HR) in reading and way-finding stage: Normal Group (left) and Fire Group (right)

3.2   Escape Behavior

The completion time of Normal Group and Fire Group was displayed in Figure 5. On average, participants in Fire Group spent more time in the way-finding stage than that in Normal Group, and Fire Group showed a higher variation in completion time. The t-test result showed that the difference of completion time between these two groups was significant (Normal Group: mean=122.0, s.d.=34.4; Fire Group: mean=166.0, s.d.= 63.6; p-value=0.076). Participants in Fire Group encountered more difficulties during the way-finding process, and cost more time.

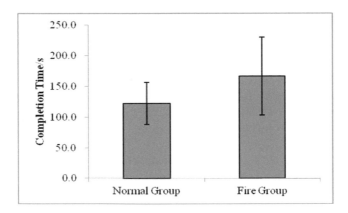

Figure 5 Average completion time of Normal Group and Fire Group

Figure 6 showed the comparison of travel distance between Normal Group and Fire Group. The average travel distance in way-finding stage for Fire Group was 489.8 m (s.d=106.0), which was longer than that of Normal Group (mean=375.3, s.d.= 40.1). The t-test result showed the difference was significant (p = 0.009).

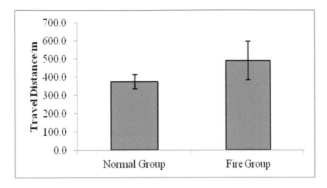

Figure 6 Average travel distance of Normal Group and Fire Group

3.3 Fixation Duration

In the way-finding stage, a visual point with the fixation duration more than 80 ms was treated as a fixation point. The average fixation duration of Normal Group and Fire Group was displayed in Figure 7. On average, participants in fire emergency showed shorter fixation duration (mean= 310.9 ms, s.d.=65.3) than that in normal condition (mean = 344.8 ms, s.d. =36.6), which indicated that participants in fire emergency may search quickly because of shorter available time and higher psychological stress. However, the t-test results didn't show a significant difference (p=0.174). This result may be explained by the great sample variance in Fire Group, whose standard deviation was nearly twice that of Normal Group.

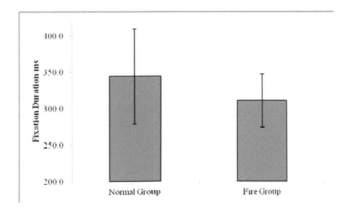

Figure 7 Average fixation duration of Normal Group and Fire Group

4 CONCLUSION

The present study used virtual reality system to investigate way-finding behavior in fire emergency. The comparison results between Normal Group and Fire Group showed that participants in fire emergency had significant higher SC and HR than that under normal condition, which proved that the virtual environment of fire emergency was effective to induce psychological stress during the escape process, and it was valid to conduct fire emergency studies. In addition, this study provided more knowledge about the behavior difference between normal condition and fire emergency. At the same location, participants needed more time to escape and traveled longer distance to find the exit in fire emergency than under normal condition. It may be explained by that psychological stress induced by fire distorted the decision making process and poor visibility in fire and smoke increased difficulties to find the right way. Participants showed shorter average fixation duration in fire emergency though the difference didn't pass significance test. In conclusion, the sample size in this study was relatively small, and great variance in sample may polluted the experiments results. Thus, more samples will be added to the present study and other variables, such as saccade duration of eye movement, the accuracy of time estimation, and the subjective evaluation will also included in the future study.

ACKNOWLEDGEMENT

The authors would like to acknowledge the support of Foxconn "Selected Talent" Research Program and Tsinghua University Research Funding (2009THZ01).

REFERENCES

Benthorn, L. and H. Frantzich. 1999. Fire alarm in a public building: how do people evaluate information and choose an evacuation exit? *Fire and materials* 23(6): 311-315.

Bosco, A., L. Picucci, A. O. Caffo, et al. 2008. Assessing human reorientation ability inside virtual reality environments: the effects of retention interval and landmark characteristics. *Cognitive processing* 9(4): 299-309.

Cole, H. P., C. Vaught, W. J. Wiehagen, et al. 1998. Decision making during a simulated mine fire escape. *Engineering Management, IEEE Transactions on* 45(2): 153-162.

Evans, G. W., C. Smith and K. Pezdek. 1982. Cognitive maps and urban form. *Journal of the American Planning Association* 48(2): 232-244.

Gluck, M. 1991. Making sense of human wayfinding: review of cognitive and linguistic knowledge for personal navigation with a new research direction. *Cognitive and linguistic aspects of geographic space, Series D: Behavioural and Social Sciences* 63: 117-135.

Golledge, R. G. 1999. *Wayfinding behavior: Cognitive mapping and other spatial processes*: Johns Hopkins Univ Pr.

Hancock, P. and J. Weaver. 2005. On time distortion under stress. *Theoretical issues in ergonomics science* 6(2): 193-211.

Janis, I. L. and L. Mann 1977. *Decision making: A psychological analysis of conflict, choice, and commitment*. New York: Free Press.

Kobes, M., I. Helsloot, B. de Vries, et al. 2010. Building safety and human behaviour in fire: A literature review. *Fire safety journal* 45(1): 1-11.

Lawton, C. A. and J. Kallai. 2002. Gender differences in wayfinding strategies and anxiety about wayfinding: A cross-cultural comparison. *Sex Roles* 47(9): 389-401.

Lynch, K. 1960. *The image of the city*. Cambridge, Mass: MIT and Harvard University press.

Meng, F., W. Zhang and R. Yang 2011. Development of Panorama Manifestation System and Its Comparison with Desk-Top System, *International Conference on Virtual Reality and Visualization (ICVRV'11)*.

Miller, J. G. 1960. Information input overload and psychopathology. *The American Journal of Psychiatry* 116(2): 695-704.

Nilsson, D., M. Johansson and H. Frantzich. 2009. Evacuation experiment in a road tunnel: A study of human behaviour and technical installations. *Fire safety journal* 44(4): 458-468.

O'Neill, M. J. 1991. Evaluation of a conceptual model of architectural legibility. *Environment and Behavior* 23(3): 259-284.

O'Neill, M. J. 1992. Effects of familiarity and plan complexity on wayfinding in simulated buildings. *Journal of Environmental Psychology* 12(4): 319-327.

Ozel, F. 2001. Time pressure and stress as a factor during emergency egress. *Safety science* 38(2): 95-107.

Paulsen, R. 1984. Human behavior and fires: An introduction. *Fire technology* 20(2): 15-27.

Ren, A., C. Chen, J. Shi, et al. 2006. Application of virtual reality technology to evacuation simulation in fire disaster. *Proceedings of the 2006 International Conference on Computer Graphics and Virtual Reality*, CSREA Press, USA: 15–21.

Shih, N. J., C. Y. Lin and C. H. Yang. 2000. A virtual-reality-based feasibility study of evacuation time compared to the traditional calculation method. *Fire safety journal* 34(4): 377-391.

Sime, J. D. 1985. Movement toward the Familiar. *Environment and Behavior* 17(6): 697-724.

Skorupka, A. 2009. Comparing Human Wayfinding Behavior in Real and Virtual Environment. *Proceedings of the 7th International Space Syntax Symposium*, Stockholm.

Smith, S. and E. Ericson. 2009. Using immersive game-based virtual reality to teach fire-safety skills to children. *Virtual Reality* 13(2): 87-99.

Smith, S. P. and D. Trenholme. 2009. Rapid prototyping a virtual fire drill environment using computer game technology. *Fire safety journal* 44(4): 559-569.

Tang, C. H., W. T. Wu and C. Y. Lin. 2009. Using virtual reality to determine how emergency signs facilitate way-finding. *Applied Ergonomics* 40(4): 722-730.

CHAPTER 20

Using Thinking Process as a Thinking Tool for the Problem of the Policy of Chronic Illness Refill Slip

Ting-Yu Wu[1], Chao-Wen Tseng[1], Wei Zhang[2] and Sheng-Hung Chang[1] *

[1]Department of Industrial Engineering and Management
Minghsin University of Science and Technology
Hsinchu, Taiwan(R.O.C.)
[2]Department of Industrial Engineering
Tsinghua University
Beijing, China
*shchang@must.edu.tw

ABSTRACT

National health insurance (NHI) system is an important part of Taiwan's social security system and now facing the predicament of financial imbalance. The policy makers have been practiced different policies to improve the predicament. However, the problems the NHI system facing are complicate and dynamic. In this research, we use Thinking Processes proposed by Dr. Goldratt as a prescribe model of human thinking to help people think over the problems of the Policy of Chronic Illness Refill Slip of NHI. We apply the Thinking Processes to national health insurance system. We practiced Step1 (What to change?) and Step 2 (What to change to?) of logic thinking in Thinking Processes to develop the strategic injection (SI). Following that, we analyzed data from the database of NHI of Hypertension to verify if the SI works. With the Thinking Processes we got the following findings: the core conflict for the Policy of Chronic Illness Refill Slip is between "both doctors and patients don't have the willingness to use the refill slip" and "trying to raise refill slip approved ratio". The strategic injection which might solve this problem is "to fix patients to a specific medical institute for outpatient services". Analyzed the data, we found that when the number of medical institutes a patient goes increased the average number of outpatient service increased and the number of medical institutes a patient goes increased the average ratio for taking refill slip decreased. If we can fix the patients to a medical institute for outpatient services, we can improve the excess expenditure for NHI system.

Keywords: Theory of constraints, Thinking processes, Strategic injection , Chronic disease , refill prescription

1. INTRODUCTION

There are different decision making with causal relations but you can't tell during daily life. The "Dynamic Decision Making" means decisions with implicit causal relations (Lerch and Harter, 2001). There are 3 main characteristics in "Dynamic Decision Making" (Edwards, 1962). First, there is a series decisions, not a sole decision. To cope with the changing environment, the decision maker need to make a series decision with different adjustments. Second, there are mutual dependences between decisions. The decision maker made a decision heading the changing of the system status. To cope with that status change, the decision maker needs to adjust his next decision. That causes a circular causality with that series decisions. Finally, there are unpredictable factors in the environment. The system status will change by not only the decision but some other factors in the environment. The decision makers also need to cope with environment factors and adjust their decision. Therefore as well as in a dynamic decision making situation includes decision makers, decision tasks, the interaction between the decision makers and the decision task and the environment. The decision task is usually to operate a system with many variables. To operate, the decision makers follow the system goal and adjust the status of some variables. In different researches (ex, Brehmer, 1987, 1990, Sterman, 1989) we found that people do poor performance in a dynamic decision task. To help people improve their performance in a dynamic decision task becomes a main research issue. National health insurance (NHI) system is an important part of social security system in Taiwan, and NHI can be treated as a dynamic decision system. But NHI is in a struggle situation with financial imbalance. With this struggle situation, it becomes an important operation goal to the NHI authority to make NHI sustainable. The NHI authority has been implemented different policies to correct this financial imbalance situation, such as Global Budget System has been implemented in July 1998 to increase peer review. But the imbalance situation still goes with duplicated prescription and unnecessary outpatient services (NHI Medical Expenditure Negotiation Committee, 2008).

With the changing environment, the expenditure of NHI for chronic illness from 1993 to 1998 is about 40% of the total expenditure of NHI in western medicine and the number is increasing (Bureau of NHI, 2008). Also, the ratio of elders above 65 years old in our population is increasing (Ministry of the Interior, 2009) and the prevalence of chronic illness for the elders is 73.93% (Liang, 2004). To cope with this trend, "Policy of Chronic Illness Prescription Refill Slip" has been practiced by

NHI. With the Refill Slip, the patients can get their prescribed medicine for three times and do not need to see their doctors. Except that if the patients use Refill Slips, their copayment will be exempted. But not as expected, the practice of the Chronic Illness Prescription Refill Slip has no significant effect on the balance of NHI.

One research about the policy of Chronic Illness Prescription Refill Slip has showed that, if Chronic Illness Prescription Refill Slip can go with outpatient drug account, it can save at least 4 -5 thousand million New Taiwan dollars in expenses in transportation to medical care institutions, registry fee and copayments. Even more, it can save 1-1.5 billion in pharmaceutical price gap between hospitals and the National Health Insurance Bureau. It also showed that people don't know about "Policy of Chronic Illness Prescription Refill Slip" and they are afraid that they can't get the original brand drug (Chen and Wu, 2004). On the other hand, the physician won't prescribe the refill slip because they are afraid that the condition of patient became worse, the patients didn't follow orders, and miss the scheduled outpatient service (Kung, Lu and Tsai, 2007). From all these researches, we found that NHI needs a more systemic view for the Policy of Chronic Illness Prescription Refill Slip to make this policy success and the important issue is how to help the NHI authority to make good policy decision with the dynamic characters in NHI system.

There are three different perspectives in the study of people's thinking and deciding. The descriptive perspective is to describe people's thinking and deciding behaviors; the normative perspective concerns the criteria to evaluate people's thinking and deciding; the prescriptive perspective concerns on helping people's thinking and deciding (Baron 2008). In this research, we adopt the prescriptive perspective and use the Thinking Process (TP) that based on Theory of Constrain (TOC) to help the NHI authority thinking and turn out to make a better decision.

The Theory of Constrain (TOC) is proposed by Dr. Eliyahu M. Goldratt in 1986. Based on TOC, Thinking Process (TP) asks three main questions, the first is "what to change", the second is "what to change to", and the third is "How to cause the change". TP develops 5 logic tools to help people thinking about the three main questions, Current Reality Tree (CRT), Evaporating Cloud (EC), Future Reality Tree (FRT), Prerequisite Tree (PRT) and Transition Tree (TrT) (Kim, Mabin et al. 2008). TP can help managers to find out the problem in their system and how to improve the performance of the system (Goldratt, 1994).

The first question is "What to Change"-- to find out the core conflict.

The practice steps for the first question are as follow: 1.to list different Undesirable Effects (UDEs); 2.to pick up 3 main issues in all the listed UDEs to construct 3 EC; 3.to integrate these 3 EC into a core conflict cloud; 4.to use CRT to verify if it is the real core conflict.

The second question is "What to Change To" – to construct complete solution.

The practice steps for the second question are as follow: 1.to find a way (called Strategic Injection, SI) to break out of the Core Conflict once and for all; 2.to define the Desired Effects (DEs) of the solution. 3.to construct a complete solution (FRT) that resolves all of the UDEs. If all DEs can be connected with a causal relation that means the SI can improve all the UDEs. We can also add Tactical Objectives (TO) to complete our SI.

The third question is "How to Cause the Change"—to construct action plan.

The practice steps for the third question are 1.to use PRT to organize all the Intermediate Objectives (IO) that are the obstructions for each TO; 2. to use TrT to change PRT into a detail project or action plan for changing each TO.

TP based on TOC has been applied in different profit and non-profit organizations in different topics such as production management, logistic management, marketing, project management and information technology (ex. Goldratt and Cox, 1986; Goldratt, 1994; Cooper and Loe, 2000; Tanner and Honeycutt,1996; Polito, Watson et al., 2006 Shi, Chang et al. , 2010). The operation of NHI is very complicated. There are different factors affecting the operation performance. It is not easy to tell what the causal relations between different decisions that the NHI authority made. Because of the character of the TP that emphasis on the causal relation, in this research we applied the TP to the policy of Chronic Illness Prescription Refill Slip. First, we use CRT to describe the current reality of the policy of Chronic Illness Prescription Refill Slip. Second, we construct a FRT for the policy of Chronic Illness Prescription Refill Slip. Finally, we use the historic data of NHI to verify the SI really a good solution for the policy of Chronic Illness Prescription Refill Slip for NHI.

2. RESEARCH METHOD

In this research, we use two main questions of TP as a main tool to help people think and analyze historic data of NHI to verify our solution.

For the first questions, "What to Change", the practice steps as below:
(1)   To collect all the UDEs by reviewing literature about the policy of Chronic Illness Prescription Refill Slip;
(2)   To pick up three main issues that most concerned and construct 3 EC for each issue.
(3)   To integrate these 3 EC to find out core conflict.
(4)   To connect all the UDEs with the core conflict and construct CRT.

For the second question, "What to change to", the practice steps as below:
(5)   To find a way (SI) to break out of the Core Conflict once and for all.
(6)   To define the Desired Effects (DEs) of the solution.
(7)   To construct a complete solution (FRT) that resolves all of the UDEs.

We verify our SI by analyzing data about hypertension patients' behavior of seeking medical help and data from medical institutes applied insurance payment from 2006 to 2007 to find out if the constructed SI can introduce real improvement for the financial imbalance of NHI. The data is from the database of Northern branch of Bureau of NHI.

3. RESULTS

3.1 What to Change

After reviewing literatures about Chronic Illness Prescription Refill Slip, we have listed 9 UDEs. There are 3 main conflicts for the policy of Chronic Illness Prescription Refill Slip. For patients, patients won't change their behavior patterns (UDE 3). For medical service institutes, it is hard to control the changing situation for chronic illness patients (UDE 4). For both the patients and medical service institutes, they both doubt the service quality of the pharmacists in drug store (UDE 8). We pick up these 3 UDEs to construct 3 ECs. After that, we integrate these 3 EC to find out the core conflict. We found the core conflict is because we have to raise the ratio that Chronic Illness Prescription Refill Slip use but the patients and medical institutes are not welling to use this Refill Slip. Also, we construct CRT to verify the core conflict that we found.

3.2 What to change to

To solve the core conflict, we proposed a SI that fixes the patient to a specific medical institute for treatment. After that we change each UDE to a DE and connect each DE and SI into FRT.

3.3 To verify the if our SI can take effect

We verify our SI by analyzing data about hypertension patients' behavior of seeking medical help and data from medical institutes applied insurance payment from 2006 to 2007 to find out if the constructed SI can introduce real improvement for the financial imbalance of NHI. Total numbers of medical record for hypertension are 2453520 from the database of Northern branch of Bureau of NHI, and the total numbers of man-count for hypertension taking outpatient service are 318846. The results of the number of medical institute a patient goes, average number of outpatient service and Average ratio for taking refill slip (%) are as the Table 1 showed.

Table 1. Number of medical institute a patient goes, average number of outpatient service and Average ratio for taking refill slip (%)

number of medical institute a patient goes	1	2	3	4	5	6	7
average number of outpatient service	8.74	9.96	12.22	14.48	16.66	18.78	20.62
Average ratio for taking refill slip (%)	37.10	29.11	24.17	20.63	19.75	19.46	16.62

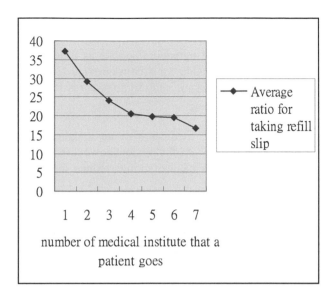

Figure 1. The relation between the number of medical institute a patient goes, and Average ratio for taking refill slip (%)

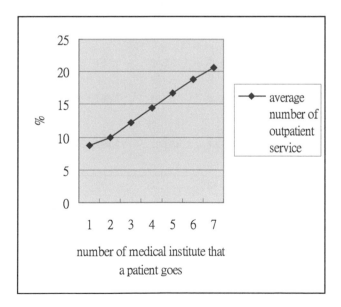

Figure 2. The relation between the number of medical institute a patient goes, and average number of outpatient service.

As these two figures show, when the number of medical institutes a patient goes increased the average number of outpatient service increased, too. On the other hand, when the number of medical institutes a patient goes increased the average ratio for taking refill slip decreased.

We use the average number of outpatient service as a dependent variable (Y) and both the number of medical institute a patient goes (X_1), and Average ratio for taking refill slip (X_2) as independent variables to do a regression analysis.

The regression model is

$$Y = 3.995 + 2.25\,X_1 + 0.063\,X_2$$

We found that the number of medical institute a patient goes is statistical significant for the regression model. As the model showed, if we want to reduce the average number of outpatient service, we should reduce the number of medical institute a patient goes.

If we fixed the patients to one medical institute for outpatient service, is there room for our SI to make improvement for the Policy of Chronic Illness Prescription Refill Slip? We found from the data that there are total 866 clinics and 63 hospitals have patients fixed to one medical institute. Table 2 tabulate different ratios that Refill Slip been prescribed and the number of medical institutes.

Table 2. Ratio of Refill Slip that the patients fixed to one medical institute for outpatient service

Ratio of Refill Slip (X%)	0	0<X≤10	10<X≤20	20<X≤30	30<X≤40	40<X≤50	50<X
Clinics	307	135	47	46	44	32	255
Ratio to all clinics (%)	35.45	15.59	5.43	5.31	5.08	3.70	29.45
Number of patients go to clinics	16,885	19,015	6,101	7,861	3,621	3,921	35,499
Frequency of patients go to clinics	182,244	181,668	48,692	73,982	28,098	30,161	225,344
Hospitals	5	12	12	12	8	4	10
Ratio to all Hospitals (%)	7.94	19.05	19.05	19.05	12.70	6.35	15.87
Number of patients go to Hospitals	73	5,638	11,735	10,334	9,657	7,128	40,326
Frequency of patients go to Hospitals	352	71,905	132,642	105,321	92,044	66,202	314,493

35.45% of the clinics and 7.94% of the hospitals didn't prescribe any refill slip. Only 29.45% of the clinics and 15.87% of the hospitals use refill slip over 50%. That shows we still have rooms for improvement. The number of patients that go medical institutes use refill slip under 50% is 101969. One refill slip reduces 2 times outpatient service. If we raise the ratio of refill slip for 1%, we can reduce 2000 times outpatient service. The insurance payment that NHI pay for outpatient service is 200 NT$ for hospitals and 400 NT$ for clinics. If we raise the ratio of refill slip for 1%, we can save about 300000 NT$ for NHI.

4. CONCLUSION

In this research, we apply Thinking Process based on TOC to find the goal of a policy making, and verify the goal by analyzing the insurance data of NHI. We found the core conflict for the Policy of Chronic Illness Prescription Refill Slip is because we have to raise the ratio that Chronic Illness Prescription Refill Slip use but the patients and medical institutes are not welling to use this Refill Slip. To solve the core conflict, we proposed a SI that fixing the patient to a specific medical institute for treatment. We analyzed the insurance data from NHI and found that if we can fix the patient to a medical institute, we could make improvement for the excess insurance payment of NHI.

For using the Thinking Process based on TOC as a prescribe model for human thinking behavior, we found that:

(8)   Using the Thinking Process can help finding the core problem in policy making.

(9)   Tools of Thinking Process can be used as an assistant tool to form solution for a problem.

(10)   Using TOC can help people understand a dynamic system and provide systematic perspectives for a dynamic system.

National health insurance (NHI) system is an important part of Taiwan's social security system and now facing the predicament of financial imbalance. In order to maintain its operation, the health insurance system needs strategies with systematic perspectives to improve its predicament. In this study, we use Thinking Processes to find out the restrictions that impede a system to achieve its goals. We apply the Thinking Processes to national health insurance system and established a reference model for constantly improving strategies for the Policy of Chronic Illness Refill Slip by analyzing data from the database of NHI.

ACKNOWLEDGEMENTS

The authors would like to thank the National Science Council of the Republic of China for financially supporting this research under Contract NSC 100-2221-E-159 -019.

REFERENCES

Baron, J. (2008) Thinking and Deciding (4th edition) New York: Cambridge University Press.

Brehmer, B. (1990) Strategies in Real Time Dynamic Decision Making. in R. M. Hogarth (Ed.) Insights in Decision Making, The University of Chicago Press, p. 262-279.

Brehmer, B. (1987) Systems Design and the Psychology of Complex Systems, in Rasmussen J.and Zunde P. (Eds.) Empirical Foundations of Information and Software Science III, New York: Plenum Press.

Bureau of National Health Insurance (2008) Medical expenditure report: No. 141, http://www.nhi.gov.tw/webdata/webdata.asp?menu=1&menu_id=4&webdata_id=815&WD_ID=20.

Chen, H.-Y. and Wu, Y.-C.(2004) A Survey Research on the Release of Prescription of Chronic Disease-Patient's Perspective, technical report, No.DOH93-TD-D-113-004, China Medical University, Taiwan.

Cooper, M. J., and Loe, T. W. (2000) Using the Theory of Constraints' Thinking Processes to Improve Problem-Solving Skills in Marketing. Journal of Marketing Education 22(2): 137-146.

Edwards, W. (1962) Dynamic Decision Theory and Probabilistic Information Processing, Human Factors, 4, 59-73.

Goldratt, E. M. (1994) It's Not Luck. Gower, England.

Goldratt, E. M. and Cox, J. (1986) The goal: a process of ongoing improvement. New York, NY: North River Press.

Kim, S., Mabin, V. J. and Davies, J. (2008) The theory of constraints thinking processes: Retrospect and prospect. International Journal of Operations and Production Management, 28(2), 155-184

Kung, P.-T., Lu, C.-H. and Tsai, W.-H. (2007) Willingness of Clinic Physicians to Approve Prescription Refills for Chronic Disease, Taiwan Journal of Public Health, Vol. 26 No. 1, 26-37.

Lerch, F. J. and Harter, D. E.(2001) Cognitive support for real-time dynamic decision making. Information Systems Research, Vol. 12, No. 1, 63–82.

Liang, Y.-C. (2004) Prevalence, Utilization, and Expenditure of Multiple Chronic Conditions in Taiwan, unpublished master thesis, National Yang-Ming University, Taipei, Taiwan.

Ministry of the Interior (2009) The forth week Statistical report for the Ministry of the Interior Taiwan of 2009, http://www.moi.gov.tw/stat/news_content.aspx?sn=2024

NHI Medical Expenditure Negotiation Committee (2008) "Q&A of Global Budget System", http://www.nhi.gov.tw/webdata/webdata.asp?menu=3&menu_id=470&webdata_id=2157&WD_ID=470.

Polito, T., Watson, K. and Vokurka, R. J. (2006) Using the Theory of Constraints to Improve Competitiveness: An airline case study, Competitiveness Review 16(1): 44-50.

Shi, G., Chang, S.-H. and Zhang, W.(2010) Challenges of Beijing road transportation system: An extended application of TOC think processes, The 3rd International Conference on Applied Human Factors and Ergonomics (AHFE). Miami, Florida UAS.

Sterman, J. D. (1989) Modeling Managerial Behavior: Misperceptions of Feedback in a Dynamic Decision Making Experiment, Management Science, 35(3), p.321-339.

Tanner, J. F., and Honeycutt, E. D. (1996) Reengineering using the theory of constraints: A case analysis of Moore Business Forms, Industrial Marketing Management 25(4): 311-319.

CHAPTER 21

A Cellular Automata Based Model as a Tool of the Organizational Culture Change Analysis

Agnieszka Kowalska-Styczeń

Silesian University of Technology
Zabrze, Poland
agnieszka.kowalska@polsl.pl

ABSTRACT

The article presents the possibility of using cellular automata to model the processes of the organizational culture change. Since the presence of leaders seems to be a key factor in the success of attitude change, the way of communication by the use of 'word of mouth' mechanism is a crucial tool in this sphere. Cellular automata allow to analyze the dynamics of changes in views and attitudes in social groups based on local interactions between people in small groups of friends, family members etc. The proposed paper shows the possibility of modeling the dynamics of cultural change in the company, if the basic assumption of this process is the presence of leaders and the their impact on possible changes.

The dependence of the model dynamics according to the changes in model parameters is shown. In particular, the simulation results reflecting changes in attitudes, depending on company size, number and characteristics of leaders and the size of informal groups of friends / acquaintances are discussed. Possible directions of the proposed approach modification and the potential areas of application in the macroergonomic analysis are proposed.

Keywords: complex system, cellular automata, change agents, macroergonomics

1   INTRODUCTION

Macroergonomic activities are directed primarily to the design of socio – technical systems, in which the "total optimization" includes the design of optimal human - organization system. Hendrick and Kleiner (2001) believe that the most appropriate moment in such an activity is effective is the general (major) change in a given system (eg. change in technology, equipment, etc.). Another favorable situation is the general change of the organization's objectives, the scope or direction of development (eg. introducing a new product or new concept of production – for example from mass to customized). Cited authors provide an increasing role of the macroergonomics in such situations. This is often expressed by the leading role of 'change agents' - leaders, specialists with extensive knowledge of ergonomics, the science of organization and socio – technical systems.

These opinion interacts with the general trend of the Leadership Complex Theory - presented exhaustively in the work of Lichtenstein et al. (2006).

The   mentioned authors propose an analysis of leadership and leadership in contemporary organizations from the perspective of complex adaptive systems (CAS). In this perspective, the role of leaders is not the issue - as in traditional approaches - of hierarchical dependence but rather a set of interacting 'agents' in the networks (agents). These interactions are largely informal but explain many of the organizational processes - such as learning in organization, innovation, diffusion and adaptation processes (Lichtenstein et al, 2006). This view of the organization appeared more than a decade earlier. Stacey (1995) proposed the inclusion of the perspective of complex systems as a strategic change process description. The cited author showed among others, interpretations of previous studies on properties of Boolean networks and cellular automata (CA), which show that the specific nature of the relationship network in social organizations implies a certain behavior and the processes occurring in them. An intriguing example of this phenomenon is shown by Kaufman (1993) based on Boolean network dependence. He showed that processes are more stable for networks with fewer connections and the possibility of falling into the transient and unpredictable states in the structures of the multiple and dense agents network connections. Interestingly, the impact of the rule itself had no effect on the relevant conduct throughout the network. Similar results were presented by Wolfram (1983, 1984, 2002), although the simulations were based on the concept of cellular automata (Kowalska - Styczeń 2007)

Stacey (1995) argues that these theoretical studies indicate that stable organizations must be characterized by a small amount of informal relationships both inside and outside the organization. Variability and dynamics of the organization is the domain of the rich connections of its members, other and / or the external environment. Interestingly, these conclusions to some extent contradict results of the empirical research presented e.g. by Gronowetter (1973) who analyzes organizations with high stability as compared to organizations with high degree of dynamics observed that the former are characterized by strong links and interaction between participants (they are heavily dependent emotionally, they spend much time together, etc.) and the latter are characterized by relatively loose relationships.

So not so much the quantity but the strength of ties affect the stability of the organization. Perhaps, however, a theoretical simulation studies show only one of many factors influencing the processes within societies with different structural characteristics.

In this paper, attempts to model the relevant aspects of the processes of change are taking place at the company level (of various sizes occurring in the economic real world) with CA technologies. The adoption of this technology, meant of course, to accept the rules and limitations described in a particular paradigm, assuming that the complex system changes can be explained by the dynamic analysis of the interaction between the elementary system parts. From the standpoint macroergonomic modeling, it is an attempt to model an influence of change agents on the working community (particularly the 'positive' change of organizational culture). Such an approach (from the perspective of complex systems science) can be viewed as the 'bottom - up' type analysis and differs from the traditional macroergonomic view proposed by Hendrick and Kleiner (2001) where the dominant role was played by the hierarchical approach of the 'top - down' type.

As proposed in the framework of complex systems science - leadership is understood as the effects of interactions between members of the organization in the informal interaction. The basic mechanism of this interaction is the 'word - of – mouth' (w-o-m) that is exchanging thoughts, ideas, opinions, etc., in informal talks within groups of colleagues, friends etc - also mostly of an informal nature.

The model enabling studies of properties of this mechanism (w-o-m) is the subject of the approach proposed here. As a tool for describing and testing processes in the organizational community the concept of cellular automata (CA) is proposed. In this approach, analyzed bonds are all treated as strong ones – following empirical study results of Brown and Reingen (1987) in which they showed such bonds as a main w-o-m referral information source.

The following sections discuss firstly the characteristics of the agent based modeling convention approach focused on cellular automata and their parameters, and then secondly formulated a model that allows simulate influence of the leaders of change (agents) on the process of changes in the views of members of the organization. The results of simulation experiments are presented and discussed. Also future works are proposed in the discussed field.

2   THE MODEL

CA is a mathematical object consisting of a network of cells in D-dimensional space, a finite set of states of a single cell and a rule, which determines the state of the cell at time $t + 1$ depending on the state at time t of the cell and cells that surround it. Surrounding cells are the neighborhood of the i-th cell.

It is then a dynamic mathematical model of processes occurring in time. D – dimensional space where the cellular automaton evolution takes place is divided into identical cells, each can assume one of the states (the number of possible states is finite). The neighborhood is the same for each cell. The cell state changes in time according to the rule F and depends on its previous state and on its neighbors' state, that is to say on the neighborhood. Thus, important parameters for a cellular automaton are: the network dimension D, the number of states of a single cell. A frequently used parameter is also the radius r, whose length depends on the form of the neighborhood of a cellular automaton. If the neighborhood consists of the nearest neighbors of the i-th cell then r = 1.

In the model presented in this article, a two-dimensional CA is proposed with von Neumann neighborhood of radius r = 1 (neighborhood 4-element, fig. 1), r = 2 (neighborhood 12-element, fig. 2) and Moore neighborhood of radius r = 1 (neighborhood 12-element, fig. 3).

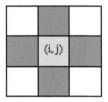

Figure 1   Von Neumann neighborhood of a two-dimensional automaton cell (i, j) with r = 1 (neighborhood 4-element)

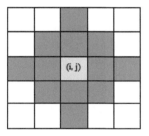

Figure 2   Von Neumann neighborhood of a two-dimensional automaton cell (i, j) with r = 2 (neighborhood 12-element)

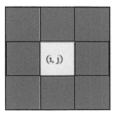

Figure 3   Moore neighborhood of a two-dimensional automaton cell (i, j) with r = 1 (neighborhood 8-element)

Our research concerns the behavior of individuals forming an organization / company. We study the environment of agents so that the model is presented as a square network of size L x L. In the calculation of the successive states of the cells, at the edges of the network, they are treated as if the whole lattice was placed on the surface of the cylinder (ie, neighbors lying on the edge of the cell are respectively opposite side of the lattice). Besides the usual agents-some of them shall be the leaders (or agents of change – disccused by Hendrick and Kleiner (2001)), who have greater power of persuasion with respect to the other members of the organization.

A single i-th cell / agent ($i = 1, 2, \dots L2$) can be in one of three states: ($s_i = 1$) occupied by agents of type A, ($s_i = 1^*$) occupied by the leaders of type A and ($s_i = -1$) seized by agents of type B. In our simulations, we examine a situation in which agents support option B, and leaders only for option A.

Denote by n the number of agents in the network, then:

$$n = L^2$$

Let the n_a - is the number of agents for option A, then $n - n_A$ is the number of agents for option B. Let $c = n_A / n$, it will be the concentration of A - agents, so $1 - c$, it will be the concentration of B - agents . In this notation, the number of agents for option:

$$n_A = cL^2$$

Let the n_{A*} - indicates the number of leaders for option A, $f = n_{A*} / n_A$ is a fraction of the leaders of option A. In this notation, the number of leaders for option A:

$$n_{A*} = fcL^2$$

Due to the presence of agents and agents-leaders in the model, and their different role, different rules of behavior are proposed for them.
Rules for Agents:
- Agent checks the preferences of their environment and changes its preference for the dominant in the environment (ie., if more than 50% of its neighbors has a different opinion than him, then he changes his opinion on the dominant one).
- Agent also changes his preference when in its environment there is at least one leader
- If more than 50% have the same opinion as the test agent or a lack of leadership, of course, nothing changes.

The rules for leaders
- If the agent is a leader even though he is in the middle cell, he does not change under the influence of environment.

Defining this way usually used cellular automaton we take into account effects of interpersonal influence and persuasion power of leaders. Thus, simulate the transmission of knowledge 'word of mouth' (w-o-m). In order to make simulation experiments one should establish the size of L (which determine number of society members), r – the size of the environment. The network can be fulfilled by the agents and leaders in appropriate proportions.

3   SIMULATION EXPERIMENTS

3.1   Design of experiments

In our simulations, the agents have a choice of two organizational culture types (A and B). Furthermore, it was assumed that at the beginning of the simulation all 'normal' agents prefer B agents - leaders prefer A, but the number of leaders preferring A is c (ie. the beginning of the simulation $f = c$), and the agents preferring B is $1 - c$

First, the change of L was examined - that is, network size affects the efficiency of the leaders of the different environments. We analyzed the average number for the option of the leaders of the 1000 simulations. Figures 4, 5 and 6 shows the dynamics of preferences, depending on the percentage of leaders at the beginning of the simulation experiment, for different network sizes and the 3 environment types defined above.Therefore, the simulation experiment was designed and performed for the different concentrations of leaders (5% – 55 %), different sizes of lattices - L = 10 (100 agents), L = 20 (400 agents), L = 30 (900 agents). These dimensions are defined to accord with the OECD (Organization for Economic Co-operation and Development) definition of the equivalent number of employees in small, medium and large enterprises appropriately.

Also an equivalent groups / teams of employees of different sizes were adopted in the model, respectively, 4-piece, 8-piece and 12-element environment.

3.2   Simulation Results

In our simulations, the agents have a choice of two organizational culture types (A and B). Furthermore, it was assumed that at the beginning of the simulation all 'normal' agents prefered B agents - leaders prefered A, but the number of leaders preferring A was c (ie the beginning of the simulation $f = c$), and the agents preferring B was $1 - c$.

First, the change of L it was examined - that is, network size affects the efficiency of the leaders of the different environments. We analyzed the average number for the option of the leaders of the 1000 simulations. Figures 4, 5 and 6 show the dynamics of preferences, depending on the percentage of leaders at the beginning of the simulation experiment, for different network sizes and the 3 environment types defined above.

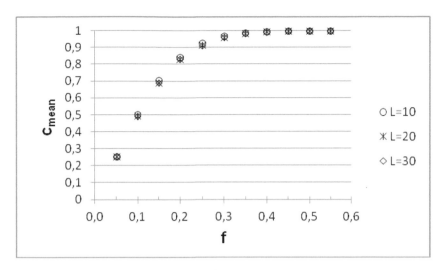

Figure 4 The change of the average concentration of agents c_{mean} for leading option, depending on the fraction f of leaders at the beginning of simulations. The results are obtained after 1000 experiments for 4-elements neighborhood

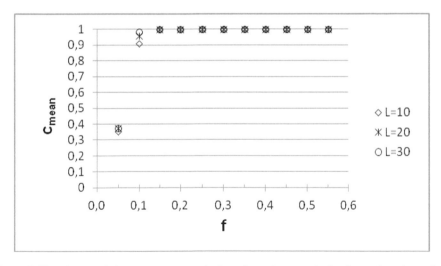

Figure 5 The change of the average concentration of agents c_{mean} for leading option, depending on the fraction f of leaders at the beginning of simulations. The results are obtained after 1000 experiments for 8-elements neighborhood.

198

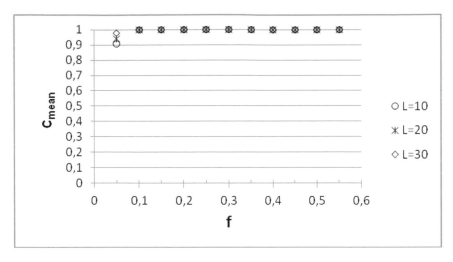

Figure 6 The change of the average concentration of agents c_{mean} for leading option, depending on the fraction f of leaders at the beginning of simulations. The results are obtained after 1000 experiments for 12-elements neighborhood

The results indicate a significant fact that changes in community opinion as a whole are independent on its size - simulated as a size of the lattice. The dynamics of the opinion change depending on the average size of cells in the neighborhood (agents) that reflects the number of (average) informal contacts - and thus the size of informal groups processes occurring in a simulated w-o-m process. A common feature of the results is the possibility to gain full confidence in the new ideas (ideas, culture) involving a smaller number of change leaders (change agents) for larger informal groups. The practical consequence may be the fact that about 30% of leaders (but with considerable authority) can convince any community (having small number of strong informal contacts) to new solutions (ideas, concepts, organizational changes, etc.).

Fig. 7, 8 and 9 show relationship between the mean number of simulation steps needed to obtain the stable state (in which there are no changes possible). These relationships are showed with dependence on the lattice dimension and neighborhood types.

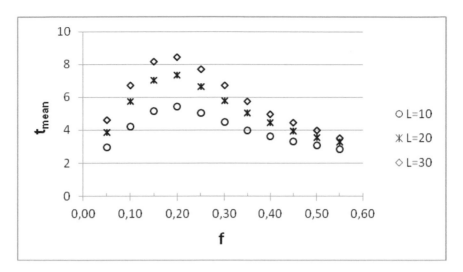

Figure 7 The change of the average number of simulation steps- t_mean needed to achieve steady state, depending on the fraction f of leaders at the beginning of the simulation, for the different lattice sizes and 4-element neighborhood after 1000 simulations

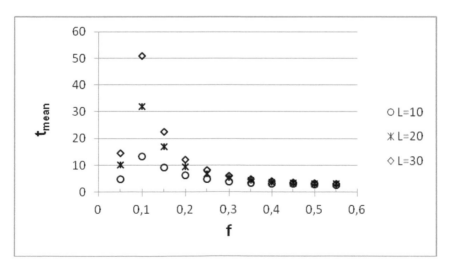

Figure 8 The change of the average number of simulation steps- t_mean needed to achieve steady state, depending on the fraction f of leaders at the beginning of the simulation, for the different lattice sizes and 8-element neighborhood after 1000 simulations

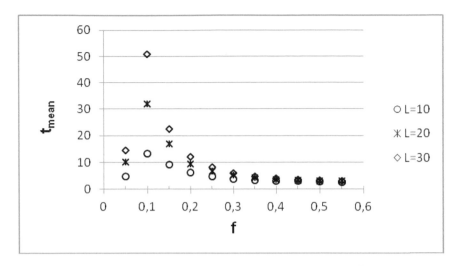

Figure 9 The change of the average number of simulation steps- t$_{mean}$ needed to achieve steady state, depending on the fraction f of leaders at the beginning of the simulation, for the different lattice sizes and 12-element neighborhood after 1000 simulations

The presented graphs show that the time needed to achieve the desired change of views (for these represented by the leaders of change) - represented in the model by simulating the number of steps to reach a steady state - depends on community size and the number of informal contacts. These features are represented in the model by appropriate lattice size and the size of the neighborhood. In all experiments, the time to achieve a steady state is the shortest for small, and longest for large neighborhoods. The relationship's character however, changes according to Figure 7 – 9. For a small and medium neighborhood (Fig. 7, 8) process of changing views reached maximal time respectively for 20 and 10 percent concentration of the leaders in the population and they are appeared not with the smallest of the tested concentrations. In the case of a large (12 element) neighborhood, the longest time appeared just for the smallest neighborhood. Difficult to predict was the fact that the maximum number of steps necessary to achieve stability in this model occurred for the average sized neighborhood (Fig. 8).

2   SUMMARY AND CONCLUSIONS

The proposed model is - like any other model - a simplification of reality. In the analyzed experiments, however, it was shown that results obtained by such simplifications on the one hand confirm the intuitive belief. For example, the greater the number of contacts creates the greater likelihood of an informal meeting of change leaders - hence the smaller number of such agents need to change the view of the whole community. On the other hand, it was difficult to predict in advance how many people of high authority (here the assumption is probably even unrealistic - one contact with the leader makes the adoption of his view) is needed to get the desired changes such as for example appropriate organizational culture. Just as in those of the above relationships shown on the basis of complex systems science (Kauffman 1993) in experiments with a simple model proposed here, some results are unexpected but suggests that the results obtained are universal effect of the regulations governing the dynamics of complex systems with the w-o-m mechanism.

Experimental studies using cellular automata open an interesting research field for the macroergonomics. Postulated role of change agents in the design of socio – technical systems (Hendricks and Kleiner 2001) can be explored using appropriate-designed study behavior of cellular automata.. An interesting direction of research seems to be adopting into the model the level of leader persuasive strength and differentiation of agent's 'degree of resistance' for innovation and change.

ACKNOWLEDGMENTS

I want to express my gratitude to Prof. Jerzy Grobelny (from Wroclaw University of Technology) for his suggestions of the idea of the role of leaders in macroergonomics and possibility of using cellular automata in this area

REFERENCES

Lichtenstein B. B., Uhl-Bien M, Marion R., Seers A., Orton J. D., and Schreiber C. (2006), Complexity leadership theory: An interactive perspective on leading in complex adaptive systems E:CO Issue Vol. 8 No. 4 2006 pp. 2–12

Brown J. J. and Reingen P. H. (1987), Social Ties and Word-of-Mouth Referral Behaviour, Journal of Consumer research. Vol. 14. 350-362

Granovetter. M.S. (1973), The Strength of Weak Ties, American Journal of Sociology. 78 (May). 1360-1380.

Hendrick H. W and Kleiner B.M. (2001), Macroergonomics. An Introduction to Work System Design. Human Factors and Ergonomics Society.

Kauffman, S. (1993), The Origins of Order: Self-Organization and Selection in Evolution, New York: OxfordUniversity Press

Kowalska – Styczeń A. (2007), Simulation of complicated economic processes with the use of cellular automatons. Wydawnictwo Politechniki Śląskiej, Gliwice (in Polish)

Stacey R.D. ((1995), The science of complexity: an alternative perspective for strategic change processes, Strategic Management Journal, Vol. 16, 477-495

Wolfram S. (1983), Statistical mechanics of cellular automata, Rev. Mod. Phys. 55, 601-644

Wolfram S. (1984), Universality and complexity in cellular automata, Physica D 10, 1-35

Wolfram S. (2002), A New Kind of Science, Wolfram Media, Inc.

CHAPTER 22

How Usage Context Shapes Evaluation and Adoption in Different Technologies

Martina Ziefle[1], Simon Himmel[1], Andreas Holzinger[2]

[1]Communication Science, Human-Computer Interaction Center
RWTH Aachen University
Ziefle@humtec.rwth-aachen.de

[2]Institute for Medical Informatics, Statistics and Documentation
Medical University of Graz, Austria

ABSTRACT

Although lots of technical devices are quite indispensable, the willingness to use the technologies is not necessarily given in all users – some devices are regarded as helpful, others are not perceived as trustful. Technical progress proceeds in every type of technology and new devices have to meet the needs of multiple users. Therefore user diversity is a key factor. This paper examines the effects of gender and age on technical interest in specific technology fields, their effects on purchase criteria for car-, medical-, and ICT and their influence on motivation for usage of these technologies. 92 respondents (21-80 years) participated in an exploratory survey revealing that general interest and interest in specific technology branches is significantly influenced by gender and age. While purchase criteria and motivation for usage differ with the technology context (medical, automobile and ICT), user diversity (gender, age) plays a minor role for adoption and evaluation criteria.

Keywords: technology context, adoption criteria, technology interest, gender, aging, technology acceptance

1   INTRODUCTION

In our society more and more people need technologies – such as medical and automobile technologies – to live an independent and autonomous life in old age.

These technologies must meet older peoples' specific demands – otherwise they neither are useful nor usable. For the development of well-accepted future technologies, we have to find out, which factors are relevant for the evaluation and acceptance of different technologies. Technical applications and electronic services can only be successfully applied if two general conditions are fulfilled: Technology must be fully accepted by the group of older people and, what is still more important, technology must meet the specific demands of the older group. So far, studies on technology acceptance consider mainly acceptance issues in younger people. Also, technology acceptance is mainly examined within computers and information and communication technologies and is connected to two major factors: the ease of use and the perceived usefulness (Davis, 1989). It is though reasonable to assume that the extent of technology acceptance depends on many more factors, especially in the older group (Seock, and Bailey, 2008; Williams and Slama 1995). Technology type, using context, age or technology generation may also be relevant for the extent of acceptance and the willingness of older people to actually use technologies (Arning and Ziefle, 2009; Gaul & Ziefle, 2009)

During the last decade, a research has made significant gains in understanding technology acceptance of ICT (e.g. Arning, Gaul & Ziefle, 2010). Though, the knowledge about determinants and situational aspects is still limited. Due to the increasing diversity of users, technical systems and usage contexts (fun and entertainment, medical, office, mobility), more aspects are relevant in understanding users' acceptance – beyond the ease of using a system and the perceived usefulness. In addition, the majority of studies had been directed to technology acceptance and adoption behaviors of young, experienced and technology-prone persons. It is though obvious that persons of different ages might have different acceptance criteria and requirements for technology, especially when considering that, for example, devices in the medical context must meet different demands than technology in the automobile context. Moreover, the gender factor is crucial in the context of technology acceptance and adoption. It had been reported that female users often show lower self-efficacy and higher technology anxiety (e.g., Busch, 1995, Davies, 1994; Downing et al., 2005), what not only affects womens' acceptance for technology (Wilkowska, Gaul & Ziefle, 2010, Ziefle & Schaar, 2011), but also their individual adoption and purchase criteria (Zhou and Xu, 2007).

1.5   Questions addressed

This exploratory study examines conditions of older users' technology acceptance taking different usage contexts into account. Empirically, perceived advantages and barriers within three different technology domains and contexts were assessed: Automobile technology, medical technology and ICT. First we explore users' technical interest in general as well as in different technology branches with a specific focus on gender and age differences. In a second step, different adoption criteria are examined in the different usage contexts. Finally, specific usage motives and acceptance conditions are investigated, again, taking usage context, gender and age into account.

2   METHODOLOGY

92 participants, aged between 21 and 80 years, took part in this study (M=55,71, SD=16,54). 45.7% of the sample was female, 54.3% male. To investigate if technical interest, purchase criteria and usage motivation differ depending on age and technology generation, the sample is split in age groups referring to three different technology generations: The young group is aged between 21 and 39 years (n=20, M=28.5, SD=5.3, 35% female/ 65% male), the middle-aged between 40 and 65 years (n=40, M=57.3, SD=6.8, 52.5% female/ 47.5% male), and the old between 65 and 82 years (n=32, M=70.8, SD=3.6, 44% female/ 56% male).

2.2   Questionnaire

In order to get a deeper insight in users' attitude towards technology in general as well as in specific technologies and their purchase criteria, we chose the questionnaire method to collect comprehensive opinions of prospective users, as well as a large number of respondents. The questionnaire was organized in four sections: The first part consisted of a query of demographic data with respect to age and gender. Section two asked for technical interest in general as well as interest in specific technology areas: household, entertainment, computer, mobile, car, medical, tools, and farming, using a four point Likert scale (no, rather no, rather yes, yes). The third part was directed to adoption criteria for three key technologies: car, medical, and ICT. Eight criteria had to be ranked by importance on a six point Likert scale: price, usability, safety, dependency, design, brand, latest state of technology, and unobtrusiveness. Finally, the usage motivation was explored. Participants had to confirm 16 statements for each of the two technology types, giving potential circumstances under which they would to use these technologies. The statements regard different dimensions and were taken from prior focus groups with older adults (Wilkowska and Ziefle, 2009, 2010; Ziefle and Schaar, 2011).

2.3   Research Variables

Respondents' age and gender were considered as independent variables. Three age groups were formed in order to detect age-specific technology preferences and to compare three different technology generations (computer, household and early technical generation, (Wilkowska and Ziefle, 2010, and Ziefle and Wilkowska, 2011). Gender is also regarded as a main factor in order to understand whether women and men have different usage motivations in different kinds of technologies (Williams and Slama, 1995, Seock and Bailey, 2008). As dependent variables, technical interest, purchase criteria and usage motivation were measured.

3.   RESULTS

Data were statistically analyzed by MANOVA procedures and ANOVAs respecting age and gender effects. Significance level was set at 5%.

3.1   Interest in Technology

First, we report on the reported interest in technology in general as well as the interest in the specific technology branches, comparing gender and age groups. For the general interest in technology, there were significant effects of gender (F $(1,9)=4.3$; $p<0.05$, Figure 1). Female respondents- independently of their age, reported to have lower interest in technology in general.

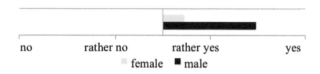

| no | rather no | rather yes | yes |

female ■ male

Figure 1 General interest in technology, effects of gender.

Also, significant age effects ($F(1,9)=2.7$; $p<0.05$) were revealed. Younger persons have a considerable higher interest compared to older persons (Figure 2).

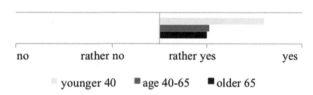

| no | rather no | rather yes | yes |

younger 40   ■ age 40-65   ■ older 65

Figure 2 General interest in technology, effects of age.

Beyond the general interest in technology, participants were asked to rate their interest in specific technology branches. Figure 3 shows the descriptive outcomes.

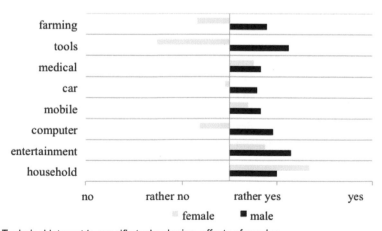

Figure 3 Technical interest in specific technologies, effects of gender.

Again, gender effects were found. Men showed a higher interest in computer technology (F(1,77)=7.7; p<0.05), tool technology (F(1,77)=30.4; p<0.00) and farming technology (F(1,77)=9.6; p<0.05). Only for household technology, female showed a higher interest in comparison to men (F(1,77)=3.1; p<0.05). When looking at ageing effects, age groups differed with respect to their interest in household technologies (F(2,83)=4.3; p<0.00), with the oldest group having the highest interest (Figure 4).

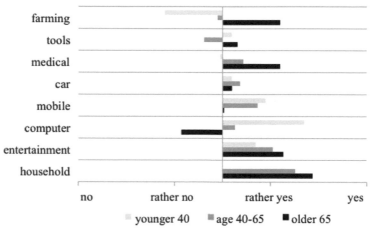

Figure 4 Technical interest in specific technologies, effects of age.

Also, age groups' interest in computer technology was different (F(2,83)=10.9; p<0.00), with a considerable higher interest in the youngest and the lowest interest in computer technology in the oldest age group. Another significant age effect was found for farming technology (F(2,83)=4.3; p<0.00), which was high in the oldest and low in the youngest age group.

3.2 Adoption Criteria

In this section we explore if peoples' adoption criteria differ for the three different contexts, comparing ICT, medical and automobile technology. Descriptive outcomes are depicted in Figure 5.

On a first sight, we see that there are criteria that are more important than others: The latest state of technology, price, usability, safety and the reliability of technology, while brand, design and unobtrusiveness seem to be of lower importance. When focusing on the different technical branches, we see different evaluation criteria though. For medical and automobile technology safety is more important than for ICT, and the design and brand is less important in medical technology than in both other technical fields (automobile technology and ICT). Interestingly, usability is equally important for all technology fields, thus can be regarded as a very generic adoption criterion. When looking at user diversity and potential gender and age effects, no significant outcomes were detected.

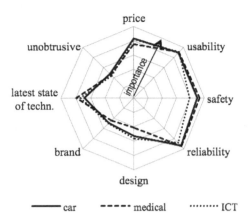

Figure 5 Adoption criteria of specific technologies

3.3   Usage motives

In this section, usage motives in two selected technology fields are examined: automobile (Figure 6) and medical technology (Figure 7). Descriptive outcomes are ranked according to the reported importance.

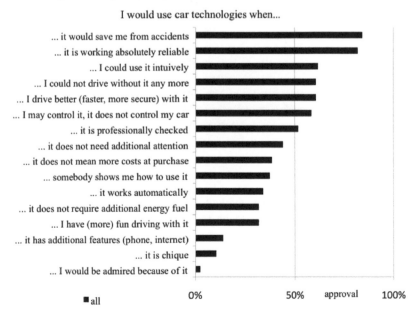

Figure 6 Usage criteria of car technologies.

With respect to technologies in the automobile context, participants attach greatest importance to "safety", "reliability" and "usability" of automobile technology. In contrast, comparatively low importance is attached to fun and design features and the social desirability of having such technologies. Interestingly, the rankings did not differ for women and men, showing that usage motives for car technologies are not differentially influenced by gender stereotypes and roles.

Age, in contrast, did impact the reported importance of usage motives in the automobile context. The younger group attached a significantly higher importance to the hedonic characteristics, thus design, fun, additional features as well as to the social desirability of automobile technologies. Interestingly, even though safety reasons revealed to be the most important for all age groups, younger adults reached the highest rankings possible (100 %).

Now the reported usage criteria for medical technology are reported (Figure 7).

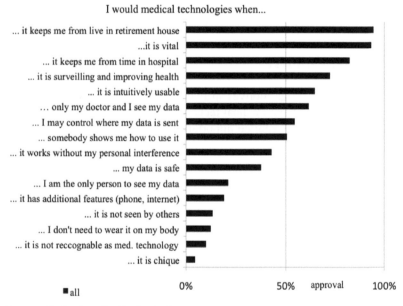

Figure 7 Usage criteria of medical technologies

From Figure 7 it becomes evident that the "independency motive" is most important. Participants indicate to be willing to use the medical technology as it might prevent them from moving to senior homes and from hospital stays. Medical technology is only accepted when it is vital (and no alternative is given). Also, "usability" and "data privacy" are core usage motives, while "unobtrusiveness", "design" and "additional features" are of minor impact for acceptance.

For the acceptance of medical technology, gender differences were prevalent. Men – in contrast to women – attach significantly larger importance to the conditional usage criterion that they would use medical technology only if it is indispensable. Also, men stress that "data control" and "data privacy", "unobtrusiveness" and " device design" as crucial for them. Women would also use medical technology for health monitoring, not only in the case of serious illness. For older persons, the most important criteria are "unobtrusiveness" and "invisibility" of medical technology. For the younger group, "data security", "privacy" and "data protection" are prominent usage motives, in contrast to the older users, which rated these conditional acceptance criteria as less important.

3.4. Correlation analyses

A first analysis regarded the relations between interests in different technology fields. In order to understand if the general interest in technology is impacted to a stronger extent by the type of technology or, rather, by the individual technology interest profile of persons (technology-prone vs. technology reluctant), we analyzed correlations (Spearman) between technology interest ratings across the different technical branches. In Table 1, outcomes are summarized.

Table 1 Intercorrelations of reported interest in different technology fields.
Bold values are significant (* p<0,05; ** p<0,01).

	house-hold	enter-tainment	Com-puter	mo-bile	car	medi-cal	tools	far-ming
household	1,0	**0,36****	-0,03	**0,28****	0,11	**,24***	0,02	0,22
entertain-ment		1,0	**0,25***	**0,51****	**0,28***	0,2	**0,32****	**0,45****
computer			1,0	**0,43****	**0,37****	0,08	**0,36****	-0,04
mobile				1,0	**0,46****	0,08	0,18	**0,22***
car					1,0	**0,35****	**0,36****	**0,23***
medical						1,0	**0,3****	**0,26***
tools							1,0	**0,62****
farming								1,0

Apparently, there is no 1:1 correlation across technical branches, according to which persons, who are interested in one type of technical field, would also be interested in another. But of course, technology-prone persons (the younger and/or the males) show higher intercorrelations of technical interest across branches compared to persons with a lower technical affinity (older persons and/or females).

A second focus was laid on the question if the general interest of technology is correlated with the rated importance of purchase criteria (Table 2).

Table 2 Correlation of technical interest and purchase criteria
Bold values are significant (* p<0,05; ** p<0,01, Spearman-Rho).

	Price	usa-bility	safety	depen-dency	design	brand	state of techn.	unob-trusive
Car	0,00	-0,14	0,72	0,25	-0,12	0,01	**-0,27***	0,02
medical	0,12	0,15	0,07	-0,05	-0,14	**-0,22***	**-0,23***	**-0,32****

For automotive technology, the general interest in technology only correlates significantly with the purchase criterion that automobile technology should represent the "latest status of technology", while all other purchase criteria are not significantly driven by technology interest. When looking at medical technology, technical interest revealed to be significantly with the "brand", the "unobtrusiveness" of the device and the "latest state of technology".

A third focus was directed to the question if the rated importance of purchase criteria is different for the technology fields (Table 3).

Table 3 Correlation of purchase criteria across technical fields

Bold values are significant (* p<0,05; ** p<0,01, Spearman-Rho).

	price	usa-bility	Safety	depen-dency	design	brand	state of techn.	unob-trusive
car/ medic.	0,63**	0,6**	0,49**	0,35**	0,54**	0,63**	0,59**	0,18

Persons attaching importance to one criterion (e.g. price) in one context (e.g. automobile) also attach importance to the very same criterion (price) in the other context (medical technology). Thus, the usage context does to a lesser extent specify the profile of required purchase criterions. Rather it is vice versa: The type of purchase criterion is more decisive independently of the technological field.

4   CONCLUSION

This study examined the interrelation of usage context (automobile, medical and IC technology) and user diversity (age, gender) on technical interest (general and in specific branches), purchase criteria and usage motives. It was found that both factors do play a major role. Gender and age significantly impacted the extent of general interest in technology. Also, technical branches were influenced by gender and age. Apparently, even though ICT and mobile technology is ubiquitous in our societies, technical interest profiles did not change over time and are - still - gendered. Regarding the specific adoption criteria, we found that "latest state of technology", "price", "usability", "safety" and "reliability" are more important than "brand", "design" and "unobtrusiveness" of technology. It is an insightful finding that adoption criteria and their specificity for technical fields are quite generic, not differentiating across gender and age groups. In contrast, the usage motives do differ with respect to technological fields, gender and age.

The findings contribute to a human-centered understanding of technology development and requirement engineering.

Acknowledgments

Thanks to Oliver Sack for his research assistance. This work was funded by the Excellence Initiative of German state and federal governments.

REFERENCES

Arning, K., & Ziefle, M. 2009. Different Perspectives on Technology Acceptance: The Role of Technology Type and Age. In A. Holzinger and K. Miesenberger (eds.). Human – Computer Interaction for eInclusion. LNCS 5889 (pp. 20-41). Berlin: Springer.

Arning, K., Gaul, S., Ziefle, M. (2010). "Same same but different". How service contexts of mobile technologies shape usage motives and barriers. In G. Leitner et al. HCI in Work & Learning, Life & Leisure. LNCS 6389 (pp. 34-54). Berlin: Springer.

Busch, T., 1995. Gender differences in self-efficacy and attitudes toward computers. Journal of EducationalComputing Research, 12, 147–158.

Davis, F. D. (1989). Perceived usefulness, perceived ease of use, and user acceptance of information technology. MIS Quarterly, 13, 318-340.

Gaul, S. & Ziefle, M. 2009. Smart Home Technologies: Insights into Generation-Specific Acceptance Motives. In A. Holzinger & K. Miesenberger (eds.) HCI and Usability for e-Inclusion. LNCS 5889 (pp. 312-332). Berlin, Heidelberg: Springer.

Gaul, S., Wilkowska, W., & Ziefle, M. 2010. Accounting for user diversity in the acceptance of medical assistive technologies. Full paper at the 3rd International ICST Conference on Electronic Healthcare for the 21st century (eHealth 2010). CD ROM.

Rodger, J.A. and Pendharkar, P.C., 2004. A field study of the impact of gender and user's technical experience on the performance of voice-activated medical tracking application. International Journal of Human-Computer Studies, 60, 529–544.

Seock, Y., Bailey, L.R. 2008. The influence of college students' shopping orientations and gender differences on online information searches and purchase behaviours. International Journal of Consumer Studies 32, 113–121.

Wilkowska, W. and Ziefle, M. 2011. "User diversity as a challenge for the integration of medical technology into future home environments," in Human-Centered Design of eHealth Technologies: Concepts, Methods and Applications, M. Ziefle and C. Röcker, Eds. Hershey, PA, USA: IGI Global, 2011, pp. 95–126.

Williams, T.G. and Slama, M.E. 1995. "Market mavens' purchase decision evaluative criteria: implications for brand and store promotion efforts," Journal of Consumer Marketing, vol. 12, no. 3, pp. 4–21.

Wilkowska, W., Gaul, S., Ziefle, M. 2010. "A Small but Significant Difference – the Role of Gender on the Acceptance of Medical Assistive Technologies". In G. Leitner et al. HCI in Work & Learning, Life & Leisure, LNCS 6389 (pp. 82 - 100). Berlin: Springer.

Ziefle, M. and Wilkowska, W. 2010. "Technology acceptability for medical assistance", in Proceedings of the 4th ICST/IEEE Conference on Pervasive Computing Technologies for Healthcare (pp.1-9). DOI 10.4108/ICST.PERVASIVEHEALTH2010.8859.

Ziefle, M. & Schaar, A.K. (2011). "Gender differences in acceptance and attitudes towards an invasive medical stent". electronic Journal of Health Informatics, 6 (2), e13, pp. 1-18.

Ziefle, M., Himmel, S. & Wilkowska, W. 2011. "When your living space knows what you do: Acceptance of medical home monitoring by different technologies". In H. Holzinger & K.-M. Simonic (Eds.). Human-Computer Interaction: Information Quality in eHealth. LNCS 7058, pp. 607–624.Berlin, Heidelberg: Springer.

Zhou G, Xu J. 2007. Adoption of educational technology: How does gender matter. International Journal of Teaching and Learning in Higher Education. 19(2):140–53.

CHAPTER 23

Biometric Measurements and Data Processing Using CAPTIV Software

Bronislaw KAPITANIAK*, Hossein ARABI**

*Ergonomist, Ex-Head of Ergonomics Unit UPMC
**Ergonomist PSA Peugeot Citroën

ABSTRACT

The operations of stapling covers on the foam present very significant working constraints assessed at the maximum level of hardness and risk, measured with the internal screening tool based on the RULA method. A physical and physiological measurement campaign was launched so as to define the level of hardness during concerned operations (stapling covers on foam during car-seat production). The measurements carried out in real field conditions concerned the forces applied on the stapler, the pressure exerted on the stapler by the operator's hand, the bioelectric muscle activities (EMG) and the biomechanical parameters of joint angulation. Biometrics system was used for the measurements and data processing was carried out with CAPTIV software. The results allowed us to observe that the forces applied on the stapler are excessive. On the other hand, the repetitiveness and the postural constraints should be considered as moderate, according to European standards EN1005.

Keywords: biomechanics, physical workload, CAPTIV

1. MATERIAL AND METHODS

In order to assess the level of hardness during operations of stapling covers on the foam a campaign of physical and physiological measurements was launched.

The measurements carried out in real field conditions concerned:
- forces applied on the stapler,
- the bioelectric muscle activities (EMG),
- the biomechanical parameters of joint angulation.

The forces were measured with a strain gauge SML-100 MFG (302422) of an approximate capacity of 50daN and with pressure sensors (AMP FSR).

The bioelectric muscle activity was measured with the Biometrics Ltd Datalog electromyographic system with EMG electrodes (Sx230 calibrated for 1000 Hz).

The exerted activities were recorded on digital video (Sony Handycam DCR-TRV620E).

The joint positions were assessed with the BIOVECT method. They were defined with freeze frames obtained from the video recording of the stapling exercise. Angle and vector analysis was carried out on the chosen frames.

CAPTIV software allowing to associate measurements with the video recorded during the measurements was used for processing the data.

The measurements were carried out on a sample of operators including different levels of anthropometric dimensions and muscle forces, as well as different levels of professional experience.

The forces exerted during the stapling activity were measured by attaching a strain gauge to the tool (the stapler) at one end and to a fixed point at the other. Then the operator carried out his task as closely to the real activity as possible. These measurements were preceded by measurements of maximal muscle forces so as to establish the reference in order to calculate the relative forces in percentage of maximal force.

The forces used by the forefinger on the stapler were measured with a pressure sensor stuck on the finger that pushed on the stapler command.

The bioelectric muscle activity was measured on the 4 muscle groups on the side of the upper limb that holds the stapler. The chosen muscles were: supinator longus, biceps brachii, deltoid, trapezius (fig.1).

The measurements were carried out with all the equipment and then synchronized with the video recorded during the measurements with CAPTIV software.

The results of the measurements were then processed with CAPTIV software allowing us to synchronize all the values measured with the video and to obtain a quantitative treatment of the results. Part of the calculation of the muscle activity and of the forces was carried out with EXCEL.

Fig. 1. Placing the EMG electrodes

The direct observations and the video were carried out during the stapling activity during one hour periods.

2. RESULTS AND DISCUSSION

2.1. Observed population

The observed operators' characteristics are presented in Table 1.

Table 1 Operators' characteristics

	age	height	weight	BMI	experience
Operator 1	26	172	73	24.7	pro
Operator 2	42	175	85	27.8	pro
Operator 3	33	176	75	24.2	beg
Operator 4	48	164	58	21.6	pro
Operator 5	41	187	79	22.6	beg
Operator 6	42	185	72	21.0	pro
Operator 7	34	180	72	22.2	pro
Operator 8	54	172	64	21.6	pro
Operator 9	47	164	58	21.6	pro
Average	**40.8**	**175.0**	**70.7**	**23.0**	

The average age of the sample represents the average age of the workers in the factory. Most operators are experienced, 2 are beginners (in stapling activity). All operators are right-handed.

2.2. Observing real activity

The results of the punctual observations of stapling activity show that the operating modes used by the operators largely depend on the type of seat that is assembled.

Indeed, in the first observation we distinguished seats that were stapled by the operator and those that were only inspected. During one hour, the operator assembled 33 seats and inspected 14. The operating mode consisted in fitting 4 staples and from 12 to 18 clips.

The break time was 9 min per 1 hour period, which corresponded only to technological pauses (waiting). No difference in the length of the cycle was observed between stapling and clipping.

A video analysis of gesture sequences was carried out on 5 recordings of approximately 15 min of the activity on a standard assembly line. The average duration of a cycle and the average frequency of stapling gestures were calculated. The results are presented in Table 2.

Table 2 Video analysis of gesture sequences

	cycle duration (s)	stapling frequency (actions/min)
Operator 6 1	66.0	3.6
Operator 7	46.8	13.9
Operator 6 2	47.6	12.9
Operator 8	68.4	10.8
Operator 9	60.4	10.2
M	57.8	10.3
E-T	10.1	4.0
CV	20%	40%

The results show that the differences in cycle duration are not very significant (variation coefficient of 20%).

The stapling frequencies vary more significantly according to assembled references.

2.3. Measurements of forces and of muscle activity

The maximal forces were recorded asking the operators to pull on the stapler in the usual working axis as strongly as they could.

The results of the measurements of maximal muscle forces and of the absolute and relative working forces are presented in Table 3.

Table 3 Measurements of maximal muscle forces and of the absolute and relative working forces

	Pushing on staple			Pushing with the forefinger		
	Fmax	Fwork	Frel	Fmax	Fwork	Frel
Operator 1	39.3	6.5	16.5%	4	1.7	42.5%
Operator 2	35.1	6.2	17.7%	3.5	2.1	60.0%
Operator 3	38.5	4.7	12.2%	3.1	0.6	19.4%
Operator 4	34.8	6.2	17.8%	3	0.7	23.3%
Operator 5	39.3	6.7	17.0%	4.3	1.3	30.2%
M	37.4	6.1	16.3%	3.6	1.3	35.1%
E-T	2.3	0.8	2.3%	0.6	0.6	16.5%
CV	6%	13%	14%	16%	50%	47%

Although the number of subjects is too small to allow a statistical analysis, we decided to calculate the variation coefficients in order to assess the interindividual variability among the operators. Of course, those coefficients have no real statistical value, but an indicative one, allowing us better interpretation of the results. This remark applies to all the calculations of forces and EMG activity, where the number is insufficient for reliable statistical calculation.

2.3.1. Forces pushing on the stapler

The maximal forces pushing on the stapler are quite similar in the 5 operators, as shown by the very low standard deviation and the variation coefficient of 6%. The forces exerted during work vary more (variation coefficient of 13%) which shows that there are individual differences in muscle strategies used by the operators according to their professional experience. No difference appears between the 2 novice operators and the 3 experienced operators. The working forces reach on average 16% of the maximal force, which must be interpreted as acceptable, given the dynamic character of the stapling task. This dynamic character is confirmed by all the video observations which show that the operators do not hold the stapler steady but accomplish a full movement, released immediately after the staple is fitted.

2.3.2. Forces pushing with the forefinger

The forces pushing on the stapler trigger with the forefinger were measured with a pressure sensor. The maximal forces vary according to the operators (variation coefficient of 16%). The forces applied during work vary even more (variation coefficient of 50%) which suggests a multitude of personalized muscle strategies. The calculation of relative forces shows that the level reached is around 35% of their maximal forces, which must be interpreted as verging on excessive forces. The observations of recorded videos allow us to confirm the dynamic character of pushing on the trigger.

2.3.3. The results of EMG measurements

The results of maximal instantaneous bioelectric activity measurements (EMG) are presented in Table 4.

Table 4 Maximal instantaneous bioelectric activity measurements (EMG)

	trapezius	deltoid	biceps	sup long
Operator 1	39.3%	33.9%	20.8%	32.5%
Operator 2	42.2%	22.8%	39.5%	43.6%
Operator 3	72.1%	21.2%	49.9%	33.3%
Operator 4	55.2%	43.5%	35.8%	39.8%
Operator 5	24.3%	25.0%	58.3%	25.4%
M	46.6%	29.3%	40.9%	34.9%
E-T	18.0%	9.3%	14.3%	7.0%
CV	38.6%	31.9%	34.9%	20.2%

The results presented in Table 4 were calculated as percentage of maximal activity observed during the exercise of the maximal force. Given that the device used for the field applications (Biometrics Datalog) does not allow recording the raw EMG signal but only an already integrated and averaged signal (avg RMS), this calculating method is most useful and brings the most relevant results.

The presented values were marked on the recording while realizing the stapling gestures and correspond to instantaneous maximal peaks of bioelectric activity. They thus represent maximal activity peaks observed during dynamic movements. The results show that for the 4 muscle groups the measured values range from 30 to 50% of the maximal activity, which corresponds to a range between "moderate" and "heavy" tasks.

To refine the analysis we calculated the mean values for each stapling cycle. The results are presented in Table 5.

Table 5 Mean values of EMG expressed on % maximal activity for each stapling cycle

	trapezius	deltoid	biceps	sup long
Operator 1	14.4%	12.5%	4.4%	8.3%
Operator 2	19.7%	7.0%	11.9%	25.7%
Operator 3	28.4%	6.3%	22.3%	33.3%
Operator 4	27.8%	13.0%	12.3%	19.1%
Operator 5	5.6%	8.5%	16.0%	8.8%
M	19.2%	9.5%	13.4%	19.0%
E-T	9.6%	3.1%	6.5%	10.8%
CV	49.9%	32.9%	48.8%	56.8%

The stapling cycle for a seat varies according to the type of seat from 30 to 60s. The results of the calculations presented in Table 4 concern cycles closer to 60s. The obtained values are lower than 20% of maximal activity, which should be interpreted as acceptable. However, considering the repetitive character of the gestures, the "acceptable" zone should be limited to 15%. This amount to say that part of the muscle activity is situated in a zone defined in the standardization as "acceptable but not recommended." This judgment is even more valid when one takes into account a very high individual variability (variation coefficient of about 50%) and the small sample. On the other hand, several authors mention the difficulties in correctly measuring the maximal forces in fieldwork conditions, arguing that the results of such measurements must be considered as under-estimated.

2.4. Gesture analysis of stapling activity

The biomechanical vector analysis was carried out with BIOVECT method on 10 frames chosen according to the postural variability observed. The BIOVECT method allows comparing the measured angles with reference comfort angles. This reference was compiled from data collected in the literature (Monod and Kapitaniak, Péninou and Kapitaniak).

The results of the angle analysis show that the posture adopted fir stapling must be considered as very slightly uncomfortable, notably because of a flexion of the back. On the other hand, the positions of the upper limbs are almost always in the zone of postural comfort. The results of the vector analysis show that the slightly bent over position does not generate excessive constraints at the lumbar level. According to an abundant bibliographic study by Mathiassen one must expect the bias of estimating the posture form short period observations to be significant and to reach 30% of error in overestimating unfavorable postures.

3. CONCLUSIONS

Although the muscle forces of pushing the stapler are in the higher limit of the maximal acceptable force, they are clearly higher when related to the maximal acceptable force as established by the NF EN 1005-3 standard. The ratio between the recommended force and the measured force ranges from 1.75 to 2.8, which indicates a health hazard two or three times as significant as the acceptable hazard. The results of the measurements show that the forefinger works in a zone higher than 30% of the maximal force, which must be considered as excessive. The results of the bioelectric activity measurements of muscles involved in the stapling action also suggest an excessive effort (from 30 to 45% of maximal efforts for instant actions, from 10 to 20% for the averages measured over cycle periods of about 1 min, and from 5 to 10% for the prolonged periods of about 30 min) and confirm this evaluation.

The results of postural analysis indicate that the main source of hardness consists in a slightly bent over position of the trunk. It is relatively little expressed by the operators because the workstations are provided with height adjustment and correctly adjusted. The stapler, however handled, does not generate joint positions of the upper limbs outside the comfort zone.

According to the global results of the present ergonomic study, the task of stapling covers on the foam in the process of producing car seats should be considered as involving a health hazard of a moderate level. This hazard is estimated by the standards as not recommended but within acceptable limits.

The results show that CAPTIV software makes it possible to carry out an accurate field analysis, very relevant from the ergonomic point of view.

BIBLIOGRAPHY

AFNOR Recueil de normes françaises, Ergonomie des postes et lieux de travail, 2008, éd. AFNOR, CD-ROM.

BAE S, WEI ZHO, ARMSTRONG T., Finger Motions in Reach and Grasp Work Elements, XVIII IEA Congress, Pékin, 2009

COLOMBINI D., GRIECO A., OCCHIPINTI E. Occupational musculoskeletal disorders of the upper limbs due to mechanical overload, Ergonomics, 1998, vol.41, n°9

FORCIER L., KUORINKA I., Work-related musculoskeletal disorders: overview, International Encyclopaedia of Ergonomics and Human Factors, 2ème edition, 2006, vol. 2, 1625-1632

KAPITANIAK B., Static load, International Encyclopaedia of Ergonomics and Human Factors, 2ème edition, 2006, vol. 1, 580-583

KAPITANIAK B., PENINOU G., HEUSCH F., Software "BIOVECT" help for postural analysis, XI Annual International Occupational Ergonomics and Safety Conference, Zurich 1996

MATHIASSEN S.E., WULFF SVENDSEN S., Systematic and random errors in posture percentiles assessed from limited exposure samples, XVIII IEA Congress, Pékin, 2009

MONOD H., KAPITANIAK B., Ergonomie. Masson, 2ème éd., 2003, 1 vol., 286 p.

MOTMANS R., ADRIAENSEN T., HERMANS V, Muscle activity during repetitive work, XVIII IEA Congress, Pékin, 2009

OCCHIPINTI E., COLOMBINI D., DE VITO G., MOLTENI G., Exposure assessment of upper limb repetitive movements: criteria for health surveillance, International Encyclopaedia of Ergonomics and Human Factors, 2ème edition, 2006, vol. 1, 1507-1509

PÉNINOU G., KAPITANIAK B., Vectorial analysis; ergonomics applied method. International Encyclopaedia of Ergonomics and Human Factors, 2ème edition, 2006, vol. 3, 3018-3022

MONOD H., KAPITANIAK B., Ergonomie. Masson, 2003, 1 vol., 120 p.

SILVERSTEIN B., Work-related musculoskeletal disorders: general issues, International Encyclopaedia of Ergonomics and Human Factors, 2ème edition, 2006, vol. 2, 1621-1624

YOUNG J., ARMSTRONG T., ASHTON-MILLER J., Active and Passive Forces in Hand/Work-Object Coupling, XVIII IEA Congress, Pékin, 2009

China's Ministry of Minor Staff Based Measurement and Application

Hu Shou-zhong Zhou Xiang Li Guang-ting

Shanghai University of Engineering Science
Shanghai, China
hushzh@126.com

ABSTRACT

The daily life of a minor size appliance only in a perfect match with the human body in order to achieve the best results. China in the national body size measurement of minors is a blank, hand size is one of them. Photographic measurement method using minors 4000 (4 ~ 17 years) of hand size measurement, based on the "long hand" and "hand width" as the benchmark, the Ministry of number of young staff-type measurements to classify research, define, and sub-file, boys and girls to develop hand-type numbers. Finally, the Lectra CAD system, plate, push version and layout designed for three different uses of minors protective gloves.

Keywords: minors, hand length, hand width, hand sizing system, gloves

1   INTRODUCTION

Since the founding of new China, body sizes of national juvenile have not been measured in our country [1], and the hand size is one of them. As a result of long-term lack of accurate juvenile body size data, leading to minor supplies less satisfaction than adults. One of directions in the field of ergonomics research is Chinese ergonomics standards of scientific data collection and its application [2]. Hand measurement and its measuring data contain important significance for planning and choice of hand types. International Organization for Standardization Technical Committee - Ergonomics anthropometric and biomechanical Committee (

ISO / TC 159 / SC3), has repeatedly made about human body measurement resolution, called on all countries to carry out human body measurement, in order to establish the International Human Dimensions Basic data.

Hand measurement is the necessary premise of divided type; the significance of dividing type is that based on the model a variety of protective equipments (such as gloves) could be produced, at the same time choose suitable protective equipment according to the type number. The formulation of minor hand size, make gloves produced on the basis of the formulation more fitting minor hand, and styles will expand to be applicable to all fields. At the same time, these shapes can reflect the shape characteristics of Chinese. On the basis of hand measurements about hand length and hand wide of Chinese minor from 4 to 17 years, type classification research could be carried out. At the same time three minor protective gloves have been designed, using Rick CAD system plate, pushing version and typesetting, to strengthen the ergonomics research.

2   THE CONTENTS OF MEASUREMENT

2.1   Data collected by photographic measurement

In China National Standard "adult hand shapes", hand length and width is defined by: hand length, connecting the styloid process of radius and ulnar styloid point palm side connection point to the tip of the middle finger point line distance; hand wide, lateral metacarpal to ulnar metacarpal point of the straight line distance. The boys and girls are divided into two categories, each number statistics. Use of photogrammetry, 4000 minors (4 ~ 17 age is minor, no longer below illustrate) hand pictures into the computer, to Photo format disk. The use of Photoshop graphics editing software measurement hand length and hand width, see figure 1.

Figure 1 the Measure of Length and Width

Measurement of those tested clothes cuffs rolled up, exposing the wrist, hand and wrist remained level. To the middle vertical line as the benchmark to measure the hand length and hand width. When the tone measured by hand photo, not all of those tested are perpendicular to the middle finger. Measurement photoshop software measurement interface to pull out a vertical line as a benchmark, picture rotation to the middle and perpendicular to parallel measurements. Marked out a middle finger to the wrist point distance as a hand; marked two points the distance from left to right hand wide.

2.2 Ministry of minor staffing definitions and sub-files

Refer to the Chinese adult hand and type classification standard GB16252 [3] similar to typing. - Refers to the hand length, hand length size; type - refers to the width of the hand to hand width dimensions, said shot the size of the Ministry of size too fat, both in mm. Hand shape representation: number and type between the values separated by a slash, and indicate the men (M) or female (F).

Minors in a vigorous stage of growth and development, the shape of the hand will be many. 2mm points file, hand length 5mm sub-file, hand width, hand long 95mm hand width 50mm for the first classification. Hand length in 95 ~ 100mm, hand width, 50 ~ 55mm, boys and girls 95/50 type. Each number very few number of coverage is less than 5% due to the limited number of measurements, such a classification lead, so I decided to hand length 10mm sub-file, hand width by 5mm sub-file.

Hand No. tranches. Underage boys and girls, hand length, hand width data from small to large, to find out in line with the Type number. Each number type belongs to the number of records in an Excel table, see Table 1 and Table 2.

Table 1 Hand measurement data of boys (1807)

Sequence Number	Hand Length (mm)	Hand width (mm)									
		50	55	60	65	70	75	80	85	90	95
1	100	5									
2	110	6	40	2							
3	120	1	48	55							
4	130		11	93	53	3					
5	140			34	149	38					
6	150			4	73	130	35	1			
7	160				7	91	110	14			
8	170					15	155	123	13		
9	180					1	64	201	70		
10	190					1	9	66	64	4	
11	200							7	8	2	1

Table 2 Hand measurement data of girls (1700)

Sequence Number	Hand Length (mm)	Hand width (mm)						
		50	55	60	65	70	75	80
1	100	5	1					
2	110	36	29	1				
3	120	1	80	46	2			
4	130		25	107	24	1		
5	140		1	64	90	11		
6	150			11	145	158	11	
7	160				61	312	101	4
8	170				12	157	135	13
9	180				1	13	33	4
10	190						2	3

Coverage obtained by the measurement data in Tables 1 and Table 2. The data belong to a number type data statistics, draw and type of coverage in all of those tested are shown in Table 3 and Table 4, male and female hand coverage.

Table 3 Boys hand coverage (‰)

Sequence Number	Hand Length (mm)	Hand width (mm)									
		50	55	60	65	70	75	80	85	90	95
1	100	2.77									
2	110	3.32	22.13	1.11							
3	120	0.55	26.56	30.44							
4	130		6.09	51.47	29.33	1.66					
5	140			18.82	82.46	21.03					
6	150			2.21	40.4	71.94	19.37	0.55			
7	160				3.87	50.36	60.87	7.75			
8	170					8.3	85.78	68.07	7.19		
9	180					0.55	35.42	111.23	38.74		
10	190					0.55	4.98	36.52	35.42	2.21	
11	200							3.87	4.43	1.11	0.55

Table 4 Girls hand coverage (‰)

Sequence Number	Hand length(mm)	Hand width(mm)						
		50	55	60	65	70	75	80
1	100	2.94	0.59					
2	110	21.18	17.06	0.59				
3	120	0.59	47.06	27.06	1.18			
4	130		14.71	62.94	14.12	0.59		
5	140		0.59	37.65	52.94	6.47		
6	150			6.47	85.29	92.94	6.47	
7	160				35.88	183.53	59.41	2.35
8	170				7.06	92.35	79.41	7.65
9	180				0.59	7.65	19.41	2.35
10	190						1.18	1.76

The coverage reflects the measured hand shapes in a given proportion in the number of (1807 boys and 1700 girls). Boys hand coverage look-up table shows: the shape of the hand with boys hand length 130mm with hand width 60mm, the proportion was 51.47 ‰, show that the hand length from 125mm to 130mm (125mm) range, while the hand width 57.5mm to 62.5mm (57.5mm) underage boys, the proportion of 1807 people in the measurement of 51.47 ‰.

According to Table 3 and Table 4 boys and girls hand and type of coverage, the coverage is greater than equal to 5% (the converse is not set the number type) and set the Type of principle, to work out hand shapes for boys and girls, see tables 5 and 6.

Table 5 Boys hand-type formulation

Sequence Number	Height (mm)	Circumference(mm)						
1	110	55						
2	120	55	60					
3	130	55	60	65				
4	140		60	65	70			
5	150			65	70	75		
6	160				70	75	80	
7	170				70	75	80	85
8	180					75	80	85
9	190						80	85

Table 6 Girls hand-type formulation

Sequence Number	Height (mm)	Circumference(mm)						
1	110	50	55					
2	120		55	60				
3	130		55	60	65			
4	140			60	65	70		
5	150			60	65	70	75	
6	160				65	70	75	
7	170				65	70	75	80
8	180					70	75	

3. CONCLUSION

3.1 Hand the growth trend

Derived from Tables 5 and 6 the number-average of the hand of each age group, analysis of growth trends and characteristics of the boys and girls hand. Exclude special hand data (such as a finger is too long, hand width is too large, etc.), 3507 Ministry of minor staffing data, find the average of each age hand length and hand width data, draw a minor staffing the Department of growth trend of linear map, see Figure 2 and Figure 3.

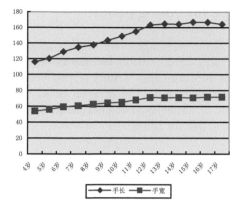

Figure2 The growth trend of girls'hand

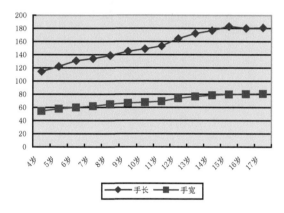

Figure 3 The growth trend of boys'hand

As is shown in the graph, compared to the increase of hand length, the growth of hand width is proportional to the growth of age just as hand length does, however, the former increase markedly while the latter increase by a small scale. According to our obeservation and measurement, the growth of the finger contribute greatly to that of the whole hand, As people are getting old, the growth of finger account for a great proportion of that of the hand length.

From the line chart, man betwent 15 and 17 keep a stable growth, the hand is growing with a diminishing rate and ever remain unchanged which demonstrate that growth for men's hand is in their first 15 years, after which the growth is rather small. In contrast, the primary time for the growth of women is in their 12 years. By comparing hands of young people with that of adults, we can find hands of Boys between 15 and 17 and Girls between 12 and 17 is similar to that of adults.

3.2 The hand type for young people

As young people in the same ages vary in their hand shapes, different people have different type of hands. From calculation, the average value of hand length and width of young men is 163.2mm and 73.4mm respectively with the type 165/75M, the number for young women is 156.7mm/68.2mm with the type 155/70F. Consequently, the type of hands for young people following 165/75 M and 155/70 F approximately equal to that for adults.

REFERENCE

[1] Zhou Xiang, Xie Hong, Chen Jing, etc. the Measurement and Analysis of the Figure for Young People Based on Fuzzy Recognition[J]. Electrica Automation.
[2] Fang Fang, Zhang Weiyuan, Zhang Wenbin, Lang Jun, etc. Research on the standards for body measuring[J]. AcademicJournal of DongHua University
[3] Xian Liwen. the Standards and Application of Body Dimension of cChinese Adults[J]. China Standards Review.

CHAPTER 24

A Study on Systematizing GUI Design Patterns for the Embedded System Products

Ichiro Hirata[1], Kenshiro Mitsutani[2], Toshiki Yamaoka[2]

1) Hyogo prefectural institute of technology
2) Wakayama university
Hyogo, JAPAN
ichiro@hyogo-kg.go.jp

ABSTRACT

The purpose of this study is to develop a graphical user interface (GUI) design efficient. We propose a method that is based on the Human Design Technology to design the screen. GUI design patterns are defined as "general operation and expression". A data mining technique of Formal Concept Analysis (FCA) was adopted to systematize the GUI design patterns.

To apply the systematized GUI design patterns to screen design, the mapping relationship between GUI design patterns and user interface (UI) design items were established. The problem of how to select GUI design patterns can be fixed using UI design items. Each of group, GUI design patterns and UI design items, were systematized focusing on class hierarchy by using FCA.

Keywords: GUI, design pattern, user interface, Human Design Technology

1   INTRODUCTION

Since Human Centered Design (HCD) has been recognized as a powerful tool for user interface, large enterprises have been introducing HCD method and facilities. But, it is hard to introduce present HCD methods to middle/small enterprises because introducing HCD method is rather expensive. Thus, we have studied a design method to introduce HCD into middle/small enterprises.

In this paper, we suggest a method of screen design with the GUI design patterns. This method makes it possible to introduce HCD into middle/small enterprises.

2   DESIGN PROCESS

Proposed GUI method is based on Human Design Technology (HDT). HDT is a logical product development and Human-centered Design method easily accessible to anyone. HDT's design process is as follows (Fig 1).

1) Gather user requirements
   User requirements are extracted to product problems. Extract problems using group interviews, observation and task analysis.
2) Grasp current circumstances
   Investigate how users perceive a target product in the market using correspondence analysis.
3) Formulate structured concepts
   Construct structured concepts based on user requirements.
4) Design (synthesis)
   Visualize a product based on the structured concepts. HDT requires that the design be based on the seventy predetermined design items.
5) Evaluate the design
   The design idea is evaluated by user test. A protocol analysis, questionnaires were made on 5 people.

HDT design process is the method for product design. So, HDT needed to customize for GUI design. When product was designed in HDT process, designer visualized using 70 design items. Proposed method is visualized using 2 items as follows.

1.   Concept target table
2.   GUI design patterns

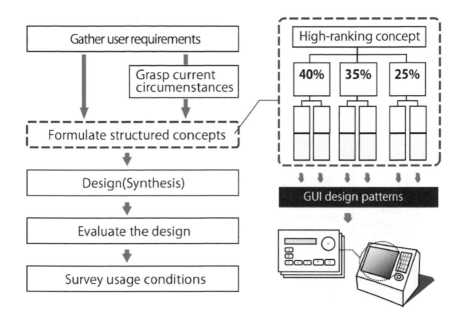

Figure 1 Design process of Human Design Technology (HDT)

2.1 Concept Target table

GUI design required information to system spec and user spec. The concept target table could define target system and users clearly (Table 1).
Contents of target system are included as follows.

System: Function, Device, Space, Hours, and Implementation
System's element: User decided 3-4 elements among the 18 elements.
User Interface
Task

Contents of users are included as follows.

Attribute: Age, Sex, Occupation, and Earn
User level: Experience, Education level, Similarity experience, Life style
User's mental model: Functional model, Structural model

Table 1 Concept target table

Clear target system	System	Function	Exclusive	
		Device	Touch pad	
		Space	Bank, Public space	
		Hours	9:00–17:00	
		Implementation	・Staff use when system is broken ・Receptionist guides user if user don't operate	
	System's element	Safety	Convenience	Modern	Functional
		Security	Efficiency	Surprise	Legible
		Confidence	Economical	Entertainment	Aesthetic
		Reliability	Tolerance	Achievement	
		Usability	Conservative	Emotional	
	User interface	User need not get new knowledge of operation		
	Task	Withdrawals, Remittance, Contributed, Money received		

Clear target user	Attribute	Age	18 – 65
		Sex	Man, Woman
		Occupation	General
		Earn	General
	User level	Experience	User has been able to use the ATM some time
		Education level	User is able to read the Japanese text
		Similarity experience	User has been able to use the station ticket reservation system some time
		Life style	Various
	User's mental model	Functional model	Model to understand How-to-use it
		Structural model	Model to understand How-it-works

2.2 GUI design patterns

GUI design patterns are defined as "general operation and expression". Some GUI design patterns of 128 case studies were investigated, and extracted patterns were classified.

As a result of this classification, GUI design patterns were grasped 84 patterns. 84 patterns were classified into 2 groups below.

1. Design: Display and layout, Access to information
2. Function: how to do the operation, touch area

GUI design patterns based on Function classification are shown in table2. GUI design patterns based on Design classification are shown in table3.

Table 2 GUI design patterns based on Function classification

GUI design pattern of Function (45items)		
Display zoon out/in	Rearrangements of items	Move items using tab key
Undo	Item editing	Drop-down menu
Cloze items	Menu bar	Tree structure
Structural input	Adjust text size	Guidance of input area
Direct operation	Continuous filtering	Aggregation structure
Default indicate	All select by one click	Drag
Gray display	Searching	Procedural operation
Right down enter	Keyword sorting	Hub structure
Metaphor selection	Shortcut key	Gradual addition
Free input	One click enter	Visible / Invisible
Tool menu	Scroll bar	Auto complete
Round button	Re-layout	Guidance
Setting	All process display	Tab button
Double click	Cancel along the way	Enter after selection
Resize	List operation	Top navigation

Table 3 GUI design patterns based on Design classification

GUI design pattern of Design (39 items)		
Icon	Hierarchical information	Waiting time indicate
Various icons	Rectangle	Emphasis
Consistency color	Information of top class	Model display
Color information	One display	Modal dialog
Picture information	Simple line	Two windows
Current position	Center information	Three windows
Footer information	Flexible screen	End of task
Common header	Dual display	Tab
Complement of change	Stripe background	Base display
Pop-up window	Emphasis of selection	1/()
Stand in	Proximity	Ruled line
Top with impact	Look flow	Check selection
Support of users' memory load	Blank effect	Breadcrumbs

3 SYSTEMATIZING GUI DESIGN PATTERNS

3.1 User Interface (UI) design items

User interface (UI) design items are prepared expert designer's know-how by precedence research. The problem of how to select GUI design patterns can be fixed using UI design items. UI design items had 29 items are as follows.
1) Flexibility, 2) Customization for different user levels, 3) User protection, 4) Accessibility, 5) Application to different cultures, 6) Provision of user enjoyment, 7) Provision of sense of accomplishment, 8) The user's leadership, 9) Reliability, 10) Clue, 11) Simplicity, 12) Ease of information retrieval, 13) At a glance interface,

14) Mapping, 15) Identification, 16) Consistency, 17) Mental model, 18) Presentation of various information, 19) Term/Message, 20) Minimization of users' memory load, 21) Minimization of physical load, 22) Sense of operation, 23) Efficiency of operation, 24) Emphasis, 25) Affordance, 26) Metaphor, 27) System Structure, 28) Feedback, 29) Help

29 UI design items were classified like GUI design pattern's classification.

Table 4 Classified UI design items

Function (9)	1) Flexibility
	2) Customization for different user levels
	8) The user's leadership
	9) Reliability
	20) Minimization of users' memory load
	21) Minimization of physical load
	23) Efficiency of operation
	28) Feedback
	29) Help
Design (12)	6) Provision of user enjoyment
	7) Provision of sense of accomplishment
	10) Clue
	11) Simplicity
	14) Mapping
	15) Identification
	18) Presentation of various information
	19) Term/Message
	24) Emphasis
	25) Affordance
	26) Metaphor
	27) System Structure
Function and Design (8)	3) User protection
	4) Accessibility
	5) Application to different cultures
	12) Ease of information retrieval
	13) At a glance interface
	16) Consistency
	17) Mental model
	22) Sense of operation

3.2   Formal Concept Analysis

Formal Concept Analysis (FCA) is a method to derive relationships from objects (GUI design patterns) and their properties (UI design items). As a result of FCA, analyzer can find user's behavior tendency in the group and reason of their behavior, and finally group's peculiarity could be grasped.

Result of FCA is a concept lattice witch is drawn as a lattice diagram. Relationships between GUI design patterns and UI design items appear on concept lattice. We analyzed 4 Groups as follows.

1) Function (UI design items) and Function (GUI design patterns)
2) Design (UI design items) and Design (GUI design patterns)
3) Both sides (UI design items) and Function (GUI design patterns)
4) Both sides (UI design items) and Design (GUI design patterns)

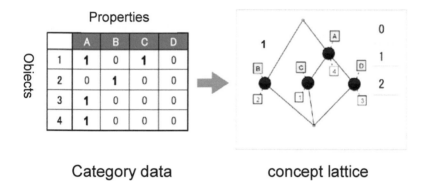

Category data   concept lattice

Figure 2 Example of FCA

4   RESULT

We analyzed each groups using FCA. Here is an example, Function (UI design items) and Function (GUI design patterns). As example, concept lattice of "Function (UI design items) and Function (GUI design patterns)" appears in figure 3. Relationships between GUI design patterns (Function) and UI design items (Function) appear on this chart.

"Efficiency of operation (UI design item)" is at summit. It means that many GUI design patterns fall under "Efficiency of operation". In short, "many GUI design patterns are efficiency of operation" is given as a peculiarity of this group.

On the other hand, "Undo (GUI design pattern)" is at the lowest layer. Concept at the lowest layer means, "no one falls under it". So, it means that "Undo" is unique pattern.

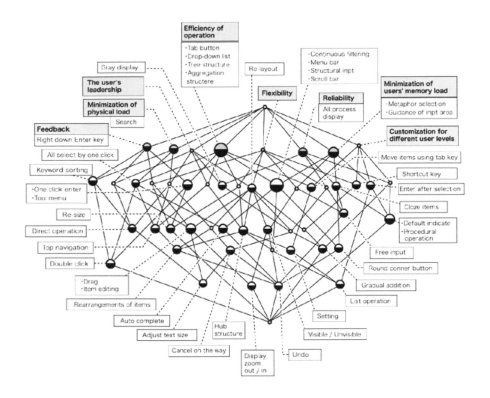

Figure 3 Relationship with UI design items and GUI design patterns (Function)

4.1   Applying GUI design patterns to design process

GUI design patterns are defined as "general operation and expression". GUI design patterns are applied to presented GUI method. The procedure to apply GUI design patterns on presented method is as follows.

(1) Constructing the structured concept: The structured concept is constructed based on user requirements.

(2) Making the concept target table: The concept target table is able to define target system and target users clearly.

(3) Select UI design items: UI design items are selected by deciding each subordinate position of the structured concept.

(4) Select GUI design patterns (Function): GUI design patterns based on function classification are selected by deciding UI design items.

(5) Select GUI design patterns (Design): GUI design patterns based on design classification are selected by deciding GUI design patterns based on function classification.

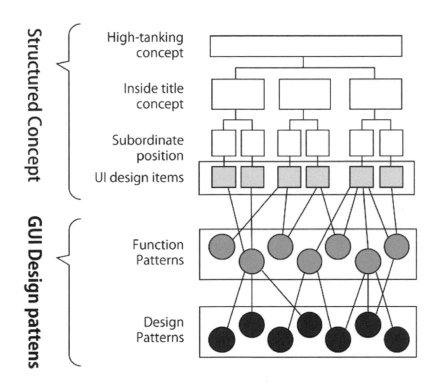

Figure 4 Relationship between structured concept and GUI design patterns

5   CONCLUSIONS

In this paper, a method of screen design with the GUI design patterns is discussed. The design process for GUI design was described based on HDT and the GUI design patterns. Logical design approach is needed for middle/small enterprises. As the usual traditional design method depends on designer's intuition or skill, it takes a lot of time to achieve design. But, beginner designers don't know GUI design items systematically. When they started GUI design, they tried to collect suitable GUI design items taking a lot of time. When they know the structure of the systematizing GUI design patterns, they collect suitable GUI design items quickly. In addition, design representation could be described quite clearly.

The methodology used in GUI design pattern can be as follows.

1. GUI could be designed quickly.
2. GUI design representation could be described quite clearly
3. GUI design process could be clearly
4. Designer's ability is little influence

ACKNOWLEDGMENTS

This work was supported by Grant-in-Aid for Scientific Research(C: 23611055).

REFERENCES

Yamaoka, T. 2003. Human-Centered Design Using Human Design Technology - Applications to Universal Design and so on–,Plenary Speech, 13th Triennial Congress of the International Ergonomics Association, 4.

Johnson-Laird, Philip N. Mental Models. 1983. Toward a Cognitive Science of Language, Inference and Consciousness. Harvard University Press

Norman.D.A, 1983. Some observations on Mental Models, Mental Models, Lawrence Erlbaum Associates, Publishers, 7-14,

Tullis. T. S, 1984. Predicting the Usability of Alphanumeric Displays. Doctoral dissertation, Rice University, Houston

Donald. A. Norman, 1986. Stephen W. Draper, User centered system design: New perspectives on human-computer interaction. Lawrence Erlbaum Associates

Erich Gamma, Richard Helm, Ralph Johnson, John Vissides, 1995. Design Patterns: Elements of Reusable Object-Oriented Software

Christopher Alexander, 1979, The Timeless Way of Building

Christopher Alexander, 1977, A Pattern Language: Towns, Buildings, Construction

Toshiki Yamaoka, Ichiro Hirata, Akio Fujiwara, Sachie Yamamoto, Daijirou Yamaguchi, Mayuko Yoshida, Rie Tutui, 2009, A Proposal of New Interface Based on Natural Phenomena and so on (1), Human-Computer-Interaction International 2009, Universal Access in HCI, Part II, Proceedings, 613-620

Ichiro Hirata, Toshiki Yamaoka, Akio Fujiwara, Sachie Yamamoto, Daijirou Yamaguchi, Mayuko Yoshida, Rie Tutui, 2009, A Proposal of New Interface Based on Natural Phenomena and so on (2), Human-Computer-Interaction International 2009, Universal Access in HCI, Part II, Proceedings, 528-534

Ichiro Hirata, Shohei Yoshida, Toshiki Yamaoka. 2009. Elderly People's Operational Image Using New Electrical Appliances, Human-Computer-Interaction International 2009, 984-988

Ichiro Hirata, Yasunori Goto, Takayuki Ueda, Teizo Nanbu, Toshiki Yamaoka. 2008. Usability Improvement of Setting a Timer- Case Study on the Heater's Remote Controller -, The 8th Asia-Pacific Conference on Computer-Human Interaction, ISBN 978-89-961324, 114-115

Section II

Theoretical Issues in Usability

CHAPTER 26

Using Modeling and Simulation to Enhance Vehicle Crewmember Survivability

Lamar Garrett, Rick Kozycki

U.S. Army Research Laboratory
Aberdeen Proving Ground, Maryland
lamar.garrett.civ@mail.mil

ABSTRACT

Incidents involving tactical vehicles are one of the leading causes of non-combat fatalities and serious injuries in Iraq and Afghanistan. According to data provided by the US Army Combat Readiness/Safety Center, tactical vehicles accounted for nearly one-third of all non-hostile fatalities (17 fatalities from November 2007 to June 2010). Significant efforts have been undertaken to improve survivability of these vehicles, especially against the constant threats of improvised explosive devices. Among these efforts has been a greater emphasis on passage ways, escape routes and hatches to ensure that they provide sufficient space for vehicle personnel to safely perform emergency egress measures. However, a vehicle fire or rollover may block access to vehicle doors or hatches and require crewmembers to use an emergency access tunnel to evacuate through the front or rear of the vehicle. Ensuring that all vehicle personnel can successfully utilize these emergency access tunnels or passageway then becomes a key safety requirement.

With regard to US Army weapon systems and platforms, digital human figure modeling has become a key tool to help ensure that Manpower and Personnel Integration (MANPRINT, AR 602-2) requirements are met. Application of human figure modeling tools and techniques has proven to be a valuable tool in the effort to examine man-machine interface problems through the evaluation of three dimensional (3D) Computer Aided Design (CAD) models of these systems. Anthropometrically and biomechanically accurate human figure models or body manikins can be used to analyze the geometric relationship between the human body and the physical environment. These analyses include reach zones, field of view, visible ground intercepts, simulation of dynamic body motions for various operator and maintenance tasks, as well as the joint torques and physical strength required to perform them. Additionally, human figure modeling can also be used to investigate the space or clearance issues associated with ingress and egress and examine the posture that one might have to assume in order to traverse an egress route or passageway. However, formidable challenges still exist to the effective application of these tools especially with regard to military platforms. For example, any accommodation analysis of these systems must include not only the physical dimensions of the target Soldier population but also the specialized mission clothing and equipment such as body armor, hydration packs, extreme cold weather gear and chemical protective equipment to name just a few. In fact, the effective use of human figure modeling (HFM) for workspace analysis requires a multi-faceted approach for a complete and accurate design assessment.

This paper highlights the methodology used to evaluate emergency egress issues for a tactical vehicle and suggest design changes for improved survivability.

Keywords: emergency, egress, human figure, modeling, simulation, survivability, tactical vehicle, MANPRINT

1   INTRODUCTION

In order to meet the threat on today's battlefields, the role of the tactical wheel vehicle has evolved significantly from its origin as a simple unprotected motorized transport to a protected system capable of operating in a full-spectrum, non-linear battlefield. However, an issue for these tactical wheeled vehicles is that they simply are not design to withstand the impact from a vehicle rollover which can occur as a result of exposure to a blast event. A vehicle rollover caused by blast forces from an anti tank (AT) mine or an improvised explosive device (IED) poses a serious threat to the crew's ability to use the vehicle's primary entry and exit points (Eisele, A, 2005). Recent efforts have been undertaken to improve the survivability of these vehicles, by placing a greater emphasis on hatches (see Figure 1), passage ways and escape routes such as tunnels to make them more accessible and ensure that they provide sufficient space for crewmembers to rapidly escape in an emergency.

Figure 1. Exterior view of roof mounted emergency egress hatches on an armored tactical wheeled vehicle.

One example of a tactical wheeled vehicle employed by the U.S. Army is the M93A1P2 "Fox", originally produced as the "TPz1 Fuchs" for use by the German military. The U.S. requirement for the Fox system was generated in the late 1980s in response to a perceived need to quickly field a chemical reconnaissance vehicle for U.S. forces in Europe. The M93A1P2 Fox (see Figure 2.) is a Chemical, Biological, Radiological, and Nuclear (CBRN) Reconnaissance System equipped with an up-armored kit and fully integrated reconnaissance system. The vehicle is intended to improve the survivability and mobility of the Army ground forces by providing increased situational awareness and information superiority to headquarters and combat maneuver elements.

Figure 2. Illustration of the Upgraded armored M93A1P2 "Fox" CBRN Reconnaissance System.

Recently, the Product Manager (PM) requested the U.S. Army Research Laboratory's (ARL's) Human Research and Engineering Directorate (HRED) perform an evaluation of the M93A1P2 Fox to ensure that human factors issues were addressed (MIL-STD-1472G, 2012). Foremost among these issues investigated were the emergency egress hatches, passage ways and escape routes. The M93A1P2 Fox vehicle has two front doors, two rear doors and a rear ramp door as well as one front crew compartment roof mounted hatch and two rear crewstation roof-mounted hatches. In addition to these egress and ingress routes, the M93A1P2 Fox also incorporates a tunnel on the left hand side of the vehicle. This tunnel connects the rear crewstation to an area just in back of the front right seat (Commander's crewstation shown in Figure 3) and can be a critical emergency egress route in the event of a rear crewstation fire or a vehicle rollover.

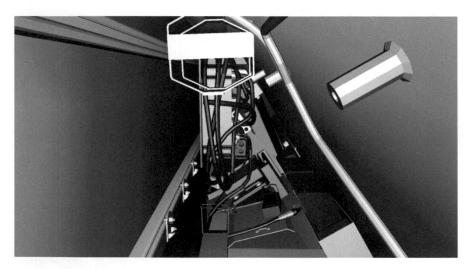

Figure 3. View of M93A1P2 Fox emergency egress tunnel located on vehicle right side. View of the tunnel is looking from the rear to the front crewstation.

2   METHODOLOGY

One approach for conducting a complete and accurate design assessment is through the use of Army Manpower and Personnel Integration (MANPRINT) to insure that key Army requirements are met (MANPRINT, AR 602-2). MANPRINT is the Army's human system integration program that seeks to optimize total system performance by ensuring a systematic consideration of the impact of materiel design on the Warfighter and their mission requirements (Garrett, 2011). For this analysis, a combination of modeling and egress testing with human subjects was used to assess the emergency egress routes. Each of these analysis methods has their advantages and disadvantages. For example, most military vehicles typically have an objective time requirement for emergency egress that might be something like "All personnel must be able to safely egress from the vehicle in less than thirty seconds." There are many variables that could factor into performing the egress. Seated passengers may experience difficulty in unfastening a seat restraint or could become entangled in the harness. A trip or snag hazards can also degrade egress performance and may be difficult to spot or predict with modeling. Finally, personnel may have to enter or exit through an emergency hatch or through a narrow interior passageway. Even if the analyst can conclude that sufficient space exist to pass through the passageway, hatch or emergency exit, it is difficult to accurately predict the time that it would take. For these reasons, evaluation of emergency ingress and egress time requirements should be performed with human subjects with the actual equipment or a prototype of the design.

2.1   Human Figure Modeling

Human figure modeling is commonly used to examine the space or clearance issues associated with ingress and egress. The analyst can also use human figure models to examine the posture that one might have to assume in order to traverse an egress route or passageway. However, any use of human figure models to examine ingress or egress clearance issues must also include models of the mission essential clothing and equipment. Items such as body armor, load-bearing vests, and protective clothing can add significant bulk to personnel that have to perform emergency ingress or egress procedures. Without models of these clothing and equipment items, it's difficult to properly assess ingress and egress issues using human figure modeling (Lockett et al., 2005). Figure 4 shows an example of human figure modeling use to examine an emergency egress scenario.

Figure 4.   Human figure model used to examine an emergency egress scenario through a restricted area inside a helicopter.

Modeling also gives the analyst the rapid ability to make virtual changes in the design and examine the impact of those changes without having to "bend metal" or prototype the design changes. In this instance, the use of modeling can be a cost effective tool that saves time as well.

Human figure modeling can also be used to examine egress for a vehicle in a rollover condition. Consideration must be given for emergency egress in rollover situations because the vehicle or aircraft is not in its normal upright state of operation. If a vehicle has rolled over on to the roof, all emergency escape hatches may be blocked. The path to the other emergency exits would certainly be different than when the vehicle was upright. Unfortunately, it may be difficult and hazardous to simulate rollover with an actual vehicle or prototype design with live subjects. However, the analyst can easily rotate the CAD model of the vehicle design and use human figure modeling to look at the egress paths that may or may not be available for this condition and examine what obstacles for egress might be present.

The modeling software used for this analysis was Jack V6.0 which is an interactive tool for modeling, manipulating, and analyzing human and other 3-D articulated geometric figures (Badler, Phillips, & Webber, 1993). The software allows the analyst to tailor the models to represent a specific user population for whom the equipment design is targeted along with models of mission essential clothing and equipment.

3   Tunnel Analysis

The modeling effort focused on the emergency escape tunnel within the Fox vehicle. It has a trapezoidal shaped area that is just 22 inches wide at the base and between 28 ½ to 32 inches tall with the top of the tunnel narrowing to only 7 inches in width. The right tunnel wall slopes upward at a 60 degree angle towards the roof forming the trapezoidal configuration. Several protuberances on both sides decrease the available space and also act as snag points during emergency egress use. The tunnel also contained a fixed mounted display, electrical junction boxes, cables and other components that also severely restricted the available space. Due to these components, the available width within some portions near the base of the tunnel was narrowed to just 6.5 inches. With this tunnel configuration, it was difficult for even the smallest of Amy personnel to safely utilize this tunnel for emergency egress (see Figure 5). For example, the 5th percentile female soldier body dimensions (Gordon, C. et al, 1988) for chest depth and buttock depth are 8.21 and 7.71 inches respectively, which exceeded available space in those restricted portions of the tunnel.

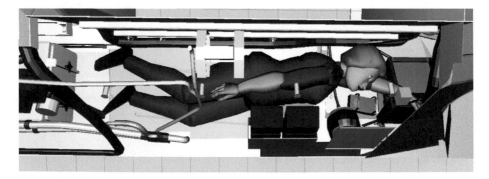

Figure 5. Small female figure shown in tunnel with restricted space due to installed components.

These body dimensions however do not include any mission essential clothing or equipment (see figure 6).

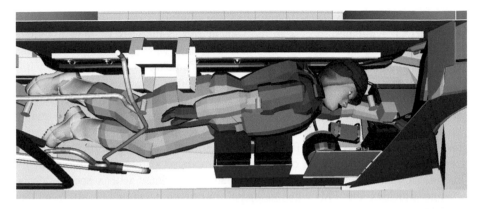

Figure 6. Small female figure equipped with models of tactical clothing and equipment inside the tunnel.

The tactical equipment worn by the Warfighter today is much larger and bulkier than the load-carrying equipment used during the Cold War and 1990s (Reed, 2009). In fact, with the additional bulk of this equipment, analyst are finding that hatch openings, pathways and tunnels on many vehicles no longer provide sufficient space to safely perform emergency egress. Also, when vehicle platforms are updated to include newer communication equipment, displays and countermeasure devices, the interior space becomes even more restricted and can have an impact on emergency egress as well. This was the case for the tunnel in the Fox vehicle, as a controller and and other components were added to this area. By using modeling to test options, relocating these components and their associated cabling and junction boxes to other sections of the vehicle removed some of the bottlenecks in the tunnel. This opened up the space restricted area of the tunnel from 6.5 inches to 13.7 inches and provided sufficient space for 95th percentile male chest depth and buttock depth dimensions of 11.04 and 11.19 inches respectively. However, it is still very likely that larger male Soldiers would probably have to remove tactical equipment before attempting emergency egress through the tunnel. Other recommendations based on human figure modeling to improve egress through the tunnel included rerouting other cabling and conduit, removing or covering threaded standoffs, redesign a protruding armored window crank handle, change design of a metal storage cage to a collapsable design and add low profile emergency lighting.

4   Conclusions

As mentioned earlier in this paper, a combination of modeling and trials with human subjects were used for this MANPRINT evaluation of the M93A1P2 Fox vehicle. The key to using this combination effectively is knowing the respective strengths and weaknesses for each method. Circumstances for each evaluation may also play a role in the methodology used. In this case, the time to perform the MANPRINT evaluation, access to the vehicle and availability of human subjects was limited. The advantage to using modeling here was that a number of design changes within the tunnel could be examined very quickly to provide the PM with a variety of options and tradeoffs. Some of the options presented could be readily implemented, but in other instances, the suggested changes could have provided additional tunnel space, but are costly to implement. Nonetheless, it is important to be able to present these possible design changes so that decision makers can weigh all options. Also, the human figure modeling software can help to determine what percentage of the population could be accommodated within a specific area of a vehicle design. The change in accommodation associated with a specific design change provides the PM with additional information on which to base their decisions.

The information provided in this paper was intended to provide some insight into the application of human figure modeling as a tool to help evaluate emergency egress issues. However, whether modeling tools are used to evaluate emergency egress testing in a simulated environment or actual subjects are used, the key factor remains the ability to identify potential shortfalls, which subsequently can make emergency egress unfeasible to execute under certain conditions. An important aspect to remember with any design evaluation effort is that the focus should be on developing a methodology to reduce risk associated with emergency egress (i.e. egress trainers); costs associated with accident prevention, and developing a more feasible approach to improve emergency egress. However, it's always critical to insure that human system integration is taken into consideration prior to resourcing a requirement.

ACKNOWLEDGMENTS

The authors would like to acknowledge Mr. Wayne Hammer, VIZIT, inc., for 3D modeling assistance.

REFERENCES

Badler, N.; Phillips, C.; Webber B. *Simulating Humans, Computer Graphics Animation Control*. Oxford University Press, New York, 1993.

Department of Defense (2012). "Department of Defense Design Criteria Standard: Human Engineering". MIL-STD-1472G.

Eisele, A. "Improved Explosives Becoming More Deadly in Iraq," *The Hill*, March 28, 2005, [http://www.hillnews.com/thehill/export/TheHill/News/Iraq/explosives1.html].

Garrett, L. The Right Fit? Observations and Insights from Recent Civilian Deployments to Iraq – Part 3: Validating Capability Gaps & Repurposing Existing Technology. XXIIIrd Annual International Occupational Ergonomics and Safety Conference, 2011.

Gordon, C.; Bradtmiller, B.; Churchhill, T.; Clauser, C.; McConville, J.; Tebbetts, I.; Walker, R. *1988 Anthropometry Survey of U.S. Army Personnel: Methods and Summary Statistics*; Technical Report Natick/TR-89/044; U.S. Army Natick Research, Development, and Engineering Center: Natick, MA, 1989.

Lockett, J. F., Kozycki, R.W., Gordon, C.C., & Bellandi, E., Proposed Integrated Human Figure Modeling Analysis Approach for the Army's Future Combat Systems, Military Vehicle Technology, SAE International, Warrendale, PA (2005) (SP-1962).

Reed, M.P.: Simulating Crew Ingress and Egress for Ground Vehicles. In: Ground Vehicle Systems Engineering and Technology Symposium, 2009.

CHAPTER 27

The Impact of Services Quality on Patient Trust and Satisfaction for Aged Patients

Julie Y.C. Liu,   Y.H. Wang

Department of Information Management,
Yuan Ze University,
Tau-Yuan, Taiwan
imyuchih@saturn.yzu.edu.tw; s989206@mail.yzu.edu.tw

ABSTRACT

Patient trust is one of the important predictors of health-related behavior, and has a significant effect on patient satisfaction and health outcomes. However, predictors of patient trust have not been comprehensively studied. This study aims to empirically examine the influence of service quality and patient perceived value on patient trust for aged patients. The analysis results indicate service quality has a significant impact on patient trust, which in turn leads to patient satisfaction. Moreover, only few dimensions of service quality play key roles on determining patient trust. The findings build a broader understanding of patient trust and satisfaction to hospitals. Future studies attempting to examine the determinants of patient trust should incorporate the concept of service quality. Health care practitioners who attempt to promote patient trust need to provide the training on raising the quality of health care service.

Keywords: service quality, patient trust, satisfaction, health care

1   INTRODUCTION

Patient trust is one of the important predictors of health-related behavior, such as maintaining continuity of care and adhering to treatment recommendations (Trachtenberg et al., 2005). In relative asymmetry of information between most patients and their physicians or care providers, patient trust is desirable in patients' own right because it less the patient questioning or enhance physician authority. Literature shows that patient trust in physicians affects patient satisfaction and health outcomes (Fiscella et al., 2004; Thom et al., 2004). Despite the known importance of patient trust, factors that influence patient trust have not been comprehensively studied. Most attention on the determinants of patient trust has been given to patient characteristics or technical skills (e.g., thoroughness, carefulness, competence) of physicians (Thom, 2001; Bendapudi et al., 2006). Until recently, there was little discussed about the role of service quality of hospitals in patient trust.

Service quality has been among the siginificant factors of patient satisfaction (Bowers, et al, 1994; Taylor and Cronin, 1994), and drawn attention from researchers in health care management (Berry and Bendapudi, 2007; Rashid and Jusoff, 2009). Researchers in marketing generally agree that service quality leads to higher levels of customer trust and customer satisfaction (Parasuraman, 1997; Wang et al., 2004; Kuo et al., 2005). Growing evidence shows that service quality in health care influences patients' behaviors (Andaleeb, 2001). Therefore, it seems rational to presume that healthcare service quality will increase patient trust and patient satisfaction. However, health care services have many characteristics that make them unique to other services. For example, the quality of healthcare service affects patients more significantly and dreadfully, compare to other services. Inappropriate medical treatments would even lead to a fatal outcome for patients. Moreover, patients and customers in other service industries are extremely different. Patients usually have some illness, pain, anxiety, fear or no willingness (Berry and Bendapudi, 2007), and come to healthcare service for their "need", rather than their "want" (Bendapudi et al., 2006). The distinct characteristics of health care service might weaken the effect of healthcare service quality on patient trust.

The purpose of this study is to empirically examine the relationship between service quality and patient trust. In addition to the well-known link between the service quality and satisfaction, we pose the following research questions: Does patients' perceived quality of service delivered by health care lead to higher levels of their trust and satisfaction? Which key dimensions of service quality dominate the effect of service quality on patient trust and patient satisfaction? To answer these questions, we conducted a survey on aged patients from different hospitals in Taiwan. The results may build a broader understanding of the determinants of patient trust and satisfaction to hospitals, and which allow service managers to plan training on increasing the quality of specific service components to improve patient trust and satisfaction. It is important to practitioners in a rapidly ageing society.

2   LITERATURE REVIEW AND HYPOTHESIS

Trust is defined as the willingness of an individual to be vulnerable to the actions of a party based on his expectation that the party will perform certain desired behaviors (Mayer et al. 1995). Patient trust represents that patients believe that their physician or care provider will act in their best interest. Patient trust has drawn much attention of researchers in health care over the past decades, and is regarded as one of the key construct for determining patient attitudes and behaviors (Thom and Campbell 1997; Hall et al., 2002; Thom et al., 2004; Trachtenberg et al., 2005). Previous research indicates that patients with trust are more willing to maintain continuity of care and adhere to treatment recommendations (Fiscella et al., 2004; Trachtenberg et al., 2005). It is generally believed that patient trust will reduce patients' negative emotion (e.g., fear, depression and anxiety) and improve patient satisfaction and treatment outcome (Thom, 1997, 2001; Berry and Bendapudi, 2007). Thom (2000) emphasizes that patient trust is determined by physician behavior and the relationship between patients and their physicians. Much effort is devoted to explore the critical physician behaviors to patient trust and develop training methods for the physician on building good relationship with their patients (Thom, 2001; Bendapudi et al., 2006). Until recently, few study notice and provide evidence in health care for the link between service quality and patient trust.

Service quality is defined as a consumer's overall assessment of product or service attributes. In service literature, the SERVQUAL metric is a measuring device of service quality (Parasuraman et al., 1988). Overall, SERVQUAL metric consists of five components: tangible, reliability, responsiveness, assurance, and empathy, and which are measured with a different score between customer expectations and customer perception of service delivery. The SERVQUAL instrument is most commonly used by researchers and practitioners as a diagnostic tool for strengths and shortfalls of service delivery in different service environments (Brown et al., 1993; Teas, 1993; Wisniewski, 2001; Jiang et al., 2002; Kettinger and Lee, 2005). Research in medical treatment and healthcare also examined and applied a variety of SERVQUAL instrument as a measure for service quality (Babakus and Mangold, 1992; Bowers et al., 1994; Lam, 1997; Taner and Antony, 2006; Rashid and Jusoff, 2009; Irfan and Ijaz, 2011).

Babakus and Mangold (1992) are pioneers of using SERVQUAL in medical service. The empirical study examined the reliability and validity of the instrument. The analysis result showed the support for the significant correlation of service quality and patients' continuity of care in the same hospitals. Taner and Antony (2006) extended the indicators of SERVQUAL to examine the difference between the service quality of public and private hospitals. They concluded that the quality of service delivered by private hospitals outperforms that by public hospitals because of better communication, assurance and patient trust. The results are similar to the findings of Andaleeb (2000) that a greater proportion of the population towards private hospitals in Bangladesh, which have better service quality than public hospitals there. Patients' perceived quality of health care services has a great influence on patient behaviors, including satisfaction, referrals, choice, usage, etc. (Andaleeb, 2000). Rashid and Jusoff (2009) argued that the service delivery in health care is more difficult than other industries because of the weakness or illness of patients and the variety of their demand. They indicated the several benefits of using SERVQUAL instrument for improving health care service.

Satisfaction is regarded as an attitude-like judgment based on one or more service interactions (Yi, 1990). Numerous evidences demonstrate that patients' perception of service quality influence patients' satisfaction and behaviors (Andaleeb, 2000; Taner and Antony, 2006). Although previous research used the different score of SERVQUAL as an indicator of service quality, some researchers demonstrate that patient perception of service quality alone has more power to predict patient satisfaction than the different score (Andaleeb, 2001; Rashid and Jusoff, 2009). This study adopts patient perception as the indicator of service quality for predicting patient trust and patient satisfaction. Literature indicates the elderly could have more experiences, which influence their attitudes and satisfaction on service (Callahan, 1992). Based on the above discussion, we propose the following hypotheses:

Hypothesis 1: The perceived quality of health care service is positively associated with patient trust.

Hypothesis 2: Patient trust is positively associated with patient satisfaction in health care.

Hypothesis 3: Only few of service quality dimensions are significant drivers of patient trust.

Hypothesis 4: Only few of service quality dimensions are significant drivers of satisfaction in health care.

3   RESEARCH METHODS

This study employed a cross-sectional survey, including samples of aged patients, greater than 50 years old, from the hospitals in Taiwan. Survey instruments were handed out to patients as they were waiting in the hospitals. A total of 34 valid questionnaires were obtained. Table 1 summarizes the demographic information of the final sample. The parameters of demographic includes, gender, education, occupation, marriage and frequency.

Table I Demographics (n = 34)

Items	Choices	number	%
Gender	Male	17	50.0
	Female	17	50.0
Education	Under senior high school	8	23.5
	Senior high school	12	35.3
	University	5	14.7
	Master	8	23.5
	No response	1	2.9
Occupation	Working Professionals	17	50.0
	Enterprise owner	1	2.9
	Housewife	10	29.4
	Un-employee	2	5.9

	Others	3	8.8
Marriage	Married	28	82.4
	Single	6	17.6
Average Frequency	One time / week	4	11.8
of visiting hospitals	One time / 2 weeks	6	17.6
	One time / month	4	11.8
	One time / 2 months	12	35.3
	One time / 3-6 mo	8	23.5

All constructs were adopted from the literature. Each questionnaire item was scored using a seven-point Likert scale ranging from "total disagreement" (1) to "total agreement" (7). The construct of trust was adopted from Harris and Goode (2004). Satisfaction was measured using questionnaire items of Carr (2007) via asking patients how satisfied they feel with the health care that they received. Service quality was evaluated with the underlying the 22-items of five dimensions in SERVQUAL (Parasuraman, Zeithaml, & Berry, 1988). The dimensions include: (1) Tangibles: the appearance of physical facilities, equipment, and personnel; (2) Reliability: the ability to perform the promised service dependably and accurately; (3) Responsiveness: the willingness to help customers and provide prompt service; (4) Assurance: the knowledge and courtesy of employees and their ability to inspire trust and confidence; and (5) Empathy: providing the caring and individualized attention to customers.

This study used Partial Least Squares (PLS), a regression-based structural equation modeling (SEM) technique, to examine the proposed model because it has strength in handling the sample of small sizes, distributional assumptions and few indicators of a construct (Chin, 1998). We adopted PLS Graph (v.3) to test the reliability of indicators and convergent validity via a confirmatory factor analysis. Reliability was examined by factor loading. Convergent validity was assessed by reliability of constructs, composite reliability of constructs, and average variance extracted (AVE) by constructs (Fornell and Larcker, 1981).

4   DATA ANALYSIS AND HYPOTHESES TESTING

According to the analysis results, all item loadings are above .72 (>.7 recommended) (Fornell and Larcker, 1981), except one indicator of tangibility, and two indicators of trust. The former indicator tangibility is about the up-to-date equipment, of which information might be limited to patients. The latter two are negative indicators. Eliminating the latter two extremely low loadings, the AVEs of all construct are higher than .56(> .5 recommended), composite reliability is above 0.79 (>.7 recommended). Table 2 lists the statistical results, which show an acceptable level of reliability and validity of collected data.

Table 2 Descriptive statistics

Constructs	Min	Max.	mean	S.D	CR	AVE
Tangibility	3.25	6.50	4.73	.86	.86	.62
Reliability	3.20	6.40	4.68	.89	.89	.63
Responsiveness	3.00	6.33	4.42	.85	.79	.56
Assurance	2.50	6.50	4.62	1.04	.95	.83
Empathy	2.00	6.67	4.44	1.05	.93	.81
Trust	3.40	6.40	4.62	.80	.90	.67
Satisfaction	3.00	7.00	4.82	.94	.94	.79

Figure 1 illustrates the path coefficients of the proposed model, where star symbol stands for the t-statistics of paths with significance at .05 level. The analysis results indicate that the only two dimensions of service quality, empathy and responsiveness, serve as important predictors of patient trust and of patient satisfaction. The path coefficient between responsiveness and trust is .371, and between empathy and trust is .473, between responsiveness and satisfaction is .588, and between empathy and satisfaction is .273. There is no significant relation between trust and satisfaction. Accordingly, hypotheses 1, 3 and 4 were supported, but hypothesis 2 was not. The total variance explained of trust and of satisfaction in the examined model were .743 and .715, respectively.

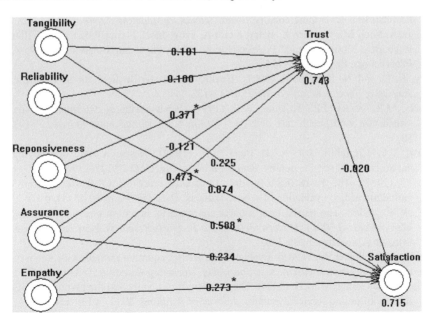

Figure 1 The result of the proposed model (Note: * indicates significance at p< 0.05 level)

5 CONCLUSION

The survey includes samples of aged patients from hospitals in Taiwan. The analysis results confirm some well-established relationships suggested in the current literature: (a) service quality and trust, (b) service quality and satisfaction. However, only few dimensions of service quality contribute to trust and satisfaction for aged patients. Specifically, the quality of empathy and responsiveness has the most significant effect on patient trust and satisfaction among all service components. It is also worth to note that for aged patients, their trust dose not influence their satisfaction to the service delivered by health care. The findings provide useful and practical suggestions for researchers and managers in improving service quality, creating and delivering superior patient trust, and achieving high patient satisfaction. As the competition of health care in Taiwan has become intense in these two decades due to the growing number of hospitals, the practitioners should pay more attention to understanding aged patients to maintain competitive advantages in an ageing society.

REFERENCES

Andaleeb, S. S. (2000). Public and private hospitals in Bangladesh: service quality and predictors of hospital choice. *Health Policy and Planning.* 15 (1), p95-102.

Andaleeb, S. S. (2001). Service quality perceptions and patient satisfaction: a study of hospitals in a developing country. *Social Science & Medicine.* 52(9), p1359-1370.

Bendapudi, Neeli M., Leonard L. Berry, Keith A. Frey, Janet Turner Parish, and William L. Rayburn. (2006). Patients' Perspectives on Ideal Physician Behaviors. *Mayo Clinic Proceedings.* 81 (3), p338-344.

Berry, L. L. and N. Bendapudi. (2007). Health care: A fertile field for service research. *Journal of Service Research.* 10 (2), p111-122.

Bowers, M.R., Swan, J.E., Koehler, W.F. (1994). What attributes determine quality and satisfaction with health care delivery?. *Health Care Management Review.* 19 (4),p49-58.

Brown, T.J., Churchill, Jr.G.A. & Peter, J.P. (1993). Research note: Improving the measurement of service quality. *Journal of Retailing.* 69 (1), p127-139.

Carr, C. (2007). The FAIRSERV model: Consumer reactions to services based on a multidimensional evaluation of service fairness. *Decision Sciences.* 38 (1), p107-130.

Chin, W.W. (1998). The partial least squares approach to structural equation modeling. In: *Marcoulides, GA(Ed.), Modern Methods for Business Research.* New Jersey: Lawrence Erlbaum Associates. p295-336.

Fornell, C. & Larcker, D. (1981). Evaluating structural equation models with unobservable variables and measurement error. *Journal of Marketing Research.* 18 (3), p39-50.

Harris, L.C. & Goode, M.M.H. (2004). The four levels of loyalty and the pivotal role of trust: A study of online service dynamics. *Journal of Retailing.* 80 (2), p139-158.

Irfan, M. & A. Ijaz. (2011). Comparison of service quality between private and public hospitals: Empirical evidences from Pakistan. *Journal of Quality and Technology Management.* 7 (1), p1-22.

Jiang, J.J., Klein, G. & Carr, C.L. (2002). Measuring information system service quality: SERVQUAL from the other side. *MIS Quarterly.* 26 (2), p145-166.

Kettinger, W.J. & Lee, C.C. (2005). Zones of tolerance: Alternative scales for measuring information systems service quality. *MIS Quarterly*. 29 (4), p607-623.

Kevin Fiscella, Sean Meldrum, Peter Franks, Cleveland G. Shields, Paul Duberstein, Susan H. McDaniel, Ronald M. (2004). Patient Trust: Is It Related to Patient-Centered Behavior of Primary Care Physicians?. *Medical Care*. 42 (11), p1049-1055.

Kuo, T., Lu, I.Y., Huang, C.H. & Wu, G.C. (2005). Measuring users' perceived service quality: An empirical study. *Total Quality Management*. 16 (3), p309-320.

Mark, A. Hall., Fabian Camacho, Elizabeth Dugan, Rajesh Balkrishnan. (2002). Trust in the medical profession: conceptual and measurement issues. *Health Services Research*. 37 (5), p1419-1439.

Mayer, R.C., Davis, J.H. & Schoorman, F.D. (1995). An integrative model of organizational trust. *Academy of Management Review*. 20 (3), p709-734.

Parasuraman, A., Zeithaml, V.A. & Berry, L.L. (1988). SERVQUAL: A multi-item scale for measuring consumer perceptions of the service quality. *Journal of Retailing*. 64 (1), p12-40.

Parasuraman, A. (1997). Reflections on gaining competitive advantage through customer value. *Journal of the Academy of Marketing Science*. 25 (2), p154-161.

Rashid, W. E. W. and H. K. Jusoff. (2009). Service quality in health care setting. *International Journal of Health Care Quality Assurance*. 22 (5), p471-482.

Teas, R.K. (1993). Expectations, performance evaluation, and consumers' perceptions of quality. *Journal of Marketing*. 57 (4), p18-34.

Thom, D. H. (2000). Training physicians to increase patient trust. *Journal of Evaluation in Clinical Practice*. 6 (3), p245-253.

Thom, D. H. (2001). Physician behaviors that predict patient trust. *Journal of Family Practice*. 50 (4), p323-328.

Thom, D. H. and B. Campbell. (1997). Patient-physician trust: an exploratory study. *The Journal of Family Practice*. 44 (2), p169-176.

Thom, D.H., Hall, M.A. and Pawlson, L.G. (2004). Measuring patients' trust in physicians when assessing quality of care. *Health Affairs*. 23 (4), p124-132.

Trachtenberg, F., Dugan, E., Hall, M.A. (2005). How patients' trust relates to their involvement in medical care. *The Journal of Family Practice*. 54 (4), p344-352.

Taylor, S. A., & Cronin Jr., J. J. (1994). Modeling patient satisfaction and service quality. *Journal of Health Care Marketing*. 14 (1), p34-44.

Wisniewski, M. (2001). Using SERVQUAL to assess customer satisfaction with public sector services. *Managing Service Quality*. 11 (6), p380-388.

Wang, Y., Lo H. & Yang, Y. (2004). An integrated framework for service quality, customer value, satisfaction: Evidence from china's telecommunication industry. *Information Systems Frontiers*. 6(4), p325-340.

Yi, Y. (1990). A critical review of consumer satisfaction. In: Zeithaml V.A *Review of Marketing*. Chicago: American Marketing Association.

CHAPTER 28

Syntax and Sequencing of Assembly Instructions

Peter Thorvald, Gunnar Bäckstrand, Dan Högberg & Keith Case

University of Skövde
Skövde, Sweden
peter.thorvald@his.se

Swerea IVF
Stockholm, Sweden

Loughborough University
Loughborough, UK

ABSTRACT

Minimalism of design is a concept often found in Human-computer interaction (HCI). It is a concept that emphasizes the presentation of as little information as possible to reduce the perceptual strain and visual search of the subject. However, in a manufacturing context, such as in manual assembly, state of the art information presentation is rarely minimalistic. Rather, organizations tend to push out as much information as possible without necessarily concerning themselves with how this information is presented to, or perceived by, the worker. This leads to a situation that is far from ideal from an HCI perspective, likely to reduce human performance and wellbeing, in turn negatively affecting overall production system performance. Obviously, there are several potential ways of addressing this issue. Perhaps the most evident way is to simply reduce the amount of information that is presented and only present the essentials. This paper will investigate and discuss how information presentation can be minimized without reducing the information content through information syntax and layout.

INTRODUCTION

This paper will investigate the potential use of alternate syntax and alternate sequencing of data in information presentation for manual assembly. Two hypotheses have been formulated based on previous work as well as experience from the automotive assembly industry (Bäckstrand, 2010; Thorvald, 2011):

Hypothesis 1 Using unsequenced data and thus minimizing the amount of presented information reduces errors and assembly time.

Hypothesis 2 The use of symbols as opposed to article numbers reduces errors and assembly time.

Hypothesis 1

In the investigation of using sequenced and unsequenced information, hypothesis 1 suggests that using unsequenced, batched information as opposed to sequenced information reduces assembly time (productivity) and errors (quality). This is argued to be due to the reduction of data presented to the worker.

Part	Quantity	Description
Part A	1	Assemble part
Part B	1	Assemble part
Part A	1	Assemble part

Figure 1. Sequenced information presentation.

As can be seen in Figure 1, sequenced information presentation presents parts in the order that they are to be assembled. While this might be preferred when presenting information to novices, such as when assembling furniture from IKEA, the expertise in a manufacturing assembly environment can be expected to be higher. Consequently, this research proposes and explores the use of unsequenced, batched information presentation so that the information presented in Figure 1 may look like Figure 2:

Part	Quantity	Description
Part A	2	Assemble part
Part B	1	Assemble part

Figure 2. Unsequenced, batched information presentation.

It is argued that presenting unsequenced, batched information as opposed to sequenced information saves space on the information medium, reduces cognitive strain related to information search, and saves in the non value adding activity of information search (Sawhney et al., 2009). Naturally, simple cases like the ones in the figures above, where the reduction of data is limited to the elimination of one row, would not be considered a major improvement but applying this to normal work instructions with 15-20 different parts might induce an improvement in both quality and productivity.

Hypothesis 2

The suggestion of using symbols as opposed to article numbers in information presentation for assembly is based on the idea that symbols carry semantic content about themselves that article numbers typically do not, and therefore they are easier to percept, process and recall. It is evident that it is most likely to be easier to process symbol representation over article numbers as the symbols are shorter, consisting of one character, whereas article numbers usually consist of several characters. However, as mentioned above, the number of characters within a representing element is not argued to be the reason for this difference. Rather, any difference found between the two is believed to be a result of the semantic content that the two representational modes include. An article number, "564163", has no connection to the long term memory of the user and is therefore subject to the risk of short term memory limitations. A symbol, Ω", on the other hand, is most likely established in the user's long term memory as "Omega". It is very likely to have personal meaning to the user, for example, the user might associate it to Greece and the Greek alphabet, it might be associated to electrical engineering as Ω is used as the symbol for ohm, a unit of electrical resistance, etc. The associative possibilities are great and this is believed to result in better recognition, recall and matching with the same symbol on the parts shelf.

Symbols are recognized in long term memory and are easily available for recognition, recall and memory. Article numbers are not connected to long term memory and thus will need to be kept in short term memory while the task is carried out.

Experimental design

To test hypotheses 1 and 2, an independent design experiment was set up where actual assembly workers assembled simple components according to different types of assembly information. All participants were given the same tasks but with different syntax to the information sources. The different modes of information design (states of the independent variable) were:

- Sequenced data presented with article numbers (IV_1)
- Unsequenced data presented with article numbers (IV_2)
- Unsequenced data presented with symbols (IV_3)

Through the different states of the independent variable defined above, the hypotheses can be rejected or confirmed through the following analyses:

- Comparisons between independent variable states IV_1 and IV_2 will confirm or reject hypothesis 1
- Comparisons between IV_2 and IV_3 will confirm or reject hypothesis 2

The risk of a confounding of the data appears when there are multiple potential sources for differences between test groups. This is why IV_1 and IV_3 cannot be compared, because they differ in two instances (Sequenced-unsequenced and article number-symbol). A comparison of these two would be meaningless as it is impossible to say what the result was caused by. However, IV_1 can be compared with IV_2 and IV_2 can in turn be compared to IV_3 since, in these comparisons the independent variables only differ in one instance for each pair.

METHOD

A total of 30 participants, randomly assigned to three independent groups took part in the study. All participants were experienced assembly workers with an age range of 20-42 years old. All participants had at least 2 years of experience in engine assembly and they were also deemed highly skilled assembly workers by Volvo Powertrain who loaned them to the study. Approximately 17% of the participants were female, a number that corresponds fairly accurately with gender diversity in the automotive assembly industry. The three states of the independent variable were tested on three separate days following each other. Variable state 1 was tested on day one, variable state 2 on day two and variable state 3 on day three. Ten subjects were assigned to each day.

Before the test, the subjects were briefed on how the test was to be carried out and how the assembly was to be completed. The task was to produce a predetermined pattern of LEGO® pieces with varying colours. Subjects were educated on the final shape of the product and informed that this would always be the same. Only the colours of the independent LEGO® pieces would differ. Figure 3 shows the final shape of the finished product.

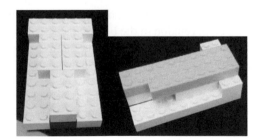

Figure 3. The final shape of the finished product shown in two states for clarity.

Figure 4, shows the building instructions that the subjects were presented with before the actual testing took place. Subjects were also offered a few pieces to test-build the final product as part of the training. The building instructions were not available to the subjects during the test; only product information was available at that point.

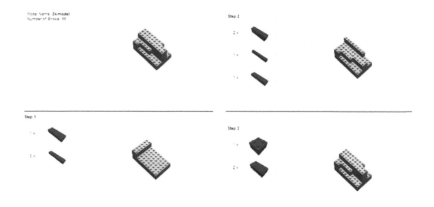

Figure 4. Building instructions. Top left frame is a description of the finished product. Each of the following frames describes the required parts and how they fit together.

Equipment

The test was carried out in a closed laboratory with the subject and two researchers present in the room. The task was carried out on a standing-height table with a material rack directly in front of it with material within reach of the subject. Information was presented to the subject via a computer laptop, placed on the table to the left of the subject and in front of the material rack used to supply LEGO® pieces to the experiment. The rack contained 13 boxes with different types of LEGO® pieces in them. Each type of LEGO® piece was also given a unique article number and a symbol. Article numbers and symbols were presented on the front edge of the material rack in relation to the boxes.

Three sets of instructions, based on the three states of the independent variable were used during the experiment. The instructions were identical in that they described the exact same sequence of the parts to be assembled, i.e. they contained the same information. The only thing that differed between the instructions was the way the product information was presented. For IV_1, tested on day 1, the assembly instructions were sequenced so that the parts of the product that were to be assembled first were presented first. Multiple quantities of articles only appeared in this state if the two articles were assembled together. The articles were presented through an article number and the quantity of the piece to be picked and assembled.

The sequence of products to be assembled was the same for all IV states and was divided into one high volume product (76% of the products) and six variants where a few of the pieces were exchanged from the high volume product. Each variant type had a frequency of about 4-5% with a total variant frequency of 24%. The variants and high volume products were randomly distributed in the sequence of assembly. Figure 5 shows the actual assembly instructions for IV_1.

Rubrik	Antal	Instruktionstext
• 56109282	2	
• 55724016	2	
• 56109282	1	
• 57046203	1	
• 55724016	1	
• 57046203	1	
• 60668495	1	
• 56109282	1	

Figure 5. Actual assembly information as presented to the subjects on day 1. Headlines were presented in Swedish and read from left to right; Headline, Quantity and Description.

On day 2, IV_2 was tested and the subjects were presented with unsequenced assembly information as can be seen in Figure 6. The information value and the syntax is still the same as in day 1 but the layout and the sequencing of the products has been manipulated.

Rubrik	Antal	Instruktionstext
• 56109282	4	
• 57046203	2	
• 55724016	3	
• 60668495	1	

Figure 6. Unsequenced assembly instructions as presented on day 2.

On day 3, testing of IV_3 included the use of unsequenced symbols as opposed to day 2's use of unsequenced article numbers (Figure 7)

Rubrik	Antal	Instruktionstext
• ☻	4	
• ♂	2	
• ☺	3	
• ♫	1	

Figure 7. Assembly information through symbols as used on day 3.

Implementation

The subjects were asked to finish as many assemblies as possible within 20 minutes. As all the subjects were experienced assembly workers they were asked to work at a normal work pace and try to assemble each product as well as possible. The number of products each subject was able to finish within the 20 minute limit varied from around 10 to 30 assemblies and for the analysis a mean for each subject was calculated. This way each group has 10 computable measurements.

RESULTS

An analysis of the difference between volume and variant products was made for all states of the independent variables. Surprisingly, only IV_2 showed a significant difference between these suggesting that the sequencing of information has greater impact than the choice of syntax ($p=0,012$).

So, in fact, only variable state 2 showed a significant difference between variant and volume products, and actually, for IV_1, variants were assembled faster than volume products according to the means (volume=53,17s, variants=52,01s). However, since the result is not significant ($p=0,81$) variants and volume products are statistically the same. Still, the indication that volume and variant products take the same time to assemble is very surprising as variants are generally believed to require more time than volume products and since the results are scattered between all IV's, a conclusion of this cannot be made. However, it could be argued that IV_2 and IV_3 both hit a floor effect in productivity in the sense that neither volume nor variant products could be assembled faster than around 25-30 seconds where means for both groups were found.

Result of hypothesis 1

The first hypothesis of the experiment was as follows:

- Using unsequenced data and thus minimizing the amount of presented information reduces errors and assembly time.

To test the above hypothesis, two sets of information sources were created. One where sequenced information was presented and one where the information was unsequenced and parts were batched together with other parts of the same kind.

Analysis of the means shows a difference between the groups and also a notably large effect size ($ES=0,9$), which is probably the reason why a significant difference between the groups could be found despite the rather small sample size ($p=0,01$).

Table 1 shows the descriptive statistics for the comparisons of IV_1 and IV_2. Both through simple analysis of the means and further significance analysis, it becomes evident that using unsequenced information presentation significantly improves productivity in manual assembly.

Table 1. Descriptive statistics for comparison between IV_1 and IV_2

	N	Minimum	Maximum	Mean	Std. Deviation
IV_1	10	28,56	96,78	52,5907	16,26455
IV_2	10	28,21	43,80	35,6690	4,51559
Valid N (listwise)	20				

Result of hypothesis 2

The second hypothesis of the experiment was as follows:

- Using symbols as opposed to traditional article numbers reduces errors and assembly time.

Just as the quality parameter arguably hit a floor effect due to the task being too easy, the productivity parameter for hypothesis 2 might have done the same. Analysis of the means in Table 2 shows a slight difference but deeper analysis showed no significant difference between the two groups ($p=0,231$).

Table 2. Descriptive statistics for IV_2 and IV_3

Groups	N	Mean	Std. Deviation	Std. Error Mean
IV_2	10	34,890741	3,7408088	1,1829476
IV_3	10	31,986535	6,3983693	2,0233420

DISCUSSION AND CONCLUSIONS

The results of the study described in this paper were:

- No quality differences between or within groups
- Significant difference in productivity between volume and variant products for IV_2
- Significant difference in productivity between IV_1 and IV_2
- No significant difference in productivity between IV_2 and IV_3
- Variants disturb assembly of the following volume product

The results for the experiment were quite surprising, as a difference in quality between the independent variable states could not be found. Actually, each independent variable showed only 0-2 errors totally, which in no way indicates a difference in the quality parameter. This is easily attributed to a probable floor effect as a result of the assembly probably being too easy. However, the productivity parameter did show differences between groups for hypothesis 1 but not for hypothesis 2.

While the absence of quality differences between the groups can arguably be assigned to a floor effect, it is still unsure whether this is the whole truth. Since IV_1 most likely did not hit a floor effect for the productivity parameter, it is somewhat uncertain whether it then can be affected by a floor effect for the quality parameter as quality and productivity very much go hand in hand. It could easily be argued that if a worker decides to work faster, this should have some kind of negative effect on quality. This issue is discussed in more depth in Thorvald et al. (2010). The quality issue will be investigated again in a follow up study to this one, again investigating the use of symbols and article numbers.

The significant difference, or absence thereof, between volume and variant products is to a large extent an enigma. As already mentioned there are several potential explanations to this such as floor effects and difficulties in understanding the instructions for IV_1, but none of the data can support or reject any of these. It might be time to challenge the generally accepted opinion that variants take longer to assemble than do volume products.

Analysis of productivity between the three information interfaces did show significant differences between IV_1 and IV_2, as expected. It becomes evident that unsequenced, batched information is better than sequenced, un-batched information. This could be attributed to the minimalistic information design that it entails. Less information on the screen gives the worker a better overview of what is to be assembled and it is even plausible that the subject uses pattern recognition to help identify the type of product to assemble. The pattern recognition that humans are so adept at (Bouma & De Voogd, 1974; Garzia & Sesma, 1993; Rookes & Willson, 2000) becomes much easier as the amount of data is reduced so drastically. A quick look at the unsequenced information can determine if everything is as it usually is. However, in the case of the sequenced information, a quick look will not be sufficient to determine what is to be assembled. In this case, subjects will most likely be forced to investigate the information source more thoroughly.

Productivity analysis of IV_2 and IV_3 did not show significant differences between the groups. However, post-hoc analysis of the data showed that the statistical power for the comparison of IV_2 and IV_3 was a mere 0,338. This means that the experiment had only a 33,8% chance of finding a significant difference between the groups, had there been one. The low statistical power is thought to be a result of the relatively small sample size and a weaker than expected effect size. As mentioned above, the issue of using article numbers or using symbols to identify parts will be further investigated in coming studies.

To sum up, the experiment reached the following conclusions.

- The experiment hit a probable floor effect and thus no quality differences between the groups were found.
- No significant difference in productivity between volume and variant products could possibly challenge the view that variant products take longer to assemble.

- Significant difference in productivity between IV_1 and IV_2 but not between IV_2 and IV_3 indicates that the sequencing of information has greater impact on productivity than does the syntax used.
- This could also be attributed to a floor effect for IV_2 and IV_3.

The study reported in this paper has been both successful and unsuccessful. Successful in that it found productivity improvements when using an unsequenced, batched information interface and ramp up points after variant products. It was unsuccessful in that it could not give a conclusive answer to the use of article numbers or symbols in information presentation. This is largely due to the floor effect that is probably present in the experiment and the weak statistical power of the experiment. These two issues might have been nullified had the experiment had access to more subjects and a more difficult task. The subjects were provided through collaboration with a major manufacturer to ensure actual assembly workers in the experiment.

REFERENCES

Bouma, H. & De Voogd, A. (1974). On the control of eye saccades in reading. Vision Research, 14 (4), 273-284.

Bäckstrand, G. (2010). Information Flow and Product Quality in Human Based Assembly. PhD Thesis. Loughborough University. Loughborough, UK. Wolfson School of Mechanical and Manufacturing Engineering.

Garzia, R. P. & Sesma, M. (1993). Vision and reading. Journal of Opthalmic Vision Development (24), 4-51.

Rookes, P. & Willson, J. (2000). Perception: Theory, Development and Organisation. London: Routledge.

Sawhney, R., Kannan, S. & Li, X. (2009). Developing a value stream map to evaluate breakdown maintenance operations. International Journal of Industrial and Systems Engineering, 4 (3), 229-240.

Thorvald, P. (2011). Presenting Information in Manual Assembly. PhD-Thesis. Loughborough University. Loughborough, UK. Wolfson School of Mechanical and Manufacturing Engineering.

Thorvald, P., Brolin, A., Högberg, D. & Case, K. (2010). Using Mobile Information Sources to Increase Productivity and Quality. In Proceedings of Applied Human Factors and Ergonomics, Miami, Florida, July, 2010.

CHAPTER 29

The Influence of Design in the Development of Exergames: A Practical Study in the Control of Obesity

Marina Barros, André Neves, Walter Correia, Marcelo Soares

Universidade Federal de Pernambuco, Design Post Graduation Program
CAC, Recife-PE / Brazil
marinalnbarros@terra.com.br

ABSTRACT

The using of virtual technologies for immersion of users are spreading in several areas of knowledge. In games, the quote is, resources are being used to work cognitive skills, attention, memory, among other factors beside the children. Currently, at the expense of usability, low cost, virtual technologies of perception and interaction, as Nintendo Wii, X-Box 360, among others, arises therefore a new class of games called Exergames, which is the union of physical activity the game. This fact gives users the development of sensory and motor skills through virtual reality mechanisms suitable for certain needs. This paper aimed to provide an overview of the current context of Exergames, especially in use for controlling childhood obesity, with some features, applications and possibilities for use.

Keywords: Exergame, Design, Ergonomics, Control Obesity

1   INTRODUCTION

Over the years the weight of the child population has increased alarmingly, and this can be observed in a survey conducted by the Centers for Disease Control and Prevention (CDC), U.S., where it shows that this population has tripled in 30 years, until 2008 . In Brazil, the data presented by the Brazilian Institute of Geography and Statistics (IBGE), show the growing obesity, where between 1974 and 1975, 2.9% of boys and 1.8% of girls aged 5-9 years were obese, as in 2008, these numbers increase to 16.6% of boys and 11.8% of girls the same age. In the U.S., can also be noted that rising obesity, the CDC reported that there was an increase of 6.5% in 1980 to 19.6% in 2008 among children aged 6-11. (IBGE 2010, CDC, 2010).

According to researcher Senninger (2009), obese children may be part of those risk groups most likely to suffer from various disorders in adulthood such as hypertension, respiratory diseases, heart disorders, diabetes, among others.

This article aims to present the exergame jointly developed by GDRLab (Game Desing Research Laboratory) and the present authors, as well as some results of the experiment with users, which is part of the project developed in academic master of design, demonstrating the importance new technologies and new approaches to try to help to reduce the problem of obesity.

2   STATE OF THE ART

The emergence of a new type of interaction mode of experience have stirred the community of HCI. Called game effort, or Exergames, which are interactive games that use movements of the players, joins an old idea, where the games are used, in addition to social benefits and mental attributes, physical attributes. (Sinclair, Hingston, Masek, 2007).

According to project director of Exergame Unlocked, Barbara Chamberlin, in order to deconstruct the image created from digital games that encourage the practice in sedentary young audience, several companies have been working in digital games design a new class of games digital take based on the physical movement of the players, currently known as exergames. Backed by different technological bases, such games have been launched by manufacturers of weight on the world stage, such as Sony and Nintendo (Hunicke, Leblanc and Zubek, 2009).

2.1 Childhood Obesity

The obesity is a chronic multifactorial and may be a consequence of eating more energy than necessary, which may be associated with multiple complications and develops through the interaction of genotype and the environment from various factors such as social behavioral, cultural, physiological, metabolic and genetic, is characterized by excessive accumulation of body fat to such an extent that compromises the health of the individual, whether child, adult or elderly, regardless of sex (National Institutes of Health, 1998).

For this reason, there is a clear tendency among members of a family having a body mass index (BMI) was similar. This is currently the main factor measurement of obesity. BMI is calculated from the relationship between the individual's weight and height squared (weight / height^2), where, for example, if the BMI is "less than 16.5, "is considered "Extreme thinness", and if we have "greater than 40", is "Morbid obesity". In the case of children, does not apply directly to this scheming, the National Center for Health Statistics (NCHS) indicates to the age of seven to nine years in the table below (Table 1).

Table 1 - BMI reference for children

Age	Gender	Normal BMI
7 to 9 years	Masculine	from 14 to 24
7 to 9 years	Feminine	from 13 to 22

Source: Adapted from Reynolds and Spruijt-Metz (2006)

The age group chosen for this study were children from six to twelve years old by the fact that, according to authors such as Piaget and Inhelder (2006) and Vygotsky (2007), this period of childhood the child acquires greater autonomy for a number issues of everyday life, how to choose clothes that will wear, choose their own foods and define the activities that will accomplish, such as watching television, playing digital games or play with friends in the playground of the building or in the backyard of his house.

2.2 Exergames

There is no way to talk about games without actually mentioning in virtual reality. This type of "reality" appeared in 1950, with flight simulators for testing, however, the potential of this instrument for training, according to Rosenblum et al (1995), expanded the application of technology to various fields such as medicine Engineering, Architecture, Psychology and Education. This is not a simple adaptation to new technologies: the coupling with the computers should be understood by cognitive links that can be produced, as the development of strategic thinking, reasoning and perception (Kastrup, 2004).

After several years of research on games, a new class of digital games based on the physical movement of the players has been developed by different manufacturers in the segment of digital games. Known as "exergames" or "activegames" have as a principle the use of new forms of interaction between player and game.

However, the games developed for the use of webcam are focused primarily on the casual gamer, and is strongly oriented to entertainment. Despite the great potential, because it is a new technology, little has been explored in the market there is a successful product.

These make the player/user to work on a series of aerobic exercise as a way to respond to events triggered in the game. Some exergames are used explicitly to exercise and others have an indirect focus on exercise, are games that have more focus on entertainment and social integration. Regardless of the focus of exergame, the player will experience the benefits of physical activity in any way. In fact, the games are just "away" or "mask" that make this focus more success with players of digital games (Hunicke, Leblanc and Zubek, 2009).

2.3. Physiological aspects of physical activity

The importance of the activity gains strength as it becomes crucial to the treatment and prevention of chronic degenerative diseases and other types of disease in general, since there is evidence that such activity helps improve the immune practitioner. In addition, physical activity is a major contributor to socialization through sports centers, spaces for walking, bringing health both physical standpoint, as the social and behavioral (Wilmore, Costill and Kenney, 2009).

There are three types of training, the strength, anaerobic and aerobic. Strength training aims to seek practitioner gain strength without necessarily gaining muscle hypertrophy. The anaerobic exercises are performed for the physical training vigorous activities which are carried out in a short time, high intensity, such as racing speed. This system uses as a form of energy the ATP-PCr (Adenosine triphosphate - phosphocreatine) and anaerobic degradation of muscle glycogen (glucose). Aerobic exercise, also called resistance training, cardio, training is conducted in endurance activities such as hiking and long distance running, as a form of energy uses ATP, and promotes transport and better use of oxygen, and this type of training aids in improved metabolism of fat, as during submaximal exercise, leads to an increased oxidation of fatty acids, present in adipose tissues, for capitation energy preserving, muscle glycogen.

3. THE GAME

This game was called "PEGGO," on account of its main objective, to "get" (in Portuguese) things on the screen while in use. The game consists of six mini games that can be played individually or in pairs depending on space available, contains between two and three stages, adding 30 minutes of play, where each stage represents a group of exercises that you want to achieve, not being mutually exclusive exercises. This one aims to get green icons on the screen, and avoid touching red icons. In the character of use as motor physical activity, it has an educational character, which is widely discussed topics in the news, such as sexuality, recycling, separate collection, among others.

This is characterized as a exergame, ie a game exercise, combining entertainment and applying activity exercise program engines, which also involve aspects of cognition. It uses a very simple hardware, which used to be just a computer with an onboard video card and a webcam attached to interaction and movement within the game to meet the same objectives.

Its target public school children 6-12 years. It can also be played by adults and other age groups, however, some concepts that are connected with the formation of knowledge about the topics covered in the game may be underestimated by the participants above the age required by the game. The following figures (figure 1 and 2) show some screens in the game while in using.

Figures 1 and 2: Screen adjustment and Game mode screen. Source: The Authors

4 STUDY WITH THE USER

The study had an experimental character, with a scientific nature, where pre-tests were performed to analyze the relevance of PEGGO agent control childhood obesity. Because it is a computer game, aided by a webcam, these requirements were met through the use of a standard PC with such a requirement, and when in use at the residence of users, we used a notebook. The explanation to all users on how to use the game through body movements for its operation, objective, name of the game, etc.., was done by means of a checklist for each user, and it was necessary to be done only once with each, demonstrating the ease of use and learning. The main goal of the game and reach the children and youth. The participants performed the tests three times a week for four consecutive weeks, and the following parameters were adopted as follows:

1. 30 minutes of play with 85% MHR, ie submaximal HR;
2. Follow-up 10 HR in 10 min;
3. Measure the waist circumference at the navel on the 1st and last test day;
4. Measurement of weight before and after exercise;
5. Calorie expenditure to year-end.

4.1 Selection of participants and the experiment

The selection of participants was done for convenience, where children of both sexes could participate in the study, adding to a total of six. It was used the following inclusion criteria: be between six and twelve years, whether or not it part of obesity, and having signed the term sheet had been signed. And as exclusion criteria: have a cardiovascular disease, and not accept the terms of the research. Of the six children only two participants had no frame of obesity.

The experiment was performed both at the Federal University of Pernambuco, as the home of some of the participants. Lasted four weeks, adding 11 days of testing, where six children, of both genders participated in the same, with the consent of their guardians. From the first day to last day of the test were measured Heart Rate (HR), calculating 85% of the same, Weight, Height and Abdominal Circumference in order to have parameters of comparison. During the tests, were always measured initial HR, HR during the test, with 10 HR and 20 minutes and final, initial and final weight and calorific final (Kcal).

4.2.1 Analysis of results

When we start the experiment, were selected six participants, aged between 06 and 12 years, where these two (33%) were with BMI - Body Mass Index - normal, and four (67%) fit to as overweight or obese, as shown in Tables 2 and 3.

Table 2: Calculation of BMI by Age of participants

ID	Name	Sex	Date of birth	Date of measur.	Height (cm)	Weight (kg)	BMI	BMI %ile
1	U.G.R.L	M	07/07/2005	26/09/2011	120	24,6	17,1	85,1
2	M.L.B.R	M	18/12/1999	26/09/2011	159	50,3	19,9	78,7
3	A.L.S	F	07/03/2004	26/09/2011	146	52	24,4	98,9
4	R.B.S	F	23/08/2003	26/09/2011	147	35,1	16,2	58,0
5	V.S.C.	M	03/05/2000	26/09/2011	150	46,3	20,6	85,5
6	M.E.B.G	F	23/04/2000	26/09/2011	152	51,3	22,2	89,3

Source: Adapted from de CDC

CDC - Centers for Disease Control and Prevention defines overweight the child is one that presents a percentile equal to or greater than 85% and less than 95% and obesity, one that presents a percentile greater than or equal to 95% (CDC , 2010).

Table 3: Percentage of BMI and BMI x sex x total number of participants

Summary of Children's BMI-for-Age			
	Boys	Girls	Total
Number of children assessed:	3	3	6
Underweight (< 5th %ile)	0%	0%	0%
Normal BMI (5th - 85th %ile)	33%	33%	33%
Overweight or obese (≥ 85th %ile)*	67%	67%	67%
Obese (≥ 95th %ile)	0%	33%	17%

*Terminology based on: Barlow SE and the Expert Committee. Expert committee recommendations regarding the prevention, assessment, and treatment of child and adolescent overweight and obesity: summary report. Pediatrics. 2007;120 (suppl 4):s164-92.

Source: Adapted from de CDC

According to this classification, it's possible to observe in Table 3 and Figure 3 that the four children with overweight and obesity, only one (17%) were female, was obese, and it is noteworthy that this was the only one who participated in a specific program of weight control, this program conducted at the University Hospital of Pernambuco, UFPE together a team of pediatric endocrinology.

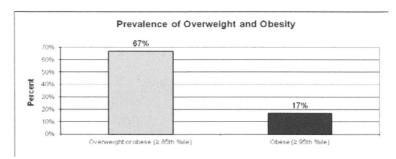

Figure 3: Percentage of overweight and obese participants. Source: Adapted from CDC

The study participants underwent weight during all 11 days of experiment, as shown in the figure 6.

They were heavy, and after weighing the test performed in the game, which lasted 30 minutes, where they monitored their heart rates from 10 to 10 minutes at the end of 30 minutes were weighed again, it is noted that the youngest participant males failed to complete 30 minutes of exercise, reaching only 20 minutes of activity for reporting fatigue. It may be noted in Figure 4, the end of the experiment there was a weight loss in all study participants, where the smallest loss was greater than 400g and 900g.

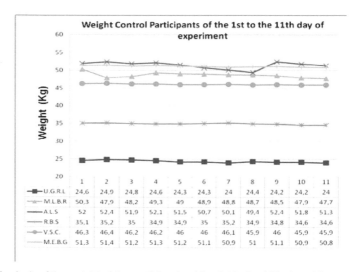

	1	2	3	4	5	6	7	8	9	10	11
U.G.R.L	24,6	24,9	24,8	24,6	24,3	24,3	24	24,4	24,2	24,2	24
M.L.B.R	50,3	47,9	48,2	49,3	49	48,9	48,8	48,7	48,5	47,9	47,7
A.L.S	52	52,4	51,9	52,1	51,5	50,7	50,1	49,4	52,4	51,8	51,3
R.B.S	35,1	35,2	35	34,9	34,9	35	35,2	34,9	34,8	34,6	34,6
V.S.C.	46,3	46,4	46,2	46,2	46	46	46,1	45,9	46	45,9	45,9
M.E.B.G	51,3	51,4	51,2	51,3	51,2	51,1	50,9	51	51,1	50,9	50,8

Figure 4: Analysis of the weight of the participants of the 1st to the 11th day of the experiment. Source: The Authors

At the end of the experiment, the participants reassessed the data to see if there was any change in the relationship between initial BMI and BMI final. And as shown in Tables 4 and 5 in the graph of Figure 5 (on the next page), there was a reduced body mass index of all participants, so that all representatives were male with normal BMI and females, showed a reduction not as significant as boys.

Table 4: Calculation of final BMI by Age of participants

ID	Name	Sex	Date of birth	Date of measure.	Height (cm)	Weight (kg)	BMI	BMI %ile
1	U.G.R.L	M	07/07/2005	26/09/2011	120	24	16,7	79,5
2	M.L.B.R	M	18/12/1999	26/09/2011	159	47,7	18,9	68,2
3	A.L.S	F	07/03/2004	26/09/2011	146	51,3	24,1	98,8
4	R.B.S	F	23/08/2003	26/09/2011	147	34,6	16,0	53,4
5	V.S.C.	M	03/05/2000	26/09/2011	150	45,9	20,4	84,4
6	M.E.B.G	F	23/04/2000	26/09/2011	152	50,8	22,0	88,5

Source: Adapted from CDC.

It is important do say the participants who achieved a significant reduction in BMI, as to normalize it, did not participate in any program of dietary control. And that despite the BMI of the participants with the highest rate had not decreased so significant, it may be noted that they showed weight loss and reduced their waist measurement significant for the period in question.

Table 5: Percentage of BMI and BMI x sex x total number of participants

Summary of Children's BMI-for-Age	Boys	Girls	Total
Number of children assessed:	3	3	6
Underweight (< 5th %ile)	0%	0%	0%
Normal BMI (5th - 85th %ile)	100%	33%	67%
Overweight or obese (≥ 85th %ile)*	0%	67%	33%
Obese (≥ 95th %ile)	0%	33%	17%

*Terminology based on: Barlow SE and the Expert Committee. Expert committee recommendations regarding the prevention, assessment, and treatment of child and adolescent overweight and obesity: summary report. Pediatrics. 2007;120 (suppl 4):s164-92.

Source: Adapted from CDC.

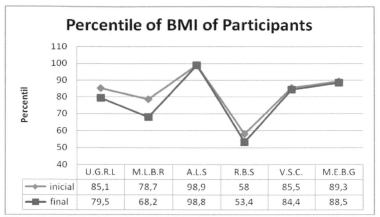

Figure 5: Analysis of the percentile BMI of participants starting and ending. Source: The Authors

5   FINAL CONSIDERATION

It can be seen that currently obesity is not just a problem of the aesthetic point of view, which causes discomfort circumstantial in nature. Being overweight can cause the emergence of several health problems, as stated earlier, such as diabetes, skeletal malformations, heart problems, among others. Approximately 16% of children and 9% of adolescents in Brazil suffer from obesity, and a given that brings with it several consequences is the fact that about eight out of ten teenagers lead to obesity in adulthood. These data indicate an increase of more than five times in the last 20 years (McArdle, Katch and Katch, 2010).

The physical exercise has been one of the most effective and most widely used for the control and treatment of obesity, whether child or adult. Physical inactivity is a risk factor and contributes greatly to the development of this evil, moreover, the absence or lack of physical activity can make the chances of having heart problems in adulthood are increased by almost twice.

During the proceedings in the experiment with PEGGO, can be used in aerobic training heart rate as an indicator of exercise intensity, and at no time noticed an excessive tiredness on the part of participating children. There was, as expected and initially rising body mass loss, and reduced BMI.

Thus, this study may help explicit, both from a scientific standpoint, as the technological point of view, where the contribution under the first point was easily perceived by the results obtained in the experiment, both with respect to products already existing product and implemented under the project, PEGGO, which validates the hypothesis of using this type of digital game as a support to control childhood obesity.

Exergames as PEGGO, whose purpose is to exercise children without a stated requirement, and still seeks entertainment and knowledge of these, in a fun and cheap (considering the hardware demand), is a simple and effective way of doing this. The number of players in the phase between 5 and 12 years who are joining the "wave" in the world of games in a row grows giddy, and not worth even a stage for this steady growth over the next ten years.

As demonstrated throughout the study, the benefits of this type of game, exergame goes beyond the physical aspects, reaching also the cognitive and social side. Addressing these issues is not merely a market issue, but of respect to human and public health.

REFERENCES

CDC - CENTERS FOR DESEASE CONTROL AND PREVENTION (2010). Childhood Obesity Facts: Health Effects of Childhood Obesity. Disponível em: http://www.cdc.gov/healthyyouth/obesity/index.htm. Acessado em 23/08/2011.

HUNICKE, R., LEBLANC, M. e ZUBEK, R. (2009) MDA: A Formal Approach to Game Design and Game Research. Design Issues, MIT Press. 4(9) p. 234 - 246.

IBGE – INSTITUTO BRASILEIRO DE GEOGRAFIA E ESTATÍSTICA (2010) POF 2008-2009: desnutrição cai e peso das crianças brasileiras ultrapassa padrão internacional. Disponível em http://www.ibge.gov.br/home/presidencia/noticias/noticia_visualiza.php?id_noticia=1699&id_pagina=1. Acessado em 23/08/2011.

KASTRUP, V. (2004). A aprendizagem da atenção na cognição inventiva. Psicol. Soc., 16, 7-16.

MCARDLE, W., KATCH, F. I. e KATCH, V. L. (2010) Fisiologia do Exercício: Energia Nutrição e Desempenho Humano. São Paulo, Ed. Guanabara.

NATIONAL INSTITUTES OF HEALTH (1998) Clinical Guidelines on the Identification, Evaluation, and Treatment of Overweight and Obesity in Adults: The Evidence Report. NIH publication 98-4083, September.

PIAGET, J. e INHELDER, B. (2006) A psicologia da criança, 2ª Ed. Rio de Janeiro: Difel.

REYNOLDS, K. D. e SPRUIJT-METZ, D. (2006) Translational research in childhood obesity prevention, Eval Health Prof, June 1; 29(2): 219 - 245.

ROSENBAUM, E.; KLOPFER, E. e PERRY, J. (2007). On Location Learning: Authentic Applied

SENNINGER, F. (2009) Obesidade Infantil: Como Fazer a Criança Emagrecer Suavemente. São Paulo, Ed. Vozes.

SINCLAIR, J; HINGSTON, P; MASEK, M. (2007) Considerations for the design of exergames. ACM. v.1 n. 4.

VYGOTSKY, L. S. (2007) A formação social da mente: o desenvolvimento dos processos psicológicos superiores, 7ª Ed. São Paulo: Martins Fontes.

WILMORE, J. H., COSTILL, D. L; e KENNEY, W. L. (2009) Fisiologia do Esporte e do Exercício.. 4ª ed. São Paulo, Ed. Manole.

276

The Health of the Pilot of Modern Machines and Their Inclusion in a Risk Society

2 MARTINS, Edgard[1] ; SOARES, Marcelo[2;] MARTINS, Laura[3]
[1,2,3]Universidade Federal de Pernambuco- Depto de Design- CAA – Recife-
PERNAMBUCO- BRASIL
edgard@upe.poli.br, *marcelo2@nlink.com.br*, laurabm@folha.rec.br

ABSTRACT

The flaws in the commitment of decision-making in emergency situations and the lack of perception related to all elements associated with a given situation in a short space of time indicate, often, lack of situational awareness. Automation always surprises the crews and often prevents them from understanding the extent of this technology that is very common in aircraft units with a high degree of automation. These facts are discussed in a subtle way by aircraft drivers who can not do it openly, as it might create an impression of professional self-worthlessness (self-deprecation). This leads to common questions like: What is happening now? What will be the next step of automated systems? This type of doubt would be inadmissible in older aircraft because the pilot of those machines works as an extension of the plane. This scenario contributes to emotional disorders and a growing hidden problem in the aeronautical field. These unexpected automation surprises reflect a complete misunderstanding or even the misinformation of the users. It also reveals their inability and limitations to overcome these new situations that were not foreseen by the aircraft designers. Our studies showed a different scenario when the accident is correlated with systemic variables. It has identified the problems or errors that contribute to the fact that drivers are unable to act properly. These vectors, when they come together, may generate eventually a temporary incompetence of the pilot due to limited capacity or lack of training in the appropriateness of automation in aircraft or even, the worst alternative, due to a personal not visible and not detectable non-adaptation to automation. We must also consider in the analysis the inadequate training and many other reasons, so that we can put in right proportion the effective participation or culpability of the pilot in accidents. Our doctoral thesis presents statistical studies that allow us to assert that the emotional and cognitive overload are being increased with automation widely applied in the cockpits of modern aircraft, and also that these new projects do not go hand in hand with the desired cognitive and ergonomic principles.

.**Keywords**: Human error, Cognitive overload, New technologies, Automation

1   INTRODUCTION

The emotional stability and physical health of workers on board aircraft are faced with the factors and conditions that enable professionals to carry out their activities and develop normally, despite the fact that these conditions may present themselves to professionals in adverse conditions (Eugenio, 2011). The modern history of aviation with its great technological complexity has pilots as redundant components that integrate embedded controls in modern aircraft. This leads us to say that the value of the worker as a permanent social group in society does not receive, currently, the proper priority. In research on the health of the pilot, there are three major perspectives that have been investigated that influence his stability, as well as the mental and emotional development of the modern airline pilot (Henriqson,2010): The previous life of the individual directly tied to experience, age, genetic and physiological vectors, The social environment, cultural environment and formal education leading to the final result, manifested by the ability, personality, strength and character and The verifiable standards of quality and quantity of life desired, ambition and achievements and its effects.

The Digital technology advances, has changed the shape and size of instruments used for navigation and communication. This has changed the actions of pilots, especially in relation to emergency procedures. There are few studies that correlate the reduction of accidents with the cognitive and technological changes. The increased cognitive load relates to these changes and requires assessment. The benefits presented by new technologies do not erase the mental models built, with hard work, during times of initial training of the aircraft career pilots in flying schools.

The public must be heeded when an aircraft incident or accident becomes part of the news. In search of who or what to blame, the pilot is guilty and immediately appointed as the underlying factors that involve real evidence of the fact they are neglected.The reading of the *Black-Boxes* notes that 70% to 80% of accidents happen due to human error, or to a string of failures that were related to the human factor (FAA, 2010). We can mention stress and the failure to fully understand the new procedures related to technological innovations linked to automation. Complex automation interfaces always promote a wide difference in philosophy and procedures for implementation of these types of aircraft, including aircraft that are different even manufactured by the same manufacturer. In this case, we frequently can identify inadequate training that contributes to the difficulty in understanding procedures by the crews. Accident investigations concluded that the ideal would be to include, in the pilot training, a psychological stage, giving to him the opportunity of self-knowledge, identifying possible "psychological breakdowns" that his biological machine can present that endangers the safety of flight. Would be given, thus, more humane and scientific support to the crew and to everyone else involved with the aerial activity, minimizing factors that cause incidents and accidents. Accident investigators concluded that the ideal situation for pilot training should include a psychological phase (Dekker , 2003), giving him or her, the opportunity of self-knowledge, identifying possible "psychological breakdowns" that biological features can present and can endanger the safety of flight. It should be given, thus, more humane and scientific support to the crew and everyone else involved with the aerial activity, reducing factors that can cause incidents and accidents. Accidents do not just happen. They have complex causes that can take days, weeks or even years to develop (Reason , 1990). However, when lack of attention and / or neglect take place resulting in a crash, we can be most certain there was a series of interactions between the user and the system that created the conditions for that to happen (Rasmussen, 1982). We understand that human variability and system failures are an integral part of the main sources of human error, causing incidents and accidents. The great human effort required managing and performing actions with the interface as the task of monitoring, the precision in the application of command and maintaining a permanent mental model consistent with the innovations in automation make it vulnerable to many human situations where errors can occur.

The human variability in aviation is a possible component of human error and we can see the consequences of these errors leading to serious damage to aircraft and people. It is not easy, in new aviation, to convey the ability to read the instruments displays. This can conduct to the deficiency and the misunderstanding in monitoring and performing control tasks: lack of motivation, the fact that it is stressful and tiring, and generate failures in control (scope, format and activation), poor training and instructions that are wrong or ambiguous. The mind of the pilot is influenced by cognition and communication components during flight, especially if we observe all information processed and are very critical considering that one is constantly getting this information through their instruments. There is information about altitude, speed and position of one's aircraft and the operation of its hydraulic power systems. If any problem occurs, several lights will light up and warning sounds emerge increasing the volume and type of man-machine communication which can diminish the perception of detail in information that must be processed and administered by the pilot. All this information must be processed by one's brain at the same time as it decides the necessary action in a context of very limited time. There is a limit of information that the brain can deal with which is part of natural human limitation. It can lead to the unusual situation in which, although the mind is operating normally, the volume of data makes it operate in overload, which may lead to failures and mistakes if we consider this man as a biological machine.

All situations in which a planned sequence of mental or physical activities fails to achieve its desired outputs are considered as errors (Reason, 2008). Thus, it is necessary that steps be taken toward reducing the likelihood of occurrence of

situations which could cause a problem. The flight safety depends on a significant amount of interpretations made by the pilot in the specific conditions in every moment of the flight. Accidents do not only occur due to pilot error, but also as a result of a poor design of the transmission of information from the external environment, equipment, their instruments, their signs, sounds and different messages. In these considerations, the human agent will always be subject to fatality, which is a factor that can not be neglected
. Because of human complexity, it is difficult to convince, in a generic way, people with merely causal explanations. Further analysis of the problem will always end with the identification of a human error, which was probably originated in the design phase, at the manufacturing stage, or given simply as a result of an "act of God". Aeronautical activities, designing human-machine systems becomes very necessary to characterize and classify human error. Human activities have always been confronted with the cognitive system.

Fig. 1 shows the human-machine interaction where difficulties with cognitive and operational perspectives needs and also physical and emotional aspects take place in a human being during the occurrence of system-level of flight.

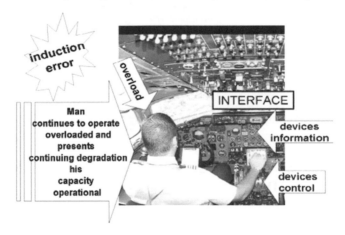

Figure 1 - Diagram of the interaction between man and machine

II. SCENARIO

On the result of the causality of accidents, we must consider the human contribution to accidents, distinguishing between active failures and latent failures due to the immediate adverse effect of the system aspect. The main feature of this component is that it is present within the process of construction of an accident long before declaring the event like an accident, being introduced by higher hierarchical levels as designers, responsible for maintenance and personnel management. We can always guarantee, with respect to organizational accidents, that the layers of defenses, that are the protective barriers, were constructed to prevent the occurrence of natural or man-made disasters. This statement is derived from the design philosophy that treats the defense in depth. In Fig. 2, based on the model "Swiss Cheese" (Reason , 1990). a fail of defense of an accident may occur as a Swiss cheese with "holes", which mean "latent failures" that sometimes began the construction of an accident long before the event. In certain circumstances, such failures (holes) can align themselves and then, the accident happens. An accident is a succession of failures. When these barriers are destroyed or are flawed or become vulnerable, the accident occurs. In this fact, that is called latent failure.

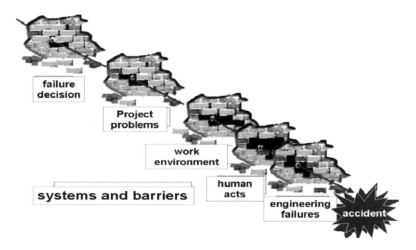

Figure 2- Latent failures, based on the "Swiss Cheese Model" of Reason

III. ACTIVITIES PURSUED BY THE PILOT AND THE INCREASED COGNITIVE LOAD

The following factors are an integral part of cognitive activity in the pilot: fatigue, body rhythm and rest, sleep and its disorders, the circadian cycle and its changes, the G-force and acceleration of gravity, the physiological demands in high-altitude, night-time take-offs and the problem of false illusion of climbing. But, other physiological demands are placed by the aviators. It is suggested that specific studies must be made for each type of aircraft and workplace, with the aim of contributing to the reduction of incidents arising from causes so predictable, yet so little studied. We must also give priority to airmen scientists that have produced these studies in physiology and occupational medicine, since the literature is scarce about indicating the need for further work in this direction. Human cognition refers to mental processes involved in thinking and their use. It is a multidisciplinary area of interest includes cognitive psychology, psychobiology, philosophy, anthropology, linguistics and artificial intelligence as a means to better understand how people perceive, learn, remember and how people think, because will lead to a much broader understanding of human behavior.

Cognition is not presented as an isolated entity, being composed of a number of other components, such as mental imagery, attention, consciousness, perception, memory, language, problem solving, creativity, decision making, reasoning, cognitive changes during development throughout life, human intelligence, artificial intelligence and various other aspects of human thought (Sternberg, 2000).

The procedures of flying an aircraft involve observation and reaction to events that take place inside the cabin of flight and the environment outside the aircraft (Green et al, 1993). The pilot is required to use information that is perceived in order to take decisions and actions to ensure the safe path of the aircraft all the time. Thus, full use of the cognitive processes becomes dominant so that a pilot can achieve full success with the task of flying the "heavier than air."

With the advent of automated inclusion of artifacts in the cabin of flight that assist the pilot in charge of controlling the aircraft, provide a great load of information that must be processed in a very short space of time, when we consider the rapidity with which changes occur, an approach that cover the human being as an individual is strongly need. Rather, the approach should include their cognition in relation to all these artifacts and other workers who share that workspace (FAA, 2005).

IV. THE DEPLOYMENT OF THE TASKS LEADING TO ACCIDENTS.

A strong component that creates stress and fatigue of pilots, referred to the design of protection, detection and effective handling of fire coming from electrical short circuit on board, is sometimes encountered as tragically happened on the Swissair Airlines flight 111, near Nova Scotia on September 2, 1998. The staff of the Federal Aviation Administration (FAA), responsible for human factors research and modern automated interfaces (FAA, 2005), reports a situation exacerbated by the widespread use an electrical product and a potentially dangerous wire on aircrafts, called "Kapton".

If a person has to deal with an outbreak of fire, coming from an electrical source at home, the first thing he would do is disconnect the electrical power switch for the fuses. But this option is not available on aircraft like the Boeing B777 and new Airbus. The aviation industry is not adequately addressing the problem of electrical fire in flight and is trying to deal recklessly (FAA, 2005). The high rate of procedural error associated with cognitive errors, in the automation age, suggests that the projects in aviation have ergonomic flaws. In addiction, is has been related that the current generation of jet transport aircraft, used on airlines, like the Airbus A320, A330, A340, Boeing B777, MD11 and the new A380, that are virtually "not flyable" without electricity. We can mention an older generation, such as the Douglas DC9 and the Boeing 737.

Another factor in pushing the pilots that causes emotional fatigue and stress is the reduction of the cockpit crew to just two. The next generation of large transport planes four engines (600 passengers) shows a relatively complex operation and has only two humans in the cockpit. The flight operation is performed by these two pilots, including emergency procedures, which should be monitored or re-checked. This is only possible in a three-crew cockpit or cockpit of a very simple operation. According to the FAA, the only cockpit with two pilots that meets these criteria is the cabin of the old DC9-30 and the MD11 series. The current generation of aircraft from Boeing and Airbus do not fit these criteria, particularly with respect to engine fire during the flight and in-flight electrical fire.

The science of combining humans with machines requires close attention to the interfaces that will put these components (human-machine) working properly. The deep study of humans shows their ability to instinctively assess and treat a situation in a dynamic scenario. A good ergonomic design project recognizes that humans are fallible and not very suitable for monitoring tasks. A properly designed machine (such as a computer) can be excellent in monitoring tasks. This work of monitoring and the increasing the amount of information invariably creates a cognitive and emotional overload and can result in fatigue and stress.

According to a group of ergonomic studies from FAA (FAA, 2005) in the United States this scenario is hardly considered by the management of aviation companies and, more seriously the manufacturers, gradually, introduce further informations on the displays of Glass cockpits. These new projects always determine some physiological, emotional and cognitive impact on the pilots.

The accident records of official institutes such as the NTSB (National Transportation Safety Bureau, USA) and CENIPA (Central Research and Prevention of Accidents, Brazil) show that some difficulties in the operation, maintenance or training aircraft, which could affect flight safety are not being rapidly and systematically passed on to crews worldwide. These professionals of aviation may also not be unaware of the particular circumstances involved in relevant accidents and incidents, which makes the dissemination of experiences very precarious.

One of the myths about the impact of automation on human performance: "while investment in automation increases, less investment is needed in human skill" (FAA, 2002). In fact, many experiments showed that the progressive automation creates new demands for knowledge, and greater, skills in humans. Investigations of the FAA (NTSB, 2011), announced that aviation companies have reported institutional problems existing in the nature and the complexity of automated flight platforms. This results in additional knowledge requirements for pilots on how to work subsystems and automated methods differently. Studies showed the industry of aviation introduced the complexities of automated platforms flight inducing pilots to develop mental models about overly simplified or erroneous system operation. This applies, particularly, on the logic of the transition from manual operation mode to operation in automatic mode (NTSB, 2011). The process of performing normal training teaches only how to control the automated systems in normal but do not teach entirely how to manage different situations that the pilots will eventually be able to find. This is a very serious situation that can proved through many aviation investigation reports that registered the pilots not knowing what to do, after some computers decisions taken, in emergences situations (NTSB, 2011). VARIG (Brazilian Air lines), for example, until recently, had no Boeing 777 simulators where pilots could simulate the emergence loss of automated systems what should be done, at list, twice a month, following the example of Singapore Airlines.

According to FAA (Federal Aviation Administration, 2010), investigations showed incidents where pilots have had trouble to perform, successfully, a particular level of automation. The pilots, in some of these situations, took long delays in trying to accomplish the task through automation, rather than trying to, alternatively, find other means to accomplish their flight management objectives. Under these circumstances, that the new system is more vulnerable to sustaining the performance and the confidence. This is shaking the binomial Human-Automation compounded with a progression of confusion and misunderstanding. The qualification program presumes it is important for crews to be prepared to deal with normal situations, to deal with success and with the probable. The history of aviation shows and teaches that a specific emergency situation, if it has not happen, will certainly happen.

V. FUTURE WORK FOR PROCEED AN SYSTEMIC EVALUATION IN THE PERFORMANCE OF THE PILOTS

Evaluating performance errors, and crew training qualifications, procedures, operations, and regulations, allows them to understand the components that contribute to errors. At first sight, the errors of the pilots can easily be identified, and it can be postulated that many of these errors are predictable and are induced by one or more factors related to the project, training, procedures, policies, or the job. The most difficult task is centered on these errors and promoting a corrective action before the occurrence of a potentially dangerous situation.

The FAA team, which deals with human factors (Federal Aviation Administration , 2010), believes it is necessary to improve the ability of aircraft manufacturers and aviation companies in detecting and eliminating the features of a project, that create predictable errors. The regulations and criteria for approval today do not include the detailed project evaluation from a flight deck in order to contribute in reducing pilot errors and performance problems that lead to human errors and accidents. Neither the appropriate criteria nor the methods or tools exist for designers or for those responsible for regulations to use them to conduct such assessments. Changes must be made in the criteria, standards, methods, processes and tools used in the design and certification. Accidents like the crash of the Airbus A320 of the AirInter (a France aviation company) near Strasbourg provide evidence of deficiencies in the project.

This accident highlights the weaknesses in several areas, particularly when the potential for seemingly minor features has a significant role in an accident. In this example, inadvertently setting an improper vertical speed may have been an important factor in the accident because of the similarities in the flight path angle and the vertical speed in the way as are registered in the FCU (Flight Control Unit).

This issue was raised during the approval process of certification and it was believed that the warnings of the flight mode and the PFD (Primary Flight Display-

display basic flight information) would compensate for any confusion caused by exposure of the FCU, and that pilots would use appropriate procedures to monitor the path of the vertical plane, away from land, and energy state. This assessment was incorrect. Under current standards, assessments of cognitive load of pilots to develop potential errors and their consequences are not evaluated. Besides, the FAA seeks to analyze the errors of pilots, a means of identifying and removing preventively future design errors that lead to problems and their consequences. This posture is essential for future evaluations of jobs in aircraft crews.

Identify projects that could lead to pilot error, prematurely, in the stages of manufacture and certification process will allow corrective actions in stages that have viable cost to correct or modify with lower impact on the production schedule. Additionally, looking at the human side, this reduces unnecessary loss of life.

VI. CONCLUSION

We developed a study focusing on the guilt of pilots in accidents when preparing our thesis. In fact, the official records of aircraft accidents blame the participation of the pilots like a large contributive factor in these events.

Modifying this scenario is very difficult in the short term, but we can see as the results of our study, which the root causes of human participation, the possibility of changing this situation. The cognitive factor has high participation in the origins of the problems (42% of all accidents found on our search). If we consider other factors, such as lack of usability applied to the ergonomics products, choise of inappropriate materials and poor design, for example, this percentage is even higher.

Time is a factor to consider. This generates a substantial change in the statistical findings of contributive factors and culpability on accidents. The last consideration on this process, as relevant and true, somewhat later, must be visible solutions. In aviation, these processes came very slowly, because everything is wildly tested and involves many people and institutions. The criteria adopted by the official organizations responsible for investigation in aviation accidents do not provide alternatives that allow a clearer view of the problems that are consequence of cognitive or other problems that have originate from ergonomic factors. We must also consider that some of these criteria cause the possibility of bringing impotence of the pilot to act on certain circumstances. The immediate result is a streamlining of the culpability in the accident that invariably falls on the human factor as a single cause or a contributing factor. Many errors are classified as only "pilot incapacitation" or "navigational error". Our research shows that there is a misunderstanding and a need to distinguish disability and pilot incapacitation (because of inadequate training) or even navigational error.

Our thesis has produced a comprehensive list of accidents and a database that allows extracting the ergonomic, systemic and emotional factors that contribute to aircraft accidents. These records do not correlate nor fall into stereotypes or patterns. These patterns are structured by the system itself as the accident records are being deployed. We developed a computer system to build a way for managing a database called the Aviation Accident Database. The data collected for implementing the database were from the main international entities for registration and prevention of aircraft accidents as the NTSB (USA), CAA (Canada), ZAA (New Zealand) and CENIPA (Brazil). This system analyses each accident and determines the direction and the convergence of its group focused, instantly deployed according to their characteristics, assigning it as a default, if the conditions already exist prior to grouping. Otherwise, the system starts formatting a new profile of an accident (MARTINS, 2007).

This feature allows the system to determine a second type of group, reporting details of the accident, which could help point to evidence of origin of the errors. Especially for those accidents that have relation with a cognitive vector. Our study showed different scenarios when the accidents are correlated with multiple variables. This possibility, of course, is due to the ability of Aviation DataBase system, which allows the referred type of analysis. It is necessary to identify accurately the problems or errors that contribute to the pilots making it impossible to act properly. These problems could point, eventually, to an temporary incompetence of the pilot due to limited capacity or lack of training appropriateness of automation in aircraft. We must also consider many other reasons that can alleviate the effective participation or culpability of the pilot. Addressing these problems to a systemic view expands the frontiers of research and prevention of aircraft accidents.

This system has the purpose of correlating a large number of variables. In this case, the data collected converges to the casualties of accidents involving aircraft, and so, can greatly aid the realization of scientific cognitive studies or applications on training aviation schools or even in aviation companies (MARTINS, 2010). This large database could be used in the prevention of aircraft accidents allowing reaching other conclusions that would result in equally important ways to improve air safety and save lives.

REFERENCES

Eugenio,C.- Automação no cockpit das aeronaves: um precioso auxílio à operação aérea ou um fator de aumento de complexidade no ambiente profissional dos pilotos, 1rd ed., vol. 1. São Paulo: Brasil, 2011, pp. 34-35.

Henriqson,E. "Coordination as a Distributed Cognitive Phenomena Situated in Aircraft Cockpits- Aviation in Focus," "A Coordenação Como Um Fenômeno Cognitivo Distribuído e Situado em Cockpits de Aeronaves," UFRGS edit., Porto Alegre, vol. 12, pp. 58 –76, Dez. 2010.

FAA- Federal Aviation Administration, "DOT/FAA/AM-10/13, Office of Aerospace Medicine, Causes of General Aviation Accidents and Incidents: Analysis Using NASA Aviation," Safety Reporting System Data, Washington DC, U.S. Department of Transportation press, Sept. 2010.

Dekker, S. "Illusions of explanation- A critical essay on error classification," The International Journal of Aviation Psychology, New Jersey, vol. 13, pp. 95-106, Sept. 2003.

Reason, J.- Human Error, 2nd ed., Cambridge: U.K., University Press, 1990, pp.92-93.

Rasmussen, J.- "A taxonomy for describing human malfunction in industrial installations," The Journal of Occupational Accidents, vol. 4, Jul. 1982, pp. 311-333.

J. Reason, J.- The Smart Human, 5rd ed., Cambridge: U.K., University Press, 2008, p. 347.

Sternberg, R.- Cognitive psychology, Porto Alegre: Brasil, Ed Artmed, 2000.

Green, R.G. and Frenbard, M. Human Factors for Pilots. Avebury: England, Technical Aldershot, 1993.

FAA- Federal Aviation Administration- FAA, "The Interface Between Flightcrews and Modern Flight Deck Systems," Federal Aviation, Administration Human Factors Team Report, May 2005. pp-34-35.

FAA- Federal Aviation Administration, "Human Error Analysis of Asrts Reports: Altitude Deviations in Advanced Technology Aircraft," Federal Aviation Administration, Human Factors Team Report, 1992, 1996, 2002.

NTSB- National Transportation Safety Board, "Airline Service Quality Performance- User Manual- ASQP," Washington, DC, Bureau of Transportation Statistics, U.S. Department of Transportation press, Oct. 2011, pp. 75-77.

NTSB- National Transportation Safety Board, "Airline Service Quality Performance- User Manual- ASQP," Washington, DC, Bureau of Transportation Statistics, U.S. Department of Transportation press, Oct. 2011, pp. 123-128.

NTSB-National Transportation Safety Board, "C.F.R.- 234 Airline Service Quality Performance Reports," Washington, DC, Bureau of Transportation Statistics, Research and Innovative Technology Administration (RITA), U.S. Department of Transportation press, Oct. 2011, pp. 45-61.

FAA- Federal Aviation Administration- FAA, "Human Error Analysis of Accidents Report," Federal Aviation Administration- Human Factors Team Report, 2010, pp. 201-206.

FAA- Federal Aviation Administration- FAA, "Human Error Analysis of Accidents Report," Federal Aviation Administration- Human Factors Team Report, 2010, pp. 108-115.

Martins, Edgard.-"Ergonomics in Aviation: A critical study of the causal responsibility of pilots in accidents," "Ergonomia na Aviação: Um estudo crítico da responsabilidade dos pilotos na causalidade dos acidentes," Msc. Monography, Universidade Federal Pernambuco, Pernambuco: Brasil, Mar. 2007, pp. 285-298

Martins, Edgard, "Study of the implications for health and work in the operationalization and the aeronaut embedded in modern aircraft in the man-machines interactive process complex," "Estudo das implicações na saúde e na operacionalização e no trabalho do aeronauta embarcado em modernas aeronaves no processo interativo homem-máquinas complexas," thesis, Centro de Pesquisas Aggeu Magalhães, Fundação Oswaldo Cruz, Pernambuco:Brasil, Aug. 2010, pp. 567-612

CHAPTER 31

Ergonomics, Usability and Virtual Reality: A Review Applied to Consumer Product

Christianne Soares Falcão, Marcelo Marcio Soares

Federal University of Pernambuco
Recife, Pernambuco, Brazil
christiannevas@hotmail.com

ABSTRACT

While it is recognized that introducing products that meet the needs of recent consumption is important, what is crucial to the success of the final product is to evaluate its usability during the early stages of the design process. Currently there are few analytical tools that can foster better support to the entire process of product design. Virtual Reality has emerged as a useful aspect of new technology that gives efficient and effective support to the design of products. Therefore, this article presents a conceptual approach to usability and its relationship with Ergonomics, and points to the importance of using Virtual Reality throughout the product development life-cycle as its use offers a more efficient approach to usability.

Keywords: Usability of consumer product, Product Design Development Process, Virtual Reality

1   INTRODUCTION

Users of consumer products have benefited from companies' growing interest in seeking technological innovation to improve their competitiveness and business performance. Industry is constantly looking for new and better products that have satisfactory sales in the market. To this end, there is a need to invest in innovative products bearing in mind that globalization makes a vast range of new products and features available, this has led to more demanding consumers and increased market competitiveness.

In order to introduce products that are compatible with current market requirements and needs, the designer should take into account product users' best needs in terms of usability and of what gives them pleasure, what appeals to their emotions, and to demonstrate that the product is safe and tailor-made for them (Seva et al, 2011). Keeping the user interested in the product requires competence from designers and ergonomists at the interface who have invested in streamlining the design process so as to manage knowledge, skills and the stock of technologies.

At the same time, according to Mahdjoub et al (2010), during the design process, engineers underestimate such factors as Ergonomics. As a result, many products around us have not been conceived to meet end-users' expectations, including the need for them to be usable. For this and many other design factors, companies have invested heavily in usability and user experience testing labs to test their products before their release to the market.

Starting with the concept of complexity which is attributed to the technology of products, Ergonomics and usability have come to play an important role in developing products that better match users' wants and needs. Usability is about the product ease of use so it is necessary first and foremost to know who is the user/ consumer, their needs and requirements so as to serve them efficiently. Human Factors & Ergonomics, in these circumstances, presents itself as an essential discipline when analyzing the interface between the product and the user.

When evaluating products, there is a need for evaluating product through several stages during the design process to ensure the final product is successful. However, many assessment tools are limited as they do not provide the means for detecting problems that may occur in real world usage because laboratory tests are only simulated use cases of the product. Therefore, extensive research is needed to investigate the efficiency and effectiveness of assessment tools based on up-to-date advanced computer technologies in order to provide better support for product design throughout the development process (Ye et al, 2007).

To meet this need, virtual reality (VR) has emerged as an option to reconcile the advantages of field and laboratory studies because it enables the user to interact with the product in a context similar to the real situation while allowing the researcher to have full control over the variables and safety conditions (Rebelo et al, 2011). Besides this advantage, VR can easily be included in participatory design methodology as VT displays design solutions effectively (Reich et al, 1996).

Thus, studies show VR as a potential tool for the design of human-centered products, especially when it is integrated with Ergonomics. However, Mahdjoub et al (2010), point out the need to integrate these technologies with a designer's knowledge in order to be more efficient in promoting the usability of the product even at the early stages of the design process.

Driven by this challenging statement, this article gives an overview of applying VR technology to evaluate the usability of consumer products. The first section sets out the theoretical concept that underpins the study and describes usability and its correlation with Ergonomics. The second section explains the concepts of virtual reality and their application in product design, and does so in order to provide the basic information needed to understand its advantages and how this is integrated with the Product Design Development Process.

The purpose of addressing this theme is to underpin a research study that sets out to investigate the appropriateness of using virtual reality technology in assessing the usability of consumer products, by making a comparative analysis of virtual prototypes and the real product. The intention is to use the results of the analysis to set parameters for exploiting VR in product design.

2   USABILITY

Usability is well-known and well-defined as to the approach of Human-Computer Interaction (HCI), the concepts of which are applied to enhance the software-user interface (Nielsen, 1993). Basing themselves on these concepts, many studies seek to apply them to fostering usability in consumer products (Jordan, 1997; Logan, 1994, Han et al, 2001).

Starting from an informal concept, usability was originally defined as the ability of a product or system to be used in an effective, efficient and enjoyable way by a specific range of users for tasks that need specific tools within a given environment. ISO 9241-11 (International Standard Organization, 2006) sets out the most classic and recognized concept of usability, "the extent to which a product can be used by specified users to achieve specified goals with effectiveness, efficiency and satisfaction in a specified context of use ".

Effectiveness refers to the extent to which an objective or task is reached. Effectiveness measures the relationship between the results obtained and the desired goals, i.e. to be effective is to manage to achieve a given objective.

Efficiency refers to the amount of effort required to achieve an objective. The lower stress, the higher the efficiency.

Satisfaction refers to the level of comfort that the user feels when using a product and to what extent the product is acceptable to the user in relation to achieving his/ her objectives. As it is more closely linked to subjective factors, this aspect can be more difficult to measure than effectiveness and efficiency (Jordan, 1998).

Krug (2000), characterizes usability from a simple perspective, as the certainty that something works well, and that a person with average (or even below average) skill and experience can use something that it is intended he/she should use, whether it is a website, a hunter jet or a revolving door, without getting irretrievably frustrated.

Based on the definitions of usability presented, and other contexts not covered, it can be observed that they share the following issues:

- There is a user;
- who owns an activity;
- using a product or system.

According to Tullis & Albert (2008), usability can have a huge impact on society, regarding access to goods and services for different user populations, such as the elderly, people with disabilities or those with language or literacy difficulties. Usability impacts everyone, every day, and crosses cultures, gender, age and economic class.

For Santos and Maciel (2010), usability is gradually being incorporated into product development and is no longer being seen as a separate activity. If a participatory design process is used, usability methods bring designers closer to users by providing a better match of products to the real user needs.

2.1 Ergonomics and Usability

Ergonomics can be understood as a discipline focused on the nature of the interactions of humans with artifacts, which starts from a unified perspective of science, engineering, design, technology and management of human-system compatibility, including a variety of products, processes, and natural and artificial environments (Karwowski, 2005). Ergonomics deals with a wide range of problems relevant to design when evaluating work systems, consumer products and workplaces, in which human-machine interactions affect human performance and the usability of products. According to Karwowski, Soares and Stanton (2011), ergonomics uses a holistic approach, centred on the user, which considers within its range of action physical, cognitive, social, organizational, environmental and other relevant factors, using appropriate methodologies, to improve design. Thus, Ergonomics is involved in the various decision-making stages when a product is being developed which leads to its having a key role in improving the performance of products.

In this context, usability becomes part of Ergonomics so that products and systems may be adjusted to the user, taking into account his/ her needs, abilities and limitations. It presents itself as the border area between the user and the parts of the system with which the user is interacting. At this threshold, Ergonomics and usability contribute perform this objective in order to maximize user satisfaction.

2.2 Attributes of usability

When analyzing usability, some principles of product design should be observed in order to avoid mistakes and difficulties in using the product. If each of these principles are complied with and if the design is for a defined and identified target audience, it will probably be possible to maintain good product usability and highest user satisfaction. Jordan (1998) lays down ten principles of usability and describes why and how each affects usability:

- *Consistency* - Similar functions and tasks are performed similarly.
- *Compatibility* - The conduct of a task using a given product is compatible with the user's expectations, based on other similar products.

- *Considerations of user resources* – These are seen to have been included when the conduct of a task takes into account the demand on the user's available physical and mental resources.
- *Feedback* – This takes place when a product has a significant response to the actions taken by users in performing a task and these are received and acknowledged.
- *Error prevention and recovery* - when performing a task, the possibility of error is minimized and should the error occur, it can be quickly and easily corrected.
- *User control* - where the user has control over the actions taken within the task without the overall process being automated.
- *Visual clarity* – when the layout of information can be easily read and interpreted without causing confusion or misinterpretation when the task is being performed.
- *Prioritization of functionality and information* - when the most important functions and information in the tasks can be very easily viewed and accessed by the user.
- *Appropriate transfer of technology* - when a technology developed to perform a certain task or for a certain product is correctly transferred to another context by appropriately performing a different task.
- *Explicitness* - where a product's functionality and method of use are explicit.

Norman (2006) defines two principles of design that are fundamental for usability: providing a good conceptual model and making things visible.

Nielsen (1993) stresses that usability is not a unique, one-dimensional property of an interface but rather it has multiple components and is traditionally associated with five attributes: Learnability, Efficiency, Memorability, Errors and Satisfaction.

Rebelo *et al* (2011) propose a division of the attributes of usability of products into those that are objective and those that are subjective. The objective attributes of usability include: *effectiveness, learnability, flexibility, understandability, memorability,* and reliability. A subjective attribute of usability includes the *attractiveness of the product*, which affects the positive attitude toward the product.

Such usability criteria should be studied systematically, using a well-developed design with well-established steps. The discipline of Usability Engineering was created as a result of this need. It sets out to conduct in a systematic way the activities needed to draft the interface throughout the life cycle of product or system development (Nielsen, 1993).

Thus, the designer determines which attributes are key components of the product for usability, and he/ she should define what each one really means for the product and how this will be measured by using usability methods to find out how each characteristic of usability can be measured.

3   VIRTUAL REALITY

During product design process, design team need to have the ability to think creatively about alternative design paths and to make decisions based on product use simulations. Meeting such needs has been getting better due to technological developments. From the second half of last century, tools that help to build up our capacity for simulation have been under development. These are the technologies in Virtual Reality (VR).

Virtual Reality technology is capable of providing an immersive work environment. By using various peripheral devices such as motion capture systems and haptic interfaces, different ways for interaction between the user and the VR system are supplied. Buck (1998) used VR technology and DHM (Digital Human Modeling) tools to enable people to take part in the design of the industrial system. VR has also been used in ergonomic applications to evaluate different aspects of the operations of human movement (Whitman *et al,* 2004; Jayaram *et al,* 2006; Wang *et al,* 2007).

Under a broader definition, virtual reality is one way to transport a person to a reality that is not physically present but is perceived as if it were. According to the definition given in Encyclopedia Britannica, virtual reality is "*the use of computer modeling and simulation that enables a person to interact with an artificial three-dimensional visual or other looked sensory environment. VR applications immerse the user in a computer-generated environment that simulates reality through the use of interactive devices, which send and receive information and goggles, headsets, gloves, or body suits are worn. In a typical VR format, a user wearing a helmet with a stereoscopic screen views animated images of a simulated environment*" ("Virtual Reality", 2010).

According to Thalmann (1997), VR refers to a technology capable of transferring a subject to a different environment without the need to move him/ her physically. To this end, the user's sensory organs are manipulated in such a way that his/ her perception is associated with what the virtual environment and not with the actual or physical environment desires. The manipulation process is controlled by a computer model based on the physical descriptions of the virtual environment.

Also according to Thalman, immersion is a key issue in VR systems, because it is fundamental for the paradigm where the user becomes part of the simulated world instead of the simulated world being a characteristic of the real world. The first "immersive VR system" was the flight simulator, where immersion is achieved through a subtle blend of real hardware and virtual image.

For a better understanding of the user's physical and psychological experience in a virtual environment, there is a need to be familiar with the concepts of immersion, presence, interaction and involvement.

The types of immersion are characterized according to the physical configuration of the interface with the user, fully-immersive (using Head Mounted Display), semi-immersive (large projection screens) or non-immersive (desktop based VR). The level of immersion is measured by the ability of users to interact and communicate with the object in virtual reality in a way similar to how he/ she interacts and communicates with objects in the real world. Thus, the less the perception of the real world (see, hear, touch), the greater the classification of immersion in VR will be (Whitman *et al*, 2004; Gutierrez *et al*, 2008).

Presence corresponds to the subjective concept that the user perceives there is in a Virtual Environment (VE). The sense of the user's presence happens when he/ she is aware of "being there" in the environment based on the technology founded on an immersive base (Slater, 1994, Gutiérrez et al, 2008).

Interaction is connected with communication between user and VR system. The capacity of detecting user motions and actions (user inputs) and refreshing VE, in accordance with these inputs, defines interaction. Involvement is related to concentration on VE. Thus, any factor distracting the user can affect his/ her involvement (Rebelo *et al*, 2011).

In order to visualize products in a VE, various devices are used. A head-mounted display (HMD) is used to visualizing the VE in an immersive manner, since it visually isolates participants from the real world. The motion tracker can be associated with the HMD in order to measure the position/ orientation of the user's head.

To capture users' movements and actions (user inputs) devices are used such as motion trackers and sensing gloves. These devices have built-in sensors that enable the computer to measure the position of the user's hand in real-time and to record the flexion of the fingers to enable natural gestures to be recognized. Gloves are necessary especially in tasks involving the manipulation of objects.

3.1   Virtual reality applied to consumer product design

Recent research has demonstrated the advantages and benefits of virtual reality (VR) in many applications, including but not limited to the automotive industry, military research, health systems, entertainment, construction, oil & gas industries etc. However, the use of VR to support consumer products design is still not widespread (Rebelo *et al*, 2011).

In practice during the product design phase itself, the product can be evaluated with the aid of CAD tools (computer aided design), in three-dimensional models, as well as of physical prototypes (functional or non-functional mock-ups). Physical prototypes, due to their costs and the time taken to construct them, can lead to negative consequences for the competitiveness of the product, which is why virtual prototypes are more commonly used.

It is in this context of substituting or complementing physical with virtual environments and taken this as the starting point that VR technology can support the design of products efficiently and effectively from the early stages of product development (e.g. conceptual design) to the entire product lifecycle activities, in various degrees of immersion, which may be chosen according to the requirements of the design (Ye, J., *et al*, 2007). Thus, by using VR, the development time and the costs of product design can be reduced as a result of having the possibility of replacing physical mock-ups with virtual prototypes.

The use of digital prototypes in various design alternatives can be viewed immediately, thus enabling users to provide feedback about them and how they are being used. Furthermore, solutions can be altered interactively and more easily so than if the object were a physical one, which means that more design alternatives in the form of prototypes can be tested, and this is more viable financially.

In addition, virtual environments (VE) enable those taking part in the research to interact safely with all types of products and in any environmental conditions, even when exposed to critical conditions that may put them at risk. VEsl also provide for data collection from the early stages of the design process, with high precision and ecological validity, which is not readily available in the real world (Teixeira *et al,* 2010).

VR, like any new and emerging technology, - also has limitations- which are related to the display devices, such as limited field of view and/ or low image resolution offered by some HMDs, as well as to interaction devices, such as the limited tactile feedback and haptic (Ye , J., *et al*, 2007). However, it is understood that the severity of these limitations is related to the quality of the technology acquired, since different devices are commercially available with varying degrees of technological sophistication. Haptic modalities involving the weight sensation, volume and shape adds restrictions that designers can overcome through recent advances in technology.

In the near future, product design based on a VR environment may provide better visualization of the product, thus enabling the designer to coexist in the same virtual space and providing a better appreciation of the geometry and aesthetics of the product. In addition, VR can support the process of participatory and collaborative planning taking place simultaneously, therefore benefiting from the possibility of having several experts and users working simultaneously on the same product and in the same environment, as seen in Mahdjoub *et al* (2010).

4 CONCLUSION

Setting out from the principle that a definition of usability had been drawn up from the views of several authors, it is essential for a product to meet the usability principles identified and described in this paper and thus to determine whether it does or does not demonstrate usability. To this end, Virtual Reality technology can be used as an important tool support the earliest stages of product development, thus helping the team of designers to develop the first concepts of the product, or even later, during the prototyping stage and more detailed studies. This factor represents an improvement in product quality in ergonomic terms.

Incorporating VR into the process of product design, together with the analysis of usability, leads to the reduced cost of evaluating solutions, thus anticipating issues that, if identified later, could jeopardize the success of the product development, as well as users' safety and well-being.

Thus, using VR, based on the application of appropriate usability method, becomes compatible with traditional approaches that have been used when developing products. As the potential of this technology allows a more natural interaction with virtual models, as noted in section 3.1, there has been an increase in the impetus to develop better support from computing resources for the process of project development. It is believed that the continuing exploration of new technologies and their integration with their application to design will result in the further evolution of product design evaluation systems that are more compatible with the needs of designers and users.

The research that has been developed by the authors intends to conduct an usability evaluation starting from comparative analysis of virtual prototypes of a consumer product, built from the VR, with real physical prototypes. Thus, the intention is to contribute to the area of the usability evaluation and to the application of advanced technologies, more specifically the application of VR in the analysis of the interface between the user and consumer products.

REFERENCES

Buck, M. 1998. Immersive user interaction within industrial virtual environment. In. *Virtual Reality for Industrial Applications*. Springer-Verlag, Berlin, 39-60.

Gutiérrez, M. A.; Vexo, F. and Thalmann, D. 2008. *Stepping Into Virtual Reality*. Lausanne: Springer.

Han, Sung; Yun, Myung; Kwahk, Jiyoung, Hong, Sang. 2001. Usability of consumer electronic products. *International Journal of Industrial Ergonomics* 28: 143-151.

ISO 9241-110. 2006. Ergonomics of human-system interaction - Part 110: Dialogue principals. Standard.

Jayaram, U.; Jayaram, S.; Shaikh, I. 2006. Introducing quantitative analysis methods into virtual environments for real-tiem and continuous ergonomic evaluations. *Computers in Industry* 57(3)2: 283-296.

Jordan, P. W. 1997. The four pleasures: Taking human factors beyond usability. *Proceedings of the 13th Triennial Congress of the International Ergonomics Association*, Vol. 2. Tampere, Finland: 364-366.

Jordan, P.W. 1998. *An introduction to usability*. London: Taylor & Francis.

Karwowski, W. 2005. Ergonomics and Human Factors: The Paradigms for Science, Engineering, Design, technology, and Management of Human – Compatible Systems. *Ergonomics* 48(5): 436-463.

Karwowski, W., Soares, M. and Stanton, N., eds, 2011. Human Factors and Ergonomics in Consumer Products. Boca Raton: CRC Press.

Krug, Steve. 2000. *Don't make me think! A common sense approach to web usability*. Indiana-Polis: New Riders Press.

Logan, R. J., 1994. Behavorial and emotional usability: Thomson consumer electronics. In: *Usability in Practice*. ed. M. E. Wiklund. AP Professional, New York, 59-82.

Mahdjoub, M.; Monticolo, D.; Gomes, S., Sagot, J. 2010. A collaborative Design for Usability approach supported by Virtual Reality and a Multi-agent System embedded in a PLM environment. *Computer-Aided Design* 42: 402-413.

Nielsen, Jakob. 1993. *Usability Engineering*. Boston: Academic Press.

Norman, Donald A. 2006. *O design do dia-a-dia*. Rio de Janeiro: Rocco.

Rebelo, Francisco; Duarte, Emilia; Noriega, Paulo and Soares, Marcelo. 2011. Virtual Reality in Consumer Product Design: Methods and Applications. In. *Human Factors and Ergonomics in Consumer Product Design: Methods and Techniques.* eds. W. Karwowski, M. Soares and N. Stanton. Vol. I, CRC Press, 381-402.

Reich, Y.; Konda, S.; Monarch, I.; Levy, S and Subrahmanian, E. 1996. Varieties and issues of participation and design. *Design Studies* 17(2): 165-180.

Santos, R.; Maciel. F. 2010. Usabilidade nos trópicos: desafios e perspectivas de um laboratório de usabilidade no Amazonas. *Proceedings of 10° Congresso Internacional de Ergonomia e Usabilidade de Interfaces Humano-Computador.* Rio de Janeiro: PUC-Rio.

Schamorrow, D. 2009. Why virtual?. *Theorical Issues in Ergonomics Science 10(3)*: 279-282.

Seva, R.; Gosiaco, K.; Santos, Ma. C.; Pangilinan, D.2011. Product design enhancement using apparent usability and affective quality. *Applied Ergonomics* 42: 511-517.

Slater, M.; Usoh, M. 1994. Body Centred Interaction in Immersive Virtual Environments. In: *Artificial Life and Virtual Reality.* eds. Magnenat Thalmann and D. Thalmann. John Wiley.

Teixeira, L.; Vilar, E.; Duarte, E.; Rebelo, F. 2010. ErgoVR - An approach for automatic data collection for Ergonomics in Design Studies. *Proceedings of the International Symposium on Occupational Safety and Hygiene, SHO2010,* Vol. 1. Guimarães, Portugal: 505-509.

Thalmann, Daniel. 1997. Introduction to Virtual Environments. *Proceedings of The International Conference on Multimedia Modeling.* Accessed January, 2010, http://vrlab.epfl.ch/~thalmann/VR/Intro.pdf.

Tullis, Tom; Albert, Bill. 2008. *Measuring the user experience: collecting, analyzing and presenting usability metrics.* USA: Elsevier Inc.

Virtual Reality. 2010. Virtual Reality. In: *Encyclopedia Britannica.* Accessed April 15, 2010, http://www.britannica.com/EBchecked/topic/630181/virtual-reality.

Wang, y.; Liao, k.; Guo, Y.; Zhang, W.; Wu, S. 2007. Development and application of integrated human machine interaction simulation system. *Journal of System Simulation* 19(11), 2492-2495.

Whitman, L. E.; Jorgensen, M.; Hathiyari, K. and Malzahn, D. 2004. Virtual reality: Its usefulness for ergonomic analysis. In *2004 Winter Simulation Conference,* eds. R. Ingalls, M. D. Rossetti, J. S. Smith and B. A. Peters, 1740-45. Washington: Winter Simulation Conference.

Ye, Jilin; Badiyani, Saurin; Raja, Vinesh and Schlegel, Thomas 2007. Applications of Virtual Reality in Product Design Evaluation. In. *Human-Computer Interaction.* ed. J. Jacko. Part IV, HCI2007, Springer-Verlag Berlin Heidelberg, 1190-1199.

CHAPTER 32

A Preliminary Study on Applying UD Principles for Signage Assessment in Hospital Environment

Chun-Ming Yang, Ching-Han Kao, Chen-Hui Lin

Dept. of Industrial Design, Ming Chi University of Technology
Taipei, TAIWAN
*E-mail: cmyang@mail.mcut.edu.tw

ABSTRACT

Due to the recent growth of aging population, foreign care-takers, and foreign labors in Taiwan, it is inevitable that a great amount of diverse population will seek medical needs or even work in the local hospitals. The signage system in the hospital environment needs to take this into consideration in order to provide universally comprehensible information for people. To this end, the initiatives of Universal Design (UD) can be introduced to help assess and evaluate UD performance of hospital signage in Taiwan and a universally comprehensible signage can be designed and planned based on the initiatives. In this study, a literature review is conducted to survey UD principles and their guidelines and design principles of hospital signage. A UD assessment, based on UD principles, is modified to accommodate signage related features. The refined UD assessment can be employed to assess UD performance of signage in a hospital and help identify ways to redesign the signage. A universally comprehensible signage in the hospital can then be promoted and help provide the efficient and effective information for visitors in hospital.

Keywords: universal design, hospital signage, signage assessment

1 INTRODUCTION

Taiwan's demographic structure has exhibited an ageing trend and an increase in foreign caretakers. In 1993, Taiwan became an aging society. By the end of 2010, senior citizens above the age of 65 constituted 10.9% of the total population, with an ageing index of 72.2%. Over the past three years, the number of senior citizens has increased by 10.7% (Ministry of the Interior, 2012), which indicates that the need for medical caretakers is continuously growing. The number of foreign workers in Taiwan in 2011 increased by 15% from that in 2008, and the number of worker in the medical care sector rose by 19% from 2008, accounting for the largest proportion of foreign workers (Bureau of Employment and Vocational Training, 2012). In the future, the population seeking medical care or employment in the medical industry is expected to increase in diversity.

Lin (1998) discovered that a person's cultural background significantly influenced their signage preference. Hence, how to satisfy users from various cultural backgrounds is an issue that requires greater attention. Additionally, the large-scale setting of the hospital often causes the patients and visitors to become stressed and disoriented, regardless of the nature of their visit or their familiarity with the hospital. Wayfinding problems are also time-consuming, especially when anger and frustration increase over time, adding to the stress level. Service personnel and hospital staff are not guaranteed to have a comprehensive knowledge of the hospital setting or provide easy-to-follow wayfinding guidance. Thus, efforts to increase the resources invested in wayfinding services cannot guarantee better outcomes; instead, they may influence the perceptions of patients and visitors regarding the quality of medical services.

Signage designs must be universally understood to satisfy every type of patient and visitor to a hospital. Only then can the users promptly understand the interior layout of building and successfully reach their destination, improving wayfinding service quality and simultaneously reducing labor cost. This study endeavors to construct universal design performance measures for the interior environment of hospitals. These measures can serve as a reference for design personnel to apply the universal design principles in signage design.

2 LITERATURE REVIEW

2.1 SIGNAGE

Socioeconomic developments have contributed to people's interest in signage. With economic progress since World War II, national infrastructure and transportation has developed rapidly, increasing people's mobility and leading to the evolution of signage design, which is receiving greater attention than ever before. Lynch (1960) contended that wayfinding is a necessary behavior caused by the interaction between people and the environment. In the nineteenth century, the fields of architecture and graphic design were merged to become environmental graphic design (Calori, 2007; Berger, 2009). In 1973, the international, professional, non-profit design organization of the Society for Environmental Graphic Design (SEGD) was founded to promote public awareness and the professional development of environmental graphic design (SEGD, 1993). Berger (2009) suggested that wayfinding is an auxiliary behavior of signage. Therefore, we can conclude that signage is a vital medium enabling communication between people and the environment.

Because hospital and community settings must accommodate increasingly diverse populations, it is vital for signage designs to be universally understood (Hölscher et al., 2005; Berger, 2009; Rousek and Hallbeck, 2011). However, the location of signs, especially those in hospitals, are often considered inappropriate (Rooke et al., 2009; Rousek and Hallbeck, 2011), causing a stressful wayfinding experience (Passini, 1996). Wayfinding problems in hospitals are costly and stressful and particularly impact outpatients and visitors, who are often already stressed and disoriented and unfamiliar with the hospital (Cowgill et al., 2003; Ulrich et al., 2004). Ulrich et al. (2004) discovered that stress is associated with wayfinding problems, which can cause an increase in blood pressure, headaches, and further stress. Therefore, when a person is confused or has insufficient information regarding the directions, their resulting anger and frustration is understandable. These problems can be resolved through inquiries; however, when the users, especially those who are not native speakers, cannot seek assistance, they can easily feel isolated and become lost, increasing their stress level (Cowgill et al., 2003). Thus, the commonality of language must be considered. Furthermore, the design of a facility should consider the interests of all users, not only the disabled (Preiser, 2008). Research has found that people spend approximately eight seconds reading a sign. Additionally, they prefer to read signs and make navigational decisions while moving instead of stopping to read signs (Johnson, 1993). Thus, the signage design is crucial for wayfinding. Gibson (2009) found that most people who move rapidly through a space tend to travel at a certain distance from signs. Therefore, appropriate information display and easy-to-understand usability are important because the information display style determines the perceptuocognitive behavior of users in complex settings. Thus, information support systems should be designed based on the perceptuocognitive abilities of most people. Based on this fundamental principle, signs must be visible, legible, and readable to enhance their usability (McCormick and Sanders, 1987). A number of elements must be considered when designing; for example, the ease of visibility (Arthur and Passini, 1992) and the effect of light on the sign visibility (Rousek and Hallbeck, 2011). Additionally, the educational level of the viewer should not affect their understanding of the signs (Cowgill et al., 2003). A well-established locational relationship between signs enables them to be thoroughly identified and understood (Gibson, 2009). By applying redundant and compound codes obtained from Human Factors Engineering (HFE) (McCormick and Sanders, 1987), diverse methods are proposed. For example, information with identical meanings can be represented using alphanumeric characters, symbols, pictograms, and even colors. Passini (1996) argued that by considering users' exact requirements and increasing design standards the principles of universal design can be applied to wayfinding design to satisfy the needs of numerous people and situations. Thus, signage design must satisfy the needs of all users and meet the environment and wayfinding requirements. Commonality of signage can be achieved with using universal design principles.

2.2   UNIVERSAL DESIGN

The concept of normalization, which refers to people with disabilities being offered the same life patterns and conditions as other citizens, originated in Northern Europe and has since spread to other parts of Europe and the U.S. (Tzeng, 2003). The promotion of accessible design that considers people with disabilities, including those caused by traffic accidents, has been part of the U.S. government's policies and civil rights movement since World War II. As early as the 1980's, a number of people recognized that even when imperfectly realized, many of the environmental provisions for accessibility had substantially more beneficiaries than expected (Duncan, 2007). This is an example of accessible design based on universal design principles that allows everyone to benefit from the product and the environment (Duncan, 2007; Story, 1998). The principle of universal design is based on accessible design.

Mace (1985) introduced the term "universal design" to describe the concept of designing products and environments that are aesthetically pleasing and useful to as many people as possible, regardless of their age, ability, or social status (Center for Universal Design, 2010). The Center for Universal Design was founded by Mace in 1989. The Center for Universal Design (2003) defines universal design as the design of products and environments that are usable by everyone, regardless of age, ability, or circumstance. Achieving usability by people of all ages, abilities, and circumstances is extremely difficult but worth striving for. As universal design increases its efficacy, the usability, safety, and marketability of products is increasing targeted to everyone (The Center for Universal Design, 2003; Duncan, 2007). A group of professions, comprising architects, product designers, engineers, and environmental design researchers, collaborated to establish the Principles of Universal Design to guide a wide range of design elements, such as the environment, product, and communication (The Center for Universal Design, 1997; Story, 1998). The Center for Universal Design at North Carolina State University (NCSU) also established principles and guidelines for universal design in 1997 (Duncan, 2007; Center for Universal Design, 2010). An important aspect of the universal design approach is an emphasis on the design process. A broad and inclusive design process includes numerous perspectives and ergonomic, human factors, and social equity considerations (Duncan, 2007). Preiser (2008) reasoned that the HFE research outcomes related to human elements could be used to develop principles and guidelines that fulfill the needs of a larger group of people and situations by facilitating the implementation and assessment of a universal design agenda. Universal design research can also identify more precisely the requirements of people with physical, sensory, and cognitive disabilities. In summary, universal design and appropriate signage are mutually complementary.

3   CONSTRUCTING PRELIMINARY UNIVERSAL DESIGN ASSESSMENT FOR SIGNAGE

This study constructed universal design performance measures for signage based on the Principles of Universal Design (Version 2.0) developed by the Center for Universal Design in NCSU. Factors that impact the design of signage were identified by integrating the principles of universal design, signage, and HFE. Each guideline of the Principles of Universal Design was assessed referencing related literature and amended, deleted, merged, and refined if necessary.

We reviewed relevant literature and deleted the items of the Principles of Universal Design that were irrelevant to signage design, that is, items 1B, 1C, 2B, 3E, 4A, Principle 5, 5B, 5C, Principle 6, 6A, 6B, 6C, 6D, Principle 7, 7B, 7C, and 7D (Table 1).

Table 1 Deleted items from the Principles of Universal Design

Items	
1B.	Avoid segregating or stigmatizing any users.
1C.	Provisions for privacy, security, and safety should be equally available to all users.
2B.	Accommodate right- or left- handed access and use.
3E.	Provide effective prompting and feedback during and after task completion.
4A.	Use different modes (pictorial, verbal, tactile) for redundant presentation of essential information.
Principle 5.	Tolerance for Error The design minimizes hazards and the adverse consequences of accidental or unintended actions.
5B.	Provide warnings of hazards and errors.
5C.	Provide fail safe features.
Principle 6.	Low Physical Effort The design can be used efficiently and comfortably and with a minimum of fatigue.
6A.	Allow user to maintain a neutral body position.
6B.	Use reasonable operating forces.
6C.	Minimize repetitive actions.
6D.	Minimize sustained physical effort.
Principle 7.	Size and Space for Approach and Use Appropriate size and space is provided for approach, reach, manipulation, and use regardless of user's body size, posture, or mobility.
7B.	Make reach to all components comfortable for any seated or standing user.

7C.	Accommodate variations in hand and grip size.
7D.	Provide adequate space for the use of assistive devices or personal assistance.

The items 5A "Arrange elements to minimize hazards and errors: most used elements, most accessible; hazardous elements eliminated, isolated, or shielded." and 5D "Discourage unconscious actions in tasks that require vigilance." of Principle 5 were similar to those of Principle 4 Perceptible Information. Therefore, we incorporated items 5A and 5D into Principle 4. Similarly, item 7A "Provide a clear line of sight to important elements for any seated or standing user." of Principle 7 was incorporated into Principle 1 Equitable Use. Finally, we refined the merged principles to satisfy the principles of universal design and develop universal design performance measures (Table 2).

Table 2 Refined UD Principles for Signage Assessment

Item
Principle 1. Equitable Use
The signage design is useful to people with diverse abilities.
1A. Provide the same means of signage use for all users: identical whenever possible; equivalent when not.
1B. Make the signage design appealing to all users.
7A. Provide a clear line of sight to important elements for any seated or standing user.
Principle 2. Flexibility in Use
The signage design accommodates a wide range of individual preferences and abilities.
2A. Provide choice in methods of signage use.
2C. Facilitate the user's accuracy and precision.
2D. Provide adaptability to the user's pace.
Principle 3. Simple and Intuitive Use
Use of the signage design is easy to understand, regardless of the user's experience, knowledge, language skills, or education level.
3A. Eliminate unnecessary complexity.
3B. Be consistent with user expectations and intuition.
3C. Accommodate a wide range of literacy and language skills.
Principle 4. Perceptible Information
The signage design communicates necessary information effectively to the user, regardless of ambient conditions or the user's visual abilities.

4B.	Provide adequate contrast between information in signage and its surroundings.
4C.	Maximize "legibility" of essential information.
4D.	Differentiate elements in ways that can be described (i.e., make it easy to give instructions or directions).
4E.	Provide compatibility with a variety of techniques or devices used by people with visual limitations.
5A.	Arrange elements to minimize hazards and errors: most used elements, most accessible; hazardous elements eliminated, isolated, or shielded.
5D.	Discourage unconscious action in signage that require vigilance.

4   CONCLUSION AND FUTURE WORK

This study investigated available or newly designed principles for signage. During this process, studies related to signage and HFE were compiled to identify the influencing factors and then applied to relevant items listed in the Principles of Universal Design. Research has shown that signage design must explore the relationship between people and the environment to satisfy users' wayfinding requirements and the principle of commonality. Results of the preliminary test will be incorporated into future universal design performance measures and, using the Delphi method, incorporated into a questionnaire survey. Questionnaires designed based on the refined measures will be pretested before undergoing the Delphi evaluation process. After conducting approximately three or four surveys, we will select the appropriate principles from unanimous survey results. The final universal design performance measures should be the result of expert consensus. Subsequently, the universal design performance measures for signage will be examined using case studies at a medical institution in Taiwan to verify the feasibility of this assessment method. We hope that the results of this study can provide people revising existing or new signage designs with an example of signage design that satisfies a greater number of people and improves the signage and ease of wayfinding in hospital environments.

REFERENCES

Arthur, P. and R. Passini. 1992. *Wayfinding: people, signs, and architecture.* NY: McGrawHill, Inc.

Berger, C. (2009). *Wayfinding: designing and implementing graphic navigational systems.* SA: RotoVison.

Bureau of Employment and Vocational Training, Accessed January 3, 2012, http://www.evta.gov.tw/content/list.asp?mfunc_id=14&func_id=57

Calori, C. 2007. *Signage and wayfinding design: a complete guide to creating environmental graphic design systems*. NJ: John Wiley and Sons.

Center for Universal Design. 2010. "The Principles of Universal Design." Accessed November 10, 2011, http://www.ncsu.edu/project/design-projects/udi/center-for-universal-design/the-principles-of-universal-design/

Cowgill, J., J. Bolek, and JRC Design. 2003. Symbol usage in health care settings for people with limited English proficiency: Part one: Evaluation of use of symbol graphics in medical settings (Research Reports Part One). Arizona: Hablamos Juntos.

Duncan, R. 2007. Universal Design-Clarification and Development. Center for Universal Design, College of Design, North Carolina State University.

Gibson, D. 2009. *The wayfinding handbook: information design for public places*. NY: Princeton Architectural Press.

Hölscher, C., T. Meilinger, and G. Vrachliotis, et al,. 2005. Finding the way inside: Linking architectural design analysis and cognitive processes. In. *Spatial cognition IV Reasoning, action, interaction*, eds. J. G. Carbonell, and J. Siekmann. International Conference Spatial Cognition 2004. Frauenchiemsee, Germany.

Johnson, C. R. 1993. "SIGNAGE AND THE ADA." Accessed March 30, 2011, http://www.west.asu.edu/johnso/signage/lama93.html

Lin, R. 1998. Cultural Differences in the Preference of Public Information Symbols. *Mingchi Instituute of Technology*, Vol. 30. pp. 57-68.

Lynch, K. 1960. *The image of city*. MA: MIT Press.

McCormick, E. J. and M. S. Sanders. 1987. *Human factors in engineering and design* (6th ed.). NY: McGraw-Hill.

"Ministry of the Interior," 2012. Accessed January 20, 2012, http://www.moi.gov.tw/stat/english/monthly.asp

Passini, R. 1996. Wayfinding design: logic, application and some thoughts on universality. *Design Studies*, Vol. 17. 3. pp. 319-331.

Preiser, F. E. 2008. Universal Design: from Policy to Assessment Research and Practice. *International Journal of Architectural Research*, Vol. 2. 2. pp. 78-93.

Rooke, C. N., P. Tzortzopoulos, and L. J. Koskela, et al. 2009. "Wayfinding: embedding knowledge in hospital environments. *Improving Healthcare Infrastructures Through Innovation*. In J. Barlow, A. Wheelock, and B. Trimcev (Chair), Improving Healthcare Infrastructures Through Innovation. 2nd Annual Conference of the Health and Care Infrastructure Research and Innovation Centre (HaCIRI). Imperial College Business School, UK.

Rousek, J. B. and M. S. Hallbeck. 2011. Improving and analyzing signage within a healthcare setting. *Applied Ergonomics*, Vol. 42. pp. 771-784.

Society for Environmental Graphic Design. 1993. "The Americans with Disabilities Act White Paper." Cambridge, MA: the Society for Environmental Graphic Design.

Story, M. 1998. "The Universal Design File: Designing for People of All Ages and Abilities." Accessed December 15, 2011, http://design.ncsu.edu/openjournal/index.php/redlab/issue/current

The Center for Universal Design. 1997. "The Principles of Universal Design." Accessed December 18, 2011, http://www.ncsu.edu/project/design-projects/udi/center-for-universal-design/the-principles-of-universal-design/

The Center for Universal Design. 2003. "A Guide to Evaluating the Universal Design Performance of Products." Raleigh, NC: North Carolina State University.

Tzeng, S.-Y. 2003. From barrier-free design to universal design-comparisons the concept transition and development process of barrier-free design between America and Japan. *Journal of Design*, Vol. 8. 2. pp. 57-76.

Ulrich, R., X. Quan, and C. Zimring, et al. 2004. "The Role of the Physical Environment in the Hospital of the 21st Century: A Once-in-a-Lifetime Opportunity." Accessed June 28, 2011, http://www.healthdesign.org/chd/research/role-physical-environment-hospital-21st-century

CHAPTER 33

Approaches to the Intuitive Use in Emergency Situations

Maria Lucia L. R. Okimoto, Caio Marcio Almeida e Silva
and Cristiana Miranda
Federal University of Paraná
Curitiba, Brazil
lucia.demec@ufpr.br

ABSTRACT

The problems of misuse of a product becomes extremely serious when people are in an emergency situation, where the time for decision making is extremely short, in which individuals tend to act intuitively. The severity of effects in these cases can lead to serious consequences to human life. In case of emergency and panic you follow your instincts and intuition to try to solve the problem it faces. Are still few systematic studies on the association of intuition and objective in the interaction between people and products, especially in the context of systems critical use.

So this article attempts to discuss the application of guidelines intuitive elements in the Design of products and hence also contribute to the increased awareness of accident reduction, in making critical decisions in emergency scenarios. It aims to investigate perceptual aspects of "intuitive" that affect the experience with products in critical condition.

Keywords: intuitive use, emergency situations, usability

1   INTRODUCTION

According to Tullis and Albert (2008), Usability can mean the difference between life and death. These authors point out that even the health sector is not immune to bad usability of products, which may be affected, for example, medical devices, diagnostic procedures and equipment, etc.

The increasing of new technologies in the home environment contributes to also an increase in the consumer products without prior usability studies. Just think of the written instructions for task actions such as lighting the pilot flame of a heater or verify the operation of an electrical device without risk of life. Accidents can occur for a period of the individual or the lack of an anti-product failure. We can cite as an example some situations such as leaving access to the stove to gas (assuming that this is the energy used in most homes in Brazil), or forget without turning off an electric iron, etc.

The National Institute of Metrology of the Ministry of Development, Industry and Foreign Trade of Brazil, INMETRO (2011) presents a classification of consumer accidents reported by consumers. According to the INMETRO (2011) product categories accounted for the largest number of accidents were the consumption of children's products, representing 13.2% of the total and similar appliances and adding 10.8%. INMETRO says you can not estimate how many people die from consuming products in Brazil.

The scenarios for using products that cause accidents can be very diverse. They can also occur under circumstances where the use of the product was not tested. In the evaluations of usability, as assumed in the NBR 4291-11, the context of use is planned according to the tasks to be performed, usually occurring in laboratory controlled environment.

This study emphasizes the need to also provide the usability testing elements that allow us to evaluate the use in an emergency situation! Understanding to this article as an emergency situation, as that situation that occurred differently, not contained in the traditional usability testing, where the subjects are under severe stress. Such as: heat, stress, natural disaster, lost in a place or road, just received a bad news, he suffered an assault, had an accident, etc. As an example of this diversity occurs when sunlight falls on the cell phone screen, impairing the conditions of usability for the same individual who uses it in an environment without direct incidence of sunlight!

Thus, this study considered an emergency situation in the use of consumer products such as the use of conditions in which individuals are under high stress resulting from caring for an immediate need. So this article discusses conceptual aspects of "intuitive use" may be involved in the context of use of emergency in order to elucidate the possible elements for assessing these conditions.

2   WHAT DOES INTUITIVE USE MEAN?

Initially, to answer this question we need to know some theoretical concepts about aspects of intuition in the context of ergonomics, usability and design in their relationship and interaction human / object. Emphasizing that the understanding of this subject is very vast and it is presented by other sciences and disciplines in a very particular issues. For the discussion in this study we seek references for the application intuitive to use in the area of Design.

Naumann (2008), defines the role of the "intuitive use" suggesting the use of this term in the context of the task, the user in relation to the environment, or in a technical system. The author recommends that the term "intuitive use" should only be attributed to human-machine interaction in a given context, to achieve the objectives.

According Hurtienne and Blessing (2007), the intuitive use is understood as the subconscious application of prior knowledge by the user. He presented a definition of intuitive use as "A technical system is intuitively usable if the user's subconscious application of prior knowledge leads to effective interaction. For these author the knowledge comes from different sources (since innate until continuum knowledge), it comes from sensorimotor, culture to professional areas of expertise (specialists).

In the definition of a concept to intuition, Alho Filho (1997) provides an overview on the literature and exposes the philosophical emphasis that guided the study of aspects of human intuition.

According to the author, the term intuition would have at least three different meanings, the sensory, interior and rational. Intuition is linked directly to sensory perceptual level of the individual and according to this author "is used as a direct and immediate vision of the reality or an understanding of a truth." We believe that the intuitive aspects of use can classify it as an intuition induced by the environment, a primary level. In this case, the use of the product as a component derived from the sensory effect of immediate care for a cause. The second type of intuition according to Alho Filho (1997) is called the interior, characterized as a kind of basic knowledge that is identified from any evidence or a conviction. In this case, there is no need to have occurred prior experience with the subject. We conclude that there may be a assimilation of features present in the situation, contributing to the perception of product use.

The third type of intuition presented is rational. According to Alho Filho (1997) this intuition is a transcendent action. There is a conscious learning of things and events, working in reality. In this case there is the use of logical reasoning and deductive reasoning, creating new strategies from previous inferences.

Figure 1 shows us the relation of elements of the perception of insight into the intuitive use of products. We also believe that the resources of perception may occur simultaneously in intuitive use.

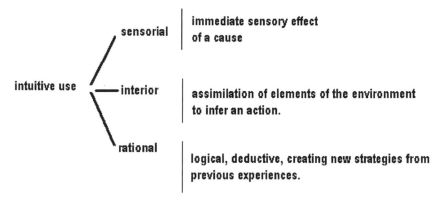

Figure 1- Elements of intuition for intuitive use (Alho Filho, 1997).

The focus of rational intuition is one of the arguments most commonly used in the literature on interfaces for interaction with the product. Blacker, Popovic e Mahar, (2003) support the argument that intuition is a cognitive process that uses the knowledge gained from previous experiences. Within this logic, intuitive use of products involves the application of knowledge gained through other products or experiences. According to the rationalist argument, people use the new products intuitively when these products have some characteristic that they already know. These authors consider the intuition part of a cognitive processing that is often unconscious and utilizes stored experiential knowledge. The authors also highlight the fact that to be evaluated aspects of intuitive use of a new product, this should be used without instruction.

Sternberg (2006) presents two approaches regarding the perception of object and shape. The first, user-centered versus subject-centered (Gestalt).The author believes that user-centered representation in the individuals acquired their shape from the object as it is perceived. The person finds the appearance of the object associated with the meaning that it transmits. In the object-centered representation the observer produces a representation of the object independent of the array processed.

The different approaches (centered on the observer or object) also refer to the investigation of the critical elements of contexts of use. Situations in which no specific use was planned, we call "non-conformities of use." These situations can cause accidents and or incidents arising from the alteration of the normal product use. In these cases we consider that there may be differences in the design and understanding of the intuitiveness of use, as the level of experience with the product.

We believe that many products are often difficult to use correctly and are often misused for a variety of reasons. This could be minimized by manufacturing products that allow interfaces from the user's intuitive strategies.

The intuitiveness is also one of the elements related to principles of Universal Design. The seven (7) Principles of Universal Design are: Principle 1, equitable use; Principle2, flexibility in use; Principle 3, simple and intuitive use, Principle 4, perceptible information; Principle 5, tolerance for error, Principle 6, decreased energy expenditure, Principle7, the appropriate space. We discuss in this article only the Design Universal principles that relate directly to aspects of intuitive use, being Principle 3, "simple and intuitive use" and Principle 4, "perceptual information" described below:

Principle 3: Simple & Intuitive Use. Use of the design is easy to understand, regardless of the user's experience, knowledge, language skills, or current concentration level

Principle 4: Perceptible Information -The design communicates necessary information effectively to the user, regardless of ambient conditions or the user's sensory abilities

We found that the intuitiveness is an established element of the principles of Universal Design. But how can we evaluate it within the concepts of usability?

Table 2 Inference of intuitive use elements' (Alho Filho, 1997) on the UD:PEC "Universal Design: Product Evaluation Countdown" and on oriented perception theory

UD/Principle	Description (UD:PEC)	Elements of the intuitive use	Oriented perception
3. Simple and Intuitive Use	3A.This product is as simple and straight forward as it can be	Sensory/ Interior/ Rational	Object/ Observer
	3B. Thos product works Just like I expected it to work	Rational	Observer
	3C.I understand the language used in this product	Interior/ Rational	Observer
	3D. The most important features of this product are the most obvious	Sensory	Object
	3E. This product lets me know that I'm using it the right way	Rational	Object
2. Flexibility in Use	2A. I can use this product in whatever way is effective for me	Sensory	Observer
	2B. I can use this product with either my right or left side (hand or foot) alone	Rational	Observer
	2C.I can use this product precisely and accurately	Rational	Observer
	2D. I can use this product at whatever pace I want (quickly or slowly)	Rational	Observer

3 SCENARIOS AND COMPONENTS IN THE ASSESSMENT OF INTUITIVE USE IN EMERGENCY SITUATIONS

The use's environment of a product in an emergency situation may be considered as a non-traditional according to the concept presented by Bernhaupt (2008).The author has defined a non-traditional environments how: "stressed, not being well, critical situations, relax, fun and operations situations". And she shows as the needs the evaluating usability in non-traditional environments and presents some considerations about of this issue. The first consideration refer to choose the right set of evaluation methods based on the context the final product will be used, the second point is done how the evaluation should take place in real usage context with real users, but this is not realistic for several domains (security, air traffic, cockpits, satellite control rooms, homes, cars, public places,etc). The third consideration must be done how the methods will be modified and developed for to allow the evaluation in non-traditional environments or how can be adapt the methods to reflect the real usage environment.

Bennett (2006) comments that some researches have noted that traditional laboratory usability studies are inappropriate for mobile applications because so many aspects of the environments in which such applications will be used are impossible to create in a laboratory setting. The author discuss how they can created a framework for conducting usability studies, design metrics and measures for these studies, deploying applications and monitoring applications in non-traditional environments. In this article, Bennett et al. (2006) reports some usability studies, and three cases were discussed. The first case presented was from researches at Indiana University with Dietary Intake Monitoring, the second by Georgia Tech Mobile devices for emergency and the third was by Researches at Carleton University about novel technologies. Some information about the test environment was presented. We can point some characteristics with these studies: the user study was small, stressful, amount of equipment was limited in the first case. In the second case the conditions were: high stress, cramped and often involve unusual equipment. And the third case was involved in defining performance criteria for novel technologies.

When there is an emergency situation, how can we test the elements of intuitiveness of the product or system?

In reviewing the literature we find an interesting study with application of usability evaluation methods for intuitive use by Hurtienne and Blessing (2007). They present us some considerations on these issues.The authors report a study correlating the Usability Measures (ISO-9241-11) and some Indicators and your relevance for Design for Intuitive Use. They concluded that "Effectiveness" is inherent in the definition of intuitive use, so these indicators will apply. In the efficiency measures they suggest that only mental workload is interesting for the field of intuitive use. They considered that "subconscious application of prior knowledge will only affect mental efficiency since it removes workload from the conscious cognitive processor. They point out that cognitive workload can be measured either by objective or by subjective means. They recommend for objectives measures (heart rate variability and under special circumstances (response times), and for subjective measures is recommended questionnaire on perception of difficult of the task. They suggest also a subjective Mental Effort Questionnaire (SMEQ) and the NASA Task Load Index (TLX). In addition they suggest that the "Satisfaction" must be measured when validating candidate theories for intuitive interaction. A questionnaire for measuring satisfaction in the context of intuitive use named Evalint by Mohs et al. The information presented by Hurtienne and Blessing (2007) is of extreme relevance for the application in the intuitive use in the emergencies.

Pozza et al (2012) present a study of usability of a touch screen mobile phone in an emergency context, the test aimed to 'make a connection. The exploratory study presented aimed at simulating emergency context for the collection of elements of intuitive use. The study was conducted with seven people. The researchers considered in assessing the emotional and behavioral aspects of the user under simulated high-stress environment, and interface characteristics of the cell phone to perform the task. The individuals were submitted during the test at a pressure environment with the addition of siren, time marker ahead to the venue of the task (in order to increase the level of stress). But to quantify the stress level of the users were obtained from different levels of heart rate before and after the tests, as shown in Table 1 below.

Table 2 Results of heart rates from initial, during and after the test

Subject	Heart rates before	Heart rates during	Heart rates after
1	85	89	76
2	86	112	97
3	93	98	95
4	95	112	99
5	85	101	94
6	94	89	90
7	85	105	103

Source: Pozza et al (2012)

Okimoto (2011) recommended for the evaluation of products with a focus on safety, usability and intuitiveness of the user context for the following emergency: the numbering of the task, the proposition of the task, the aspect(s) for intuitiveness in use that were considered in the task, evidence in the product, an indication of the success of the task, specified sensory stimulus (stimuli) for the task, the sensory stimulus (stimuli) used for the task, the success of the aspects considered for intuitive use, the success of sensory stimuli, and observations how the task was done.

4   CONCLUSIONS

The first conclusion we can draw from this study refers to the concepts generated by the authors Blacker, Popovic e Mahar, (2003) e Alho Filho (1997) e Hurtienne and Blessing (2007) on intuitive use. We realize that the concept of intuitive use is very broad and requires the approach of various evaluation methods for the identification of perceptual elements: sensory, interior and rational as described by Alho and Filho (1997), as shown in the Table 1. The table 2 shows us the synthesis of the elements discussed in this article, and a confrontation of the elements of assessment proposed in the document "Universal Design: Product Evaluation Countdown, UD:PEC"(2002), principle 3,"Simple and Intuitive Use" and principle 2 ,"Flexibility in use".

These principles were based on the concepts presented by Alho Filho (1997) refer to the classification of the elements responsible for intuitive use, and the nature of the source of the element for intuitive use, "object or observer". In principle 3 "Simple and intuitive," presented by UD: SGP we identified more strongly the application of rational elements for intuitive use, according to our judgment.

Regarding the nature of the element for intuitive use (for perception) we believe that these are distributed among all present criteria. This criterion was for us to easier identification and judgment to an emergency context.

This study allowed us to visualize and clarify the various issues that relate to the introduction of the intuitive aspects in the design phase in which it must be focused on the "perception-oriented or induced" by Design. We conclude that the topics "Usability", "intuitive use" and "Universal Design", are complementary and inter relate to each other. However, the instruments used to check each of the criteria see table 2, are not oriented to the same elements perception. We conclude that to evaluate the intuitive use for emergency context there is a need for methods suitable for the construction of context of use to meet the perceptual larger number of elements. (as visualized in exploratory studies of Bernhaupt (2008), Bennett (2006), Hurtienne and Blessing (2007), Pozza et all (2012) and Okimoto (2011).

This study presents our contribution so much questioning the possibility of inclusion of guidelines for the certification of products for the emergency context and for the product design so that it can be used in an emergency setting.

The topic may bring new insights for further exploratory studies of "User Experience", UX, in the context of an emergency. This issue points out issues for future research such as the introduction of non-traditional scenarios (covering unusual situations of use) in order to anticipate possible risks and physical and mental damage to individuals when used in high-stress situation, outside the standards normal use.

REFERENCES

Alho Filho (1997) ALHO FILHO, Joaquim Lopes. O delírio: Transtorno da intuição. São Paulo: Robe Editorial, 1997.

Blackler, A. L., Popovic, V. & Mahar, D. P. (2003) The nature of intuitive use of products : an experimental approach. *Design Studies*, *24*(6), pp. 491-506.

Bennett,G; Lindgaard,G.; Tsuji,B.; Connelly,K.H. and Siek,K.A. Reality Testing: HCI-challenges in non-traditional environments. Proceeding CHI EA '06 CHI '06 extended abstracts on Human factors in computing systems. ACM New York, NY, USA.2006.

Bernhaupt, R. IHC2008 Final Keynote1147, Porto Alegre, Brasil, 24. October, 2008.

Hurtienne, J., Blessing, L. (2007) Design for Intuitive Use - Testing Image Schema Theory for User Interface Design. Proceedings of the 16th International Conference on Engineering Design (ICED07).

INMETRO (2011), *Acidentes de consumo:Uma Visão Técnica sobre o Tema. Acesso outubro de 2011<http:// www.inmetro.gov.br/consumidor /pdf/ acidentes-consumo-uma-visao-tecnica.pdf>.*

Naumann (2008), Naumann, A., Pohlmeyer, A.E., Hußlein, S., Kindsmüller, M.C., Mohs, C., Israel, J.H.: Design for intuitive use: beyond usability. In CHI Extended Abstracts(2008).

Okimoto,M.L.L.R, unpublished Project, UFPR. Diretrizes para a certificação de dispositivos de controles de produtos com enfoque na segurança, usabilidade e intuitividade do usuário para contexto de emergência. Curitiba, Brazil (2011).

Pozza, F.; *Roncalio,V. Miranda,C.; Okimoto,M.L.* Usability evaluation of touchscreen phone in emergency context, unpublished article, will be presented in AHFE 2012. San Francisco ,USA.2012.

Sternberg, R.J. (2006). The Nature of Creativity. Creativity Research Journal .Volume 18, Issue 1, Taylor & Francis.

Tulis,T. and Albert,B. (2008) Measuring the User Experience: Collecting, Analyzing, and Presenting Usability Metrics - Morgan Kaufmann Publishers. 2008.

Universal Design: Product Evaluation Countdown , UDPEC (2002).Center for Universal Design, , http://www.ncsu.edu/ncsu/design/cud/pubs_p/docs/UDPEC.pdf

CHAPTER 34

Proposal of Human-centered Design Process Support Environment for System Design and Development

Yukiko Tanikawa, Ryosuke Okubo, Shin'ichi Fukuzumi

NEC Corporation
Tokyo, Japan
y-tanikawa@cw.jp.nec.com

ABSTRACT

This paper proposes the support environment to improve the usability of the systems efficiently in the development process. Human-centred design for interactive systems (ISO9241-210) is applied to the system development process and necessary activities from a system planning and proposal phase to a design and development phase are clarified. These help system engineers without expertise improve usability of systems. For upper process of system development (from a system planning and proposal phase to a requirement definition phase), we developed the method for clarifying and describing customer needs concerning usability. This method was used for evaluation that describes customer needs about usability. Moreover the customer needs described by the method were compared with those described by a usability expert, so that the both needs were confirmed to be almost the same. From this verification, validity and usefulness of the method were verified.

Keywords: Human-centered design, system development process, usability, customer needs concerning usability, human interface requirement

1   INTRODUCTION

In recent years, usability is an important factor influencing a customer's decision of choosing a system, as well as functions and a price. In case of business systems used especially in companies, government offices and local governments, customers more often place top priority on usability.   Due to this, redoing a system development which is caused by recognition difference between a customer and a system vendor about usability of a system has often come out.

Necessity for usability is increasing as mentioned above. On the other hand, improving usability is not easy for system engineers. We interviewed some system engineers and extracted three major issues why they feel difficulties in improving usability. (1)Engineers don't have knowledge about what they should do to improve usability. (2)There is no method to define customer's needs about usability of the system, so engineers can not set goals for the usability. (3)Because of (2), engineers can not estimate the cost for the usability of systems and thus they have risk of profit deterioration if they try to improve usability.

As one of methods for improving usability, it is effective that usability experts participate in a system development project (Goransson, Gulliksen and Boivie, 2003) (Paech and Kohler, 2003) (Memmel, Gundelsweiler and Reiterer, 2007). However, the numbers of usability experts are limited, and small projects can not afford to commission experts. In order to solve these problems, we have been developing documents and tools based on experience and know-how of usability experts, and have been providing these to system engineers so that the engineers without expertise can work on usability improvement (Hiramatsu and Fukuzumi, 2008) (Suzuki, Bellotti, Yee, et al. 2011) (Fukuzumi, Ikegami and Okada, 2009) (Fukuzumi, 2005). Though these documents and tools are useful to develop a system with high usability, engineers without expertise have a heavy workload to master the documents and tools and to use them depending on a system development process and a project situation.

Now we are working on development of process support environment to improve usability of a system efficiently in a system development process used in each project. This process support environment has two main features as follows. (Feature 1) Human-centred design for interactive systems (ISO9241-210, 2010) is applied to a system development process, and necessary tasks are clarified from a system planning and proposal phase to a design and development phase. (Feature 2) The method to clarify and describe customer's needs about usability is devised and applied to a system development process. These features are intended to help system engineers without expertise improve usability of systems.

This paper proposes human-centered design process support environment for a system design and development. For upper process of system development (from a system planning and proposal phase to a requirement definition phase), we developed the method for clarifying and describing customer needs concerning usability. We outline this method, and present the result of verification experiment about its validity and usefulness.

2   HUMAN-CENTERED DESIGN PROCESS SUPPORT ENVIRONMENT

2.1   Problems in design

Human-centered design process support environment aims to help system engineers without expertise efficiently improve usability of systems, which are business systems used especially in companies, government offices and local government.   In order to achieve this aim, it is necessary to support consensus building with customers as well as to support thinking and activity of system engineer in a system development process. We think the following four problems should specifically be solved.

[Problem 1] Specify design activities and procedures necessary to develop systems with usability, and define those application scenes (requirement definition phase and testing phase, etc.).

[Problem 2] Clarify and describe customer needs concerning usability, and specify human interface requirements (HI requirements) based on the needs.

[Problem 3] Agree on specified HI requirements including cost with customers.

[Problem 4] Adapt system design definitely to HI requirements specified.

2.2   Overview of human-centered design process support environment

Whole picture of human-centered design process support environment is presented in Figure.1.

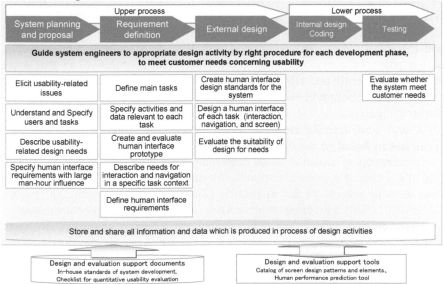

Figure 1 Human-centered design process support environment

As depicted in Figure.1, this process support environment specifies design activities necessary to develop systems with usability for each development phase. The design activities and the procedure are standardized so that system engineers without expertise are able to work on them. Moreover this environment accumulates to share all information and data generated during the design activities, as presented in Figure.1.

We focus on description of customer needs concerning usability, specification of HI requirements, and confirmation on adaptability of the requirements and system design. This process support environment provides system engineers with guide about appropriate activities in the scenes above. And it also support selecting appropriate documents or tools based on experience and know-how of usability experts for the activities.

2.3 Design guidelines of human- centered process support environment

The solution policy to four problems given by 2.1 is described as design guidelines of this process support environment as follows.

(1) Specify design activities and procedures necessary to develop systems with usability, and define those application scenes

To improve usability of systems, system engineers without expertise need to know answers to the following three questions. (a) What kind of activities should they do? And what should they aim through the activities? (b) How should they do activities? (c) In which development phase should they do activities?

In this process support environment, we specify necessary design activities and the target of the activities to improve usability. And we materialize the work item and work procedure in the design activity based on human-centred design for interactive systems (ISO9241-210, 2010) (Lutsch. 2011). These are solution to (a) and (b) above. Moreover each design activities are associated with appropriate design phase in consideration of every single phase objective so that system engineers efficiently and effectively work on design activities. It is solution to (c) above. The association between a design activity and a development process is as shown in Figure 1.

(2) Clarify and describe customer needs concerning usability, and specify HI requirements based on the needs

Specification of customer needs concerning usability means identifying usability goal of a system. It is essential to develop a system with usability. Besides it should be done in upper process as fast as possible (Zimmermann, Grotzbach. 2007) (Paech, Kohler. 2003). This is necessary to prevent redoing a system development repeatedly.

Upper process includes not only requirement definition phase but also system planning and proposal phase. Business systems are generally developed to solve some business issues, that are presented in system planning and proposal phase by customers. And the solutions to the issues often need usability of systems (Zimmermann, Grotzbach. 2007). Thus this process support environment provides assists from system planning and proposal phase.

We devise following two methods to clarify and describe customer needs concerning usability and to specify HI requirements. The methods are based on human-centred design for interactive systems (ISO9241-210, 2010) and also based on experience and know-how of usability experts.

[method A] Derive usability-related design needs of customer from user profile and their task characteristics, and associate needs with HI requirements.

[method B] Using operable sample or prototyping, embody customer needs for look and feel, operability and usage scenario that are hard to communicate by words, and associate needs with HI requirements. We integrate these methods into design activities and procedure, in the process support environment we propose.

(3) Agree on specified HI requirements including cost with customers

Customer needs concerning usability and HI requirements specified should be agreed with customers and shared (Metzker and Offergeld. 2001).

Needs, requirements for the needs, and development cost for the requirements are agreed thoroughly in process of consensus building with customers. If this process failed, both a customer and a vendor are not able to get benefit. For example, if requirements that exceed the budget frame are agreed, a vendor has a risk of profit deterioration, and a customer doesn't get expected results. Then this process support environment provides man-hour information for each HI requirement. Man-hour information has two types. One is numeric man-hour based on know-how of usability experts. The other is consideration points concerning man-hour.

(4) Adapt system design definitely to HI requirements specified

Developing a system to meet HI requirements specified and agreed in upper process is very important. Information agreed in upper process should be shared and be taken over to system engineers in charge of lower development process (Metzker and Offergeld. 2001).

A system development project needs variety of roles. They are project manager, business process designer, system designer, programmer, sales, etc. Members in charge of lower process such as internal design usually differ from those in charge of upper process such as requirement definition. According to our project support experiences, we know that requirements themselves are succeed to lower process from upper process; however customer needs causing requirements and other background information are hardly succeeded. Besides it is obvious that designing and developing a system without requirement-related information is a critical cause of redoing a system development.

This process support environment provides base to share all information and data generated during the design activities among project members. In addition, this environment provides screen design templates and elements that meet HI requirements. Moreover it provides tools that can be used depending on a system development process and verification objectives, in order to verify adaptability of the requirements and system design (Hiramatsu and Fukuzumi, 2008) (Suzuki, Bellotti, Yee, et al. 2011) (Fukuzumi, Ikegami and Okada, 2009) (Fukuzumi, 2005).

3   SUPPORT METHOD TO DESCRIBE CUSTOMER NEEDS

As mentioned in Section 2, clarifying and describing customer needs concerning usability means specifying usability goal of a system. This activity is essential to develop a system with usability. And it should properly be done in upper process of system development. Describing customer needs properly in upper process is help to prevent redoing a system development repeatedly. Besides it increases new system development effect for a customer. And it leads to enlarge customer satisfaction including not only the development system but also the development process.

Upper process includes system planning and proposal phase and requirement definition phase. Each phase has own objective and feature. Design Activities to describe customer needs concerning usability should be matched to objectives and features of each phase, so that system engineers may do the activities by themselves. We set the kind, the range, and the details degree of needs as follows.

System planning and proposal phase:

Objectives of this phase are to specify business issues or business process issues and to determine a direction of solution for the issues, and also to devise a rough investment plan. With consideration of these objectives, usability-rerated objectives are set as follows.

- Elicit what a customer solves with usability, namely business issues and business process issues being related to usability.
- Describe what kind of usability a customer demands, namely usability-related system design needs, and goals.
- Agree with a customer about usability-related system design needs and rough design cost.

Requirement definition phase:

Objectives of this phase are to concretely describe customer needs in specific tasks, to define requirements for realizing the needs, and to estimate a system development cost in rough figures. With consideration of these objectives, usability-rerated objectives are set as follows.

- Describe how a user performs a specific task using a system, namely needs for interaction between a user and system, and navigation in a specific task.
- Define HI requirements for realizing the needs above.
- Agree with a customer about usability-related system design needs, HI requirements, and usability-related design cost.

In this process support environment, we devised methods to describe customer needs concerning usability according to each development phase. Features of these methods are outlined as follows.

3.1   Activities and procedure to specify customer needs

The procedure to derive usability-related design needs of customer from user profile and their task characteristics and to associate needs with HI requirements is set based on human-centred design for interactive systems (ISO9241-210, 2010). Whole activities and procedure to specify customer needs are depicted in figure.2.

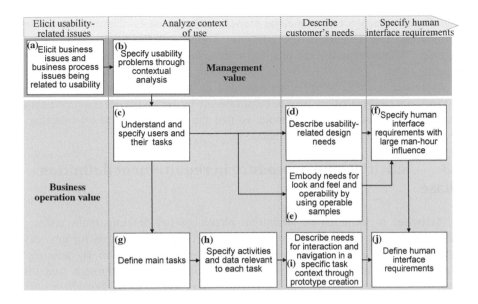

Figure.2 Activities to specify customer needs related to the usability

3.2 Activities and procedure in system planning and proposal phase

Activities in system planning and proposal phase are (a) Eliciting business issues and business process issues being related to usability, (b) Specifying usability problems through contextual analysis, (c) Understanding and specifying users and their tasks, (d) Describing usability-related design needs, (e) Embodying needs for look and feel and operability by using operable samples, and (f) Specifying HI requirements with large man-hour influence, as shown in figure.2.

(a) Eliciting customer business issues and business process issues being related to usability is the first activity. If a customer have some issues, (b) specifying usability problems through contextual analysis is necessary. For example, in a case a customer have awareness of the issues that operational mistakes often happen, it is necessary to specify where and how often and what kind of mistake happen, through field observation and its analysis. However, this phase is before the order so that context analysis is desirable to be with low load. To meet the demands, our method provides simple hearing items to check issues and to judge necessity of deep analysis in the next phase.

Next activity is (c) understanding and specifying users and their tasks. For this activity, our method offers hearing items so that system engineers without expertise do this activity as easy as possible. And, from this specification of user profile and their task characteristics, (d) usability-related design needs of a customer are derived. In addition, (f) HI requirements are specified. Requirements specified in this phase are the ones with large man-hour influence to devise a rough investment plan. For these activities, our method prepared rules about association between user profiles and task characteristics, customer needs, and HI requirements. System engineers without expertise are possible to describe needs and specify requirements by using them.

Moreover operable samples are provided to embody customer needs for look and feel, operability that are hard to communicate by verbal (e). HI requirements are associated with each operable sample, so that the needs embodied by selecting an operable sample derive HI requirements.

3.3 Activities and procedure in requirement definition phase

Activities in Requirement definition phase are (g) defining main tasks, (h) specify activities and data relevant to each task, (i) describing needs for interaction and navigation in a specific task context through prototype creation, (j) defining HI requirements, in addition to (b) specifying usability problems through contextual analysis, and (c) understanding and specifying users and their tasks.

All activities in this phase are done based on activities in previous phase. If deep contextual analysis is necessary, it should be done for (b) specifying usability problems. Then (g) defining main tasks, and next (h) specifying activities and data relevant to each task should be done. For (h) specifying activities and data, our method offers activity analysis sheets with items to check data characteristics, in order to help system engineer thinking.

Moreover (i) Describing needs for interaction and navigation in a specific task context through prototype creation is the following activity. In this activity, our method offers elements for prototyping such as screen layout patterns, transition patterns, and components. And it also offers means to support selecting elements suitable for activities and data specified beforehand. HI requirements are associated with each element, so that the needs specified by selecting elements and prototyping derive HI requirements. In this way, (j) HI requirements are defined.

4 PROCESS OF DESCRIBING CUSTOMER NEEDS APPLIED METHODS

We developed process support environment that built in the methods presented above for system planning and proposal phase. Process of describing customer needs by using this process support environment is outlined as followed.

First, system users and their tasks are specified by using analysis sheet on the screen, which is consist of items to check user profiles and task characteristics. After checking these items, process support environment derives design guidelines using rules to derive them as design needs based on checked items, as depicted in figure.3. Besides HI requirements is also derived using the same rules.

Figure.3 Screen image of HCD process support environment

At the same time, this process support environment selects and displays operable samples suitable to user profile specified as presented in figure.3. When one of operable sample is selected based on concrete customer demands, HI requirements associated with the operable sample are derived. This association between an operable sample and requirements is possible to confirm on the screen.Customer needs and HI requirements are examined and negotiated to agree with customers or among project members. The examination and negotiation are done while referring to man-hour information for each requirement. Requirement list with information deriving origin to each one is displayed or printed, so that priorities are set with customers and needs are refined as a whole.

These activities are done according to navigation by process support environment as shown in figure.3.

5   VERIFICATION AND CONSIDERATION

We verified this method. Objectives of this verification are (1) to verify validity of HI requirements derived, and (2) to evaluate usefulness of this method.

A usability expert tried activities from describing customer needs to specifying HI requirement at system planning and proposal phase in a practical project. He tried activities by two methods. One is doing activities by using the method we devised. The other is doing activities by a usability expert himself. After this trial, we compared the both results, that is, needs described and requirements specified. This was verification for objective (1) above. And we compared the both process. This was verification for objective (2) above.

About verification for objective (1), customer needs described by the method and those described by the usability expert were confirmed to be almost the same. From this verification, validity of the method was verified. About verification for objective (2), man-hour to these activities was reduced in trial using the method. Moreover the experimenter points out that quality of activities improved. From this verification, usefulness of this method was verified in this trial case.

6   CONCLUSIONS

This paper proposes human-centered design process support environment for a system design and development. For upper process of system development (from a system planning and proposal phase to a requirement definition phase), we developed the method for clarifying and describing customer needs concerning usability. This method was used for evaluation that describe customer needs about usability. Moreover the customer needs described by the method were compared with those described by a usability expert, so that the both needs were confirmed to be almost the same. From this verification, validity and usefulness of the method were verified.

REFERENCES

ISO 9241-210: Human-centred design for interactive systems.2010.

B Goransson, J Gulliksen, I Boivie.2003: The usability design process - integrating user-centered systems design in the software development process. *Software Process: Improvement and Practice Volume 8, Issue 2:* 111-131

B. Paech, K. Kohler. 2003. Usability Engineering integrated with Requirements Engineering. *ICSE'03 International Conference on Software Engineering:* 36-40

T Memmel, F Gundelsweiler, H Reiterer. 2007. Agile human-centered software engineering. *BCS-HCI '07 Proceedings of the 21st British HCI Group Annual Conference on People and Computers: HCI...but not as we know it - Volume 1:* 167-175

C Lutsch. 2011: ISO Usability Standards and Enterprise Software: A Management Perspective. *Proceedings of HCI International 2011:* 154-161

D Zimmermann, L Grotzbach. 2007. A Requirement Engineering Approach to User Centered Design. *HCI'07 Proceedings of the 12th international conference on Human-computer interaction: interaction design and usability:* 360-369

E Metzker and M Offergeld. 2001. An Interdisciplinary Approach for Successfully Integrating Human-Centered Design Methods into Development Processes Practiced by Industrial Software Development Organizations. *8th IFIP International Conference, EHCI 2001:* 19～33

T Hiramatsu, S Fukuzumi. 2008. Applying Human-Centered Design Process to SystemDirector Enterprise Development Methodology. *NEC Technical Journal Volume vol.3 no.2 (June, 2008):* 12-16

S Suzuki, V Bellotti, N Yee, et al. 2011: Variation in Importance of Time-on-Task with Familiarity with Mobile Phone Models. *CHI '11 Proceedings of the 2011 annual conference on Human factors in computing systems:* 2551-2554

S Fukuzumi, T Ikegami, H Okada. 2009. Development of Quantitative Usability Evaluation Method. *Proceedings of HCI International 2009:* 252-258

S Fukuzumi. 2005: Web Contents Accessibility Check Tool. *Proceedings of HCI International 2005*

CHAPTER 35

Ergonomic Study on the Information Capacity of Isolating Language without Kanji Characters

M. Ikeda, T.Konosu

Chiba Institute of Technology Graduate School of Social Systems Science
Chiba,Japan
s0742008PM@it-chiba.ac.jp

ABSTRACT

As for the study on text information processing, a large number of studies about agglutinative language and inflective language have been researched. The objective of this study is to clarify reading characteristics of Thai language (isolating language) by using ergonomic method. In this study, we measured the information capacity per one fixation by the eye movements of the participants reading Thai language sentences. As a result of calculating the information capacity based on syllable of Thai Language, it was suggested that information capacity per syllable was 5.9 bits in Thai language. Also, the information capacity per one fixation was 88 bits. Thai language has fewer occurrence rate of peripheral search guidance elements, compared to Japanese (isolating language with Kanji characters) or English (agglutinative language write with a space between words). Also, it can be suggested that the cognitive search guidance affects stronger than the peripheral search guidance, because of expert knowledge which is necessary for increases as it is difficult to sentence. Therefore, it can be suggested that fixation duration is relatively longer since the information capacity per syllable increase because Gaze length became relatively longer because cognitive processing takes time in understanding sentences.

Keywords: eye movements, isolating language, information capacity, search guidance

1   INTRODUCTION

1.1   Study of text information processing

Several experiments such as those on measurements of eye movements have been conducted in studies of the information capacity of humans. Other methods for studying information capacity include the stimulus reduction (in which a number of characters are presented briefly, and the characters that can be read have to be reproduced), instantaneous tachistoscopic presentation, and a method of determination in which the number of characters in a line are divided by the number of fixation points. In addition, the information capacity per one fixation can also be measured by the effective field of view. A method of measuring the effective field of view is given in a study that uses the moving window method. Osaka (1989) conducted a Japanese text-reading experiment on the basis of the effective field of view and found that the number of characters for each fixation point was 7 characters in the case of texts with hiragana and katakana and 13 characters in the case of texts with mixed kanji and kana characters. In previous research on information capacity, the languages studied were mainly inflective languages, represented by Western languages, and agglutinative languages, to which the Japanese language belongs. This study focuses on isolated non-kanji languages that have been relatively less studied and derives the information capacity in linguistic processing. The Thai language was selected for this study. Although the Chinese language would have been more appropriate considering its large number of speakers, the Thai language, which has a phonetic writing system, was considered as the most suitable, since there are already some studies that focus on the kanji system, which is an ideographic writing system. Since there is no text segmentation in the Thai language, it is considered that there is little scope for peripheral search guidance. Further, in text reading, the presence of phonological information is considered to be more important than in other languages. For this reason, there is a possibility that a cognitive process that is different from those found in other languages will be found.

1.2   Purpose

The objective of this study is to clarify the reading characteristics of the Thai language (isolating language) by using an ergonomic method. In this study, we measured the information capacity per fixation by examining the eye movements of the participants reading Thai language sentences. Thus, it became possible to consider the differences in cognitive processing of language format. In the Thai language, a consonant or a vowel does not independently represent a sound or a meaning, and sound expressions are made only in combinations. This language is not suitable for defining the information capacity per word, as Japanese or agglutinative languages are; therefore, the information capacity per syllable is calculated in this study.

2   EXPERIMENT

3.1   Overview

In the experiment, the eye tracker Free View (Takei Scientific Instrument) and a 50-inch monitor positioned with 120cm visual range, presented the sentences in 12pt font size letters. Sentences of different difficulty level were randomly presented and eye movement was measured and fixation, duration, and count were compared. The participants were 16 students from Thai-Nichi Institute of Technology.

3.2   Stimulus

Two sentences were constructed for each of the three difficulty levels while considering vocabulary level and the complexity of grammar. In total, six sentences were constructed: low difficulty A (758 characters) and B (530 characters) that were based on sentences from essays and textbooks; medium difficulty C (581 characters) and D (560 characters) at a general reading level; and high difficulty E (243 characters) and F (462 characters) that require expert knowledge for comprehension.

3   RESULTS

3.1   Information Capacity

Thai consists of the combination of consonants (48 letters), vowels (18 letters), and tone codes. A consonant or vowel does not independently represent a sound or meaning, and their sound expressions are made only in combination. Since defining the information volume per letter, as we do with Japanese or inflected languages, is not suitable for Thai, this study calculated information capacity per syllable instead. As a result of calculating the volume based on combinations of Thai, it was suggested that information capacity per syllable in Thai is 5.9 bits. Also, the information reception per gaze was 88 bits. Figure 1 shows the information capacity per one fixation at each difficulty level calculated based on information capacity. There was a significant difference at alpha = 0.01 in the information capacity per one fixation between the medium- difficulty sentences (96.0bits) and the high-difficulty sentences (46.3bits) in the information capacity per one fixation. In the comparative study of Japanese and German readings, Fukuda (2009) clarified that there is no difference between the two languages as to the information capacity per word (around 50 bits with both languages) while there is a difference in the amount of words per one fixation. Although this study's data was consistent with Fukuda's, the information capacity inclined to appear slightly larger.

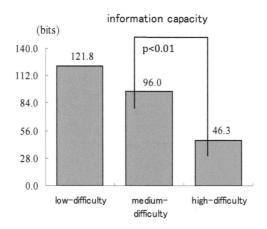

Figure 1 t-test result of average information capacity

3.2   Eye movement track

Figure 2 shows the average fixation frequency. There was a significant difference at alpha = 0.01 in average fixation frequency between the low-difficulty sentences (4.9times/100cha) and the medium-difficulty sentences (6.7times/100cha), and that there was a significant difference at alpha = 0.01 in average fixation frequency between the low-difficulty sentences (4.9times/100cha) and the high-difficulty sentences (13.5times/100cha) in the average fixation frequency. Figure 3 shows the average fixation duration. There was a significant difference at alpha = 0.01 in average fixation duration between the low-difficulty sentences (28.0msec/cha) and the medium-difficulty sentences (42.0msec/cha) in the average fixation duration. Kounosu (1998) demonstrated that the fixation frequency, not the fixation duration, decreases as the difficulty level intensifies. This study shows longer fixation duration than the study by Kounosu. Compared to Japanese (an agglutinative language of Chinese characters system) or English (inflected language with spacing between words), Thai has fewer occurrence rate of morphological search guiding elements. Also, it can be suggested that the cognitive search guidance affects stronger than the peripheral search guidance, because of expert knowledge which is necessary for increases as it is difficult to sentence. Therefore, it can be suggested that fixation duration is a relatively longer since the information capacity per syllable increase because fixation duration became relatively longer because cognitive processing takes time in understanding sentences.

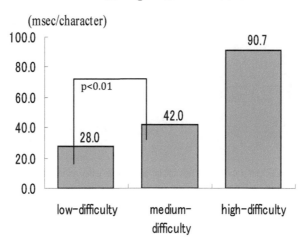

average fixation frequency
(times/character)

p<0.01

15.0 ⌐
12.0
9.0
6.0
3.0
0.0

4.9   6.7   13.7

low–difficulty   medium–   high–difficulty
                  difficulty

Figure 2 t-test result of average fixation frequency

average fixation duration

(msec/character)

100.0 ⌐
80.0
60.0
40.0
20.0
0.0

p<0.01

28.0   42.0   90.7

low–difficulty   medium–   high–difficulty
                  difficulty

Figure 3 t-test result of average information capacity

3   CONCLUSION

Thai language has fewer occurrence rate of peripheral search guidance elements, compared to Japanese (isolating language with Kanji characters) or English (agglutinative language write with a space between words). Also, it can be suggested that the cognitive search guidance affects stronger than the peripheral search guidance, because of expert knowledge which is necessary for increases as it is difficult to sentence. Therefore, it can be suggested that fixation duration is relatively longer since the information capacity per syllable increase because Gaze length became relatively longer because cognitive processing takes time in understanding sentences. Next, Figure 4 shows eye movements trajectory. It was found that participant's gaze fixated mostly at the beginning of the sentence for Japanese comprehension while their gaze fixated mostly at the end of the sentence for Thai comprehension. The reason for this difference in fixation patterns may be caused by the "do sakot." Thai syllable structure can be consonant + vowel or consonant + vowel + consonant (final consonant), and this final consonant is the do sakot. The frequency in appearance of do sakot may affect fixation during Thai comprehension, which suggests the necessity of further investigation of do sakot.

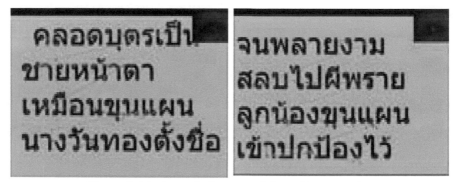

Figure 4   t-test result of average information capacity

REFERENCES

Osaka, N : Eye fixation and saccade during kana and kanji text reading, Comparison of English and Japanese text processing, Bulletin of the Psychonomic Society, No.27, pp.548-550,1989. .

Fukuda, R : Comparison of human visual information capacity in reading Japanese, German, and English, Perceptual and Motor Skills,No.108, pp.281-296,2009.

Konosu,T. Jimbo,T. Shigematsu,J. and Fukuda,T: A ergonomic study of Japanese graphical and phonological reading process. Japanese Ergonomics Society, Vol.32, No.4, 1998.

CHAPTER 36

The Fame is Afoot: Q&A Systems and the Future of HCI

Simeon Keates[1], Philip Varker[2]

[1]University of Abertay Dundee
Dundee, Scotland, UK
s.keates@abertay.ac.uk

[2]IBM T.J. Watson Research Center
Human Ability and Accessibility Center
Yorktown Heights, NY, USA
varker@us.ibm.com

ABSTRACT

In February 2011, the IBM Watson Q&A (Question and Answer) system took part in a special challenge, pitting its question and answer capability against former Jeopardy!^{TM} grand champions in a televised match. Watson emerged victorious from the challenge, demonstrating that current question answering technology has advanced to the point where it can arguably be more dependable than human experts. This new system represents a significant breakthrough in humanity's decades-long endeavour to build computers that are "more like us" – fundamentally designed to map to and support human abilities, and where the needs of people are central to the design process. The emerging Q&A systems, such as IBM's Watson system, have the potential to address longstanding challenges in the fields of cognitive and decision-making support related to flexibility, precision and accuracy. This paper foresees a future in which ever more powerful question answering systems are used ever more seamlessly to enhance human cognition in all aspects of our lives and work. We look forward to a day when technology truly adapts to people, supporting all of us in reaching our greatest human potential in creativity, productivity and ingenuity.

Keywords: HCI, Q&A systems, user-centred design

1   INTRODUCTION

Fundamentally, computers were developed as data processing machines and the ultimate purpose of data processing is to provide answers to questions. From the earliest days of the computer, it has long been envisaged that computers should be able to provide question and answer capability. A new class of Q&A computing systems, exemplified by the IBM Watson system, is poised to create a radical breakthrough in this area (Dillenberger *et al.*, 2011).

In the earliest examples of Q&A systems, the questions to be answered had to be rigidly structured according to the syntax of the stored data or knowledge and the syntax "understood" by (i.e. programmed into) the computer. Since the questions to be answered had to be structured according to the syntax of the stored data or knowledge and the syntax understood by the computer, computers have historically been very good at providing very precise answers, but not necessarily accurate ones. Watson is arguably the first of a new generation of systems that attempt to move beyond such limitations. Such systems offer radically improved question and answer ability through the application of deep analytics to unstructured "natural language" questions, i.e. questions not specifically formatted to follow a computer's syntax or that are simply broken down into the component words (as in a typical search engine).

However, Watson was the result of many years of development and was powered by a specially constructed database built over 4 years by a dedicated team of 20 specialists. The medium-term development of Watson is to build specialist databases, such as health informatics, for high-value markets (Bigus *et al.*, 2011).

The long-term development challenge is to build ever more powerful Q&A systems more quickly, for general-purpose applications and with a smaller footprint. While these engineering advances are being made, it is also essential to also consider the human perspectives of such a system. In this paper we consider the fundamental questions of what might such systems be used for and how might they be accessed. It is only by considering such questions that the full potential of this exciting new technology will be realised (Keates, Varker, Spowart, 2011).

2   THE USER CONTEXT

For the purposes of this paper, it will be assumed that the engineering challenges of making a hypothetical future Q&A system (which, for the sake of convenience, shall be called QAFUTURE in this paper) that supports full natural language processing queries of all kinds of general and specialist knowledge datasets available for everyday use will be conquered. There are projected advances in speech recognition technology that are also very relevant to future systems of this type. (Picheny *et al.*, 2011).For reasons of practicality, it will also be assumed that some versions of future Q&A systems might be able to be accessed through a simple smartphone interface, which supports either spoken or typed questions equally adeptly. It is open for speculation as to the best format for the answers. In the Jeopardy!™ challenge, Watson formulated its answers in the standard Jeopardy question response, i.e. "*What is ...?*" However, this was a very particular response for a very particular set of circumstances. In an earlier paper, the authors discussed at length the issues around the potential ambiguity in how users may phrase questions and how a system such as Watson may have to resolve such potential ambiguities (Keates, Varker, Spowart, 2011).

An example of this in practice was encountered in the Jeopardy!™ challenge. Under the category "U.S. CITIES" when presented with "*Its largest airport is named for a WWII hero; its second largest for a WWII battle,*" Watson answered: "*Toronto*". While to most readers this seems to suggest a serious limitation in the technology, in practice it arose because Watson had been trained not to be too literal in its interpretation of Jeopardy categories, which are sometimes only tangentially related to the actual questions. Further confusion could have arisen because, for example, the Toronto Blue Jays baseball team plays in the American League and there are cities called Toronto in the U.S.

In another example under the category "THE ART OF THE STEAL", Watson was flummoxed by the difficult wording of the question, "*In May 2010 5 paintings worth $125 million by Braque, Matisse & 3 others left Paris' museum of this art period.*" Watson's response was "*Picasso*" – an artist rather than an art period, although the correct answer, "*Modern Art,*" was Watson's third-highest confidence answer. There are many implicit assumptions in how people ask and answer questions and systems such as Watson either have to have these assumptions coded explicitly or have to be given machine-learning capabilities. The developers of Watson have implemented both of these approaches.

The conclusion of the earlier paper was that a system such as the hypothetical QAFUTURE should display confidence ratings in all answers provided, as well as the sources used for the answer (Keates, Varker, Spowart, 2011).

The aim of this paper is to examine how such a system, with all of the technical wrinkles ironed-out, could be of real-world benefit and application from a user perspective, i.e. identifying those areas of life endeavour that could realistically be supported by systems of this type.

3.0   QAFUTURE AND AREAS OF LIFE ENDEAVOUR

In another earlier paper focusing on general-purpose cognitive support systems, the authors identified 5 such areas of life endeavour that could be supported by new technology (Keates, Kozloski, Varker, 2009), specifically:

1.   Lifelong learning and education
2.   Workplace
3.   Real world (i.e. extended activities of daily living)
4.   Entertainment
5.   Socialising

It must be noted that this list is indicative only. It is not intended to be a complete list of all such areas of life endeavour. The remainder of this paper will examine possible uses of the speculated QAFUTURE system for each of the 5 areas identified. The examples given are indicative only and are included to stimulate thoughts and provoke discussion about possible other uses of this technology. They are not intended to describe the only possible uses and applications.

3.1   QAFUTURE and Education

It is arguably easiest to imagine the uses of QAFUTURE in a traditional classroom environment. A faultless Q&A device that answers all questions 100% correctly will probably quickly become the number one "must-have" gadget on schoolchildren's wish lists, with thoughts of never having to learn another fact again racing through their minds. It is simple to envisage how such a device could have the same effect as the introduction of calculators had on mathematics, spell-checkers on essay writing and even GPS systems on navigation. All of these inventions have had a significant impact on the lives of their users and have clearly made certain tasks very much easier, but they are not without their own issues.

For example, basic numeracy skills are still taught at elementary school level, despite the ubiquity of calculators. There are many reasons for this, not least that calculators are not always available and often it is faster to do the calculation in your head. One of the most important reasons, though, is error checking. How do you know if you accidentally press the wrong button on the calculator? The only sensible solution is to estimate the correct answer and use the calculator to provide the precise answer.

Similarly spell-checkers have improved life in the keyboard age, especially for people with conditions such as dyslexia. However, again, how do you know whether the suggested correction is the correct word without a basic idea of what the correct spelling should be? Spell-checkers have not eliminated the need to understand the basic fundamentals of how to spell.

GPS systems have become indispensible aids when driving in unfamiliar towns and cities. However, they are not infallible and they do sometimes get confused, for example by differences in heights between roads that look connected on a 2D map, but are actually at different elevations. It is also very easy to enter a slightly wrong address and end up several hundred miles away from your true destination, such as going to Stamford Bridge (just outside York in the north of England) when aiming for Stamford Bridge football ground (home of Chelsea FC in London) – a difference of 230 miles.

A quick search on the Internet and it is easy to find substantial research into the effects of such systems on our innate abilities to do basic sums, spell and even navigate. While such systems are incredibly useful, an over-reliance on them can lead to a decrease in our ability to perform the basic tasks that they are designed to assist with. Thus, introducing a system such as QAFUTURE into the classroom would have to be done carefully to not interfere with the acquisition of the basic skills of knowledge retention and processing.

However, if introduced and used with care, a system such as QAFUTURE could allow students to swiftly expand their understanding of subject matter by allowing them to acquire knowledge in the format that best suits their own knowledge acquisition processes. In effect, QAFUTURE would allow students to ask any and all questions around a topic that they can think of and to build their own mental model of the topic in hand. QAFUTURE would become their own personalized encyclopedia, providing answers that precisely address the knowledge gaps of the students. Such a device, with the infinite patience of a computer, would allow students to build a network of knowledge around a topic, effectively providing a route to the scaffolding techniques that are essential for successful knowledge acquisition and processing (Keates *et al.*, 2007). Indeed, such a system is only a few steps away from a fully automated personal tutor.

To be truly considered as fully automated tutors, Q&A systems would need additional capabilities, such as the ability to formatively create a pedagogical approach crafted to meet the individual needs of the user. Furthermore, it is not clear at this stage how Watson could disambiguate different perspectives of the same event. Two sides engaged in a conflict will have very different views and opinions about events and being able to identify and contextualize those differences will most likely take further development of QAFUTURE. This further refinement of Q&A systems' operating capabilities will be similar in size to the change required to move from the current approach to web searching into a fully semantic web search – a long-held aspiration that has yet to be accomplished.

3.2   QAFUTURE in the Workplace

Working for a large multinational company such as IBM presents many challenges, especially in terms of the sheer volume of information that is generated on a day-to-day basis. If an accessibility engineer in Austin, TX, is tasked to consider which applications to develop to support a bid for a government contract in Germany, for example, it would be extremely useful to be able to find more information about the company's sales strategy in that country, key personnel and other pertinent information quickly. A QAFUTURE system, kept fully up-to-date on all of the company's internal intranet pages and other approved communications would be able to provide very targeted answers very quickly and save the engineer to time of find a range of possible intranet pages, not all of which will be relevant to the task at hand, and having to trawl through them to build up a potentially incomplete mental model of the sales situation in Germany. QAFUTURE would offer significant improvements in both efficiency and effectiveness for the engineer in this example.

QAFUTURE could also serve as a productivity enhancement tool, for example streamlining committee management. Many workplace committees are dominated by routine paperwork and matters requiring action by other departments. QAFUTURE could allow a committee member to ask a simple question such as, "*Are there any items on the agenda that refer directly or indirectly to my department?*" The system could search through all of the supporting paperwork and highlight any issues that either affect the specific department or alternatively affect all departments. Such a filter would allow much more efficient preparation for the meeting.

Another possible application of Q&A systems technology would be in helping find expertise within the company. Asking a question such as *"Who knows about Section 508 checklists and their application to website development?"* could produce a very targeted list of potential experts who may be able to help. Such a capability could save a very large company a significant amount of time and money.

3.3 QAFUTURE in Real World Support

Real world activities considered here are the basic issues of everyday living, such as remembering appointments, taking medication, phoning the electricity company, etc. Q&A systems could easily be adapted for use as a simple reminder system. Such a use would be of significant value for users with cognitive and memory impairments in particular. There is substantial research in this area, developing specific solutions for specific situations. The advantage of a system with the power of QAFUTURE, though, is the ability to combine all of the separate solutions into a single all-encompassing one.

QAFUTURE could also assist in providing succinct answers to simple but infrequently performed queries, such as *"What is the telephone number of my medical insurance carrier for questions concerning rejected claims?"*

3.4 QAFUTURE and Entertainment

QAFUTURE offers a number of intriguing possibilities for entertainment. A very simple application would be for planning an evening out. QAFUTURE could act as a restaurant guide and movie critic, swiftly searching through online databases to recommend the best steakhouse in the area and the best action movie to follow it. Similarly, QAFUTURE could be used for the ultimate in shopping assistance, producing very fine-grain recommendations for very specific requests.

The system could also offer an opportunity as an entertainment medium in its own right. If the database being parsed by QAFUTURE contains all the information about a fictional world, such as names, locations, trophies, etc., then QAFUTURE itself could become a game. As the user asks natural language questions, they explore the virtual world in a potentially unbounded adventure version of Dungeons and Dragons[TM]-type adventures.

Other possibilities include the use of QAFUTURE for general knowledge and trivia type games, such as the Jeopardy![TM] challenge, but using Q&A systems as the arbiter and question-master, not a competitor.

3.5 QAFUTURE and Socialising

Socialising is considered here to be the promotion of direct inter-personal communication and also simply spending time with people you like.

QAFUTURE also offers exciting opportunities to support socialising activities. If the database of knowledge being parsed is based on social media information, such as Facebook profiles for example, then Q&A systems could offer a kindred-spirit finding service. A user could ask a question such as, *"Who lives within 100 miles of me who likes rock climbing and sushi?"*

Clearly there are privacy issues that need to be addressed in such a system, although no more than those facing services provided by companies such as Facebook and Google at the present moment in time.

4   CONCLUSIONS

This paper has explored wider uses of forthcoming generation of Q&A technology suggested by the IBM Watson system and its success in the Jeopardy!™ challenge. If such systems can be built to offer guaranteed responses, how will they change the nature of how people conduct their daily lives? Computers have always excelled at providing precise answers to questions. The challenge for users has often been how to best formulate questions to obtain the best possible answer. What happens to human thinking and HCI when the users are freed from such limitations? Systems like Watson represent a genuinely exciting glimpse of the future of computing.

However, the full potential of that future will only be realised if the user perspective is considered throughout the development of these new systems.

DISCLAIMER

The authors of this paper were inspired by the performance of IBM's Watson system in the Jeopardy!™ match, and are very intrigued about the potential of this technology to radically reshape how we think about people using computers. However, this paper represents "blue sky" speculation about such future possibilities, and in no sense should it be interpreted as an indication of future plans by IBM or any other organization.

REFERENCES

Bigus, J.P.; Campbell, M.; Carmeli, B.; Cefkin, M.; Chang, H.; Chen-Ritzo, C.-H.; Cody, W.F.; Ebadollahi, S.; Evfimievski, A.; Farkash, A.; Glissmann, S.; Gotz, D.; Grandison, T.W.A.; Gruhl, D.; Haas, P.J.; Hsiao, M.J.H.; Hsueh, P.-Y.S.; Hu, J.; Jasinski, J.M.; Kaufman, J.H.; Kieliszewski, C.A.; Kohn, M.S.; Knoop, S.E.; Maglio, P.P.; Mak, R.L.; Nelken, H.; Neti, C.; Neuvirth, H.; Pan, Y.; Peres, Y.; Ramakrishnan, S.; Rosen-Zvi, M.; Renly, S.; Selinger, P.; Shabo, A.; Sorrentino, R.K.; Sun, J.; Syeda-Mahmood, T.; Tan, W.-C.; Tao, Y.Y.Y.; Yaesoubi, R.; Zhu, X.; "Information technology for healthcare transformation," *IBM Journal of Research and Development*, vol. 55, no.5, pp.6:1-6:14, Sept.-Oct. 2011

Dillenberger, D.E.; Gil, D.; Nitta, S.V.; Ritter, M.B.; "Frontiers of information technology," *IBM Journal of Research and Development*, vol. 55, no.5, pp.1:1-1:13, Sept.-Oct. 2011

Keates, S.; Adams, R.; Bodine, C.; Czaja, S.; Gordon, W.; Gregor, P.; Hacker, E.; Hanson, V.; Kemp, J.; Laff, M.; Lewis, C.; Pieper, M.; Richards, J.; Rose, D.; Savidis, A.; Schultz, G.; Snayd, P.; Trewin, S.; Varker, P.; "Cognitive and learning difficulties and how they affect access to IT systems," *Universal Access in the Information Society*, 5(4), pp. 329-339, April 2007

Keates, S.; Kozloski, J.; Varker, P.: 2009. "Cognitive Impairments, HCI and Daily Living." In *Proceedings of the 5th International Conference on Universal Access in Human-Computer Interaction* (UAHCI '09), Constantine Stephanidis (Ed.). Springer-Verlag, Berlin, Heidelberg, 366-374

Keates, S.; Varker, P.; Spowart, F.; "Human-machine design considerations in advanced machine-learning systems," *IBM Journal of Research and Development*, vol. 55, no.5, pp. 4:1-4:10, Sept.-Oct. 2011

Picheny, M.; Nahamoo, D.; Goel, V.; Kingsbury, B.; Ramabhadran, B.; Rennie, S.J.; Saon, G.; "Trends and advances in speech recognition," *IBM Journal of Research and Development*, vol. 55, no.5, pp.2:1-2:18, Sept.-Oct. 2011

CHAPTER 37

Benefits in the Production Process through the Acquisition of Competence in Ergonomics, Case Study of an Automotive Industry

VARASQUIN Adriano, BALBINOTTI Giles, VIEIRA Leandro

Pontifícia Universidade Católica do Paraná - PUCPR
Curitiba, BRAZIL
avarasquin@hotmail.com

ABSTRACT

The globalized world and a highly competitive require ever greater productive efficiency by businesses, producing goods or providing services to lower cost, for that all resources should be used as best as possible at levels that maximizes profits. For these diets was the emergence of actions to streamline the work, offering better production processes, work routines and conditions to improve productivity and reduced production time. Among them comes to ergonomics, with the purpose of analyzing and studying the different situations that involve a man and work. This article presents a study related to the acquisition of skills (training) related to ergonomics and working conditions, improving processes and results of an organization. Through practical demonstrations of the tools used in their processes, actions aimed at improving conditions and job gains by the company as a reduction in accident rates, absenteeism and improving the quality of its products. Information was obtained by applying a field survey, structured questionnaires, data analysis and records of real situations that occur in day-to-day an automotive.

Keywords: Ergonomics training, manufacturer, effectiveness of ergonomics.

1. Introduction

The word training refers to the acquisition of knowledge, skills and competencies as a result of training or teaching of practical skills related to specific useful skills. Besides training the technological advancements and competitiveness of the modern world require that workers continually update their skills throughout their working lives. In the early studies related to training some experts, according CHIAVENATO (1999), considered a training environment to suit each person to the office and develop the workforce of the organization from the positions held. Currently this line of thinking has extended this concept, where the training is considered a process for preparing people to perform more efficiently their tasks related to his office busy developing their skills to become more productive, contributing to the achievement of objectives of an organization. To identify a training need as comments CHIAVENATO (1999), it is necessary to find a lack of professional training in an area where an individual or group to improve or increase their efficiency, effectiveness and productivity at work, becoming a beneficial for employees and their organization.

2. Training Procedure

Learning can be defined as a process where we acquire experiences that enhance our ability to perform certain actions. Training is a process with the procedures and actions that assist in learning and skills development. According CHIAVENATTO (1999) the training process has four steps:
1. Diagnosis: It is a survey of training needs or future.
2. Design: Development of the training program.
3. Implementation: Implementation and conduct of the training process.
4. Rating: Verification of training results.
Another important tool for the identification of training needs are the indicators of the company as low productivity, waste materials, poor use of space, lack of motivation of employees, etc..
According BISCARO (1994) there are several factors involved in the learning process (training), which determine the methods and actions that should be used in learning. As an example of a quality problem, we based the identification of sector or faulty mechanism of the process, to determine the tools to be used during training

2.1 Ergonomics

The studies related to ergonomics emerged after World War II, as a result of interdisciplinary work done by various professionals of the time. Initially linked to industrial activities involving man and machine. According IIDA (2005), ergonomics expanded today, covering all types of human activities mainly in the service sector such as health, education and transportation. Including human activities such as posture, movement, flow of information, jobs, cognition, controls, forms of work organization, human factors, among others.

Another important aspect studied by ergonomics as BALBINOTTI (2003), is the concern improvements in the jobs and the demand put them in the best conditions possible to avoid the accident or excessive fatigue and improve performance. Ergonomics is now an aspect increasingly approached within companies, to be connected to virtually all areas directly related to health and welfare of employees, and the quality of products and services offered by the organization.

2.2 Ergonomic Training

Ergonomic issues may have a role in improving processes and results of an organization. Being a science applied to the development of machines, systems and tasks, with the aim of proposing greater safety, health and comforts the general conditions of work. Also improving the quality of life of employees, by fighting the agents of evil in the physical and mental health. According MÁSCULO and VIDAL (2011) manly and ergonomics is the science that studies and makes the reality of working people seeking productivity gains through its main component, human comfort.

With the increasing complexity of work environments, there is a need for a detailed study to identify risk situations and effects of accidents to workers' health. Training ergonomic add knowledge to the individual, helping to fit your work environment, as comments such actions IIDA (2005) reduce fatigue and monotony, or eliminate highly repetitive work and a possible lack of motivation caused by low employee participation in achieving their activities. Ergonomic training also assists in the analysis of general environmental conditions such as temperature, noise, vibration, toxic gases and lighting. These can be recognized by certain symptoms such as high level of errors, accidents, illness, and absenteeism and employee turnover. Behind this evidence may be experiencing a mismatch of machinery, failure in work organization, which cause muscle aches and tension among workers as comments IIDA (2005), with the spread of ergonomics and activities linked to it as training, we can develop a greater awareness of the workers within the organization in relation to the effects of work environment and health. Ergonomics is an excellent tool for accident prevention, because it is a science all scenarios involving the execution of an activity, whether manual or mechanical thereby assisting in reducing absenteeism and possible quality problems linked to jobs with ergonomics difficult.

3. Methodology

The research presented was applied to an automotive industry from Paraná-Brazil, where ergonomic aspects are addressed in daily business. Ergonomic training within the system helps the organization in achieving results, improving working conditions by reducing absenteeism and accidents, improving the quality of their products as shown in the description of the results.

The company offers various internal training related to ergonomics, among them those that stand out are:

JOB observation: Tool used to discuss situations of obvious improvements in the job, seeking greater efficiency, shorten cycle times, movements made by the workers, reactivity, and check the speed of processes.

Kaizen: The tool set aimed at continuous improvement of processes, improvements in employment and also the working conditions of employees. According BALBINOTTI (2003), kaizen tool can reduce the physical effort, through actions such as the correct way to use tools and improved posture.

Considered by experts one of the main tools for improving ergonomic aspects.

Form analysis of safety and ergonomics: Training designed to line supervisors for making diagnoses in order to make actions in the workplace safer and with less complexity. Encouraging actions to improve the ergonomics of workers, supervisors, technicians and experts, identifying positions classified by the color green stand with satisfactory rate, yellow posts with one or two restrictions observed and identified by red posts with important restrictions.

Dexterity: Operation of the job, allowing the training of gestures and steps reducing the time of execution of activities related to the post, improving quality, fluidity in movement, and helps correct posture of the workers. The field of skills reduces the dispersion of performance, and therefore the risks of non-quality, promoting the optimization of operations by improving the fluidity of movement, ergonomics and safety in the workplace. By applying the four rules of economy of motion training participants perceive the search for quality, speed, flow and safety.

Regulatory Standard 33 (Standard Brazilian regulatory work in order to establish minimum requirements for the identification of confined spaces and recognition, evaluation, monitoring and control of risks in order to permanently guarantee the safety and health of workers who interact directly or indirectly in such spaces. Mandatory training with a workload of 40 hours for employees who have activities related to employment confined area or environment not designed for continuous human occupancy, having limited means of entry and exit, which is insufficient ventilation to remove contaminants or where it can there is a deficiency or oxygen enrichment.

Regulatory Standard 05: Ergonomics is now an aspect increasingly approached within the Brazilian standard regulatory work related to the Internal Commission for Accident Prevention, commissioned with the aim of preventing accidents and illnesses resulting from work. Identifying risks of work processes, mapping of risks and preventive measures for solution of problems of safety and health at work. All internal commission for accident prevention shall receive a 20-hour course for the feasibility of it.

Regulatory Standard 10: Standard Brazilian regulatory work in order to plan preventive controls and systems to ensure the safety and health of workers whose duties are directly and indirectly linked to activities in electrical installations and services with electricity. For the control of risks in these activities, identifying the best materials and equipment to perform implementation and maintenance of electrical installations and work done in its vicinity. Also a 40 hours training required for employees who have their activities directly or indirectly related to electrical installations.

3. Discussion of Results

This research was conducted in an automotive industry, to analyze the company's results were used statistical data and the application of a questionnaire to identify ergonomic training is most effective in improving outcomes and the importance of training at the point of view of employees. The questionnaire has nine questions, with three closed and six open applied to 25 employees in five departments.
Below the results obtained in the research represented graphically and comments made by participants.

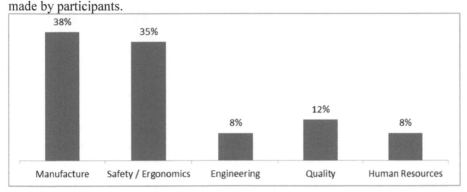

Graphic 1 Sectors of research participants.

The graph 1 represents the areas of research participants, represented mostly by employees in the manufacturing and safety / ergonomics, followed by quality, engineering and human resources. The research was performed mostly in manufacturing, being the area most related to this study.

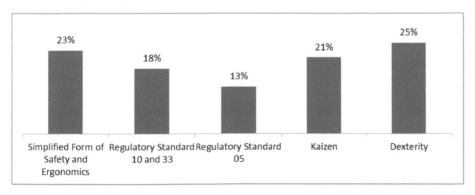

Graphic 2 Training in ergonomics.

344

We can see from the graph 2, the training Dexterity, Form Simplified Safety and Ergonomics and Kaizen are the main tools for improving ergonomic aspects identified in the survey. Certainly the three tools are more related to ergonomics, by being directed process improvement, training movements and a key aspect of ergonomics that is observation improvement of ergonomics.

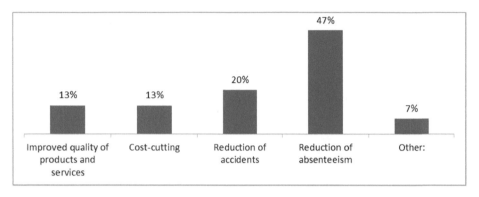

Graphic 3 Aspects that ergonomics is most effective with in the company.

The graph 3 shows that the main benefit of ergonomics to the company's effectiveness in improving the rate of absenteeism and accident reduction in the company. Below some of the main comments from employees regarding the improvement of ergonomics in the workplace, indicators, and benefits in relation to ergonomic training offered:
- Surely, when the company invests in an ergonomic program the level of productivity increases, as the company works for the welfare of employees and satisfied if the employee works and contend the quality of products increases.
- When a company implements a program of applied ergonomics and training related to ergonomics, it shows the concern that the employee has. And even the official is concerned about their health within the company.
With this the company increases productivity, reduces absenteeism (missing work), reduces accidents and incidents, reduce distant and restricted.
- By having good ergonomic conditions the operator is concerned with making the process better.
- Reduced absenteeism, fewer rejected by physical problems, less stress among team members, etc ... And as a result, increased productivity with quality.
- Improving the cognitive part is evident from the reduction of quality problems. A workstation ergonomically structured submit an operator's less wear, better organization (with 5S which also focuses on the ergonomic aspects) of a generating station improves the products presented.
- Not only the training (include more than one. 5S). The systematic monitoring and animations are extremely important to strengthen the training aspects passed.
- For safety in the workplace, employees need in 1st place to have a safe attitude and if he is trained he will be more aware of the risks and prevention of these risks, so the chances of an accident at work are smaller.

- Better ergonomics ensures that is comfortable doing the task correctly. Therefore, the chance of quality problems because the operator is trying to do the easier it is much smaller.
- The quality of a product according to improve good ergonomics of the job as the operator will not become tired which could degrade the product.
- The best training is Dexterity, it is the first and must be well detailed, so the operator can develop attitudes for other training such as Kaizen.
- The health benefits are directly on the operator, but the company gains by not having employees restricted or removed without damaging the posts.
- I believe that all training should be more comprehensive ergonomic and disseminated so that they can benefit and believe that training is beneficial to the health of the operator and not only for the measurements.
- The company is clearly concerned with culture and ergonomics, but even with the existing culture, we still have jobs with conditions to be improved.
- The skill, in particular, is essential for continuous improvement in the workplace, not only reached the ergonomics, but also increasing productivity, quality and reducing costs.
- Several benefits, but in particular, reduces the rate of absenteeism in that the ergonomic conditions improve, reduce the clearances.
- Since the operator can interact with the group for these improvements because it is the main actor of any change in his surroundings.
- The main objective of any organization is one, financial, and the training she hopes to return as soon as possible.
For it invests to reduce restrictions on jobs, increase productivity and product quality.
- Best results in ergonomics in the organization are those where there is involvement of all shifts, for if there is global involvement the risk of rejection is high.

Based on some secondary data from company and field research, it was found that the company has a real concern for the improvement of ergonomics, by offering training and other activities proposed under the representation of some indicators that show the effectiveness of these actions:

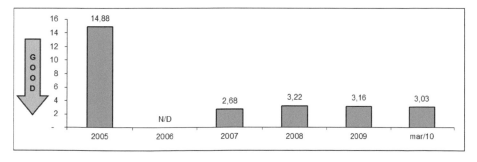

Graphic 4 Index Absenteeism.

As can be seen in Graph 4 rates of absenteeism (absence of employees on the job) had a significant reduction from 2005 to 2010, after many lawsuits filed by the company as ergonomics training related absenteeism can be caused by several factors such as motivation and pathological diseases, affecting profits and results of a business, reducing productivity, influencing the quality of the product or service and also in production costs.

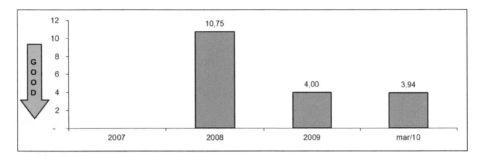

Graphic 5 Index Accident.

The accident rate shown above in graph 5, were also reduced after ergonomic actions. The index is calculated as the proportion of accidents in respect of hours worked by employees, encompassing any type of accident, according IIDA (2005) some organizational decisions may create conditions that favor the occurrence of accidents. These are called latent failures and include activities such as product design, design work, planning and control of production, investment in automation equipment. Often by financial constraints and limited time, companies use poor quality materials in their facilities, therefore, are not always guaranteed the best conditions for carrying out the work, creating latent failures, such as purchase of materials and inadequate equipment or poorly maintained, poorly designed jobs, and other excessive loads.

Any action to improve the ergonomics and conditions of employment are to prevent and avoid accidents, thereby reducing downtime on production lines, thus reducing costs.

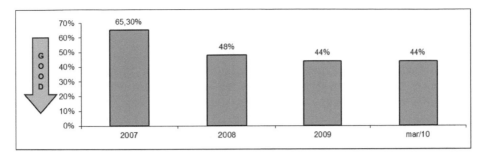

Graphic 6 Critical Jobs

The jobs in the ergonomics program proposed by the company is divided into two groups, critical jobs where the employee has medium or high percentage of accidents and get muscle problems, and less critical where the low level of risk for accidents and muscular problems. The critical posts as noted in graph 6, were reduced Identifying these posts can be aided by training Sheet Simplified Safety and Ergonomics, previously mentioned in the description of the training. According BALBINOTTI (2003), ergonomics seeks not only to prevent workers the jobs stressful and / or dangerous, but seeks to put them in the best working conditions possible to avoid the accident or excessive fatigue and improve yield.

Classified as critical positions within the company, also represent potential quality problems during or after product manufacture. As noted in Chart 7, 42% of quality problems come from the ranks with hard classification.

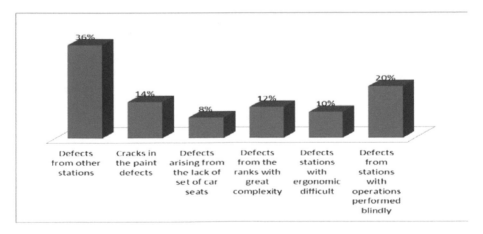

Graphic 7 Relationship with the quality of ergonomics.

The results presented demonstrate the effectiveness of ergonomics to combat absenteeism, reduce accidents and improve the quality of processes, products and services, through small actions and relatively low cost to the contrary that many organizations think, or simply unknown to implement the improvement its processes.

4. Conclusions

With the increasingly competitive scenario, the search for profit and larger slices of the market (Market Share), somehow end up harming the smooth running of processes and living within a few by all organizations that interact in this environment. Through this research it was concluded that issues such as quality of life, health and wellness have become matters of considerable attention within the company, an employee motivated by favorable working environment for the realization of his work is much more efficient, thus providing lower costs to company and also improving the quality of its products and services. These questions can be provided with training related to ergonomics, such as the Kaizen tool that seeks continuous improvement in the process of reducing Dexterity dispersions of execution, favoring the improvement of operations in the fluidity of movement, the tool sheet Simplified Safety and Ergonomics applied to supervisors to identify manufacturing jobs with higher risks of accidents among others.

With other field research training were identified by employees as effective in improving processes and ergonomic tool 5S which also focuses on improving the organization of ergonomic work environment. Another issue much discussed in the study was the quality of products and services and the level of productivity related to ergonomics, many employees believe that investment in ergonomics programs reduce absenteeism and rework, improving productivity, these issues were confirmed by results presented by company, which reduced considerably after the implementation of various actions ergonomic.

With the modern world, few workers depend on physical strength, the image of the worker is becoming reduced, portrayed by only one person sitting in front of a computer, focused on decision-making is thus possible to conclude that the improvement of productivity and results many, has a direct relationship with the company's ergonomic program. Ergonomics in the modern world has become increasingly important in improving the cognitive aspects, such as our capacity for learning and memory, mainly assisting in achieving the main objectives of an organization.

REFERENCES

BALBINOTTI, Giles 2003 A Ergonomia como Princípio e Prática nas Empresas.

BÍSCARO, A.W 1994. Métodos e técnicas em T&D.

MÁSCULO, Mario e VIDAL, Cesar 2001, Ergonomia – Trabalho adequado e eficiente.

CHIAVENATO Idanlberto 1999, Gestão de Pessoas: O novo papel dos recursos humanos nas organizações.

IIDA, Itiro 2005 Ergonomia - Projeto e Produção.

REGULATORY STANDARDS - Standards for Safety and Health at Work, Ministry of Labour. Available in http://www.normaregulamentadora.com.br/. Accessed on October 12, 2011.

ABERGO - ASSOCIAÇÃO BRASILEIRA DE ERGONOMIA, Available in http://www.abergo.com.br/. Accessed on Dezember, 20, 2011.

Documentation of the Automobile Industry, 2011 material.

CHAPTER **38**

Correlation between OCRA Exposition Index, Posture and Work Cycle Time

Romero Cardoso Oliveira, Francisco Soares Másculo*,*

* Production and Systems Department, School of Engineering, Paraíba Federal University ,Brazil

romero.c.oliveira@gmail.com; masculo@gmail.com

ABSTRACT

The objective of the study was to verify the relationship among the use of inadequate postures through the method Occupational Repetitive Actions, OCRA, and the generation of Repetitive Strain Injuries, RSI, in workstations of repetitive work. The proposed study is of exploratory and descriptive nature, with quali-quantitative approach and was done in a company of the shoe sector. It was verified the quantitative of RSI, as well as, the situation of the repetitiveness in the workstation through the method OCRA, collecting data and deepening the knowledge on the influence and importance of the repetitiveness in this company. The results show the discrepancy among the data picked by OCRA and the numbers of RSI registered in the medical department of the appraised company. It is known about the importance of OCRA as a validated method and globally used, however it is difficult the compilation of theoretical data for the generation of an index that detects the danger of a specific work in provoking lesions.

Keywords: Method OCRA, inadequate postures, Repetitive Strain Injury

1   INTRODUCTION

The work environment is a space that includes ergonomic variables, in the areas physical, cognitive or organizational, that if not observed can influence the performance and the worker's labor capacity directly. The posture or the positioning of the parts of the body are of fundamental importance for the biomechanics of the movements, the body positioning when very adjusted will provoke the execution of correct postures without overload in the body structures. When not positioned in a correct way it may causes the execution of forced postures, that will take interference in the movements during the accomplishment of the task, worsening the action of the harmful aspects on the body parts.

In the last years the discussions on diseases related to the work took impulse, because these can lead to the worker's incapacity temporary or permanent, not just bringing negative consequences for this, but also for the company or institution where they work (NECKEL; FERRETO, 2006). Many problems are related to the generation of RSI, and one of the principal is the high repetitiveness level and execution of the goals which the workers are exposed.

However, some companies didn't still go back their concerns to the workers, putting for the society the consequences of an atmosphere of work harmful and exempting themselves of the improvement obligation in the productive processes, in the organization of the work and in the relationship with the worker, that would be possible solutions for the decrease of the occupational risks (PASTRE, 2001). Salim (2003) affirms that in spite of being countless the factors that influence the genesis of the occupational disturbances, "its determination, ultimately, surpasses the social structure, relating, above all, with the changes in course in the organization of the work and secondarily with the peculiar technological innovations to the productive restructuring".

In this manner, we see the relevance of the analysis of the repetitiveness of the work as primordial factor in the prevention of RSI, and as a main point in the improvement of the work conditions, and consequently, influential in the productivity of the company. For evaluation of the work conditions in the companies, or in any it labor workplace, a lot of methods of analysis of the rhythm and work load are available in the ergonomics area. The method of evaluation OCRA is an important instrument and validated and recognized for evaluation of the level of repetitiveness of the tasks, generating an index of risk level for acquisition of occupational diseases.

In this sense it is noticed the importance of the analysis of the work with respect to the generation of occupational damages, and methods that aid are of great importance. In the practical experience using the method OCRA was noticed variability in the importance or weight of some aspects on other in the final index, however it is plausible that some factors provoke more physical stress than others on the body structures. Then it is looked for inside in this work the observance of the method OCRA of these variability and if these are in agreement with the real factors of generation of RSI that will be analyzed in the research.

This study contributes to systematize the generating aspects of RSI, what facilitates the prevention and the combat to these occupational damages, as well as it aids to the construction of workstations proactive that avoid the use of these harmful factors during the execution of the tasks. The portal of Brazilian Social Welfare shows the concern with the great number of cases of RSI, which increased 126% between 2006 and 2007, from 9.845 cases in 2006 to 22.217 cases in 2007. It is considered that, in 2009, the government spent around 2 billion of reals with the cases of RSI in Brazil.

For the observance of RSI in a worker is known that the genesis can come from several aspects. This research will focus in the use of forced or inadequate postures for the execution of the tasks, which is an item of fundamental importance, because the posture has direct influence on the incidence of vectors of force harmful on the body structures, and that consequently, can generate occupational damages.

This study tries to answer the following inquiry: which is the existent relationship among harmful postures observe through OCRA in the work place and the diagnosis of RSI, in these work places?

The object of research of this work was the shoe industry, where in which important characteristics were observed, as the accomplishment of a lot of tasks in handmade way, development of activities in cells of production with specialized tasks, accomplishment of the activity in high-speed and small cycles. All these data and numbers show the relevance of the approached subject and the need of studies that look for the progress of the diagnosis and characterization of the occupational diseases, for consequently they aid in the decrease of the number of these lesions.

The objective of the study was to verify the relationship among the use of inadequate postures through OCRA and the generation of RSI in work repetitive.

2   Method

The proposed study is of exploratory and descriptive nature, with quali-quantitative approach and was done in a shoe company, it was verify the quantitative of RSI, as well as, the situation of the repetitiveness in the workplace through the method OCRA, collecting data and deepening the knowledge on the influence and importance of the repetitiveness in this company.

The universe of the research was this company, it was included in the study two branches in the Paraíba State of the company, where samples of the repetitive workstations were analyzed, of the productive sections of the company. The selection of the sample it was non probabilistic, looking for to assist the pré-established criterion in the research, that it was the visualization, through the accompaniment of the activity, of a repetitive cycle in the execution of the task.

The collection of data was through the analysis of the workstations of repetitive works of the company, as well as the collection of the cases of RSI of this referred company. It is understood as workstation with repetitive character all those activities with a cyclical characteristic in the execution of the activity, that can be observed by the accompaniment of the activity and for the measurement of the activity cycle.

The contact with the company was through a consulting contract where the data were picked in this period, for two researchers in a total of 3 months. The parameters designated for the collection of data of the research brought us the analysis of 73 workstations with a defined task cycle. This filter of the workstations was done through the observation of each task in the amount of employees of the company, of the assembly sections, cut, sews, finishing, among other, observing or not the presence of a defined cycle in the accomplishment of the work.

The method Occupational Repetitive Actions (OCRA), developed by researchers Enrico Occhipinti, Daniela Colombini and Michele Fanti, it is an evaluation method and analysis of the risk factors associated to repetitive movements of superior members, through the calculation of a quantitative index (PAVANI, 2007; ANTONIO, 2003). This characteristic of the execution of a certain cycle is essential for the application of OCRA, and important condition in the emergence of RSI. And for the research of the data of RSI, the medical department of the company was contacted, being picked the data of work related musculoskeletal disorders (WMSD), through the report of notification of pathologies of superior members, related as lesions for repetitive efforts.

The method OCRA is a method of quantitative evaluation of the repetitiveness in work, it uses analysis of several points, as postures of superior members adopted during the task, force exercised in certain movements, duration of the cycle, number of actions accomplished by each superior member, among others. Each approached aspect is equal to a punctuation, which will be counted in the end through the observed technical actions (OTA) and recommended technical actions (RTA), whose division will give the final index that will indicate the risk presence for the generation of occupational diseases.

2.1   Data Analysis

The analysis of the data of the OCRA method, as well as of the RSIs data was described in a quali-quantitative way, describing the characteristics of the lesions and analyzing the behavior of these pathologies in the company.

In this sense the application of the method was done in workstations with repetitive characteristics, confirming or not the risk verified in that place. The data of the index posture of the instrument were analyzed verifying the quantitative of the presence of the forced/inadequate movements, for each articulation.

It was also picked data of RSI of the company, that served as reference for the analysis of the posture item of the method, where it was verified if the body structures of the superior members that more influences the value of OCRA are the really the more observed.

3   Results and Discussion

The study analyzed 73 workstations using the method OCRA. Just in a post the index multiplication posture had the value 1, which it is the best punctuation, in other words, only one workstation presented the execution of postures in forced amplitude that had no influenced the conditions in the index OCRA, and consequently, for the method, it doesn't influence the generation of RTIs.

As the analyses are accomplished differentiating the members, left and right, we will have 146 results of values of factor posture, where: 2 members obtained value 0,7; 41 members the value 0,6; 33 members with the value 0,5; 31 members inside of the punctuation range that generates the factor 0,33 and, finally, 16 members are with the factor in the value of 0,1, as can be seen in Table 1.

Table 1 OCRA final index, factor posture of the OCRA and cicle life values

	OCRA FINAL INDEX		VALUE FACTOR "POSTURE" OF THE OCRA		CICLE LIFE (min)
Jobs	MD	ME	MD	ME	
1	3,35	1,13	0,50	0,70	0,20
2	1,69	2,15	0,60	0,50	0,58
3	1,92	1,12	0,60	0,33	0,30
4	3,19	3,72	0,60	0,60	0,12
5	2,29	6,40	0,70	0,50	0,25
6	14,36	6,05	0,50	0,60	0,58
7	5,60	2,07	0,33	0,33	0,66
8	3,99	2,41	0,50	0,60	0,07
9	7,18	3,69	0,60	0,70	0,17
10	0,48	0,56	0,70	0,60	0,32
11	5,14	15,07	0,33	0,10	0,15
12	2,39	1,03	0,60	0,70	0,13
13	26,85	6,78	0,10	0,33	0,42
14	9,96	1,79	0,60	0,60	0,20
15	4,36	24,90	0,50	0,10	0,05
16	3,45	3,45	0,50	0,50	0,13
17	46,70	38,91	0,60	0,60	0,13
18	15,96	8,87	0,60	0,60	0,13
19	2,16	2,69	0,60	0,6	0,60
20	15,96	16,90	0,10	0,10	0,08
21	19,78	13,45	0,10	0,10	0,92
22	0,88	1,34	0,50	0,33	1,13
23	0,99	0,99	0,60	0,60	0,05
24	2,74	2,74	0,70	0,70	0,23
25	2,14	2,14	0,70	0,70	0,43
26	35,43	28,99	0,10	0,10	0,18
27	3,92	2,35	0,60	0,70	0,27
28	10,36	17,77	0,33	0,10	0,75

	OCRA FINAL INDEX		VALUE FACTOR "POSTURE" OF THE OCRA		CICLE LIFE (min)
Jobs	MD	ME	MD	ME	
29	13,01	1,53	0,60	0,60	0,20
30	5,55	5,55	0,50	0,50	0,25
31	10,32	10,32	0,33	0,33	0,75
32	5,67	8,32	0,33	0,10	0,20
33	6,10	5,23	0,60	0,70	0,66
34	6,43	2,51	0,50	0,60	0,93
35	3,84	1,54	0,33	0,70	0,17
36	7,37	2,61	0,33	0,70	0,28
37	2,77	3,10	0,70	0,50	0,30
38	15,56	13,69	0,50	0,60	0,21
39	3,12	2,65	0,50	0,50	0,93
40	3,65	1,46	0,33	0,70	0,58
41	2,22	1,92	0,50	0,50	0,15
42	9,62	8,25	0,33	0,33	0,12
43	14,77	12,92	0,50	0,50	0,46
44	6,64	3,35	0,33	0,50	0,20
45	4,12	2,94	0,50	0,70	0,32
46	0,59	0,59	1,00	1,00	0,13
47	6,26	4,47	0,50	0,60	0,33
48	2,83	3,40	0,60	0,50	0,17
49	5,59	5,59	0,60	0,60	0,50
50	5,75	5,48	0,33	0,33	0,65
51	3,94	13,00	0,33	0,10	0,13
52	3,88	2,54	0,50	0,60	0,20
53	5,75	1,08	0,50	0,60	0,40
54	8,87	11,29	0,33	0,33	0,67
55	24,74	10,58	0,10	0,10	0,53
56	3,35	2,36	0,50	0,60	1,20
57	7,01	9,71	0,33	0,33	1,07
58	3,18	2,81	0,33	0,33	0,30
59	2,87	1,20	0,50	0,60	0,20

	OCRA FINAL INDEX		VALUE FACTOR "POSTURE" OF THE OCRA		CICLE LIFE (min)
60	7,68	6,15	0,33	0,33	0,10
61	4,88	2,33	0,60	0,60	0,75
62	15,60	2,26	0,10	0,33	0,13
63	2,31	1,85	0,70	0,70	0,13
64	2,77	2,77	0,70	0,70	0,15
65	3,66	2,51	0,60	0,70	0,28
66	2,03	2,03	0,50	0,60	0,08
67	2,49	13,45	0,60	0,10	0,18
68	6,40	4,12	0,50	0,70	0,23
69	9,84	9,74	0,33	0,50	0,08
70	3,12	6,29	0,33	0,50	0,52
71	2,13	1,81	0,60	0,60	0,83
72	0,72	0,17	0,33	0,70	0,33
73	3,40	1,41	0,50	0,60	0,17

On the members with result 0,7, it was observed that 3 only had the final index of OCRA in the range of high risk the generation of RSI, what shows the influence of the posture inside of the value OCRA, which only represented 13,04%. Among the 41 members with factor multiplier 0,6, 17 were very high with the risk of obtaining of RSI, that elevates the percentile value, in comparison with the previous factor, for 41,46% of the members.

Continuing the analysis for the verification of the value 0,5; 17 members of the 33 were shown to be in the range of high risk, revealing a percentile of 51,51%, confirming the tendency of influence of the factor posture in the final index OCRA. The factor 0,33 display clearly this tendency, since of the 31 members with this value, 23 are in the range of high risk, in other words, 74,19% of the superior members analyzed had serious risk of obtaining a lesion for repetitive effort.

The worst level found in the research, and that corroborates with the previous data, are of the members that are with the multiplication value in 0,1, where all the members are in the risk range, with the smallest value of the final index OCRA 8,32, that is more than the double of the inferior limit to be in this risk group.

Through this initial analysis it was intend to demonstrate the importance of the factor posture inside of the method used for the analysis, which it is observed inside by the practice and experience of the companies.

Still, from the results, it can be accomplished the analysis of the values of the final indexes OCRA, for both members. Then, the research obtained in the total, 146 results of the exposition index (EI), being 73 of each member. For the right member it was verified the existence of only 13 workers (17,81%) with the index inside of the acceptable level of EI, being 16 (21,92%) in the level of low risk of generation of RSI, and the great majority inside of the range of high risk for acquiring RSI, with 44 workers inside of this level, in other words, 60,27% of the sample accomplished for the right member. In the left member it was counted 23 workers inside EI of the normality range, what represents 31,51% of the number of analyzed individuals. In the range of low risk, 18 workers, representing the percentile level of 24,66%, and inside of the range of high risk, where again it was most of the sample, it came 32 individuals, showing that 43,83% of the workers presented great risk of acquiring RSI.

With these results, it is noticed the largest gravity of the right member in the exhibition to factor of risk, what can be explained by most of the population to be skillful and to use this with more frequency in the activities of more precision or force. In the general analysis it was verified that out of the 146 results of the two members, 36 were in the normality range, 34 in the range of low risk and 76 in the range of high risk. In other words, 75,34% of EI for member presented risk of developing pathologies in the work for repetition, and 52,05% of the total offered high risk for the development of these diseases.

The analysis of the work cycle showed that the average of the repetitive cycle of execution of the tasks was in 0,36 minutes, or 21,6 seconds, where the smallest value of duration of the cycle was in 0,05 minutes or 3 seconds, what showed the high index of repetition of the movements that the workers were exposed in a lot of tasks, that provokes constant waste of the same body structures, and probably, lesions.

These values show the impact of the repetitiveness inside of the analyzed company, in the sense of futures problems with relation to the workers' health, and in consequence, of financial and human damages for the researched industry.

Comparing the three items studied it was found that 22 out of the 73 workers had analyzed the two members outside the area of high risk of acquiring WMSD, that is, 51 workers had at least one member with a serious risk of obtaining an occupational disease. By correlating the presence of risks with the working cycle, we it was seen that the 22 workers with low risk or acceptable situation, only six had work cycle time above the average seen above, which is 0.36 minutes. This equates to 27.27% of the workers. Among the 51 that are in the range of high risk, 18 have the above average cycle, that is, 35.29%. This shows that through the OCRA index there is no relationship between the presence of a small duty cycle and high degree of risk scale of this tool, because the presence of low cycles do not necessarily led to high rates of OCRA.

This can be explained since the presence of a short cycle does not require the execution of many movements, and thus not causing large physical exhaustion, which can happen in the opposite, in other words, a large cycle can occur in carrying out many movements, and there indeed have pronounced impacts and generate high risk.

Regarding the comparison of the OCRA index factor with posture, it was verified that the 22 workers are low risk and acceptable, only three (13.63%) had a member with a 0.33 posture factor; the others had posture factors between 0.5 and 0.7. Workers at high risk 26 (50.98%) presented postural factor less than or equal to 0.33. At this point there is the presence of a direct relationship, as to put at high risk if he saw the increased presence of harmful postures, which did not happen with low-risk group. So there is a relationship between the presentable postures in amplitude and the risk of forced purchase of MSDs, which is more plausible compared to the cycle time, as a harmful posture will necessarily overburden the body structure, influencing the development of WMSD.

4   CONCLUSIONS

The results show a direct and strong influence of the posture factor in the final late OCRA index. This shows a trend of concern about the attitude in the workplace, and consequently to the development and availability of working materials, since these are the ones who provide or require the adoption of the most awkward postures.

This point to a need for improvement in two key aspects of ergonomics: man and job. Since the deficiency lies in jobs with disabilities and workers without postural orientation.

Comparison of the final index OCRA with the work cycle, showed no influence on the results of low cycles risk of acquiring TSI´s. This is also related to the above comparison, this relation is only valid if the low cycles bring with them harmful positions, that is, repetitive cycles to be executed little movement will not have detrimental impacts on the body structure.

The results appear inside for a larger importance of the factor posture of the evaluation method, including other evaluation criteria. The movements adopted in the workstation impose overloads biomechanics that can have an influence factor in the generation of larger TSI´s than observed in the tool.

It is comprehensible that the ergonomic analysis of a workstation, in the sense of analyzing the degree of risk of a work environment, is complex and it takes in consideration several factors that difficultly will be included in only one analysis tool. Therefore the importance of OCRA as an instrument that gets to join several data, inside of analyzed parameters and tested by qualified professionals, and that supplies an important index, as parameter for an ergonomic diagnosis of the work position.

However, the need of researches is observed that deepen this theme about the generation of damages of work, and their causes and effects, which it is of great social and economical importance. This work intended to contribute for the increment of these studies, knowing that the pursuit of this diagnosis is important, with propositions of how to develop means that facilitate the posture analysis of a workstation, since this is a current theme and present in the reality of companies that are concerned about the negative influence of TSI´s.

REFERENCES

ANTONIO, Remi Lópes. *Estudo ergonômico dos riscos de ler/dort em linha de montagem: aplicando o método Occupational Repetitive Actions (OCRA) na Análise Ergonômica do Trabalho (AET).* Florianópolis, 2003. Dissertação – Mestrado do Programa de Pós-Graduação em Engenharia de Produção – UFSC.

NECKEL, Franciane; FERRETO, Lirane Elize. *Avaliação do ambiente de trabalho dos docentes da unioeste campus de Francisco Beltrão-PR.* Revista Faz Ciência, 2006, vol. 08, n° 01, pp. 183-204.

OCCHIPINTI, Enrico. *OCRA: a concise index for the assessment of exposure to repetitive movements of the upper limbs.* ERGONOMICS, Vol. 41, n°. 9, 1290-1311, 1998.

PASTRE, Tatiana Maglia. *Análise do estilo de trabalho em montagem de precisa.* Porto Alegre, 2001. Dissertação – Mestrado Profissionalizante em Engenharia – UFRGS.

PAVANI, Ronildo Aparecido. *Estudo ergonômico aplicando o método Occupational Repetitive Actions (OCRA): Uma contribuição para a gestão da saúde no trabalho.* São Paulo, 2007. Dissertação – Mestrado em Gestão Integrada em Saúde do Trabalho e Meio-ambiente – Centro Universitário – SENAC.

SALIM, Celso Amorim. *Doenças do trabalho: exclusão, segregação e relações de gênero.* São Paulo Perspec., São Paulo, v. 17, n. 1, Mar. 2003.

CHAPTER 39

A Review of Functional Evaluation Methods Used to Include People with Disabilities at Work

Laura Bezerra Martins[a], Béda Barkokébas Junior[b],
Bruno Maia de Guimarães[c]

[a]Federal University of Pernambuco, Brazil, laurabm@folha.rec.br;
[b]University of Pernambuco, Brazil. bedalsht@poli.br;
[c]Federal University of Pernambuco, Brazil. bmguimaraes@hotmail.com

ABSTRACT

23.9% of the population of Brazil, namely 45.6 million people have some type of disability. This paper discusses methods of evaluating the functional capacity of the individual, which were found in the literature, and that can be used for the process of including people with disabilities (PD) at work. The methods cited and discussed are: the Functional Independence Measure (FIM), the Work Ability Index (WAI), the AMI System (Available Motions Inventory), the Pain Disability Questionnaire (PDQ), the Short Musculoskeletal Function Assessment (SMFA) Questionnaire and the Competency Profile. Such methods demonstrate the importance of making a functional assessment of PDs so that the inclusion process ensures good interaction between human beings and their tasks or activities by lowering the barriers imposed by the environment and the organization of work, thus making people more effective.

Keywords: People with disabilities, Inclusion at work, Functional Evaluation.

1   INTRODUCTION

Despite the efforts being made to include people with disabilities (PD) in society, it is clear that their deficiencies can very often adversely affect their search for work. PDs come up against prejudice at various points as they are seen as people who generate costs and not profits.

Around 23.9% of the population of Brazil, i.e. 45.6 million people have one or more disabilities. Of this total, 78% are visually impaired, 29.09% have a motor disability, 21.3% are hearing impaired and 5.73% are mentally challenged (IBGE, 2011). It is also to be noted that according to WHO (2002), 10% of the world's population, i.e. 610 million people have some form of disability.

Some companies have hired workers with disabilities without using appropriate methods, namely without analysing accessibility factors, the demands of the jobs offered or the potential of individuals, which has resulted in problems to do with job fit, accidents and effects that have both a negative economic impact on the company and on its image and policy on social awareness besides causing PDs themselves psycho-social problems (Simonelli and Camarotto, 2008).

According to Martins, Barkokébas Junior and Guimarães (2012), the aim of comparing the demands of workplace with the functional capacity of workers with disabilities is to determine if the work makes smaller or greater demands than the worker can learn to deal with. Thus, the results of such comparisons should avoid PDs having to struggle to adapt to the work or to offer work that is far short of their professional qualifications and abilities.

Thus, an ergonomic approach to these issues is indispensable, since, the tools that Ergonomics uses identify the tasks and physical, intellectual and organizational demands of jobs and the functional capabilities of workers with disabilities, and with such information, appropriate adjustments can be made to jobs (Guimarães, Martins and Barkokébas Junior, 2012).

Therefore, there is a need to determine the functional profile of a worker with a disability so as to include him/her in the workforce. To do so, a detailed evaluation of the PD's functional capabilities must be made to establish what their abilities, needs and limitations are.

The evaluation of the functional capabilities can be used when selecting people for jobs to ensure that workers with disabilities are placed in an appropriate manner, such that their skills are matched to the demands of the job (Innes, 2006).

A battery of tests is generally used to verify that the individual is able to meet the demands of a job (King et al, 1998). When an individual is to be placed in a new job, it is believed that a more comprehensive and generic assessment is needed. A series of requirements must be tested to obtain as much information as possible in order to consider a range of employment opportunities (Lechner apud King et al, 1998).

According to Martins and Guimarães (2010), a detailed evaluation of the functional capacity of workers with disabilities in conjunction with task analysis enables an appropriate placement to be made, i.e. one which combines their skills with the requirements of the job. Associating the worker's functional profile with the requirements of the job also enables identification of the tasks of the job that need to be tailored to the individual.

According to Fadyl (2009) in order to place or relocate an individual with disabilities in a job, in addition to assessing his/her functional capacity, the following factors should be included: a psychological assessment; an evaluation of their cognitive skills; an evaluation of social and family issues; an assessment of behavior and interpersonal relationships; an assessment of the physical and social environment of the workplace.

This paper sets out to discuss methods, found in the literature, of assessing an individual's functional capacity that can be used for the process of including people with disabilities at work.

2   FUNCTIONAL EVALUATION METHODS FOR PEOPLE WITH DISABILITIES

2.1.   FUNCTIONAL INDEPENDENCE MEASURE (FIM)

The Functional Independence Measure is a tool for assessing disability in patients with functional restrictions of various origins, and was compiled by Granger et al (1986). In Brazil, it has been translated into Portuguese and the 2001 version was validated by Riberto et al (2004). This version is also regarded as being very reliable.

Its primary purpose is to evaluate quantitatively the amount of special help required by a person to perform a series of everyday motor and cognitive tasks. This evaluation is carried out by using interviews and directly observing the patient. Among the activities evaluated are: self-care (feeding, personal hygiene, bathing, dressing from the waist up, getting dressed from the waist down, using the toilet), transfers (bed, chair, wheelchair, WC, bathtub or shower), locomotion (walking or by wheelchair and stairs), sphincter control (control of urine and feces), communication (comprehension and expression) and social cognition (memory, social interaction and problem solving) . Each of these activities is evaluated and receives a score that goes from 1 (total dependence) to 7 (complete independence), so the total score ranges from 18 to 126. (Riberto et al, 2004).

The cognitive domain of the Functional Independence Measure is one of the greatest differentials of this functional assessment tool compared with others because the activities that it includes are normally only assessed in separate neuropsychological tests. The idea of testing functional independence for cognitive activities presents an innovative way to address these aspects of the higher functions of the brain, since as well as checking on the patient´s ability to complete them, such testing also enables a check to be made on to what extent this ability is recognized by family members and caregivers, who may go on to delegate such activities to someone with one or more of these disabilities (Riberto et al, 2004).

Thus, in the Functional Independence Measure there are no steps for completing the instrument which is a list with a description of the items which are evaluated and given a score of 1-7 and, in the end, the values are summed so as to obtain the total score.

2.2.   WORK ABILITY INDEX (WAI)

The Work Ability Index was compiled in Finland to be used in occupational health services for the purposes of identifying to what extent a worker has the skills to perform his/ her job and of serving as a method of assessment for health examinations and surveys in the workplace, i.e. it assesses workers' perception of their ability to carry out their work (Tuomi et al, 1997).

The Work Ability Index is a tool for measuring the capacity for work from the perspective of the workers themselves by means of ten questions summarized in seven dimensions: (1) "capacity for current job and compared with the best in their life", represented by a score of from 0 to 10 points; (2) "capacity for work in relation to the demands of the job", using two questions about the nature of work (physical, mental or mixed) and that, when weighted, provide a score of from 2-10 points; (3) "current number of self- and physician-diagnosed illnesses" obtained from a list of 51 illnesses, scored from 1 to 7; (4) "estimated lost time at work due to illness", obtained from a question with a score ranging from 1 to 6 points; (5)" absences from work due to illnesses", obtained from a question about the number of absences, categorized into five groups, scored in a range of from 1 to 5 points; (6) "own prognosis on the capacity for work", obtained from a question with a score of 1, 4 or 7 points; and (7) "mental resources" based on a score of from 1-4 score points obtained by weighting the answers to three questions. The results of the seven dimensions provide a measure of the capacity for work ranging from 7 to 49 points (Martinez et al, 2009).

According to Tuomi et al (1997), the individuals assessed can be classified according to the final score, as shown in Table 1:

Table 1 – Classification of the Index of Capacity for Work

Points	Capacity for work	Objectives of the measures
7-27	Low	To restore the capacity for work
28-36	Moderate	To improve the capacity for work
37-43	Good	To improve the capacity for work
44-49	Excellent	To maintain the capacity for work

Source: Tuomi et al (1997)

In his studies, Medeiros Neto (2004) found that the Work Ability Index is applicable to assessing a PD's functional capabilities. However, according with the same author, the instrument has not yet been validated in Brazil as to using it to assess PDs.

2.3. AMI (AVAILABLE MOTIONS INVENTORY) SYSTEM

The AMI System was originally developed to assess the residual capacity of the upper extremities of individuals with neuromuscular disabilities such as cerebral palsy, while performing light manual tasks in industrial jobs. It is applied to the functional assessment of PDs, placing PDs in jobs, quantifying functional losses due to injuries or accidents, designing and/ or modifying tasks, machines and workplaces, as well as evaluating the performance of work after modifying tasks thus determining the potential for improving productivity (Tortosa et al, 1997).

It evaluates the physical capacity of the person in 2 categories: operations with control and assembly commands, including measures of the person's strength, precision and pace of physical performance, using 71 different evaluations in each hand, where the manual tasks of the industrial sector are simulated in a cabin. Scoring and data analysis are done on a computer using specific software.

The assessment of each individual is conducted progressively using the following analyzes: raw scores of the skills and movements. The first level of the AMI System assessment are the gross evaluations which are calculated in kilograms for strength tests and correct actions per minute for time measurement tests. The second level scores capabilities, in which points are given to the activities of the individual with disabilities compared to subjects who have none. The third and final AMI level of analysis includes a classification of movement as per the part of the body that performs the task (Tortosa et al, 1997).

The 71 ways of scoring the capacity of the individual can be transformed into scoring fourteen classes of movement by means of a weighting system such that the subject can have a profile score of the class of movement for each AMI subtest: the value zero is the average score of people with no disabilities and a negative value indicates a sub-standard performance (Tortosa et al, 1997).

Thus, the importance of this tool is in assessing the skills of someone with neuromuscular disorders, in order to provide parameters to define the jobs and tasks to be performed by the worker and this forms part of the process of including PDs at work.

2.4. THE PAIN DISABILITY QUESTIONNAIRE (PDQ)

The PDQ is a quantitative method of assessing functional capacity focused on the disability and functionality of individuals with chronic musculoskeletal disorders. It was developed by Anagnostis et al (2004), using items from other methods that emphasize pain as a dysfunction and information from health professionals so as to construct the questionnaire.

The aim of the PDQ is to evaluate to what extent the presence of pain hinders an individual undertaking his/ her everyday social and professional activities.

The questionnaire is answered by the person assessed and consists of 15 questions related to their functional and psychosocial status. Each answer is included on a horizontal scale ranging from 0 to 10. The answer is about to what extent pain interferes with the item in question: 0 is given when the pain does not interfere and 10 when it impossible to undertake the activities. In the end, the points given to each question are added up to obtain the final value that represents the extent to which pain does interfere with activities of the person assessed.

The fact that the PDQ is a quantitative method makes it easier for the evaluator to understand to what extent pain interferes with the everyday social and professional activities of the individual assessed.

2.5. THE SHORT MUSCULOSKELETAL FUNCTION ASSESSMENT (SMFA) QUESTIONNAIRE

SMFA is a quantitative method that was drawn up to check to what extent the individual is affected by functional changes arising from musculoskeletal disorders, and thus to determine the physical limitations of the person assessed (Swiontkowski, 1999).

The SMFA questionnaire is answered by the individual evaluated and contains two parts: the first is the index of functional impairment and the second is the index of discomfort. The index of functional impairment contains 34 questions on functional performance, of which 25 questions evaluate the difficulty that individuals have when performing certain activities. They are grouped into four categories: everyday activities, emotional status, function of the arms and hands and mobility. The answers to each question may range from a score of 1 to 5 points, where 1 point is given for performing the activity without difficulty, 2 points for without much difficulty, 3 points for with moderate difficulty, 4 points when it is very difficult to perform the activity and 5 points if the person is unable to perform the activity. The last nine questions on functional performance assess the frequency with which individuals had difficulty performing certain activities in the previous week and the answers also range from 1 to 5 points, where 1 point is given when they had no difficulty in performing their activities, 2 points for seldom, 3 points for sometimes, 4 points for most of the time and 5 points when they always had difficulty performing their activities in the previous week.

In the second part, the discomfort index has 12 questions that enable individuals to assess to what extent they are inconvenienced by problems in their leisure, recreation, work and family activities. The answers to each question range from 1 to 5 points, where 1 point is given for no discomfort, 2 points for little discomfort, 3 points for moderate discomfort, 4 for great discomfort and 5 points for extreme discomfort (Swiontkowski, 1999).

The final SMFA score is obtained by summing the scores of questions from both parts, where the higher the score (maximum score possible 230), the lower the functional capacity of the person assessed (Swiontkowski, 1999).

Thus, the SMFA questionnaire is a method that is easy to use and quickly applied and through which the functional limitations of the individuals assessed can be discovered from their answers. Even though this method was initially drawn up to evaluate people with musculoskeletal disorders, it is believed that it can be used to determine a PD's functional profile since it contains general questions about the performance of various parts of the body which are needed to carry out everyday professional and leisure activities.

2.6.   THE COMPETENCY PROFILE

The Competency Profile was put forward by Mélennec *apud* Tortosa et al (1997) in order to facilitate the work of forensic expertise regarding a PD's disability, in order to set out a complete profile of the physical and mental attitudes of the subject examined by quantifying the global disability and defining precisely what the person is or is not capable of doing. The following form part of the evaluation:

- Profile of the subject - age, level of education, previous professional activities and causes of interruption, family and social status, medical and surgical history relevant to drawing up the current competence profile.
- Intelligence and intellectual functions – conducting special tests or consultations with other specialists, when it was not possible to evaluate these in the previous topic. Or else, with regard to elementary actions such as understanding simple sentences, reading, expressing themselves, etc.
- Psyche and psychological attitudes - will of the subject, motivation towards or interest in doing work and the state of mental health. Examination by a specialist may be requested.
- Sense organs and communication - describes the kind of change in vision, hearing, touch or smell and the consequence of this in performing activities such as moving about, reading, handling, communicating with people, with a view to highlighting the need for aids.
- Upper limbs and use of hands - ability to grasp, recognize by touch, carry and hold objects.
- Heart, lung, and capacity for effort - tolerance of effort using simple tests, such as climbing stairs, bending one's legs.
- Spine - flexibility of the spine, using the fingers-floor distance.
- Lower limbs - the ability to walk, to stand with or without aid, kneel, sit, walk up and down stairs, run, etc.
- Other information - sometimes information is sought on the digestive function, skin, kidney and urinary function, the endocrine, frequency of epileptic seizures, etc.

3   CONCLUSIONS

The methods described evaluate the functional capacity of the individual, which results in obtaining his/her profile, i.e., his/her abilities, skills and limitations, which can be used for the purpose of selecting, designing or modifying equipment, processes, places and work environments with a view to including the people with disabilities adequately at work. .

Such methods demonstrate the importance of making a functional assessment of PDs so that the inclusion process ensures good interaction between human beings and their tasks or activities by lowering the barriers imposed by the environment and the organization of work, thus making workers more effective.

Finally, it is clear that in order to make jobs suitable for people with disabilities, the worker's functional capabilities must be compared with the demands of the tasks of the job. By doing so, it is possible to make adaptations to the workplace and to the tasks so that they support the skills and needs of workers with disabilities.

REFERENCES

Anagnostis, C., Gatchel, R. J., Mayer, T. G. 2004. The Pain Disability Questionnaire: a new psychometrically sound measure for chronic musculoskeletal disorders. *Spine* 29 (20): 2290-2302.

Fadyl, J. K. 2009. *Development of a new measure of work-ability for injured workers.* Master's Degree in Health Science - School of Rehabilitation and Occupation Studies. Auckland: Auckland University of Technology.

Granger C. V., Hamilton B. B., Keith R. A., Zielezny M., Sherwin F. S. 1986. *Advances in functional assessment for rehabilitation.* In: Topics in geriatric rehabilitation. Rockville, MD: Aspen.

Guimarães, B. M., Martins, L. B., Barkokébas Junior, B. 2012. Issues Concerning Scientific Production of Including People with Disabilities at Work. *Work* 41: 4722-4728.

IBGE - Instituto Brasileiro de Geografia e Estatística. 2011. *Tabulação Avançada do Censo Demográfico 2010.* Rio de Janeiro: IBGE.

Innes, E. Reliability and Validity of Functional Capacity Evaluations: An Update. 2006. *International Journal of Disability Management Research* 1(1): 135–148.

King, M. P., Tuckwel, N., Barret, T. E. 1998. A critical review of functional capacity evaluations. *Phys. Ther.* 78 (8): 852-866.

Martinez, M. C., Latorre, M. R. D. O., Fischer, F. M. 2009. Validade e confiabilidade da versão brasileira do Índice de Capacidade para o Trabalho. *Rev. Saúde Pública* 43(3): 525-532. Jun.

Martins, L. B., Guimarães, B. M. 2010. Ergonomia e a inclusão laboral de pessoas com deficiência. *Revista Brasileira de Tradução Visual* 3(3).

Martins, L. B., Barkokébas Junior, B., Guimarães, B. M. 2012. Including people with disabilities at work: a case study of the job of bricklayer in civil construction in Brazil. *Work* 41: 4716-4721.

Medeiros Neto, C. F. de. 2004 *A Influência dos fatores ergonômicos sobre a capacidade laboral de pessoas portadoras de deficiência física no setor calçadista paraibano: um estudo de caso.* João Pessoa: UFPB (Mestrado em engenharia de produção) Programa de Pós–Graduação em Engenharia de Produção, Escola de Engenharia. Universidade Federal da Paraíba, João Pessoa.

Organização mundial de saúde – OMS. 2002. *Salud e Envejecimento*: um documento para del debate. Madri.

Riberto, M. et al. 2004. Validação da versão brasileira da medida de independência funcional. *Acta Fisiátrica* 11 (2): 72-76.

Simonelli, A. P., Camarotto, J. 2008. Analysis of industrial tasks as a tool for the inclusion of people with disabilities in the work market. *Occup. Ther. Int.* 15(3): 150–164.

Swiontkowski, M. F, Engelberg, R, Martin, D. P, Agel, J. 1999. Short Musculoskeletal Function Assessment Questionnaire: Validity, Reliability, and Responsiveness. *J Bone Joint Surg Am* 81(9): 1245-60.

Tortosa, L. et al. 1997. *Ergonomia y Discapacidad.* Madrid: Ministerio de Trabajo y Asuntos Sociales.

Tuomi K., Ilmarinen J., Jahkola A., Katajarinne L., Tulkki A. 1997. *Índice de capacidade para o trabalho.* Tradução de FM Fischer. Helsinki, Finlândia: Instituto Finlandês de Saúde Ocupacional.

CHAPTER 40

The Organizational Factors in the Causalities of Aircraft Accidents Related to Maintenance

Campos,Reginaldo.[11,] MARTINS, Edgard[2]; SOARES, Marcelo[3]

1 Universidade Federal de Pernambuco- Depart of Design- CAA – Recife-
Pernambuco- Brasil, regiscampos3@gmail.com
2UFPE- Universidade Federal de Pernambuco- Depart of Design- CAA – Caruaru -
Pernambuco- Brasil; edgardpiloto@gmail.com
3UFPE- Universidade Federal de Pernambuco- Depart of Design- CAA – Recife-
Pernambuco- Brasil; marcelo2@nlink.com[br]

ABSTRACT

According to reports on aviation accidents, most investigations the blame is always linked to human factors. Our research was able to prove clearly that the level of organizational factors, such as training systems and qualifications, financial resources and the allocation of scarce resources to technical training to perform repairs on aircraft systems, also contribute to the occurrence of aviation accidents . Cultural or value systems that permeate the organization has a direct influence, as factors contributing to errors in the maintenance area. The pressure of time, the organization becomes the present at every moment of the operation, as well as significant financial losses due to downtime of the aircraft, in some cases can lead to fines and additional costs such as transportation and accommodation of passengers, loss of image and other additional consequences.

Keywords: Guilt; Organization; Accidents

1. INTRODUCTION

Working conditions and preparation of aviation maintenance professionals can determine situations that could lead to problems, even to the most significant problems in relation to flight safety, such as minor incidents, accidents sometimes imperceptible or even catastrophic. Not all factors that contribute to the causation of aircraft accidents are considered, especially the organizational factors. Technological advances force the maintenance professional to increasingly improve their knowledge and recycle constantly, but organizations do not always prioritize these needs. The lack training raised in our research showed that 22% of the occurrences of accidents were the perceived lack of training of maintenance engineers. When filtered, our sample showed that the occurrence of accidents related to maintenance, which in 81% of cases the pilots could do nothing to correct the situation, preventing it from any action which could avoid the accident. Given these data, we conclude that the cognitive factor acts directly when the mechanic is doing its thing, sometimes the lack of training on new equipment that he is handling, which are often factors that contribute directly to the cause of aviation accidents. We must be attentive to the precepts of administrative rules and procedures, human and organizational should be taken as the focus of Organizational Ergonomics. According to Hendrick (1991), the term "Organizational ergonomics" or "macroergonomics" is defined as the optimization of system design and organizational work, considering the importance of people, technological variables, and their environmental interactions. This term was used by the author as the basis for creating rules and administrative and organizational human, signa-ling that they must be treated in the eyes of Ergonomics in large organizations.

2. CONTEXTUALIZATION

The global aviation market is growing every day and every time there is a greater need to focus on flight safety of the aircraft. In this context, we evaluated in our study areas involved, focusing on the maintenance area to enable a safe aviation. It is a measure of responsibility attributed to the aviation mechanic to keep aviation safe at all times, either in air or when it is still on the ground being prepared for the flight phase. The mechanical aircraft is often working in difficult or as a result of the pressure time, or because the environment for which it is inserted.

The analysis of the actions of operating personnel during the maintenance process involved in accidents and incidents has been the traditional method used in aviation, to assess the impact of human performance in flight safety.

When reports of the process of investigating air accidents and incidents are presented to international bodies like the NTSB, FAA, CENIPA, among others, little stands out the processes that may have led to these incidents and accidents investigated. The aircraft mechanic sometimes not received adequate training to perform their jobs and little is considered the factors that are around you that may be contributing to the occurrences. Although most investigations have a tendency to search for data for future preventative actions, not always the organizational factors, ergonomic, psychological, cognitive or considered. The environment to which this work is inserted during the execution of their task, often contributes decisively to facilitate the generation of pro-blems. The organizational factors are not always mentioned or even considered in investigations, where the lack of a policy for monitoring this work or due to time factors, are often contributors to aircraft incidents. This contribution can be exemplified by the lack of organization of spare parts, lack of access to information or the lack or inefficiency in the procedures to be followed during the execution of a task or pass their turn. There is a hidden war between the security procedures to keep an aircraft in flight conditions and the pressure that the equipment is in operation time greater than is possible. Although the safety factor should considered as a priority in aviation, many times, the mechanic is obliged to make available as soon as possible to fly this plane. This compromise between production and safety is a complex and delicate balance, because it requires that these professionals are highly effective in maintaining the implementation of mechanisms to achieve successful flight safety.

Our studies have addressed the factors that influence the performance and competence of an aircraft mechanic in performing the task in your workplace. We discussed important issues like the consequences of corporate pressure that can directly interfere with the proper execution of the task an aviation mechanic. Another issue addressed was the lack of continuous and adequate knowledge in the mechanical handling systems, referring to problems related to ineffective training. This problem is compounded by the continuing evolution of aircraft technology that requires more training of those responsible for maintenance of these artifacts.

Operational and organizational aspects can interfere with the daily work of these professionals, where the impacts of these factors affect the performance of an aircraft mechanic.

3. PROBLEMATIZATION

Maintenance is used in various types of organizations to avoid possible gaps and breaks in plant and machinery. This area is critical to provide reliable, quality improvement and cost reduction, since the planned maintenance is less costly, and essential when considering the safety aspects of aviation. There are several types of maintenance that can be employed, such as corrective maintenance, preventive and predictive. The equipment must be maintained according to the company's decision, either by the manufacturer's guidelines or policy of the organization itself, but always giving priority to flight safety.

General aviation makes maintenance procedures, more elaborate than the other segments or, as the financial and the consequences in case of fault are much greater. According to Nakajima (1988), when applied correctly the maintenance service recommended by the manufacturers, the organization fails to repair the device continuously, to trace the causes and thus to manage these maintenance. There are several reasons that lead us to implement procedures for aircraft maintenance, which are:

a) Increased reliability: good maintenance generates fewer stops without programming;
b) Improvement of quality machines and equipment with maintenance deficit, is likely to generate errors that cause poor performance and therefore cause quality problems;
c) Cost reduction: when cared for, the equipment works more efficiently;
d) Increase of life: simple care such as cleaning and lubrication, ensure the durability of the machine, reducing small problems that can cause wear or deterioration, and;
e) Improve safety: machinery and equipment and care are less likely to behave unpredictably or outside the established standard, thus reducing possible risks in the operation.

Some peculiarities of the area of maintenance of aircraft are deployed for the operation of flight safety, which is a priority in any situation in aviation. Any savings that might compromise this operation without security is not usually practiced. Investigations on the occurrence of incidents or accidents seek information that will point out the responsibilities and where such irregularities are found, the consequences can be harmful not only financially, but mainly for the company's image.

Some airlines are worrying to developing a database that aims to make a system of risk control, where the difficulty lies in the influence of the organization of this type of control.This control is only possible through the creation of the database in order to visualize the errors that occurred in maintenance procedures, but these are rarely reported, since this can result in punishment for the employee. Sometimes the organizational processes and procedures may reflect errors in the maintenance of an aircraft, such as the release of a piece that should go to repair and is delivered to the mechanic to be used in the repair of an aircraft. Any failure in these processes can cause various problems.

Often these professionals are required to perform a task outside the standard procedures in the manuals. These factors may be lack of time, lack of suitable components for replacement, lack of proper tools or even the task different from that provided in the manuals of procedures. The model based on the "Swiss Cheese" of Reason (2000), makes an analogy with the above mentioned factors that may be contributing to incidents or even accidents.

4. THE SOURCE

This analysis was derived from studies of the author to his defense of Master Campos (2011), where they were raised and several causal factors in aircraft that were involved in accidents related to maintenance, the factors for these events, such as organizational.The research was based on Aviation Database Software developed by Martins (2006) to compile the elements of instrumentation for the defense of their master's theses and doctoral degrees. This method was used and adapted to evaluate the mechanical aspects of aircraft, which are also considered other factors that contribute to the occurrence of incidents and accidents in aviation. The basis of our research reports were completed by the official investigation of aviation accidents and incidents worldwide recognition, the main factor for its occurrence was related to maintenance. We analyzed 125 reports randomly, dating from 1946-2010, and there was no distinction between countries, manufacturers and size of aircraft. Featuring the evidence related to ergonomic studies, this proved especially organizational influences as contributory factor for the occurrence of aviation accidents and incidents. The following evidence was considered:

Code	Description of evidence causes ergonomic Problem
B1	Structural (material)
B2	Project error
B3	The wrong choice aircraft
B4	Support Staff of earth
B5	Stress
B6	Control Tower (ATC procedure with - Air Traffic (Control)
B7	Fatigue and physiological problems / physiological disorders
B8	Psychological and emotional problems
B9	Error leadership
B10	Training Problems
B11	Training inappropriate / poor training
B12	Mechanical error/ maintenance error / failure of instruments
B13	Layout error
B14	Emotional overload and / or cognitive
B15	Distribution wrong task
B16	Instruments in position deficient
B17	Language Error
B18	Communication error
B19	Information error
B20	Collective error
B21	Other ergonomic problems / cognitive

Table 1 - Summary of Evidence of Ergonomic causes (Martins, 2006)

During our assessments and correlating the evidence, it is worth noting the B10 factors (training issues), B11 (Inadequate training / inefficient) that correlated with B12 (error factor for mechanical maintenance / error / failure of the instruments) brings substantial information for our work in accordance with Figure 1, which demonstrates the lack of proper training can have disastrous consequences for aviation.

Correlation between evidence ergonomic

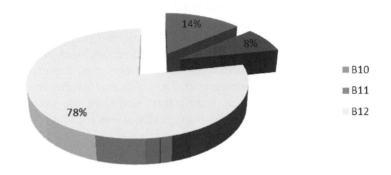

Figure 1 - Correlation of Ergonomic Evidence of Causes - Source: Campos (2011)

5. CONCLUSIONS

Although the bodies of the countries that have already been researched systems evaluation and accreditation of mechanical flight, through our research we concluded that even with accident records dating from 1946, we found that the incidence of failures and even training activities of without professional training is relatively straight forward. Following are the recommendations pertaining to this item:

a) Provide solutions to technical and organizational problems related to the maintenance area;

b) Technological advances force these professionals to specialize their knowledge and recycle constantly, but organizations do not always prioritize this perceived need, such as lack of training data have been highlights in 22% of the occurrences of accidents in our sample. These data we conclude that the cognitive factor acts directly when the mechanic is performing its task, because the lack of training, qualification and recycling equipment in which he handled and contributory factors are often the direct cause of air accidents.

c) Technological advances force these professionals to specialize their knowledge and recycle constantly, but organizations do not always prioritize this perceived need, such as lack of training data have been highlights in 22% of the occurrences of accidents in our sample. Through these data, we conclude that the cognitive factor acts directly when the mechanic is doing its thing, because the lack of training, education, and recycling the equipment in which he dealt, and contributing factors are often the direct cause of air accidents.

It should be a reflection on the organization's role in the failures that occur within the area of maintenance. It should not be released from the responsibility of maintenance technicians for their work, or shift blame away from workers to management, but to create conditions for the maintenance activities to ensure flight safety and all those involved in the operation.

We found that some factors were evident as a lack of skilled mechanics, procedures when inspecting aircraft while in transit in airports, or are in the hangars for maintenance companies. We also stress that were not found in our research reports on aviation accidents, elements indicating that the airline was in compliance with all or part of the rules and procedures for the maintenance area. We believe this is a good time of investigative bodies to determine the parameters for this type of factor was more prominent in research. We are facing an increasing use of air transport and in accordance with the CENIPA (2011), the number of accidents tends to break records in Brazil, since up to 20 June 2011 had been 74 recorded accidents involving aircraft drills, 67% around the year 2010.

REFERENCES

CAMPOS, R. M. - Ergonomics in aviation: a critical study of the liability of aircraft mechanics in the causation of accidents – Recife - Brazil: 2011 - UFPE

CENIPA- http://www.cenipa.aer.mil.br/cenipa/paginas/historico.php - Accessed 23/06/11

H. W. HENDRICK, Ergonomics in organizational design and management, Ergonomics, 34, Santa Monica, 1991, 743.

MARTINS, Edgar Thomas - Monograph Masters-A Critical Study of Responsibility of Pilot in Aircraft Accident, 2006, UFPE

NAKAJIMA, S. Total productive maintenance. Productivity Press, 1988.

REASON, J. (2000): Human error: models and management. BMJ 2000, 320:768-70

CHAPTER 41

Visual Monitoring System Applied to Team Workout

- An Inclusive Design Monitoring System

Gago da Silva, J., Moreira da Silva, F.

CIAUD – Research Center in Architecture, Urban Planning and Design
Technical University of Lisbon, Portugal

gago.silva@gmail.com
fms.fautl@gmail.com

ABSTRACT

This paper describes the development of an experimental teaching module, where students developed several scenarios to test a Vest Display System, the system allows the monitoring of physical conditions through biometric data input, providing fast visual assessment to team instructors in workout activities.

Keywords: Accessibility, Sign Design, Scenario Design

1   INTRODUCTION

The classroom based research project, focused in a physical condition monitoring system applied to a team coached activity, using a common analogy present in video games, characters' special powers and energy meters. In this project, the augmented reality concept can be used and adapted to a different and more accessible output. The referred project was conducted with a group of students, in the Digital Environments course - the Master in Graphic Design degree from ESART (www.esart.ipcb.pt), and was developed in a time span of three months.

The project was designed to be implemented in a Vest Display System (V.D.S), a device consisting in a flexible led screen installed on a vest using Input data, which displayed the heart rate and tension, retrieved from a wrist watch.)

2   MONITORING PHYSICAL ACTIVITIES

To control and alert for certain situations related with physical conditions during training activities regarding team coordination, the instructor needs to have an overall vision and surveillance. To properly coordinate the team, the instructor is compelled to be in a distant point of observation, but simultaneously maintaining a close view to each of the team members. There are certain involuntary signs of physical stress that are expressed by the person involved in the training activities. The instructor needs to have a fast assessment of the team endurance values and their response to stress factors. Different scenarios require distinct visual positions and information reading settings, but each one of those scenarios can contribute to a larger set of sign language used in the vest system display. It's crucial in a classroom project the development of these concepts, designing an alert system focused on team physical activities and a warning language system relating to an already understood sign system. One of the problems is related to this sign system, whether it is useful and easily identifiable, but also whether it is affordable, functional and understandable.

In a classroom project several scenarios can be developed for a better understanding of the project values and scopes of activity. One of the scenarios were the introductory scuba-diving classes, where divers used the vest over their equipment while performing a several actions planned by the class instructor. If the monitoring is done in a distant position, the instructor will need a better view to each individual's reactions to the exercise. This vest digital system, used by each diver, could improve safety in the planning of diving lessons, creating a useful accessibility system.

3   V.D.S

The core of this educational project is the creation of a visual warning system applied to a vest. The vest could be used in team activities focusing in managing situations of stress related with physical activities. Each trainee shows his physical condition through a visual system, using a sign system, an energy meter display, similar to those used in some video games interface. The display shows biometric data allowing a more efficient supervision by the team instructor when he is managing team efforts during training activities. Therefore the instructor will be warned to any alert information displayed by the vest, engaging in a faster response to stress situations observed in the each member of the class.

The system uses an input system, consisting on a wrist watch collecting biometric information, and an output system composed of a flexible led screen, using a conventional sign system.

In order for the sign system to be more understandable we used a cultural convention, through the use of an analogy to a video game, where different dynamic signs show the character's stamina and energy levels, providing useful information to the player. In this project the provided information is allows the instructor to manage students' effort limits in planned training sessions. The system could also facilitate a first visual recognition of any alert signal related to stress issues providing a fast response to the unfolding of the activities planned for the class, re

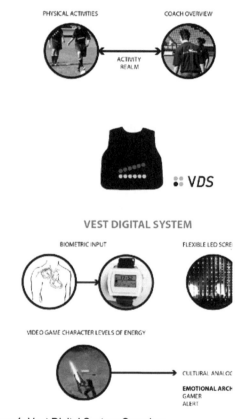

Figure 1: Vest Digital System Overview

This problem is interesting because it states the need to create a warning signal system using accessible technology on augmented reality basis. Working with a group of students of the Digital Environments course in the Master in Graphic Design, each one of the students chose a scenario and developed his visual sign system according to the chosen scenario. Four scenarios were created, all of them introductory trainings: scuba diving, fire fighting, football and team cycling.

The visual system in the V.D.S works as a simplified system of output, translating complex information to visuals, and providing a fast assessment system of response. The biometric values provides actualized information to the instructor managing the training activities.

The rapid advances in ICT (Information Communication Technology), nanotechnology, biomonitoring, mobile networks, pervasive computing, wearable systems and drug delivery approaches are transforming the health care sector and fuelling the m-health phenomenon (Olla, Tan, 2009). The M Health or mobile health is a recent term for medical and public health practice supported by mobile devices, such as mobile phones, and patient monitoring devices.

The biometric data in the V.D.S is used as prevention to plan physical training. Using the biometric parameters related with physical stress issues (fatigue, abnormal heart rate) the students involved in this project conceived several scenarios where a series of interaction actions were developed.

These scenarios were focused in a training activity, normally supervised by the team instructor. The scenarios developed in the training activities were:fire fighting, football, scuba-diving and team cycling. In each of the scenarios, the instructor uses the information of the physical intensity performed by each team member, collected through an occupational ergonomic device.

The developments in occupational ergonomics are mirrored in exercise and sport contexts. The intensity of the exercise can be monitored so that the athlete is not overloaded. The stimulus must be varied regularly to avoid habituation and boredom. Safety is a priority if the training is to be effective, since injuries are often caused by using faulty techniques. For example, the majority of weight lifting injuries occur when athletes handle loads of weight that are too heavy or adopt poor lifting techniques (Reilly, 2010).

One starting point in an ergonomics analysis is the quantification of load on an individual. The assumption is that the task imposes demands on the individual, whose responses can be used as indicators of physiological strain. Physiological criteria corresponding to exercise intensity include energy expenditure, oxygen uptake, body temperature, heart rate, and blood metabolite concentrations. Such variables have been associated with subjective perceptions of exertion, task difficulty, and thermal comfort as well as with biomechanical measurements that reflect the level of power output. (Reilly, 2010)

The ergonomic factors involved in visual perception applied to each scenario are dependent from variable barrier conditions, like visual movement and noise. All the scenarios are dependent of conditions related with movement, but in activities as scuba diving and fire fighting there are situations related with changes in the environment. A water flow creates murky images and in fire fighting those situations of lack of visibility are usually related to smoke and particles in the air. Having in mind these particular situations, the students performed several simulations using different visual accessibility conditions according to the description of each scenario.

The student also retrieved information about reading conditions. Using scuba-diving as the example there are situations related with colour and shape perception which are influenced by different materials and light conditions.

In the light conditions associated with scuba-diving there are changes in how colour is perceived. Water absorbs colours, starting with vivid ones such as red and orange, and ten feet deep in the water the red turns to black, and at thirty feet orange becomes imperceptible. (Crockett, 2002).

The transference of light in water is different because air, glass and water have different densities. The light travels at different speeds. Light travels around 25% slower in water than in air, therefore light rays will be refracted (bent) as they cross the interface between each element. As light crosses the water, glass, air boundary of a driver's mask, this causes a false impression of distance by a ratio of approximately 4:3 and magnification by a third (Jackson, 2005).

Similar difficulties can be found on environments where smoke and dust are involved. The project planning followed the accessibility factors expressed by each environment settings, When engaging the questions of accessibility there was an additional effort in creating the appropriate narrative for each scenario simulation.

When developing a learning experience on the theme of accessibility, it is crucial for the students the exercise of their capacity and ease in creating different scenarios, using for that purpose the digital technology as vehicle in channelling information through different scenarios. A crucial feature in designing interactive systems, Scenario-based design offers significant and unique leverage on some of the most characteristic and vexing challenges of design work: Scenarios are at once concrete and flexible, helping developers manage the fluidity of design situations.(Jacko, Andrews, 2003).

This effort can be accomplished by the use of narrative creation tools, following a collection of facts retrieved from scientific or technical literature from each activity area of the scenario proposed. With that purpose multiple perspectives about the problem are stimulated, in how to convey the informations provided by the V.D.S vest digital system.

The narrative is 'channelled' by scenario simulation, enabling different approaches and new appliances for the V.D.S. When engaged, the theme of accessibilities are tested using different user routes and points of access into the experience.

A cultural analogy, used as channel of communication, is provided to the students as an easy access and allocation of information in the field of simulation. Each scenario requires an exercise. Testing several variables in order to design a possible system. The use of popular cultural symbols such as the energy meters used in the 'RPG' video games provides an easy identification element, when creating a suitable reading condition for each of the scenarios created by the students.Thus creating a multi-perspective approach to the project; the cultural focus in identifying the visual elements, comprises also a system of accessibility.

The multiple perspectives are also fundamental principles when drawing in a wide variety of tools, when the design requires a test simulation as it happens in user interface design, Figure 2. Thanks to the student's feedback it is possible to outline several experimental stages of activity in the project design focused on accessibility:

- Scenario/Narrative creation
- Sign system
- Scenario simulation study

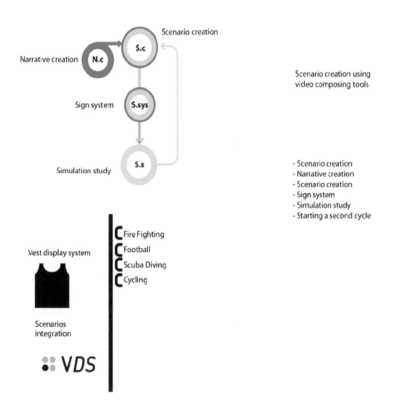

Figure 2: *Experimental sequence of activities developed in the V.D.S project.*

3 RELATED WORK

The appliance of visual systems using led technology has been found in a prototype work of David Forbe, the author of a TV vest composed of 14,400 green, red, and blue LEDs that together compose 4,667 pixels. The display image is sent from an iPod into a circuit board on the vest's left shoulder, and additional video processing converts the iPod data into signals for the LEDs, which travel along Ethernet-like cables to four separate boards.

Erick Johnson, the author of the project Light Bright (lightbright.net) developed a complex DIY fabrication of a total of 1,536 three colour (RGB) LEDs. A very flexible modular hand-made Red-Green-Blue LED display panel.

This introductory research to the problem of visual retrieval of biometric information was only possible with the help of the students of the Digital Environments course - Master in Graphic Design (ESART 2012), we also would like to acknowledge their support. Although this Introductory project aroused some interesting problems like the ergonomic factor in reading situations involving movement and visual barriers according to each scenario. We must draw a line between structural frame work and the visible results obtained from each scenario involving the visual vest system. The students scenarios aroused several questions related with the briefing structure. The accessibility factors must focus on different levels of accomplishment in the sign system design but also creating different evaluation scenarios.

4 CONCLUSION AND FURTHER WORK

The V.D.S project will be repeated with other groups of students to reach different outcomes within the project briefing. The role played by technology can only be enhanced when the right conditions in accessibility are created, for that reason simplifying the overall functionality in the system to help further developments in use and deployment.

The use of known cultural conventions will help in the system implementation. Bringing in context the technology and its use in practical scenarios will contribute to achieve and help the development of digital integrated systems such as the V.D.S system, merging the technical system (the technology) and the implementation in application scenarios (the human factor).

We envision as future work in the V.D.S system the development of more testing scenarios, the creation and analysis of other barriers to accessibility, using video game scenarios as visual metaphor, delimited in training activities were teamwork is a common practice. The short time available for the project development may decrease production quality, thus a new frame of work for the project development is contemplated.

The development of competences through time is crucial for Ludvigsen, et al. (2010) the creation of new practices, using new technologies is dependent from re-orderings and the emergence of knowledge and competences, time is also a critical factor in such interactions of knowledge and for competencies necessary to conquer new stability of practice in the use of new technologies.

The development of this project in a multiple perspective practice of teaching opens new possibilities, driven by two forces; the accessibility of visual information applied to team activities, and the use of accessible analogies based on popular culture icons. These two driving forces convey stronger outcomes when developing a visual system made for accessibility purposes.

ACKNOWLEDGMENTS

The authors would like to acknowledge the ESART (www.esart.ipcb.pt), CIAUD, the course of Digital Environments – Master in Graphic Design 11/12.

REFERENCES

Crockett, Jim. (2002). The Why-To of Scuba Diving. Aqua Quest Publications, Inc.

Jacko, Julie A., Sears, Andrews. (2003). The human-computer interaction handbook: fundamentals, evolving technologies, and emerging applications. Routledge

Jackson, Jack., 2005. Complete Diving Manual. New Holland Publishers

Ludvigsen, S. , Rasmussen, I., Lund. A., & Säljö, R., 2010. Intersecting Trajectories of Participation; Temporality and Learning. Learning across sites. Routledge.

Meckler, D (2011).You Built What Wearable Led Television.[online] Available at:http://www.popsci.com/diy/article/2011-11/you-built-what-wearable-led-television [accessed January 2012]

King, William R., 2009. Knowledge Management and Organizational Learning. Volume 4 de Annals of Information Systems. Springer

Reilly, Thomas., 2010. Ergonomics in sport and physical activity: enhancing performance and improving safety. Human Kinetics.

CHAPTER 42

Users' Awareness of Computer Security and Usability Evaluation

Ryang-Hee Kim, Kee-Sam Jeong, and Soon-Ei Bae
Dept. of Fashion Design, University of Hanseo,
360 Daegok-ri, Haemi-myeon,
Seosan-si, Chungcheongnam-do, ZIP: 356-706, South Korea,
Yanghee1003@naver.com

Young-Kuk Kwon ,Han-Soo Je, Ji-Woong Choi,
Dae-Gyun Shin, Jae-Seok Jang,
Dept. of Safety Engineering, Seoul National University of Science and Technology,
KongNung 2-Dong, NoWon-Ku, Seoul, ZIP: 139-743, South Korea,
safeman@seoultech.ac.kr

ABSTRACT

Currently, the network system (PC) has been used in supplying electric power, opening facility of dam, web-surfing, air defense weapon of military, weather prediction and national security. There has been a leaking of personal information from a major portal site by the hackers. The victims' personal information is used for crimes, such as "Illegally opening an account", or "sending Spam". The aim of this study is to develop the usability evaluation method for the personal computer security. Also, there is no difference in age groups for awareness and usability of computer security difference in age groups for awareness and usability of computer security'. In order to do this, we selected heuristic evaluation method to develop the evaluation method for the personal computer security system. Then, we collected 137 usability principles from previous literatures, including Nielsen (1994)`s checklist, screened a total of 34 usability principles among them in terms of tangible user interface main properties, and refined twenty five usability principles. The selected usability principles are classified systematically using statistical method, T-test and Principle Components Analysis (PCA) using Factor Analysis. 196 subjects (Groups: 20-40's, 40-70's) were divided into two groups: Adolescence group (N: 108, Female; 54, Male; 54), Senior-age group (N: 88, Female; 44, Male; 44). As a result, 25 usability principles are categorized into 5 groups according to their correlations and we named them respectively (1) User Interaction-support, (2) User Cognitive-support (3) User Performance-support, (4) User Satisfaction-support, (5) User Physicality principles, as usability checklist items are produced a 5-point Likert scale. Finally, the result of T-teat showed that users' awareness of computer security and usability evaluation groups were statistically significant for differences between Adolescence group and Senior-age group (**p<.01). Also, we evaluated this users' awareness of computer security system using checklists which is a one of currently available tangible user interface prototype, and could suggest 25usability items for users' awareness of computer security system for the Aged.

Keywords: Tangible Computer Security System, Ubiquitous Computing, Users' Awareness, Heuristic Evaluation Method, Usability Evaluation

1   INTRODUCTION

21st century is IT society, and IT technology will be everywhere in the foreseeable future. There is possibility of leaking one's personal information in the middle of internet if keep pursuing convenience (Hollerer, T., et al., 1999). Not so long ago, there has been an outflow of personal information on a website. The information was leaked easily by the hackers, although this website is a huge company with 35 million users. This incident shows the lack of security awareness of South Korea.

The numbers of people, accessing resources on the internet through personal computer, are continually increasing, and internet protocol itself was not structured to be safe. TCP/IP communication stack has high level risk of hacking through the network by malicious users or process since it does not contain any approved security standards. Even though internet connection has become safe recently, still, many incidents are caused which give us lessons that nothing is perfectly safe. So, it is necessary to enhance security conscious of the other country as well as South Korea where computer users put high priority on IT products' convenience. This research was progressed in order to find differences in awareness and usability of computer security by different gender and age groups. Also, we'd liked to develop users' awareness of computer security system using checklists which is a one of currently available tangible user interface prototype.

2   THEORETICAL BACKGROUND

2.1   Computer Security Practices

The convenience of high-quality IT product has inverse relationship with its security. In the movie 'War Games', a high school student hero accidently hacks Super Computer of U.S. Department of Defense to cause a threat of nuclear war. In this movie, the student used a modem (called WOPR) to connect to Department computer and plays a game with artificial intelligent software which manages nuclear missile reservoir (http://www.pcbee.co.kr).

As for the modern security practices, new millennium has come, where about 400 million people use internet globally (Computer Industry Almanac, 2004). Economical loss by 3 most perilous internet viruses is reported to be about 1.32 billion dollars for last 2 years (http://www.tscllc.com/Small-Business-Solutions/Tablet-pc-benefits.asp). In the Security Standard regard, all industry follows laws and regulations enacted by institutions such as AMA, or IEEE; it is the same with information security field.

2.2 User's Relevance of Computer Security

Computer Security could be defined by itself or anything that is related to computer, in other words, hardware, software, input and output data, handling data, computer communication, computer facilities, and its management. Types of Computer Security are as follows: 1) Network Security is used for disclosure of inside information and protection from external intrusion, 2) System Security is preventing unlawful usage of computer system by using loopholes of computer system's operating systems, applications, or server, and then 3) Data Security is for protecting system data, preventing interception or modification of the data in transferring, and safely transferring the data.

Computer vaccine means Electronics & Computer Science/Computer Science Computing a piece of software designed to detect and remove computer viruses from personal computer vaccine (virus check software) system such as "V3", Ever Green, Norton Antivirus, HaURi, and "Alyac". V3 is the first developed vaccine in Korea among the personal computer security system, whereas "Alyac" is the virus check software developed by Eastsoft etc.

On the other hand, usability is the ease of use and 'learnability' of a human-made object. The object of use can be a software application, website, book, tool, machine, processing, or anything with a human interacts. A usability study may be conducted as a primary job function by a usability analyst or as a secondary job function by designers, technical writers, marketing personnel, and others (ISO 9241-11, 1998).

It is widely used in consumer electronics, communication, and knowledge transfer objects (such as a cookbook, a document or online help) and mechanical objects such as a door handle or a hammer. The primary notion of usability is that an object designed with a generalized users' psychology and physiology in mind is, for example: 1) More efficient to use—takes less time to accomplish a particular task, 2) Easier to learn—operation can be learned by observing the object, and 3) More satisfying to use.

Table 1 PC User's Retention, Usage, and Relevance (2005, Statistics Department, %)

Division	PC utilization	Search	E-mail	Learning	Games	Chats	Shopping/finance	Others	Internet usage
Male	65.0	32.7	7.7	8.1	14.5	0.5	1.1	35.4	64.9
Female	56.6	29.8	7.4	5.8	6.6	1.3	5.5	43.8	56.4
15~19	94.3	37.4	14.1	16.2	21.3	3.8	1.2	6.0	94.3
20~29	85.6	46.0	11.5	7.8	13.2	0.8	5.9	14.7	85.5
30~39	72.2	40.0	7.8	7.6	11.0	0.5	5.1	28.2	72.1
40~49	43.4	23.5	4.9	4.4	7.4	0.5	2.3	57.0	43.2
50~59	19.3	11.4	1.6	1.7	3.5	0.0	0.7	81.2	19.0
60 and above	5.2	2.6	0.9	0.3	0.7	0.1	0.3	95.2	5.1

In accordance with the preceding studies, this research was progressed in order to find differences in age groups or genders for awareness and usability of computer security. Also, we evaluated this users' awareness of computer security system using checklists which is a one of currently available tangible user interface prototype. And then, according to the results of quantitative and qualitative measurement, we could suggest some usability items for users' awareness of computer security system.

3   METHODOLOGY

3.1   Measurement Procedure

In this research, we have used district survey method to find computer security awareness and usability of gender and age. The survey was constructed with 5-Likert scale, consisting of 13 demographic questions and 25 main usability questions. We collected of 137 usability principles from previous literatures including Nielsen (1994)'s checklist, and then we screened a total of 34 usability principles among them in terms of tangible user interface main properties, and then refined 25usability principles (Anderson, J. A., et al, 2011, Kim, M. H. and Ji, Y. G., 2007, Ji, Y. G., et al, 2007),. 196 subjects (Groups: 20-40's, 40-70's) were divided into three groups: Adolescence group (N: 108, Female; 54, Male; 54), Senior-age group (N: 88, Female; 44, Male; 44) who have been using the personal computer security participated in the test.

3.2   Statistical Analysis

The selected usability principles are classified systematically using statistical method, T-test and Principle Components Analysis (PCA) using Factor Analysis (Seong, L. K., 1998) by using SPSS Inc's PASWStatistic18.0 (Table 2).

Table 2 Research Framework

Division	Investigation Item, Method		
IndependentVariables	Age	Subject	Koreans
Number of Demographic Questions	13	Sample Size	196
Number of Usability Questions	25		
Dependent Variables	Sense of Security, Usability		
Measure, Analysis Method	5-Likert Scale	SPSS 18.0, T-test and Principle Components Analysis (PCA) using Factor Analysis	

4   RESULTS AND DISCUSSION

4.1   Correlation between Computer Security Awareness by Gender

Investigation of recognition of 'Nate-On' hacking incident by gender

In order to find correlation between computer security awareness by gender, we found how many people recognize the incident (Figure 1). Many were aware of and were interested in the incident. As a result, women lack computer security, and mostly not interested in computer security compared to men. The results of survey on 196 chosen male and female are shown in Figure 2. We have used histograms to configure values of men and women who use vaccine and computer passwords. As the result of comparison, both men and women preferred using vaccine programs than using computer passwords, and men showed more frequent use of vaccine and security programs. For questions 7 (Current security system is safe) and 11 (I'm aware of security program that I'm currently using), both men and women showed equal average value, and men and women showed 4.2 and 3.1 value respectively for question number 12 (security is related with its method).

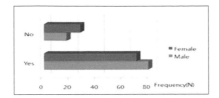

Figure 1 Relation of vaccine usage based on gender

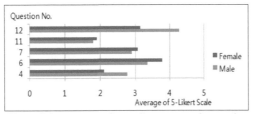

Figure 2 Awareness of security programs by gender

Satisfaction of security programs by gender

For question No. 21 (satisfied with interface of security measures of currently using PC), both men and women showed average of 3 (normal), and for question No. 22 (satisfied with design of currently using security measures). For question No. 23 (satisfaction of visual check of currently using security measures), women displayed average of 2.2 while men showed 3.0.

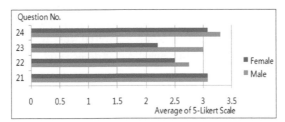

Figure 3 Satisfaction of security programs by gender

4.2 Computer Security Awareness Correlated with Age

Investigation of recognition of 'Nate-On' hacking incident by age

We have investigated 196 randomly chosen people consist of different age groups (20~40's to Adolescence group, 40~70's to Senior-aged group). In the examination of age-based security-cautious relationship with the vaccine, 77 of 108 Adolescence group, and 66 of 88 Senior-aged group showed high frequency of using the vaccine compare to not using at all.

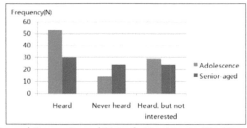

Figure 4 Recognition of 'Nate-On' hacking incident by age

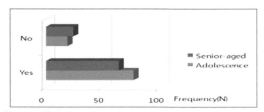

Figure 5 Relation of vaccine usage based on age groups

Awareness of Security programs by age groups

In order to find the average for awareness of security programs, we have selected 5-point Likert scale to get the average value for survey questions regarding awareness. For question number 6 (hacking has nothing to do with me), Senior-aged respondents showed average of 3.6 and adolescence respondents showed average of 3.45, showing most of Senior-aged group answered yes, while adolescence group answered normal. Other than this question, Adolescence group showed higher average value than Senior-aged group.

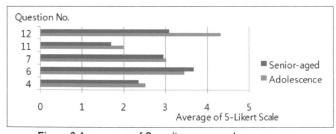

Figure 6 Awareness of Security programs by age groups

Satisfaction of Security Programs by age groups

In order to find the differences in satisfaction of a security program on different age groups through survey questions, we have measured averages for question No. 21~24. While adolescence respondents showed higher averages for question No. 21~23, Senior-aged respondents had higher average of 3.37 for question No. 24 (satisfaction of currently using PC's information security).

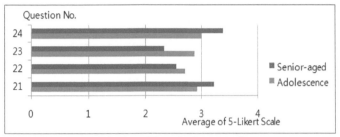

Figure 7 Satisfaction of Security Programs by age groups

4.3 Usability Evaluation of Computer Security Correlated with Age

The results of evaluation of computer security usability correlated with age are shown in Table 3. There are significant differences between Adolescence group and Senior-aged group for some awareness and usability items of computer security (Kim, W. K. 2001). Also, 25items for computer security usability evaluation are categorized into five groups according to their correlations and we named them respectively (1) Factor 1; User Interaction-support (Questionnaires No. 1, 2, 8, 24), (2) Factor 2: User Cognitive-support (Questionnaires No. 4, 6, 7, 11, 12), (3) Factor 3: User Performance-support (Questionnaires No. 5, 9, 14, 20. 25), (4) Factor 4: User Satisfaction-support (Questionnaires No. 16, 17, 21, 22, 23), (5) Factor 5: User Physicality (Questionnaires No. 3. 10, 13, 15, 18),. So, total usability items of 25checklists are developed with 5-point Likert scale. PCA reduced the 25usability measurements to five factors, and the Varimax rotation more clearly distinguished the usability components. Reliability within these factors were calculated by Cronbach's alpha, Factor 1; 0.818, Factor 2; 0.724, Factor 3; 0.620, Factor4; 0.544, Factor 5; 0.856. All of the reliability within these factors was very a high correlation respectively.

Table 3. The comparison and analysis of the computer security usability level between Adolescence group (N=108) and Senior-aged group (N=88) in some awareness and satisfaction questionnaires

	Differences of mean value		P-value (T-test)
	Adolescence group (N=108)	Senior-aged group (N=88)	
Awareness questionnaires	-28.34	-28.08	-4.61**
Satisfaction questionnaires	-6.65	-7.89	0.54

** p<.01   (paired t-test)

5 CONCLUSIONS

Modern computer facilities give human-being qualitatively simplified, uniformed, heuristic, and improved modern social life and the needs of those have been increased especially for significance of the efficient computer security. This research was carried out to find the significant differences for awareness and usability of computer security with correlated age. As using statistical analysis, the systematical results of the level of computer security usability between Adolescence group and Senior-aged group were shown significantly different in several computer security usability items, especially for the awareness and satisfaction questionnaires.

On the other hands, another aim of this research was to develop users' awareness of computer security system using checklists which is a one of currently available tangible user interface prototype. 25 items for computer security usability evaluation are categorized into five groups according to their correlations and we named them respectively (1) Factor 1; User Interaction-support, (2) Factor 2: User Cognitive-support, (3) Factor 3: User Performance-support, (4) Factor 4: User Satisfaction-support, (5) Factor 5: User Physicality. Accordingly, 25 usability checklists could be suggested with 5-point Likert scale. Ultimately, we hope that the results of this research could be expected to utilize the guidelines for designing the tangible user computer security system prototype for the Aged.

REFERENCES

Anderson, J. A., Wangner, J., Bessesen, M., and Williams, L. C. (2011), Usability Testing in the Hospital, *Journal of Human Factors and Ergonomics in Manufacturing & Service Industries*, 00 (0) 1-12 DOI: 10. 1002/hfm, Wiley Periodicals, Inc.

Hollerer, T., Feiner, S., Terauchi, T., Rashid, G. and Halaway, D. (1999), Exploring MARS: Developing Indoor and Outdoor User Interfaces to a Mobile Augmented Reality System, Computers and Graphics, 23(6), pp. 779-785.

ISO 9241-11, (1998), Ergonomic Requirements for Office Work with Visual Display Terminals (VDTs)-Part II: Guidance on usability, ISO, Geneva.

ISO 13207, (1999), Human-Centred Design Processes for Interactive Systems, ISO, Geneva.

Ji, Y. G., Jin, B. S., and Ko, S. M., (2007), Development of a AHP Model for, Telematic Interface Evaluation, *Proceedings of HCII 2007 Conference on Human compute Interaction*, LNCS 4550, Beijing, China, July.

Kim, M. H. and Ji, Y. G., (2007), The Development of usability evaluation method for tangible user interface, Thesis, University of Yonsei.

Kim, W. K. (2001), Physical disability and depression in older adults: Predictability of structural and functional aspects of social support, *Journal of the Korean Clinical Psychology*, 20(1), pp. 49-66.

Nielsen, J., (1994b), Heuristic Evaluation, In, J. Nielsen and Mark, R. L. (Eds.), Usability Inspection Methods, pp. 25-62, New York: John Wiley and Sons.

Seong, L. K., (1998), Experimental Design and Analysis, pp.194-267, Gyunggi-Do, Korea, Free-Academy Press.

-http://www.tscllc.com/Small-Business-Solutions/Tablet-pc-benefits.asp

-http://www.pcbee.co.kr/it/viewer_tx.php?content_num=38044

kostat.go.kr/portal/korea/kor_nw/.../index.board (2005)

CHAPTER **43**

An Empirical Approach for Driver-Driver Interaction Study: Attributes, Influence Factors and Framework

Yutao Ba, Wei Zhang

Department of Industrial Engineering, Tsinghua University, Beijing 100084, China
bb0438@163.com, zhangwei1968@gmail.com

ABSTRACT

During daily driving, interaction with other drivers is a complex part of driving task, where information is primarily contained in anonymous and transient driving behaviors. However, there are few theoretical frameworks and empirical evidences focused on driver-driver interaction. This study aims at elaborating the mechanism of driver-driver interaction process. We reviewed the related literatures describing elements of drivers' internal status, external behaviors which contain attributes and influence factors to interaction process between drivers. A perception-cognition-emotion-behavior framework was proposed to interpret the anger and aggression caused by defectiveness interaction. The need for further research is to empirically establish and validate the framework, based on which more efficient driver-driver communication system could be designed for the better emotions and safer behaviors.

Keywords: driver-driver interaction, driving behaviors, cognition, emotion

1. INTRODUCTION

Driving behavior is a direct consequence of the stimuli received from the road infrastructure, surrounding environment and atmosphere inside the vehicle (De Waard, 1996). During daily driving, drivers interact with traffic environment, vehicle and other road users. Interaction with other drivers is more complex than interaction with the traffic environment and vehicle, because other drivers, in contrast to stationary objects, are in motion and driver has to estimate their speed and direction to be able to avoid them (Gunilla Björklund, 2005). During this interaction process, information is primarily contained in anonymous and transient driving behaviors and simple signal (e.g. horn and lights).

To make the driving task possible, every driver has to take the intentions and behaviors of other driver into account, because human being is the weak link in the human-machine-environment system (Gunilla Björklund, 2005). Drivers always used others' behaviors as a source of information, or compare their behavior with that of others which is considered as collective expressions (Zaidel and David, 1992) Due to perceptual and attention limits, in some situation, drivers are like to misunderstand the information implicated in others' external behaviors, and lead to cognition basis, negative emotion status (anger or irritation) or stress.

Furthermore, in the next generation of vehicle, some new devices will be embedded which will provide more communication between drivers of abjection vehicles. More and more car-to-car communication device is in the development process, which enables vehicle to communicate directly with other vehicles on the road, enabling drivers to exchange information about current traffic congestion, road conditions, etc. Some multimedia communication such as emoticons can express a wide range of driver emotions, including remorse, lust, worry (slow down), and, of course, anger. Verify the efficient of interaction media between drivers is pre-requisite for design of more popular and widely used driver-driver communication system.

More and more researchers (Gunilla Björklund, 2005, Wilde and Gerald, 1976) draw their attention to the topic of interaction between drivers. But, there are still few theoretical frameworks and limited empirical evidences. Present study, therefore, is to give more comprehensive framework and empirical approach for further research. First, a review about studies of driver-driver interaction and other related researches is conducted, considering both the drivers' external behaviors and inherent status as the results of interaction process with other drivers in traffic flow. Then, we conclude the attributes and influence factors for an integrated model of the driver-driver interaction. Finally, we suggest an empirical approach in driving simulator to validate the proposed model.

2. LITERATURE REVIEW

2.1. Definition and attributes of driver-driver interaction

There is still no consensus of the definition of driver-driver interaction. On a daily driving, road users have to share the road with a great number of other road users, each of them with the major goal to move from one place to another, with a minimum of disruption and at a pleasant speed (Gunilla Björklund, 2005). So, drivers in the same traffic flow interact with each other and finally everyone's behaviors are affected by others (Nass, Clifford; Ing-Marie Jonsson; Helen Harris;

Ben Reaves; Jack Endo; Scott Brave and Leila Takayama, 2005). According to Zaidel (Zaidel and David, 1992), every individual driver is influenced by the social environment consisting of other road users. At the same time, every road user is a part in other road user's social environment. Zaidel stated four potential ways the other drivers' behaviors can influence a driver:

Others' behavior can be used as a source of information: As long as the drivers share a common driving culture, the informative value of others' behavior is high. In other cases, misunderstandings can easily occur. Observing others' behavior in a novel situation can also be a clue to the required behavior.

Others can be considered as a reference group: drivers compare their behavior with that of others. The concept "social norms" might be thought of as the summary representation of the opinion of others. To the extent that individuals accept the collective expressions of opinions as their own, they hold similar norms and may share common rules of behavior.

Imitation of others behavior: Imitation refers to following someone else's behavior without directly being instructed or forced to do so and with no directly communication between the imitator and the person whose behavior is imitated. The potential imitator must have the like ability to perform the behaviors and also some prior inclination to behave in that way or, at least, not having strong objection against it.

Communication with others: Communication in traffic can clarify intentions of the drivers and explain otherwise puzzling or offensive behavior, and in that way reduce misunderstanding, frustration, and conflicts. Improved communication may facilitate co-operative behavior among drivers, and thereby create a more positive social climate in traffic.

Some other research topics in driving safety are also related to the social interaction process between vehicle drivers:

Traffic rules: Traffic rules are usually divided into two parts, formal rules and informal rules (Gunilla Björklund, 2005). The formal rules (e.g. traffic laws) indicate the intentions and behaviors of the road users, which consider as the basis to expect others' behavior during interaction. Informal traffic rule defined as standards that are understood by members of a group, and that guide and/or constrain social behavior without the force of laws. These norms emerge out of interaction with others; they may or may not be stated explicitly, and any sanctions for deviating from them come from social networks, not the legal system. Some traffic rules are not congruent with the drivers' intention and natural behavior pattern (Cialdini, Robert B. and Melanie R. Trost eds, 1998). If behavior, which supplement or contradict formal traffic rules, become common in a specific situation or place, it is an indication that a informal traffic rule, or a social norm might take function.

Risky driving behaviors: Risky behaviors usually include aberrant behaviors, violation, conflict and aggression. Aberrant driving behaviors mention that driving style makes the individual salient (e.g. behaviors that stand out or deviate in some other way), which is departure from inform rules. The behaviors which are according with "social norms" are considered as the standard to judge which driving behaviors are aberrant. Previous research (Zaidel and David, 1992) showed aberrant behavior increase the likelihood of these persons being imitated. Violation is considered as contravention of formal traffic rules. The previous study (Parker, D. et al., 1995) showed violations have a motivational component and therefore causality should be considered within a broader social context. It has been show that drivers' attitude and intentions to commit driving violation (drinking and driving, speeding, run red light) are influenced by their beliefs concerning social expectations to perform the behaviors. Conflicts are because a road user's ability to correctly predict another road user's behavior is reduced if the other road user complies with a different rule system (Wilde and Gerald, 1976). An irritated drive is also more disposed to perform what is called aggressive behaviors (Lajunen et al., 1998, 1999), such as unsafe distance, "cutting" in front of other drivers, speeding, and running red traffic lights. Most of those behaviors are known to be associated with accidents. Other aggressive behaviors, such as headlight flashing and long time horns may not be dangerous as such, but may provoke other drivers and make them irritated and behave aggressively.

Attitude: Attitude is considered as "a psychological tendency that is expressed by evaluating a particular entity with some degree of favor or disfavor". According to the theory of planned behavior (TPB) (Icek Ajzen, 1991) , behavioral beliefs produce a favorable or unfavorable attitude toward the behavior. If attitudes toward specific behaviors are clearer, there may be a better chance for affecting change in the behavior.

Attribution: Attribution refers to the process by which individuals arrive at causal explanation for their own and others' behaviors (Berkowitz, Leonard ed., 1977). Several studies (Gunilla Björklund, 2005, Ulfarsson, 2001) have shown that drivers are subject to attribution biases when judging the behavior of other road uses. For example, a driver who perceives others to drive at excessive speeds is also more likely to drive faster than a driver who perceives others to comply with the limits (Lajunen et al., 1998, 1999).

Stress: Stress may be raised due to the complex traffic environment or other drivers' behaviors. During the interaction process, the influence of stress may uncertainty. In some researches (Westerman, S. J. and D. Haigney, 2000), high levels of driver stress were found to be associated with increased self-report of lapses, errors and violations, as a result of concretive coping and impaired control and attention. Previous study (Matthews, Gerald; Lisa Dorn and A. Ian Glendon, 1991) found lower driving speed and faster detection of hazard as a result of alertness during stress driving.

Emotion: Emotions also play a powerful, central role during our driving task. They impact our beliefs, inform our decision-making and in large measure guide how we adapt our behavior to the world around us. Our interactions with each other are a source of many of our emotions and we have developed both a range of behaviors that communicate emotional information as well as an ability to recognize the emotional arousal in others.

2.2. Influence factors

In the past several years, a large amount of studies have conducted to examine different factors influencing the driving behaviors. To better understand factors that impact response driving behavior as a result of interaction process, we classify those influence factors into two parts, individual factors (*Table 1*) and external factors (*Table 2*). The major essential individual factors include age, experience, gender, impairment and personality and the major essential external factors include traffic environment, vehicle technology, training culture and context.

Table 2. Essential individual impact factors

Factor	Literatures	Conclusions
Age	H. R. Booher (1978)	The aging process diminishes the visual capacity of drivers and can contribute to crash likelihood.
	K. Ball and C. Owsley (1991)	Selective attention and slower information processing for old driver are associated with greater crash risk.
	Brian A. Jonah, 1986 (1986)	Young drivers are at greater risk of being involved in a casualty accident than older drivers and this risk is primarily a function of their propensity to take risks.
Experience	R. R. Mourant and T. H. Rockwell (1972)	Inexperienced drivers scan the road differently than experienced drivers do.
	K. Renge (2000)	Experienced drivers could understand the signals (blinkers, headlights, hazard lamps and hand gestures) better than novice drivers. Novice drivers could understand formal signal and everyday signal better than informal signal.
Gender	T Özkan and T Lajunen (2005)	Sex (be male or be masculinity) can be predictor of the risky driving behaviors.
	H. S. Lonczak et al. (2007)	Men reported more traffic citations and injuries, but did not differ from women in reported driving anger.
Impairment	J. Yu et al. (2004)	Aggressive driving and road rage tended to be affected mostly by alcohol problems and feelings of depression.
	M. S. Huntley and Centybea.Tm (1974)	Sleep deprivation significantly decreased the effects of alcohol on coarse-steering reversal rates.
personality	Sujata M. Patila et al. (2006)	Greater risk-taking, hostility, aggression, and tolerance of deviance propensity were significant influence the risky driving behaviors.
	H. Iversen and T. Rundmo (2002)	Scored high on sensation seeking, normlessness and driver anger reported more frequent risky driving compared to those who scored low on these variables.

Table 3. Essential external factors on the interaction process

Factor	Literatures	Conclusion
Traffic environment	D. A. Hennessy and D. L. Wiesenthal (1999)	Both driver stress and aggression are greater in high than low congestion condition.
	T. Horberry et al. (2006)	Visual clutter was caused by the complex road environments, which may contribute to high mental workload and stress.
Vehicle technology	J.E.B. Tornros and A.K. Bolling (2005)	Driving speed decreases as an effect hands phone use, and may interpret this effect as a compensatory effort for the increased mental workload.
	Y. T. Ba and Z. Wei (2011)	Vehicle devices can improve driving behavior, but in some situation it will cause high mental workload.
Training	D. L. Roenker et al. (2003)	A measure of processing speed and spatial attention can be improved with training in simulator.
	T. Rundmo and H. Iversen (2004)	The respondents perceived the risk to be higher after the training than before.
Culture	M. C. Stiles and J. I. Grieshop (1999)	Car accidents or near misses due to differences between the China and American were explored. Drivers are more aggressive in China than in the US.
Context	D. Shinar and R. Compton, (2004)	The task context with time pressure as is often the case during rush hours will cause more aggression and violation.

3   FRAMEWORK AND EMPIRICAL APPROACH

During the process, the traffic rules, form and inform, can be consider as the baseline of varies driving behaviors. Rational process (e.g. attitude and attribution) and emotional process (e.g. stress and emotion) could take functions simultaneity. Rational process holds the human's rational state, which consists of some knowledge about the external world and the people themselves. This knowledge evolves by some previous experience, and its evolution depends on the processing of rational inputs from traffic environment and others driving behaviors. Emotional process holds the drivers' stress and emotional status, whose temporal evolution is governed by emotional stimuli from perceived behavior and result influence the response driving behavior. Goal setting by context is explained as a dominant factor impact the risk acceptance in motivational models (Wilde, 1989). The major goal of each driver is move from one place with a minimum of disruptions and at a pleasant speed.

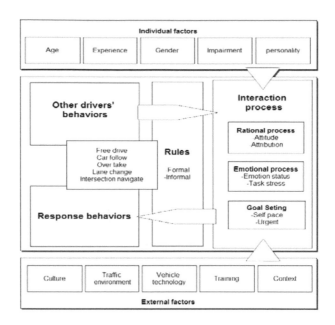

Figure 1. Framework of driver-driver interaction model

Based on the review of the literatures, we can conclude during the interaction process, the drivers' behaviors could consider as the result of cognition process about perceived behaviors from other drivers. The mechanism of driver interaction is constructed and proposed framework of driver-driver interaction model is show in *Figure 1*.

A series of experiments will be designed to validate the proposed model. In order to develop more effective empirical approach, we suggest using driving simulator as alternative method. Compare with the field study, diving simulator are safer, more controllable and repeatable. But before we conduct the experiment in virtual environment, the driving simulation approach needs to be evaluated for its validity in specified interaction scenario. For this reason, two associative experiments are designed. Aim of first experiments is to evaluate to fidelity (or face validity) of interaction process in the driving simulator. The purpose of the second experiment is to explore the relationship between perceived driving behaviors and response driving behaviors, interaction process and response driving behaviors, influence of internal and external factors and then validate the proposed framework of driver-diver interaction model.

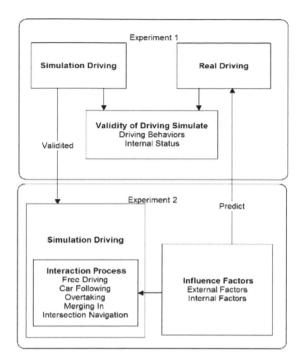

Figure 2.Empirical approach for validating driver-driver interaction model

To draw inferences confidently about real driving behaviors from driving simulation data requires establishing the validity of driving simulate (Reimer et al., 2006). In short, do the driving behaviors observed and internal status (e.g. attitude, attribute, stress and emotion) measured in high fidelity experimental simulations map onto them in real-world driving experiences? To what extent does the simulator produce driving behaviors and internal status that are comparable to real driving behaviors? The first experiment is design to answer above question. Some typical interaction scenario will be designed in pair both in the simulation driving and real driving, such as free driving, car following, overtaking, C and intersection navigation. Some configuration of simulator will be adapted until we get satisfied behavioral validity (absolute or relative). Based on the validity of driving simulate, the second experiment is to exam the potential influence factors and construct the framework of driver-diver interaction model. Subjects were required to drive in a high-fidelity simulated environment. The whole empirical approach is show in *Figure 2*.

5   CONCLUSION

In present research, we describe an empirical approach to validate the framework of driver-driver interaction model which is based on the results of previous research. In general, the interaction process between drivers is a special kind of social communication, where information is primarily contained in anonymous and transient driving behaviors. Due to perceptual and attention limits, in some situation, drivers are like to misunderstand the information implicated in others' external behaviors, and lead to cognition basis, negative emotion status (anger or irritation) and stress. Some important contributes of interaction process are concluded in previous research, such as traffic ruler, risky behaviors, attitude, attribution, stress and emotion. Various individual and external factors may have influence on driving behaviors, and they may also affect the interaction process.

In the driver-driver interaction model, cognition process is divided in two parts: relational process and emotional process. Rational process is related to knowledge about the external world and the people themselves. Emotion process is governed by emotional stimuli from perceived behaviors and their results influence the response driving behaviors.

To validate the proposed model, the proposed empirical approach suggests two steps. In the first step, compare between simulation and real-world driving is suggested to establish the validity of driving simulator. The second step explores the relationship between perceived driving behaviors, interaction process, response driving behaviors and influence of varies factor in the driving simulation.

In future research, experiment should be conducted to examine the excitation mechanism of the influence factors. For example, the participants with aggression and none-aggression personality traits will be recruited to test the significant effect of personality traits. The goal setting by context will be divided to self pace and urgency to test the significant effect of motivation. Base on those results, we may answer how to design more efficient interaction system between drivers will be given. The significant and opportunity of present research is to gain less risky behaviors, more positive emotion and maximization of social effectiveness by improving driver-driver interaction.

REFERENCES

Ajzen, Icek. 1991. "The Theory of Planned Behavior." Organizational Behavior and Human Decision Processes, 50(2): 179-211.

Ba, Y. T. and Z. Wei. 2011. "A Review of Driver Mental Workload in Driver-Vehicle-Environment System." Lecture Notes in Computer Science, 2011, 6775/2011, 125-34.

Ball, K. and C. Owsley. 1991. "Identifying Correlates of Accident Involvement for the Older Driver." Human Factors, 33(5): 583-95.

Berkowitz, Leonard ed. 1977. "The Intuitive Psychologist and His Shortcomings: Distortions in the Attribution Process." New York: Academic Press.

Björklund, Gunilla. 2005. "Driver Interaction: Informal Rules, Irritation and Aggressive Behaviour." Uppsala: Acta Universitatis Upsaliensis.

Booher, H. R. 1978. "Effects of Visual and Auditory Impairment in Driving Performance." Human Factors, 20(3): 307-19.

Bryan Reimer, Lisa A. D'Ambrosio, Joseph F. Coughlin, Michael E. Kafrissen and Joseph Biederman. 2006. "Using self-reported data to assess the validity of driving simulation data." Behavior Research Methods, 38(2): 314-324.

Cialdini, Robert B. and Melanie R. Trost eds. 1998. "Social Influence: Social Norms, Conformity and Compliance. " US: McGraw-Hill.

De Waard, D.1996. "The Measurement of Drivers' Mental Workload." Ph.D. thesis. University of Groningen, Traffic Research Centre, Haren, Netherlands.

Jonah, Brian A. 1986. "Accident Risk and Risk-Taking Behaviour among Young Drivers." Accident Analysis & Prevention, 18(4): 255-71.

Hennessy, D. A. and D. L. Wiesenthal. 1999. "Traffic Congestion, Driver Stress, and Driver Aggression." Aggressive Behavior, 25(6): 409-23.

Huntley, M. S. and Centybea.Tm. 1974. "Alcohol, Sleep-Deprivation, and Driving Speed Effects upon Control Use During Driving." Human Factors, 16(1): 19-28.

Horberry, T., J. Anderson, M. A. Regan, T. J. Triggs and J. Brown. 2006. "Driver Distraction: The Effects of Concurrent in-Vehicle Tasks, Road Environment Complexity and Age on Driving Performance." Accident Analysis and Prevention, 38(1): 185-91.

Iversen, H. and T. Rundmo. 2002. "Personality, Risky Driving and Accident Involvement among Norwegian Drivers." Personality and Individual Differences, 33(8): 1251-63.

Lajunen, Timo; Dianne Parker and Stephen G Stradling. 1998. "Dimensions of Driver Anger, Aggressive and Highway Code Violations and Their Ediation by Safety Orientation in Uk Drivers." Transportation Research Part F: Traffic Psychology and Behaviour, 1(2): 107-21.

Lajunen, Timo; Dianne Parkera and Heikki Summalab. 1999. "Does Traffic Congestion Increase Driver Aggression?" Transportation Research Part F: Traffic Psychology and Behaviour, 2(4): 225-36.

Lonczak, H. S., C. Neighbors and D. M. Donovan. 2007. "Predicting Risky and Angry Driving as a Function of Gender." Accident Analysis and Prevention, 39(3): 536-45.

Matthews, Gerald; Lisa Dorn and A. Ian Glendon. 1991. "Personality Correlates of Driver Stress." Personality and Individual Differences, 12(6): 535-49.

Mourant, R. R. and T. H. Rockwell. 1972. "Strategies of Visual Search by Novice and Experienced Drivers." Human Factors, 14(4): 325-&.

Nass, Clifford; Ing-Marie Jonsson; Helen Harris; Ben Reaves; Jack Endo; Scott Brave and Leila Takayama. 2005. "Improving Automotive Safety by Pairing Driver Emotion and Car Voice Emotion" CHI '05 Extended Abstracts on Human Factors in Computing Systems. New York.

Nass, Clifford; Ing-Marie Jonsson; Helen Harris; Ben Reaves; Jack Endo; Scott Brave and Leila Takayama. 2005. "Improving Automotive Safety by Pairing Driver Emotion and Car Voice Emotion" CHI '05 Extended Abstracts on Human Factors in Computing Systems. New York.

Özkan, T and T Lajunen. 2005. "Why Are There Sex Differences in Risky Driving? The Relationship between Sex and Gender-Role on Aggressive Driving, Traffic Offences, and Accident Involvement among Young Turkish Drivers." Aggressive Behavior, 31(6): 547-58.

Parker, D., J. T. Reason, A. S. R. Manstead and S. G. Stradling. 1995. "Driving Errors, Driving Violations and Accident Involvement." Ergonomics, 38(5): 1036-48.

Patila, Sujata M., Jean Thatcher Shopeb, Trivellore E. Raghunathanc and C. Raymond Binghamb. 2006. "The Role of Personality Characteristics in Young Adult Driving." Traffic Injury Prevention, 7(4): 328-34.

Renge, K. 2000. "Effect of Driving Experience on Drivers' Decoding Process of Roadway Interpersonal Communication." Ergonomics: 43(1): 27-39.

Roenker, D. L., G. M. Cissell, K. K. Ball, V. G. Wadley and J. D. Edwards. 2003. "Speed-of-Processing and Driving Simulator Training Result in Improved Driving Performance." Human Factors, 45(2): 218-33.

Rundmo, T. and H. Iversen. 2004. "Risk Perception and Driving Behaviour among Adolescents in Two Norwegian Counties before and after a Traffic Safety Campaign." Safety Science, 42(1): 1-21.

Stiles, M. C. and J. I. Grieshop. 1999. "Impacts of Culture on Driver Knowledge and Safety Device Usage among Hispanic Farm Workers." Accident Analysis and Prevention, 31(3): 235-41.

Shinar, D. and R. Compton. 2004. "Aggressive Driving: An Observational Study of Driver, Vehicle, and Situational Variables." Accident Analysis and Prevention, 36(3): 429-37.

Ulfarsson, G. F., V. N. Shankar, et al. 2001. Travel aid. Washington State Transportation Center (TRAC), Washington.

Westerman, S. J. and D. Haigney. 2000. "Individual Differences in Driver Stress, Error and Violation." Personality and Individual Differences, 29(5): 981-98.

Wilde, Gerald J. 1976. "Social Interaction Patterns in Driver Behavior." Human Factors, 18(5): 477-92.

Wilde, G. J. S. 1989. "Accident Countermeasures and Behavioral Compensation - the Position of Risk Homeostasis Theory." Journal of Occupational Accidents, 10(4): 267-92.

Yu, J.; P. C. Evans and L. Perfetti. 2004. "Road Aggression among Drinking Drivers: Alcohol and Non-Alcohol Effects on Aggressive Driving and Road Rage." Journal of Criminal Justice, 32(5): 421-30.

Zaidel, David M. 1992. "A Modeling Perspective on the Culture of Driving." Accident Analysis & Prevention, 24(6): 585-97.

Section III

Usability in Web Environment

CHAPTER 44

Why Should Home Networking be Complicated?

Abbas Moallem, Ph.D

NETGEAR
350 East Plumeria Drive
San Jose, CA 95134, USA
Abbas.moallem@NETGEAR.COM

ABSTRACT

In today's life, an Internet connection in a home can be considered a utility like a telephone or even to some degree electricity. Not having an efficient home network is like having just one electrical outlet in the whole house versus having outlets throughout the house and being able to plug in all your appliances at once any time you want. Networking devices from wireless routers to storage, to digital media receivers and so on, are more and more available on the market and their speed and quality of reception constantly improves. Despite these hardware improvements, the software parts that enable users to interact and manage the extended functionalities of these devices are still very complicated and out of reach for average home users.

In the competitive market of networking device manufacturing companies, limited resources are allocated to improve the ease of use of the user interfaces (UIs) where users interact with the devices on a daily basis.

This paper presents the main issues in home network software usability, based on survey results and empirical observation, and describes a new generation of tasked-based user interfaces for NETGEAR devices, enabling home networking users to easily set up and interact with their devices.

Keywords: Home Networking, Routers User Interface, Smart Appliances, Smart Network usability, NETGEAR

1   INTRODUCTION

With the speedy expansion of the Internet and Internet-based technologies in homes, an Internet connection became a basic utility, like electricity, gas or water. According to the website internetworldstats.com, 78.3% of the US population is Internet users. Users are not just satisfied with having their computer connected to the Internet. They need wireless connections at home for their computer devices, and a cross-access connection that lets them, view, modify, and control their computer devices. Users also need to control and monitor or solve problems when they are away from home. They equally want an uninterrupted Internet connection for smart phones or Wi-Fi connections from public places and through their preferred providers. Satisfying user needs depends on three main hardware factors.

- High-speed Internet connection at the provider level.
- Reliable and high-speed router to provide fast and expanded coverage.
- Device inter-connectivity and discoverability in a network.

To appropriately configure and manage the network devices, and problem-solve the issues, users need to first install hardware correctly, and use the user interfaces of each devices to configure, set, and manage their network.

Not long ago, the Internet connection was limited to one or more computers at home. Now, gradually within a "smart home" (Cheng & Kunz, 2009) a multitude of computerized devices are interconnected and interacting with each other. Users should be able to not only configure them but also monitor them daily and interact with them and more importantly, problem-solve network issues.

Despite a tremendous progress in hardware: from cable modems, router, media devices and so on, improvements in the usability of their software is still very primitive. It is commonly agreed upon among users that managing home networks is a tedious, difficult task that is out of reach for most users with limited networking experience.

An important body of research points out a variety of usability issues for home networking technologies. In a recent article Edwards et al. conclude that "Network usability problems run deep because the technology was originally developed for research labs and enterprise networks and does not account for the unique characteristics of the home: lack of professional administrators, deep heterogeneity, and expectations of privacy" (Edwards, 2011). The issues that seemed to be related to this lack of ease in this technology include: user interfaces are built for advanced users with IT backgrounds, heterogeneity in users, home networking configuration, and problem-solving of home network issues. For example, Grinter et al. (2005), note that home users are often unable to verbally articulate accurate information about their networks or even have a different mental model of their network based on their level of expertise.

Poole, E.S. et al. (2009) have found that home users drew on a variety of resources to conceptualize their home networks. This study suggests that they need to be offered drawings that exhibit properties associated with network education or training, on some schemes such the physical layout or routines of the home as methods of organizing the network. Alternately, when troubleshooting the networking problem, home users would like to pinpoint locations of malfunctioning devices.

Expansion of the smart home concept should also benefit older adults. Older adults seem to have an overall positive attitude towards this technology. Usage of health monitoring sensors, or security devices in the home to enhance their lives is a good example of how people benefit from this technology (Demiris G. et al, 2009). However the ease of use of these applications is crucial for successful usage by older adults.

Home network devices should make it easy to add computers, or any other smart appliances to the home network, and establish interaction by offering users an easy to use and intuitive method of centralized management and monitoring of the devices.

The results of a survey on home wireless usage reveals that currently 22% of people surveyed bought their wireless router based on speed, followed by 17%t for low price, and 1% for ease of use. However, when participants were asked about purchasing a new wireless router, speed remained the top factor and increased in importance to 37%, while ease of use moved up to the second priority with 17% (TMC News, 2011).

Consequently, making home networking smart is not achievable unless the usability of the devices (software and hardware) along with their configurations is extensively improved and accessible to not only expert users but all types of users.

2   DESIGNING SMART NETWORKS ACCESSIBLE TO ALL

If you have a home networking device, the chances are you have experienced a variety of home networking user interfaces. You might even have asked someone with technical knowledge to help you with installing, connecting, and configuring the device that you acquired or had problems with. This might have happened for any type of device from a Wi-Fi router to more advanced devices such as storage, security systems, and interactive TV or sound systems.

The poor usability of the user interfaces (UIs) is not observable only on one or two brands of products. A simple comparison among the most common brands on the market reveals that all brands suffer to some degrees from the same usability issues.

2.1 Terminology

Table [1] shows the 26 common terms used in configuration and setting of major Wi-Fi router brands.

WEP, WPA-PSK [TKIP], WPA-PSK WPA2-PSK [AES], WPA-PSK [TKIP] + WPA2-PSK [AES], *WPA/WPA2 Enterprise, Passphrase, Enable SSID Broadcast, Channel, Guest Network, IP Address, IP* *Subnet Mask, Gateway IP Address, DNS Servers, Primary DNS, Secondary DNS, Router MAC Address ,* *Domain Names, Internet IP Address, Static IP Address, IP Subnet Mask, Media Server, HTTP, HTTP* *(via internet), FTP (via internet),USB Settings, IPv6, Router, Modem, DSL Modem, Cable Modem*
Table 1: Labeling terms used in most router user interfaces

In a survey, I asked a group of potential users the meaning of the 26 (Table 1) most common terms in home networking. These terms are often used as field labels in router UIs. Forty-three participants (mostly college students and former graduate students in software engineering, human factors, and psychology) completed an online survey (19 male, 24 female, 45% under age 25, 41% age 25 to 45, 45% with formal education of college, 39% graduate schools, and finally 86% of participants own a router a home). The results show that a very small percentage of people know the meaning of the terms used in these devices' user interfaces (UIs). Many users do not understand even the more common terms shown in Table 2, such as DSL modem or IP address. For example, 43% of participants are not sure of the meaning of "Passphrase, "DNS Servers" 41%, WEP "48%" or even router "9%."

Term	Not sure at all	Know what it means but not sure.	Definition Provided
Router	9%	34%	27%
Modem	2%	41%	57%
DSL Modem	9%	48%	43%
Cable Modem	11%	43%	45%
WEP	48%	16%	36%
WPA-PSK [TKIP]	57%	2%	41%
Passphrase	43%	20%	36%
IP Address	9%	39%	52%
DNS Servers	41%	20%	39%
Router MAC Address	36%	27%	36%
Domain Name	11%	39%	50%

Table 2: Understandability of selected networking terminologies used in field labeling of router UI

Considering that this survey was conducted out of the context of the UI, and the number of respondents was limited (43 participants), it is hard to draw a global conclusion. However, empirical observations that support this data indicate to what extent users are unfamiliar with all terms used. Knowledge of the terms and their meaning requires education, training and reading of many pages of documentation and online help, which is another hard task for the average user. To underline this unfamiliarity with the terms, the readers of this article can check to see if they are familiar with these terms.

2.2 User Interfaces

Comparison of the user interfaces of four major brands of devices illustrated the similarity of the design among the different brands. Basic screens of router interfaces for these brands are shown in Figure 1a, b, d and d). A quick review of the screens, despite some differences in color palette or layout, reveals that the UIs are built using the same terminologies. Each UI displays a number of parameters requiring users to understand them and set or configure them without knowing the context, what the settings do, and the relationship among the parameters. All four UIs evaluated use a feature-based design approach targeted to advanced IT users.

Figures 1.a to 1.d show the main page UIs of the four router brands. The field labels and terminology seem very similar if not identical.

2.3 Design Constraints

Designing better and more user-centered interfaces for home networking devices requires a good understanding of the main existing constraints. The most problematic constraints in designing the user interfaces (UIs) seem to be the diversity of computer networking and applications that users use.

Figure 1.a: NETGEAR- Field Labels	Figure 1.b: 2:Cisco-Linksys-Field Labels
Figure 1.c: D-link- Field Labels	Figure 1.d:Belkin Modem Router

Figure 1a, 1b, and 1c, show UI screens of the four router brands evaluated. The field labels and terminology seem very similar if not identical.

I asked a group of networking experts to rate and explain the score that they give to the usability of the networking technology. Eight experts participated in this survey. On the question of, "To your opinion, and based on your knowledge of home networking technology, How would you rate the usability of your home networking product (1 lowest, 10 highest)? Usability is defined as the effectiveness, efficiency, and satisfaction with which users can achieve tasks in a particular environment of a product," experts gave an average score of 8 to the usability of networking products. When they were asked to explain the score, a variety of reasons were indicated (Table 3). This data, although not enough in terms of the number of participants, reveals that even the experts think that the usability of these devices needs more improvement.

Commentary
• Usability is very important. .. the product with rich features but with bad usability, I will still give low rating for such product.
• My rating is based on my family usage to keep wireless connections with an AP.
• To perform some functions requires making changes in multiple areas of Setup.
• I think usability is the most important part of a product.
• The current home products today could be significantly improved so that they are more intuitive to operate. The test that I use at home is to test usability of a product around a range of users including my 6 year old daughter and my 75 year old mother. In most cases today, both have difficulty working with home products. On the other hand, both can easily use Apple IOS devices.
• Had to stop using my XXXXX router when I went to Uverse. AT&T provides a gateway. The XXXX router cannot coexist on the same home network as the gateway.
• Common sense.
• There are missing features that I'd like to have and they are not easily accessible without buying a new device.
• Average Score given by participant for usability is 8/10

Table 3: Expert comments of the usability of home networking technology.

3. DESIGN CONCEPTS

In an attempt to extensively enhance usability of the home networking device user interfaces (UIs), NETGEAR aimed to make common tasks in installing, configuring, and monitoring the network easy for the average home user according to each user's needs.

To achieve this goal I classified usability needs in two categories: hardware and software usability. In terms of software usability, the following concepts were identified:

- Users' needs are different. All users don't need all the functionalities offered by devices. Consequently, the device functionalities should be scalable. The interface should not complicate the task of average users who need to complete only a few basic tasks.
- The UI should use common and easy to understand terminology and avoid complex and technical language.
- The UI should offer a tasked-based interface in terms of what users want to achieve rather than a combination of a variety of features or parameters.
- Users should be able to install, configure, and monitor their network devices in the shortest time possible.
- The UI should let users install, set and monitor the application with the least number of navigation clicks.
- The UI should be appealing and not distracting when users have the application open for long period of time.
- The UI should be built with fewest objects, simplest behavior, general rules, and terminology, and should be consistent.
- The UI should use the most common paradigm that users are familiar with to reduce learning and remembering of the design pattern.

4. USER EXPERIENCE ARCHITECTURE

The main usability problems observed in home networking might be related to the following areas:

- Selecting the proper device that matches the user's specific needs.
- Installing the hardware according the instructions provided by the manufacturer.
- Installing the software and early configuration.
- Monitoring, managing, and adjusting settings for the device.

The users' challenges start from the moment that they purchase and try to install the hardware and its related software. As an example, we can just look at the first step, meaning the sequence of the hardware connections when users have a cable provider. Most routers require users to:

- Unplug the modem (remove the modem batteries and unplug the modem's power cord).
- Wait a few minutes.
- Connect the router to the modem and computer then plug it in to an electrical outlet and wait for the wizard. (This sequence is provided on documentation or on the installation CD, which from the user perspective is the same).

Although there are valid technical reasons to follow this sequence, users get quite confused. They often try to do this task over and over and blame themselves, assuming that they have made a mistake in the sequence, while the issues might be something else. In one router brand, to prevent errors a big orange sticker was placed on the router to prevent the wrong sequencing (Figure 2). Figure 3 shows one of the early steps in configuration, the security setting, for one major router brand.

Despite all the issues in product selection and installation usability, this paper will focus on monitoring, managing and settings the device, meaning tasks that require the use of software and occur after the hardware is installed.

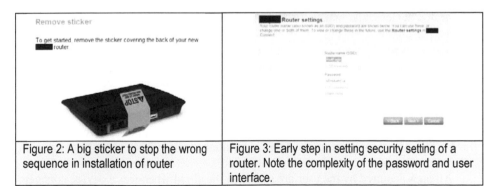

Figure 2: A big sticker to stop the wrong sequence in installation of router

Figure 3: Early step in setting security setting of a router. Note the complexity of the password and user interface.

4.1 Home Networking Devices User Interfaces

Comparison of the several home networking device brands illustrates that most, if not all of them, target advanced users with some networking IT knowledge.

Despite some effort to provide a more basic user interface for average home users (Example: NETGEAR Genie-Netgear.com (2012), it is still hard to claim that the average home users will have a smooth user experience when installing home networking appliances.

All the major brand home networking appliances use a complex feature-oriented rudimentary user interface (UI) design. Figure 4 shows the UI for a D-Link router setup page. Among other things, we notice a tremendous amount of text, very technical terminology, complex labeling and difficult navigation. The situation is not very different from one product to another among major brands.

4.2 A Different Way of Designing for Home Networking Devices: NETGEAR Smart Network Case

NETGEAR is a worldwide provider of networking products built on a variety of technologies such as wireless, Ethernet, and Powerline.

With a wide variety of devices that offer different functionalities within the networking context, usability improvement at all product levels is quite challenging. However, with customers' needs for more user-friendly products, the usability improvement at all level product levels is desired at NETGEAR.

In an effort to create a new way of approaching the networking tasks for all users, NETGEAR aimed to make home networking accessible to average home users who do not have advanced knowledge in IT. Although it is understood that the average home user might not need all the features offered in a product, reducing the features is out of question since the appliances should continue to offer advanced setup features for expert users.

To achieve this goal, the new initiative architecture was based on a unified user interface using a cloud based technology. This architecture helps users perform the basic tasks with ease while still allowing advanced users to log in to the device UI and perform advanced configuration and monitoring.

This approach, named "NETGEAR Smart Network," is an intelligent network that optimizes the applications that users run, enhances the performance of their connected devices, and makes those devices easier to connect and use. The network can be personalized and customized based on the devices that users need, what they do, and the type of network traffic they run.

After users connect and install their NETGEAR device, they are invited to create an account. Upon sign in and login, the web application recognizes all the devices owned and registered by the user and the system displays all the available information for each device. Users then can install the different apps, which are task-based entities, on each selected device. The available apps are offered through an NETGEAR Apps Store method that is familiar to users. The Smart Network also allows developers to create new applications.

All apps acquired by the user are saved on the cloud whenever a device connects to the Internet. Users can update them and change the configuration from anywhere. For example, a user can log in to NETGEAR Smart Network from anywhere with Internet connection and change or monitor activities in his/her home network, access a webcam, connect to the home network, or change printer settings.

4.3 Design Elements

When users log in, they can immediately click any device on their home network without additional navigation. All available information about the device is displayed in a device widget. Each device widget identifies a physical device. Users can also customize their owned devices with user friendly names.

In one navigational click (My Apps Tab) users can view all the applications they own. Each App Widget displays the app and the device on which it is installed including the dynamic information about the app. Clicking on the app opens the app pages where easy to understand screens and the UI help the user manage changes or perform tasks. (Figures 5 and 6)

This common way of doing the same tasks on all devices, an easy-to-understand terminologies, and consistency of the UI object provides a user-centric approach to all devices.

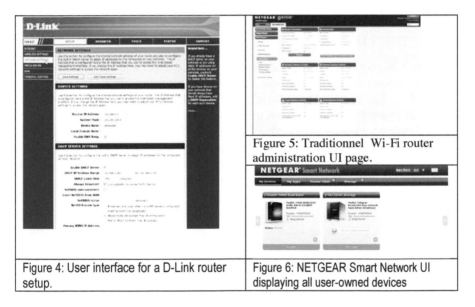

Figure 5: Traditionnel Wi-Fi router administration UI page.

Figure 4: User interface for a D-Link router setup.	Figure 6: NETGEAR Smart Network UI displaying all user-owned devices

4.4 User Interface Simplicity

With user interface simplicity in two widgets (device and app), one-click navigation, understandable terminologies, a limited number of terms used in the UI, and simple visual design, users are empowered to perform tasks with least IT skills. With advanced information visualization techniques, users can easily understand what is performed and status of the system.

Before and After

Figure 7 and 8 show the functionality offered through the NETGEAR traditional UI and the same functionality though the app in the NETGEAR Smart Network UI. The usage of appealing visualization, easy to understand, task-based terminology

416

and advanced UI design technique not only make user tasks easy, but also visually appealing and fun.

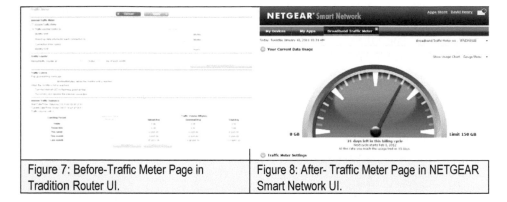

| Figure 7: Before-Traffic Meter Page in Tradition Router UI. | Figure 8: After- Traffic Meter Page in NETGEAR Smart Network UI. |

5. CONCLUSION

Today's market is not about focusing on more features and functionalities. We are going toward a situation where comparable products often offer the same functionality. The example of the mobile smart phone illustrates that despite the complex functionalities among smart phones, people prefer the phone that is easy to use and easy to operate rather than one that offers more functionalities. This philosophy is applicable to all types of technology. With the expansion of smart home appliances, success will go to the home networking enterprises that offer an easier to use product, satisfying the most common users.

ACKNOWLEDGMENTS

The author would like to acknowledge my colleagues Ye Zhang for reviewing this paper and his valuable input and Betsy Miller for editing this article.

REFERENCES

Cheng J. and Kunz Th., 2009, A Survey on Smart Home Networking Inside the smart home, Springer. Carleton University, Systems and Computer Engineering, Technical Report SCE-09-10, September 2009.

Edwards, W. K. 2011, Before building the network or its components, first understand the home and the behavior of its human inhabitants. June 2011, vol. 54, no. 6, communications of the ACM.

Grinter, R.E., Ducheneaut, N., Edwards, W.K. and Newman, M., 2005, The Work To Make The Home Network Work. ECSCW, Springer/Kluwer, Paris, France, 2005, 469-488.

Poole E.S., Chetty M., Grinter R.E. and Edwards W. K., 20XX, More Than Meets the Eye: Transforming the User Experience of Home Network Management GVU Center, School of Interactive Computing. Georgia Institute of Technology, Atlanta, GA, USA.

Demiris G., Rantz M., Aud M.D., Marek K.D., 2004, Inform H. M., Older adults' attitudes towards and perceptions of 'smart home' technologies: a pilot study. (JUNE 2004) VOL. 29, NO. 2, 87-94.

TMC News, (2011) Cisco Home Networking Issues Consumer Survey [Professional Services Close - Up], August 31, 2011, TMC News.
http://www.tmcnet.com/usubmit/2011/08/31/5741669.htm

Yang J. and Edwards W. K., (2007) ICEbox: Toward Easy-to-Use Home Networking Graphics, Visualization and Usability Center, College of Computing, Georgia Institute of Technology. Accessed on December 2011.
http://www.cc.gatech.edu/~keith/pubs/icebox-interact.pdf

MacMillan, R.: Plugged In: Wireless Networking Baffles Some Customers. Reuters news report, March 10 (2006).

NETGEART, Inc (2012), Accessed on January 10th, 2012.
http://www.netgear.com/landing/en-us/netgear-genie.aspx

CHAPTER 45

Delivering Intelligent Healthcare Web Services in a Cloud for Aging in Place

Chang-Yu Chiang[1], Chuan-Jun Su[2]

Department of Industrial Engineering & Management
Yuan Ze University
135, Yuandong Rd., Zhongli City, Taoyuan County 32003, Taiwan (R.O.C)
s978905@mail.yzu.edu.tw [1]
iecjsu@saturn.yzu.edu.tw [2]

ABSTRACT

As the elderly population has been rapidly expanding and the core tax-paying population has been shrinking, the need for adequate elderly health and housing services grows while the resources to provide services is decreasing. Delivering healthcare services in a more effective manner by using modern technology is of paramount importance. The seamless integration of such enabling technologies as ontology, intelligent agent, web services, and cloud computing is transforming the healthcare from hospital-based treatments to home-based self-care and preventive care. A ubiquitous healthcare platform based on this technological integration, which synergizes service providers with care-receivers needs to be developed to provide personalized healthcare services at the right time, in the right place, and the right manner.

This paper presents the development and overall architecture of IAServ (Intelligent Aging-in-place Healthcare Web Services Platform) for the provision of personalized healthcare service ubiquitously in a cloud computing setting to support the most desirable and most cost-efficient method of aging - aging in place. The IAServ is expected to offer pervasive, accurate and contextually-aware personal care services. Architecturally the Sensor Network-implemented IAServ leverages Web service and Cloud Computing to provide economic, scalable, and robust healthcare services over the Internet.

Keywords: Ubiquitous Computing, Healthcare, Aging-in-Place, Web Service, Cloud Computing, Ontology

1   INTRODUCTION

Rapid advances in medicine and technology are leading to marked increases in human longevity. Combined with declining birth rates, societies around the world are now facing challenges associated with rapidly aging populations. A recent report as an example from Taiwan's Department of Health shows that nearly 90% of senior citizens in Taiwan live with one chronic disease, while over half live with three or more chronic diseases. Many of these people are incapable of living independently, and the requirements for care will only continue to increase. These issues are raising broader concerns about the overall care and quality of life of the elderly.

Home Care is conceived as the integration of medical, social and familiar resources addressed to the same goal: enabling the integral care of the elders in their own residence (Valls et al., 2010). It emphasizes the concept of the "aging-in-place" which refers to living where care-receivers have lived for many years, or to living in a non-healthcare environment, and using products, services and conveniences to enable care-receivers to not have to move as circumstances change. The practice of the aging-in-place would significantly have the advantages of underpinning to elders' feelings of dignity, quality of life and independence. Today, however, the current connections between elderly health and housing are tenuous at best. As a result, the most desirable and most cost-efficient method of aging - aging in place - is difficult, even under the most ideal conditions (Lawler, 2001).

The goal of health services provision is to improve health outcomes in the population and to respond to people's expectations (personalization), while reducing inequalities in both health and responsiveness. The health care needs of the population should be met with the best possible quantity and quality of services produced at minimum costs. The use of information technology (IT) to enhancing healthcare delivery and to improving the quality of life is of paramount importance. In this research, we aim to seamlessly integrate such enabling technologies as ontology, intelligent agent, web services, and cloud computing for establishing the IASERV platform to provide ubiquitous, personalized healthcare services.

In the context of healthcare, there are virtually no cases with the same presentation, every care-receiver having a rather unique history and manifestation of a disease. A robust methodology is therefore needed for the provision of personalized healthcare services, which takes into account the health status and various context of each care-receiver (Riaño et al., 2009). To provide personalized health care services, ontology and rules of inference are utilized in IASERV to convert low-level context to high-level context. Ontological design used in IASERV furnishes an accurate formalization of healthcare goals for responding to health care providers and receivers needs by directing and addressing their current aims. The personalization in IASERV is achieved by means of compiling individual version of a personal profile, which derives the action plan of the personal agents in the multi-agent environment.

Web Services aim at building environments consisting of modular, distributable, sharable, and loosely coupled components. Web services are the software applications available on the Internet/Web which perform specific functions. Built on XML, a standard that is supported and accepted by thousands of vendors worldwide, Web services first focus on interoperability. XML is the syntax of messages, and Hypertext Transport Protocol (HTTP), the underlying protocol, is how applications send XML messages to Web services in order to communicate. Basically, Web services were born to solve three main issues in distributed programs, those are Firewall Traversal, Complexity, and Interoperability. With Web Services, users can share, integrate, and reuse all of these to create new functionalities on the Internet in a loosely-coupled and interoperable way.

Cloud computing is Internet-based computing, whereby shared data, software, platform, and infrastructure are provided to users through typically a web browser on demand, like the electricity grid. Generally, cloud computing customers do not own the physical infrastructure, instead avoiding capital expenditure by leasing usage from a cloud provider. The most comprehensive definition of cloud computing comes from The National Institute of Standards and Technology (NIST), Information Technology Laboratory: "Cloud computing is a pay-per-use model for enabling available, convenient, on-demand network access to a shared pool of configurable computing resources (e.g., networks, servers, storage, applications, services) that can be rapidly provisioned and released with minimal management effort or service provider interaction." (Mell and Grance, 2011)

In order to establish an interoperable and ubiquitously accessible IASERV with short time-to-market and low running cost, healthcare services are developed as web services and deployed in a cloud computing setting. The development of IASERV aims to improve the quality of aging-in-place, make home care services more proactive, and reduce the overall cost in the provision of care services. Ontology that provides the formalization of healthcare goals is used in the process of creating the personalized care plan for each care-receiver. The Web Services furnish the various applications which are relevant to the care activities needs. Sensor Networks are also used to allow easily collecting the environment information involving the temperature, humidity, person's location and so on, while Mobile Agents offer contextually-aware, timely and accurate information for the provision of quality care, with care and treatment data recorded automatically.

Section 2 of this paper presents the related works to discuss the development of Ambient Intelligence (AmI), the essence of personalized services, and cloud computing development. Section 3 presents the system design and architecture of the IASERV. Finally, conclusions of this research and comments on suggested future extensions are made in Section 4.

2   RELATED WORKS

2.1   Ambient Intelligence Applications in Care Services

Context-aware computing systems and methods are described. In particular embodiments, location aware systems and methods are described. In the described embodiments, hierarchical tree structures are utilized to ascertain a device context or location (Parupudi et al., 2010).Contextual awareness is a concept that has been described for some time, but technologies (e.g. wireless technologies, mobile tools, sensors, wearable instruments, intelligent artifacts, handheld devices, etc.) are only now becoming available to support the development of applications. Such technologies could help health care professionals to manage their tasks while increasing the quality of patient care (Bricon-Souf, 2007).

Su and Chen proposed an "Intelligent Community Care System (ICCS)" by applying RFID (Radio-frequency identification) and Mobile Agent technologies to enable care givers and communities to offer pervasive, accurate and contextually-aware care services (Su and Chen, 2010). RFID allows care-givers to easily locate the care-receivers, while Mobile Agents furnish contextually-aware, timely and accurate information for the provision of quality care, with care and treatment data recorded automatically. Fraile et al. presented a hybrid Multi-Agent architecture, named HoCa, for the control and supervision of dependent environments (Fraile et al., 2008). The HoCa architecture provided the basic idea in incorporating the alert management system based on SMS and MMS technologies and context control system based on Java Card and RFID technologies.

With the past researches (e.g. mentioned above), the basic idea referring to the infrastructure of developing a care services system platform equipped AmI mechanism was performed. The IASERV proposed in this paper would be built and integrated the novel ideas, (e.g. cloud computing, web services, etc.), to provide the more robust care services on-top of the past contributions.

2.2   Profiles and Personalized Services

The user profile is the data instance of a user model that is applied to adaptive interactive systems. It also acts as the key component which is adopting to provide the personalized services or information in the personalization system (Gauch et al., 2007). Golemati M. et al. showed that the generation of the user profile which is available in semantically inferring the personalized services, from user information collection, profiles construction to profiles representation (Golemati et al., 2007). Eslami MZ et al. proposed an effective service tailoring process and architecture to personalize homecare services according to the individual care-receiver's needs. In the proposed approach, the tailoring process is divided into six steps which present how to conFig the personalized services from the user profile (Eslami et al., 2010). Chang YK et al. developed a personalized service recommendation system (PSRS) in a home-care environment. The PSRS has capability of providing proper services based on the user's preferences and habits which are recorded in the user profile. Through the user profile, the system will be able to automatically launch to safety alert, recommendable services and healthcare services in the house (Chang et al., 2009).

The feasible solution, which presents how to generate the personalized services from user profile, can be summarized into two processes: generating the goal activities from the user profile and defining the services which can perform the activities. In generating the goal activities, the ontology, named "goal ontology", is created to present the relationship between the user profile and goal activities. In defining the services, the ontology, named "task ontology", is created to present the relationship between the goal activities and the services. This solution has broadly used in many researches. In this paper, this solution is also cited to support the personalized services configuration.

2.3   Cloud Computing

While the Web environment and IT services are rapidly evolving, a new development paradigm comes out. Cloud computing is recently coming into prominence as a new computing paradigm for IT services based on the Internet, and it typically involves over-the-Internet provision of dynamically scalable and often virtualized resource. The concept of cloud computing is that, instead of hosting all applications and data in the client-side, they can now be load balanced on many different servers in many different locations.

A method for integrating cloud computing systems includes establishing a connection between a cloud computing system architecture and cloud computing systems. Each of the cloud computing systems includes computing resources. The method further includes integrating the computing resources with external integration architecture by establishing a second connection between the cloud computing system architecture and the external integration architecture (Hadar et al., 2010).

IBM defined that cloud computing provides a completely Internet-based, dynamic and scalable service-oriented IT environment which can be accessed by using Web-capable device anywhere. IBM indicated that cloud computing is different from earlier on demand computing concept and said that cloud computing not only delivers scalable virtualized operating environments (infrastructure as a service, or IaaS); it can also provide application design, development and management tools (platform as a service, or PaaS), and the applications themselves (software as a service, or SaaS). Furthermore, data and software will be stored in the cloud and accessed via the web protocols, eliminating the traditional need for client-side software deployment.

NIST (National Institute of Standards and Technology, US) provides the most comprehensive definition of cloud computing as stated in the previous section (Mell, and Grance, 2011). Furthermore the essential characteristics, service models, and deployment models are also addressed in the definition.

3   IASERV DESIGN AND ARCHITECTURE

The design of IASERV centers around seamless integration of such enabling technologies as Ontology, Sensor Network, Web Services and the FIPA (Foundation for Intelligent Physical Agents) - compliant agent framework JADE (Java Agent Development Environment) (Bellifemine et al., 2007) to provide personalized and context-aware healthcare services. The services are implemented as web services and deployed in an environment of cloud computing for economic, scalable, and ubiquitously accessible service provision as illustrated in Figure 2.

Functionally, the IASERV is able to infer the appropriate care plan which defines the needed healthcare activities through the elders' profiles and relevant external web services. The caregivers or professionals are allowed to administrate the healthcare according to a medical goal. The contextually-relevant information such as the impact of recent weather changes to a person suffered from hypertension can also be accessed by the caregivers through their tablet PC, PDA, or other mobile devices. The IASERV architecture comprises three main components: 1) the repositories layer, 2) the web services running in a cloud setting, and 3) the agent environment as shown in Figure 1.

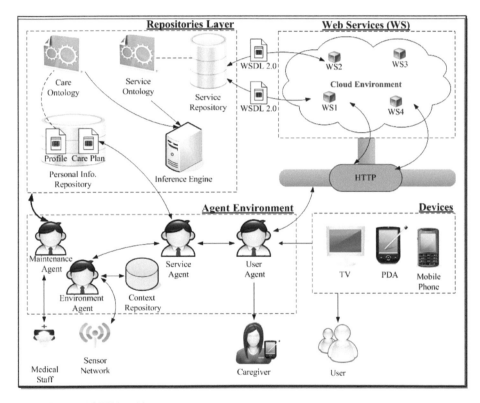

Figure 1 The IASERV architecture

3.1 Repositories Layer

The Inference engine in the Repository Layer plays a vital role in deriving personalized care plan based on a care-receiver's profile. A professional firstly compile a care-receiver's profile into ontological format through a predefined UI (User Interface). The ontological profile will be subsequently used as input to the

424

Inference Engine for generating a personalized care plan by using the Care ontology and Service ontology.

The personal profile includes the basic information (e.g. name, sex, age, characteristics, preference, interest, etc.) and the personal states (e.g. clinical states, disabilities, impairments, etc.) of an individual.

The Care Ontology encapsulates the relationship between the care-receiver's profile and the relevant care services. The classes defined in Care Ontology include User State, Goal-specification, Care Service, and Care Activity as illustrated in Figure 2. The User State class indicates the status of a care-receiver while the Goal-specification and the Care Service classes describe the intended medical goal and define the care services and tasks associated with specific sates respectively.

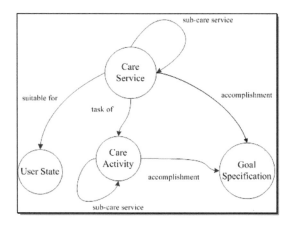

Figure 2 The partial model of Care Ontology

The Service Ontology encapsulates the relationship between the care activities and the physical web services. The classes covered in the Service Ontology include Care Activity, Web Service, Service Profile, Process Model, and Service Grounding as shown in Figure 3. The Care Activity specifies the tasks that need to be performed in a care service while the Web Service registers services that are available for executing tasks.

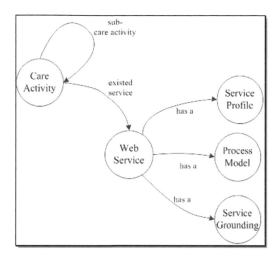

Figure 3 The partial model of Service Ontology

3.2 Agent Environment

The Agent Environment is in charge of the management and coordination of all system applications and services. The agents would perform the defined care services in a care plan for a care-recipient by dispatching the Web Services with the contextual data through the HTTP protocol. The agent environment in IASERV comprises four agents and each one of them with specific roles, capabilities and characteristics:

- **Environment Agent:** The Environment Agent is responsible for managing the environmental data (e.g. temperature, humidity, location, etc.). It works with the Context Repository and connects with the sensor networks. The Environment Agent collects the environmental data from the sensor network and stores the data into the Context Repository. When performing the contextual relevant services, the Environment Agent offers the corresponding contextual data to the Service Agent for the provision of contextually-aware services.

- **Service Agent:** The Service Agent is designed for monitoring the personalized care plan. When detecting the scheduled care plan, it will prepare the information of the web services with the needed contextual data and subsequently delegate to the User Agent for performing tasks in a care service.

- **User Agent:** The User Agent is responsible for accessing the web service and performing the care services on behalf of care-receivers when receiving the tasks from the Service Agent. According to the data from the Service Agent, the User Agent makes a request to the web service with the needed input data parameters via HTTP protocol. The resulted information will then be published to the user device by the User Agent to complete the care service.

- **Maintenance Agent:** The Maintenance Agent essentially provides the repositories management functions. The medical staff may administer care-receiver's care plan if the care services are not suitable for the care-receiver. The Maintain Agent serves as a bridge between the professionals and repositories. The professionals administer (post, update, delete) the services and profiles through the GUI provided by the Maintenance Agent.

5. CONCLUDING REMARKS AND FUTURE WORKS

In this paper, we present the design and development of IASERV based on seamless integration of such enabling technologies as Ontology, Sensor Network, Web Services and the FIPA (Foundation for Intelligent Physical Agents) - compliant agent framework JADE (Java Agent Development Environment) to provide personalized and context-aware healthcare services. The IASERV employs ontologies to improve the integration and orchestration of all the technologies involved such as web services, context-aware healthcare services, sensor network, and software agents. Rules of inference are formulated, which utilize the structural information in ontologies to derive personalized, context-aware healthcare plan for elders. The sensor network of implementation for web services and the deployment to cloud computing environments can not only shorten the development cycle but also reduce maintenance costs and enable ubiquitous access. With the proposed system design, an economic, scalable, and robust healthcare platform can be built over the Internet.

One critical issue remained to be addressed is the protection of users' privacy and personal information. While the advantage of mobility in the agent paradigm helps us in developing distributed health care systems and reaching a large section of mobile clientele, it also opens the system and makes it vulnerable to be attacked by malicious entities. In this research we adopted a JADE plug-in JADE-S (Moreno et al., 2010) that allows us to establish security characteristics such as access control, secure communications, etc. in the platform, so that it can start to be used in real environments. While JADE-S provides fundamental support for application level security, it has such shortcomings as SSL cannot be individually but globally applied to all messages, the information of agent identity cannot be accessed from the program, etc. The proposed IASERV can be further strengthened by incorporating with more sophisticated and robust security models to provide better protection for users' personal data and privacy (Wong and Ng, 2006).

REFERENCES

Bellifemine, F., G. Caire, and D. Greenwood. 2007. *Developing Multi-Agent Systems with JADE*. England: John Wiley & Sons, Ltd. Press.
Bricon-Souf, N. 2007. Newman CR. Context awareness in health care: A review. *International Journal of Medical Informatics* 76(1):2-12.

Chang, Y. K., C. L. Yang, and C. P. Chang, et al. 2009. "A Personalized Service Recommendation System in a Home-care Environment." In: Proceedings of the 15th International Conference on Distributed Multimedia Systems, DMS 2009, San Francisco, USA.

Eslami, M. Z., A. Zarghami, B. Sapkota, and M. Sinderen. 2010. "Service Tailoring: Towards Personalized Homecare Services" In: Proceedings of the 5th International Workshop on Architectures, Concepts and Technologies for Service Oriented Computing, ACT4SOC 2010. Athens, Greece.

Fraile, J. A., J. Bajo, B. P. Lancho, and E. Sanz. 2008. "HoCa Home Care Multi-agent Architecture." In: Proceedings - International Symposium on Distributed Computing and Artificial Intelligence, DCAI 2008 (pp. 52-61), 22-24 October, Salamanca, Spain.

Gauch, S, M. Speretta, A. Chandramouli, and A. Micarelli. 2007. User profiles for personalized information access. In. The adaptive web, eds. P. Brusilovsky, A. Kobsa, and W. Nejdl. Springer-Verlag Berlin, Heidelberg.

Golemati, M, A. Katifori, C. Vassilakis, G. Lepouras, and C. Halatsis. 2007. "Creating an Ontology for the User Profile: Method and Applications." In: Proceedings of the First IEEE International Conference on Research Challenges in Information Science, RCIS 2007 (pp. 407-412), Ouarzazate, Morocco.

Hadar, E., C. E. Gates, and K. H. Esfahany, et al. 2010. System, Method, and Software for Integrating Cloud Computing Systems. Patent - US20100125664A1.

Lawler, K. 2001. "Aging in Place: Coordinating Housing and Health Care Provision for America's Growing Elderly Population," Accessed February 28, 2012, http://www.jchs.harvard.edu/publications/seniors/lawler_w01-13.pdf.

Mell, P. and T. Grance. 2011. "The NIST definition of cloud computing," Accessed February 28, 2012, http://csrc.nist.gov/publications/nistpubs/800-145/SP800-145.pdf.

Moreno, A, D. Sanchez, and D. Isern. 2003, Security Measures in a Medical Multi-Agent System. Frontiers in Artificial Intelligence and Applications. IOS Press.

Parupudi, G., S. S. Evans, B. J. Holtgrewe, and E. F. Reus. 2010. Context Aware Systems and Methods Utilizing Hierarchical Tree Structures. Patent - US7743074.

Riaño, D., F. Real, and F. Campana et al. 2009. An Ontology for the Care of the Elder at Home. In. Artificial Intelligence in Medicine. eds. C. Combi, Y. Shahar, and A. Abu-Hanna. Springer Berlin, Heidelberg 235-239.

Su, C. J. and B. J. Chen, "An Intelligent Community Care System Using Network Sensors and Mobile Agent Technology." Paper presented at AHFE International Conference, jointly with 1st Advances in Human Factors and Ergonomics in Healthcare, Miami, USA, 2010.

Valls, A., K. Gibert, D. Sánchez, and M. Batet. 2010. Using ontologies for structuring organizational knowledge in Home Care assistance. International Journal of Medical Informatics 79(5):370-387.

Wong, S. W. and K. W. Ng. 2006 "Security Support for Mobile Grid Services Framework." In: Proceedings of the International Conference on Next Generation Web Services Practices, NWeSP 2006 (pp. 75-82), 25 September – 9 October, Seoul, South Korea.

CHAPTER 46

The Role of Web Layout Design Factors in Modeling the Internet User Behavior

Rafał Michalski, Jerzy Grobelny, Wojciech Lindert

Institute of Organization and Management (I23),
Faculty of Computer Science and Management (W8),
Wrocław University of Technology,
27 Wybrzeże Wyspiańskiego,
50-370 Wrocław, Poland

ABSTRACT

The presented research deals with the effects of web page design influencing the usability measured both objectively as well as subjectively. The examined factors included the background color, the number of local navigation hyperlinks, and the proportion of graphics in relation to the text on the html page. The obtained typical task execution times were first analyzed by means of the analysis of variance and later compared with the users' preferences obtained by means of the Analytic Hierarchy Process technique (Saaty, 1977).

Keywords: usability, web page design, subjective evaluation, AHP

1   INTRODUCTION

The comprehensive understanding how people search information in web sites and provision of appropriate recommendations in this regard has a huge significance not only for broadening our knowledge about the human behavior, but also is practically useful. The studies in this field give important indications for developing the guidelines in the area of human-computer interaction (HCI).

Apart from the efficiency and effectiveness of the given interactive systems, the understanding individuals' likings is recently also an important issue in this field.

The users' preferences are connected with the user satisfaction which constitutes one of the main dimensions of the usability concepts defined both by HCI researchers and practitioners (e.g. ISO 9241, 1998; ISO 9126, 1998; Dix et al., 2004; Folmer and Bosch, 2004).

Although, there was a number of research regarding the search and click tasks (e.g. Schaik and Ling, 2001; Kalbach and Bosenick, 2003; Pearson and Schaik, 2003; Grobelny et al., 2005; Michalski et al., 2006), and preferences (e.g. Tractinsky et al., 2000; Hassenzahl, 2004; Lavie and Tractinsky, 2004; Grobelny and Michalski, 2011) separately, there were few of them that concerned both objective and subjective measures that dealt with the various web site templates commonly used in practice.

Therefore, the main goal of this study is to explore the influence of the selected factors of designing the web pages on the users' operation efficiency and preference structure. For this purpose, an investigation was conducted that consisted of two stages. The first one dealt with performing the simple search and point tasks whereas the second one concerned the preference evaluation by means of the pairwise comparisons. In next passages of this work the aforementioned experiments are presented in details, analyzed and discussed.

2   METHOD

Participants

Thirty eight subjects participated in the efficiency examination. All of them were students of the Wroclaw University of Technology at the age between 19 and 25 years. They were not paid for taking part in the examination. Among the participants there were 22 males and 16 females. The great majority of the examined persons possessed own computer for more than three years and used Internet on a daily basis. Almost 70 percent of subjects used the Mozilla Firefox web browser, 16% Opera, 11% MS Internet Explorer, and only 5% Google Chrome. As many as 87% users used Microsoft Windows operating systems including Windows XP (54%) and Windows Vista (34%).

In the second part of the study concerned with the preferences assessment 65 students were investigated. Among them there were 33 women and 32 men. The age ranged from 19 to 28 with the average of 21 years. Most of the subjects (95%) declared using Microsoft operating systems. The Mozilla Firefox was the most popular web browser (49%), the second was Google Chrome (20%), next Opera (16%), and only 6% used MS Internet Explorer.

Apparatus

The examinations were conducted in an open-access self learning room in one of the Wroclaw University of Technology buildings. In order to prevent the attention disturbance a special box was used.

The performance tasks were carried out on one laptop computer with the external laser mouse. The resolution was set to 1024×764 pixels at its 17" screen. The default parameters of the screen and mouse were employed. The Windows XP Professional operating system was installed along with the Microsoft Power Point 2003 used for displaying stimuli. Free light version of the uLog (Noldus Information Technology, 2012) software was employed to record the user activity such as mouse clicks and keystrokes performed during the tests. For the preference study a different device was used. The 17'' screen computer laptop with the 1440×900 pixels resolution. Microsoft Windows Vista Business operating system was installed together with the IrfanView software in a 4.28 version (Skiljan, 2012) used for making the comparisons.

Independent variables

Four independent variables differentiating the analyzed web site structures were manipulated: background color, the number of links in the local navigation, the proportion of the web page filled by graphics, and the way the local menu items were arranged.

The number of links in the local navigation. The number of hyperlinks were set on two levels seven (L07) and 14 (L14). According to the Miller studies (1956) a human being is capable of manage about seven simple chunks of information in the working memory, hence this number should be optimal also in specifying the quantity of links in a menu. The second level was set twice as high.

The proportion of the web page filled by graphics. This factor was also examined on two levels: 15% (G15) and 30% (G30). Those values results from the heuristic recommendation that the optimal number of illustrations is three, and they should not occupy more or less from 5% to 15% of the whole web page available space (Nielsen and Tahir, 2001). In other words, devoting more than 15% of the web page layout to the images may indicate the graphics overloading. The second level was doubled.

The local menu orientation. Two types of arranging the local menu items were used. The first, horizontal (H) one, was situated in the top section of the web page, directly under the global menu, whereas in the second arrangement they were located vertically (V) in the left hand side of the screen.

The web site templates used in this study were created generally according to the recommendations provided by Ani Phyo (2003). She considers that the typical web site contains the following modules: global, local, and administrative navigations, web page specific content (e.g. headings, subheadings, text, images, video clips, music clips, animations, captions and other objects), web page title, search tools, and page footer.

In this study there were, however, some modifications introduced. Because there are many web sites which do not separate the administrative navigation module, it was not included in our design. Additionally, there are some doubts whether those objects really help users in navigation (Nielsen, 1999). The items from the administrative navigation were located either in the local menu or the page footer.

The title was also not incorporated into the research templates as this element does not matter in this investigation. Every studied web site mock-up consisted of the following six areas which were also indicated in the exemplary web sites presented in Figure 1:

(1) The global menu that contains four buttons. For the mock-up with the vertical local menu they included: registration, e-mail, forum, and chat links, while for the one with a horizontal local menu: forum, galleries, novelties, and registration. For each of the experimental condition the order of those buttons was arranged differently. This solution should prevent users from learning the buttons locations, which was not examined in this study.

(2) The local menu which included either seven or 14 elements. The smaller local menu contained: history, services, price list, picture gallery, location map, news, files to download buttons. In the case of the bigger solution the following targets were employed: science, blog, e-mail, auction, offers, games, chat, TV, help, tips, music, sport, film, and business. All of the hyperlinks were also situated at a random order for every version of the investigated web page.

(3) The content specific for the given web page – the graphics used here is different for every web page mock-up. The place devoted to the text was filled with the *Lorem Ipsum* words (Lorem ipsum, 2012). This Latin text allows the user to concentrate on its visual aspects instead of the meaning.

(4) The logo – there were two versions developed for this study: one for the web site versions with horizontal local menu and one for the vertical ones. The image was located in the same place for all variants.

(5) The search mechanism – it is the same size and shape in all the studied web page and is situated in approximately the same location.

(6) The web page footer – contains the same number of items for all experimental conditions and their order is also identical. The footer consists of the following hyperlinks: terms of use, contact, privacy policy, copyright information, security on the internet.

Figure 1 The investigated web sites prepared according to the first and second template

Dependent measures

The dependent variables being measured were twofold nature. The task efficiency was measured by registering the acquisition times. The time was computed from when the users clicked on the task execution order to when the target item was selected. The users' preferences expressed towards the web site mock-ups were obtained by means of the application of the procedure proposed by Saaty in the AHP method (1977). The temporal parameter results were gatherer by means of the uLog program (Noldus Information Technology, 2012) that records some of the user activities in the operating system.

Experimental design and procedure

In the efficiency as well as the preferences evaluation all three independent variables were used and the combination of these factors, each one on two levels, resulted in 8 different web site designs: (2 no of local links) × (2 graphics-text proportions) × (2 local menu orientations).

All of the examined persons have never seen the research web pages before the investigation. Prior to the proper examination, the subjects were informed about the goal and the scope of the study. Then they were asked to fill in the anonymous questionnaire that included questions about the: age, gender, education, computer possessing, time spent daily for surfing on the Internet, operating system, and web browser type used. Next, the proper computer based investigation took place. The within subject experimental design was applied so every participant tested all of the experimental conditions. The two parts of the study were conducted separately. The efficiency part began by displaying on the white background the task execution order. After the mouse clicking, the given web page appeared on the screen. The user had to find and click the earlier specified target.

The proper study of determining the preference structure by means of the AHP technique started by filling out the questionnaire, which consisted of 28 comparisons, each row contained one comparison. A subject answered the question "Which of the two web sites do you like more?", and indicated his/her opinion on the Likert-type scale: decidedly prettier, prettier, somewhat prettier, the same preference. Simultaneously the given pair of web sites was displayed on the computer screen. After pressing the space, next pair of web pages to compare appeared. The whole comparisons' procedure lasted approximately seven minutes.

3   RESULTS AND DISCUSSION

In the sections that follow the obtained objective and subjective results' analyses are provided. The gathered data are first investigated by means of the descriptive statistical parameters and then the analysis of variance was conducted for the described in previous sections independent factors. Next, the preference related data were depicted and analyzed by means of the similar statistical tools.

Objective results and analysis

Descriptive statistics

The average value of the task completion time for all of the experimental conditions amounted to 4.4s, with the standard deviation of 3.3s, and the mean standard error – 0.19s. The shortest time registered for the mock-ups equalled 1s whereas the longest amounted to as much as 34s. The descriptive characteristics including average acquisition times, standard deviations, mean standard errors, and extreme values for acquisition times obtained for the examined web sites are put together in Table 1. From these data, it can be easily noted that the second experimental web page mock-up was operated the fastest by the users with the mean value of 2.6s. The worst results were obtained for the eighth condition where the average acquisition time was equal 8s. The biggest dispersion of the results was observed also for the eighth layout where the standard deviation and mean standard error amounted to 5.4s and 0.88s respectively. The smallest values of these parameters were computed tor the second web page template.

Table 1 Descriptive statistics of the target acquisition times

No	Experimental condition	Mean time (s)	SD	MSE	Min	Max
1.	L07_G15_H	3.3	2.5	0.40	1	12
2.	L07_G15_V	2.6	1.3	0.20	1	9
3.	L07_G30_H	4.7	1.5	0.24	2	9
4.	L07_G30_V	3.0	2.1	0.34	1	13
5.	L14_G15_H	4.2	2.3	0.37	1	9
6.	L14_G15_V	5.6	3.4	0.54	2	23
7.	L14_G30_H	3.7	2.6	0.43	1	12
8.	L14_G30_V	8.0	5.4	0.88	2	34

Analysis of Variance

In order to verify the significance of differences in the task completion times, a standard three-way analysis of variance was employed. The calculated F statistics and respective p values for the main effects are put together in Table 2.

According to the obtained results all of the effects were statistically meaningful. Additionally, two interactions happened to be significant: the interaction between the number of local menu links with the local menu orientation, and the interaction among all of the three examined factors.

Table 2 Descriptive statistics of the target acquisition times

Factor	df	F	p
Number of links in the local navigation (NLN)	1	35	< 0.00001*
Proportion of the web page filled by graphics (PGR)	1	8.4	0.0040*
Local menu orientation (LMO)	1	6.1	0.014*
NLN × PGR	1	0.019	0.89
NLN × LMO	1	36	< 0.00001*
PGR × LMO	1	1.9	0.17
NLN × PGR × LMO	1	9.1	0.0028*

The mean task execution times along with 0.95 confidence intervals denoted by whiskers for all the statistically significant factors and interactions are presented in Figures 2–6.

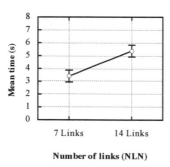

Number of links (NLN)

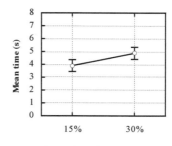

Proportion of graphics (PGR)

Figure 2 Mean task execution times depending on the number of links, $F(1, 296) = 35$, $p < 0.00001$.

Figure 3 Mean task execution times depending on the proportion of graphics on the page, $F(1, 296) = 8.4$, $p < 0.005$.

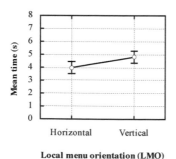

Local menu orientation (LMO)

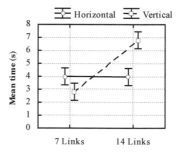

Number of links (NLN)

Figure 4 Mean task execution times depending on the local menu orientation, $F(1, 296) = 6.1$, $p < 0.05$.

Figure 5 Mean task execution times for the interaction between the number of links and the local menu orientation $F(1, 296) = 36$, $p < 0.00001$.

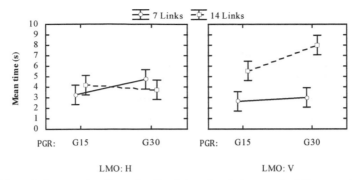

Figure 6 Mean task execution times for the interaction between all of the examined factors $F(1, 296) = 9.1$, $p < 0.005$.

These obtained outcomes for the number of links in the local menu support the theoretical expectations stemming from the Hick-Hyman law (Hick, 1952; Hyman, 1953), which links the search performance times with the number of stimuli to be processed. In this study the smaller number of items in the local menu decreased the number of possible objects to be searched. This was probably the cause of the registered considerably shorter task performance times for the seven-item web mock-ups. However, it was somewhat surprising that the times performed on web pages with the 15% of graphics share were executed meaningfully faster than in the case of the layouts with the proportion of 30%. This phenomenon could be connected with the more demanding visual processing of the graphics than the text. The decidedly better completion times for the horizontal in comparison with the vertical one orientation of the local menu were already reported in some investigations (Pearson and Schaik, 2003; Michalski et al., 2006; Michalski and Grobelny, 2008) and were rather anticipated. On the other hand, the interaction between the number of links and the local menu orientation was unexpected. This result shows the superiority of the horizontal arrangement over the vertical one only for the 14 objects included in the local menu, whereas for the menu with seven-items the vertical orientation was better operated.

Preferences' results and analysis

The results pertaining to the users' preferences towards examined web pages are presented in this section. As it was mention earlier the preferences were examined by means of the Analytic Hierarchy approach. This technique allows for assessing the comparisons' consistency level, which can be used to verify the reliability of the obtained results. The consistency ratio threshold used for this purpose was set at the level of 0.2 in this study. The application of this criterion, resulted in exclusion of 28 persons, and thus the results of 37 participants were subject to research in next sections. The relative likings are expressed as average values of the obtained AHP weights.

Descriptive statistics

The basic descriptive statistics of the obtained relative preferences for all examined web page templates are summarized in Table 3.

Table 3 Descriptive statistics of the web page weights

No	Web page	Weight	Rank	SD	MSE	Min	Max
1.	L07_G15_H	0.135	2	0.0855	0.0141	0.0271	0.419
2.	L07_G15_V	0.126	4	0.0923	0.0152	0.0241	0.344
3.	L07_G30_H	0.149	1	0.11	0.0181	0.0218	0.390
4.	L07_G30_V	0.127	3	0.09	0.0148	0.0197	0.408
5.	L14_G15_H	0.102	8	0.0858	0.0141	0.0206	0.325
6.	L14_G15_V	0.120	6	0.0864	0.0142	0.0222	0.300
7.	L14_G30_H	0.125	5	0.0958	0.0157	0.0269	0.362
8.	L14_G30_V	0.116	7	0.0760	0.0125	0.0211	0.344

The presented results show that the markedly best perceived web page was the one with horizontal local menu consisting of seven hyperlinks and the 30% share of graphics. The worst evaluated experimental condition was the one with the vertical menu including 14 objects and the 30% of space occupied by graphical elements. It can also be observed from the data that generally better rated were web pages with lower number of local navigation links.

Analysis of Variance

To test the significance of differences in the average weights computed for individual web sites, a standard three way analysis of variance was used. The obtained results revealed that only the number of links in the local navigation (NLN) was statistically significant merely at the level of 0.1. The rest of the analyzed factors along with all of the interactions were irrelevant. The detailed results of this analysis are demonstrated in Table 4.

Table 4 Descriptive statistics of the target acquisition times

Factor	df	F	p
Number of links in the local navigation (NLN)	1	3,05	0,082*
Proportion of the web page filled by graphics (PGR)	1	0,63	0,43
Local menu orientation (LMO)	1	0,30	0,58
NLN * PGR	1	0,0075	0,93
NLN * LMO	1	0,93	0,34
PGR * LMO	1	0,97	0,32

The mean weights for the effect of the number of links in the local menu are illustrated in Figure 7 and show slightly higher preferences towards web pages with smaller number of menu items.

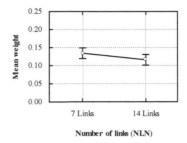

Figure 7 Average AHP weights depending on number of links in the local menu, $F(1, 288) = 3.05$, $p < 0.1$; whiskers denote 0.95 confidence intervals.

Regression analysis

The further exploration was meant to verify whether the objective results obtained in the first part of this research corresponded to the preference investigation conducted in the second phase. For this purpose the linear regression was applied and the outcome is visualized in Figure 8.

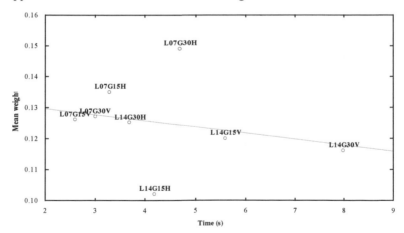

Figure 8 Mean times on slide depending on the web site, $F(3, 92) = 8.4$, $p < 0.0001$; Mean weight = 0.134 - 0.00197 × Mean time, $R^2 = 6.3\%$, $F(1, 6) = 0.404$, $p = 0.55$.

The analysis revealed no correspondence between the objective efficiency results and later subjective ratings as the R squared amounted to barely 6.3%. The lack of correlation might be possibly attributed to the fact that the subjects making the preference comparisons were not performing any tasks on the examined web pages. Therefore, their opinions expressed only the perceived attractiveness of presented layouts. In reality those preferences could be affected by the users experience in operating the given web site solution. Similar effect was observed in the study presented by Michalski (2011). It can also be noticed that in the Figure 8 there are two extreme values that probably strongly influence the regression results. If these outliers are excluded from the analysis then the linear regression happens to be significant, $F(1, 4) = 7.4$, $p = 0.05$, $R^2 = 65\%$, and takes the following form: $Mean\ weight = 0.136 - 0.00254 \times Mean\ time$.

4   CONCLUSIONS

The present research was designed in a way that should reflect to a considerable degree the natural processes of searching and clicking the objects located in the web page. Therefore, the typical web site layouts were applied and three main factors: the number of items in the local menu, proportion of graphics to text, and the local navigation orientation were analyzed both objectively and subjectively. The objective analysis included the efficiency analysis of searching and selecting the target, while the subjective approach involved the preferences' examination. Whereas the search and click investigation showed significant influence of all of the examined effects along with some interactions, the preferences were meaningfully affected (just at the level of 0.1) only with respect to number of items in the local navigation area. Such outcomes showing different impact on the studied factors depending on the objective or subjective analysis were further confirmed by the linear regression results. This may indicate that the users' ratings of the viewed web sites may have little in common with the efficiency with which the web sites could be operated and emphasizes the significance of the subjective perception.

REFERENCES

Dix, A., J. Finlay, G. D. Abowd, and R. Beale 2004. *Human Computer Interaction*. 3rd Edition, Harlow: Pearson Education.

Folmer, E. and J. Bosch 2004. Architecting for usability: a survey. *The Journal of Systems and Software* 70: 61–78.

Grobelny, J. and R. Michalski, 2011. Various approaches to a human preference analysis in a digital signage display design. *Human Factors and Ergonomics in Manufacturing & Service Industries* 21(6): 529–542.

Grobelny, J., W. Karwowski, and C. Drury 2005. Usability of Graphical icons in the design of human-computer interfaces. *International Journal of Human-Computer Interaction* 18: 167–182.

Hassenzahl, M. 2004. The interplay of beauty, goodness, and usability in interactive products. *Human–Computer Interaction*, 19, 319–349.

Hick, W. E. 1952. On the rate of gain of information. *Quarterly Journal of Experimental Psychology* 4: 11–36.

Hyman, R. 1953. Stimulus information as a determinant of reaction time. *Journal of Experimental Psychology* 45: 188–196.

ISO 9126 1998. *Software product quality, Part 1: Quality model*, International Standard.

ISO 9241 1998. *Ergonomic requirements for office work with visual display terminals (VDTs)*, Part 11: Guidance on usability, International Standard.

Kalbach, J. and T. Bosenick 2003. Web page layout: A comparison between left- and right-justified site navigation menus. *Journal of Digital Information* 4, Article No. 153, 2003-04-28. Accessed 12 February 2011, http://journals.tdl.org/jodi/article/view/94/93.

Lavie, T. and N. Tractinsky 2004. Assessing Dimensions of Perceived Visual Aesthetics of Web Sites, *International Journal of Human-Computer Studies* 60(3): 269–298.

Lorem ipsum 2012, Wikipedia, Accessed February 18, 2012, http://en.wikipedia.org/wiki/Lorem_ipsum.

Michalski, R., J. Grobelny, and W. Karwowski 2006. The effects of graphical interface design characteristics on human-computer interaction task efficiency. *International Journal of Industrial Ergonomics* 36(11): 959–977.

Michalski, R. and J. Grobelny, 2008. The role of colour preattentive processing in human-computer interaction task efficiency: a preliminary study, *International Journal of Industrial Ergonomics* 38: 321–332.

Michalski, R. 2011. Examining users' preferences towards vertical graphical toolbars in simple search and point tasks. *Computers in Human Behavior* 27(6): 2308–2321.

Miller, G. A. 1956. The magical number seven, plus or minus two: some limits on our capacity for processing information. *Psychological Review* 63(2): 81.

Nielsen, J. 1999. *Designing Web Usability* (1st Ed.). Peachpit Press.

Nielsen, J. and M. Tahir 2001. *Homepage Usability: 50 Websites Deconstructed*. New Riders Publishing.

Noldus Information Technology 2012. *uLog*, Accessed February 18, 2012, http://www.noldus.com/human-behavior-research/products/ulog,.

Pearson, R. and P. Schaik 2003. The effect of spatial layout of and link color in web pages on performance in a visual search task and an interactive search task. *International Journal of Human-Computer Studies* 59: 327–353.

Phyo, A. 2003. *Return on Design: Smarter Web Design That Works* (1st Ed.). New Riders Press.

Saaty, T. L. 1977. A scaling method for priorities in hierarchical structures. *Journal of Mathematical Psychology* 15: 234–281.

Skiljan, I. 2012. *Irfanview*, Accessed February 18, 2012, http://www.irfanview.com.

Schaik, P. and J. Ling 2001. The effects of frame layout and differential background contrast on visual search performance in Web pages, *Interacting with computers* 13: 513–525.

Tractinsky, N., A. S. Katz, and D. Ikar 2000. What is beautiful is usable. *Interacting with Computers* 13(2): 127–145.

CHAPTER 47

Web-oriented Architecture for Long-Term Health Monitoring in a Cloud – Using Balance Ability Monitoring as an Example

Chang-Yu Chiang[1], Chuan-Jun Su[2], Bernard C. Jiang[3]

Department of Industrial Engineering & Management
Yuan Ze University
135, Yuandong Rd., Zhongli City, Taoyuan County 32003, Taiwan (R.O.C)
s978905@mail.yzu.edu.tw [1]
iecjsu@saturn.yzu.edu.tw [2]
iebjiang@saturn.yzu.edu.tw [3]

ABSTRACT

The global trends in aging population lead to significantly increased demand for long-term care (LTC) monitoring. In LTC area, many researchers have significant contributions in evaluating the elders' health status via analyzing gauged data from assistant devices, such as analyzing elders' balance ability by using a force plate. The applications are independently available at very specific environment and scenarios. Those are lack of interoperability and scalability and are not sharable. While the Web-oriented Architecture (WOA) is emerging as a new architecture for developing dynamic, highly scalable, and extensively interoperable Web applications, it has been practically implemented on a global scale today for thousands of organizations and has recently become a growing grassroots phenomenon. In this paper, we propose a WOA driven information platform for enabling effective long-term health monitoring (WLTHM) platform. The WLTHM platform aims to consolidate the once-fragmented world of LTC services into a more efficient platform, enabling the evaluation functions to be sourced and built more easily, quickly, inexpensively, and at higher quality levels. In addition, in the system development section and scenarios, we present the balance ability monitoring as an example to validate the viability and feasibility of the proposed approach.

Keywords: Long-Term Care, Health Monitoring, Balance Ability Analysis, WOA, Web-oriented Architecture

1   INTRODUCTION

Rapid advances in medicine and technology are leading to marked increases in human longevity. Combined with declining birth rates, societies around the world are now facing challenges associated with rapidly aging populations. A recent report as an example from Taiwan's Department of Health shows that nearly 90% of senior citizens in Taiwan live with one chronic disease, while over half live with three or more chronic diseases. Many of these people are incapable of living independently, and the requirements for care will only continue to increase. These issues are raising broader concerns about the overall care and quality of life of the elderly.

As the elderly population has been rapidly expanding and the core tax-paying population has been shrinking, the need for adequate elderly health and housing services grows while the resources to provide services is decreasing. Delivering healthcare services in a more effective manner by using modern technology is of paramount importance. The seamless integration of such enabling technologies as ontology, intelligent agent, web services, and cloud computing is transforming the healthcare from hospital-based treatments to home-based self-care and preventive care. A ubiquitous healthcare platform based on this technological integration, which synergizes service providers with care-receivers needs to be developed to provide personalized healthcare services at the right time, in the right place, and the right manner.

The subject of long-term health care has received increasing attention both in developing and developed countries in the world, due to a belief that an ageing population will greatly swell the demand for long-term health monitoring and tracking service (Peng et al., 2010). Recent health monitoring services trends try to offer solutions in forms of information technology (IT)-based devices, which can be operated remotely and automatically, relieving the carers from the need to stay near the patient. Klingeberg and Schilling (2012) presented a wireless coupled recording device for long-term monitoring of the vital sign. The ECG, the blood pressure and the skin temperature and recorded for the determination of the movements during recording. Chen (2011) developed an embedded human pulse monitoring system with intelligent data analysis mechanism for disease detection and long-term health care. The proposed system can be applied to monitor and analyze pulse signal in daily life and also has a friendly web-based interface for medical staff to observe immediate pulse signals for remote treatment. There are also a lot of similar works proposed (Paradiso et al., 2005; Preve, 2011). However, most of them are aimed at very specific pathologies and scenarios or are programmed with very specific functions. Additionally, they are definitely not user friendly and lack usable interfaces, so IT experts are required for fine tuning the system to each patient in a process that may take days or even weeks of work.

Web-Oriented Architecture (WOA) aims at building environments consisting of modular, distributable, sharable, and loosely coupled components. WOA merges the core design of Service-Oriented Architecture (SOA) (Welke et al., 2011), the user sharing of Web 2.0 and the functionalities of mash-up and regards entire web as resources via the representational state transfer (REST). REST is a style of software architecture for distributed hypermedia systems such as the World Wide Web (Fielding and Taylor, 2002). With WOA, users can easily share, integrate, and reuse all of these to create new functionalities on the Internet in a loosely-coupled and interoperable way.

Cloud computing is Internet-based computing, whereby shared data, software, platform, and infrastructure are provided to users through typically a web browser on demand, like the electricity grid. Generally, cloud computing customers do not own the physical infrastructure, instead avoiding capital expenditure by leasing usage from a cloud provider. The most comprehensive definition of cloud computing comes from The National Institute of Standards and Technology (NIST), Information Technology Laboratory: "Cloud computing is a pay-per-use model for enabling available, convenient, on-demand network access to a shared pool of configurable computing resources (e.g., networks, servers, storage, applications, services) that can be rapidly provisioned and released with minimal management effort or service provider interaction." (Mell and Grance, 2011).

In this research, we propose a WOA driven information platform for enabling effective long-term health monitoring (WLTHM) platform in a Cloud. The WLTHM platform aims to consolidate the once-fragmented world of LTC services into a more efficient platform, enabling the evaluation functions to be sourced and built more easily, quickly, inexpensively, and at higher quality levels. In addition, in the system development section and scenarios, we present the balance ability monitoring as an example to validate the viability and feasibility of the proposed approach.

This research is organized as follows. Section 2 reviews related literature on the works in WOA integration and postural balance ability. Section 3 presents the system design of WLTHM platform and usage scenarios. Finally, Section 4 provides conclusions and suggestions for future studies.

2   LITERATURE REVIEW

2.1   WOA Approach for Integration

Both SOA and WOA aim at building environments consisting of modular, distributable, sharable, and loosely coupled components. SOA based on SOAP web services follows the path of distributed object approaches with object specific methods and remote procedure calls (operation centric approach). WOA follows the established path of the Web and promotes usage of a single uniform functional interface for all kind of objects (resource centric approach). Unlike the complex SOA, recent developments in the context of Web 2.0 show how easy it can be to link or compose IT components dynamically (Szepielak et al., 2010). At this point the idea of REST (Fielding and Taylor, 2002) is often used in a Resource-oriented Architecture (ROA), which uses the Web in a way it was originally intended: by using only the HTTP methods such as POST, GET, PUT and DELETE as actions. It can be positioned between SOA and ROA and uses technological standards from both these concepts where applicable.

REST is a web service design style which specifies a collection of architecture principles defining how data resources are represented and addressed. The key principles of REST style include the following notions. Systems that follow the REST principles are often called "RESTfull".

- Application state and functionality are abstracted into resources. Any information, which is offered by the system and can be named, is possible to be represented by a "resource". Any concept that needs to be addressed, referenced and accessed must fit within the definition of a resource.
- Resources must be uniquely identified and addressable using a universal syntax, such as a Uniform Resource Identifier (URI) used in HTTP.
- All resources must share a uniform interface for the transfer of state between client and the resource, which is hosted on the server. The set of operations, as well as supported content types, need to be well defined. At the same time, code on demand (such as JavaScript) could be optionally supported.
- The communication protocol between the resource data provider and consumer/client has to be: client-server, stateless, layered and cacheable.

Developing WOA with RESTfull web services (Fielding and Taylor, 2002) provides enhanced scalability, generality of interfaces, independent deployment, reduced interaction latency and they can encapsulate legacy systems. This approach has been broadly applied in much research areas. Thies and Vossen (2009) sharpened understanding of the concept behind a WOA and pointed out how governance and the control of a WOA can be achieved, by using a controlling software element called the Web Architecture Controller which specifies the Web-centric core elements of a WOA. Szepielak et al. (2010) illustrated how resources distributed across major information systems, (including: Engineering Data Management System, Requirement Management System, Inventory Management System, Facility Management System, and CAD system), are integrated into a single Integration Portal based on the WOA. There are also many other research fields, such as e-Learning (Dodero and Ghiglione, 2008), smart home (Belimpasakis and Moloney, 2009), etc., which show how profit from WOA.

2.2   Balance Ability – Postural Control Analysis

The pest researches indicated that attention plays an important role on postural (balance) control stability for elderly people when standing still (Amiridis et al., 2003; Jamet et al., 2004). Yang and Jiang (2010) designed a static (standing still) experiment which collect the center of pressure (COP) signals to quantify and compare the variability in complexity of postural stability in healthy young and elderly faller subjects under the conditions of standing still and with and without an attention influence   (Figure 1). Through the experiment and multi-scale entropy (MSE) analysis, the influence of an attention task is investigated for elderly subjects. In particular, in order to understand the dynamic properties of postural control, the variability in COP position time series was quantified, along the horizontal plane, in anteroposterior (AP) and mediolateral (ML) directions. In this research, we quote Yang and Jiang's work and use it as the example of balance ability monitoring in WLTHM platform.

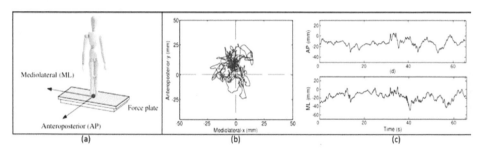

Figure 1. The COP signal measurement and trajectory. (a) is a diagram of a force plate test system; (b) is displacement of COP within the force plate; and (c) is COP displacement in the AP and ML directions over time, respectively (Yang and Jiang, 2010).

3   WLTHM PLATFORM DESIGN

WLTHM platform provides an integration environment which is designed based on WOA with RESTful web services. Such a RESTful web service can be thought of as a collection of resources. Through the Internet, users are allowed to independently access these services through HTTP methods such as POST, GET, PUT and DELETE; and to integrate and reuse them to create new services. Through WLTHM platform, services developers can create and public new services or execute various applications from the existed resources. Through WLTHM platform, various services and applications and users' practiced results can be gathered and collected.

The system framework of WLTHM platform comprises 1) a Data Access Layer which stores backend data, 2) RESTful web services, and 3) a User Interface which renders the front-end user interfaces (UIs) for each user. The multi-layer WCPD framework is illustrated in Figure 2.

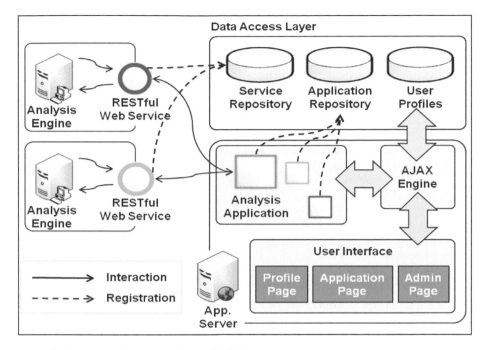

Figure 2. The system framework of WLTHM platform.

Data Access Layer (DAL)

The DAL is responsible for storing static data (e.g., User Profiles) and applications furnished by services developers. It encompasses three storage areas: 1) Services Repository, 2) Applications Repository, and 3) User Profiles:

(1) Service Repository - stores all registered resources from service developers. Developers can focus on their core skills and publish their analysis functionalities through RESTfull web service (e.g. balance ability analysis engine).

(2) Application Repository- a database that stores all health monitoring applications (i.e. Analysis Applications). The applications are usually created by using mash-up tools and remixing the RESTfull web services.

(3) User Profiles: stores the participators' profiles that encapsulate personal information and configuration, and the list of uploading applications.

RESTful Web Services in WLTHM

RESTful web services in WLTHM platform provide an abstraction for publishing analysis engine and giving remote access to the Data Access Layer. The analysis engine is means of the calculating functionality performance, such as the determination of user's health status. In this research, we use the balance ability evaluation as the example and develop a balance ability analysis application. The

446

applied method of balance ability analysis is proposed by Yang and Jiang's (2010) work as described in section 2.2. It is developed by using the commercial software, MATLAB, and published through the RESTfull web service which is implemented by using JAVA programming language. The relationships between Analysis engine, RESTfull Web Service, and Analysis Application are shown as Figure 3.

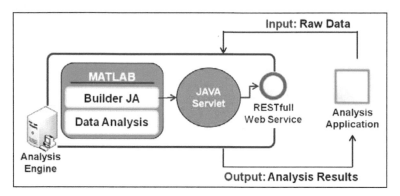

Figure 3. Relationships between Analysis engine, RESTfull Web Service, and Analysis Application.

Following the REST principles, it regards the contents of the Data Access Layer as resources, each of which is referenced with a simple URI and represented in Javascript Object Notation (JSON) format. For example, the users' practiced results can be accessible via the simple HTTP GET method with a URI: http://localhost:8080/wlthm/resources/results/, which is a RESTful web service named "results" in the "wlthm" project (Figure 4).

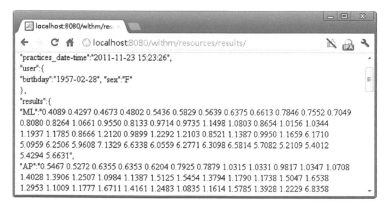

Figure 4. Accessing the users' practiced results via RESTful web service.

The User Interface Layer (UIL)

The major task of the UIL is to represent the UIs for WLTHM users and asynchronously send/respond to users' requests to/from the server side. Based on the Asynchronous JavaScript XML (AJAX) and the WLTHM framework, the UIL encompasses 1) the User Interface, which generates presentation in the browser, and 2) the Ajax Engine, which makes asynchronous requests. Figure 5 shows the example usage of balance ability analysis in WLTHM platform.

User Interface is user-facing browser page which interacts with applications, encompassing 1) Profile Page, 2) Application Page, and 3) Admin Page:

(1)   Profile Page: a page for recording the usage and personal configurations. Users may maintain their personal settings (i.e. personal information or preferred applications) through in this page.
(2)   Application Page: the interface listing and executing analysis services or applications.
(3)   Admin Page: a page which manages WLTHM platform. Hosts can manage all participants within this platform and configure the overall settings.

AJAX allows the user's interaction with the application to occur asynchronously, independent of communication with the server. The technology enables web applications to call the web-server without leaving the actual page and the process can be run in the background without notice of the user. This avoids loading the same form or page including the html markup multiple times, reduces the network traffic and increases the user acceptance.

Figure 5. The usage of balance ability analysis in WLTHM platform.

4   CONCLUSION

In this paper, we present the design and development of WLTHM platform based on seamless integration of such enabling technologies as WOA and RESTful web services to provide an integration environment for supporting long-term health monitoring in a cloud. WLTHM platform employs WOA with RESTful web services approach to build environments consisting of modular, distributable, sharable, and loosely coupled components. The health related services, applications, and static data can be sourced and accessed simply and quickly. The functionalities are also can be extended easily through the RESTful web services. The WLTHM deployment to cloud computing environments can not only shorten the development cycle but also reduce maintenance costs and enable ubiquitous access. With the proposed system design, an economic, scalable, and robust healthcare platform can be built over the Internet. The WLTHM platform consolidates the once-fragmented world of LTC services into a more efficient platform, enabling the evaluation functions to be sourced and built more easily, quickly, inexpensively, and at higher quality levels.

ACKNOWLEDGEMENT

This research was supported by the National Science Council (NSC 100-2221-E-155 -065 -MY3), Taiwan.

REFERENCES

Amiridis, I. G., V. Hatzitaki, F. Arabatzi. 2003. Age-induced modifications of static postural control in humans. *Neruroscience Letters* 350(3):137-140.

Chen, C. M. 2011. Web-based remote human pulse monitoring system with intelligent data analysis for home health care. *Expert Systems with Applications* 38(3):2011-2019

Fielding, R. T. and R. N. Taylor. 2002. Principled design of the modern Web architecture. *ACM Transactions on Internet Technology* 2(2):115-150.

Jamet, M., D. Deviterne, G. C. Gauchard, G. Vançon, and P. P. Perrin. 2004. Higher visual dependency increases balance control perturbation during cognitive task fulfillment in elderly people. *Neuroscience letters* 359(1-2):61---64.

Klingeberg, T. and M. Schilling. 2012. Mobile wearable device for long term monitoring of vital signs. *Computer Methods and Programs in Biomedicine* In Press, Available online 28 January 2012.

Mell P. and Grance T. "The NIST definition of cloud computing." Accessed February 28, 2012, http://csrc.nist.gov/publications/nistpubs/800-145/SP800-145.pdf.

Paradiso R, G. Loriga, and N. Taccini. 2005. A wearable health care system based on knitted integrated sensors. *IEEE Transactions on Information Technology in Biomedicine* 9(3):337-344

Peng, R., L. Ling, and Q. He. 2010. Self-rated health status transition and long-term care need, of the oldest Chinese. *Health Policy* 97(2-3):259-266.

Preve, N. 2011. SEGEDMA: Sensor grid enhancement data management system for HealthCare computing. *Expert Systems with Applications* 38(3): 2371-2380.

Szepielak, D., P. Tumidajewicz, and L. Hagge. 2010. "Integrating Information Systems Using Web Oriented Integration Architecture and RESTful Web Services." In: Proceedings - 2010 IEEE 6th World Congress on Services, SERVICES 2010 (pp. 598 - 605), 5-10 July, Miami, Florida, USA.

Thies, G. and G. Vossen. 2008. "Web-Oriented Architectures: On the Impact of Web 2.0 on Service-Oriented Architectures." In: Proceedings - 2008 IEEE Asia-Pacific Services Computing Conference, APSCC 2008 (pp. 1075 - 1082), 9-12 December, Yilan, Taiwan.

Thies, G. and G. Vossen. 2009. "Governance in Web-oriented Architectures." In: Proceedings - 2009 IEEE Asia-Pacific Services Computing Conference, APSCC 2009 (pp. 180 - 186), 7-11 December, Biopolis, Singapore.

Welke, R. R. Hirschheim, and A. Schwarz. 2011. Service-Oriented Architecture Maturity. *Computer* 44(2):61-67.

Yang, W. H. and B. C. Jiang, "Multi-scale entropy analysis for postural sway signals with attention influence for elderly and young subjects." Paper presented at AHFE International Conference, jointly with 1st Advances in Human Factors and Ergonomics in Healthcare, Miami, USA, 2010.

CHAPTER 48

Heuristic Evaluation for e-Commerce Web Pages Usability Assessment

Katarzyna Jach, Marcin Kuliński

Institute of Organization and Management (I23)
Faculty of Computer Science and Management (W8)
Wrocław University of Technology
Wybrzeże Wyspiańskiego 27
50-370 Wrocław, Poland

ABSTRACT

Heuristic evaluation is a popular tool for a preliminary assessment of web page usability. For e-commerce web sites however, more specific techniques are required. In the paper application software for heuristic checklist based on general heuristics created by Nielsen and Molich (with later modifications) and the set of detailed recommendations is presented. The specificity of the application called TECHA (Tool for E-Commerce Heuristic Assessment) originates in its object-oriented structure dedicated for e-commerce web page evaluation.

Keywords: heuristic evaluation (HE), effectiveness, e-commerce

1   INTRODUCTION

The average abandonment rate of shopping cart during online shopping sessions jumped from 71% to 75% within the first six months of 2011 (Ouellet 2011). While some customers do return to complete a purchase, many more remain lost for further shopping transactions. There is a lot of marketing reasons for such behavior (i.e. comparing prices by the consumers by using online carts, high or unexpected shipping costs, temporary interruptions of internet connection etc.). One of them is a poor usability of e-commerce web pages. Experiments made by Lohse and Spiller (1999) show how substantially the interface features explain variance in sales for online stores. Using a stepwise regression approach, the authors identified the set of interface design features that have had the largest influence on Internet store traffic and monthly sales.

The goal of the usability assessment is not only to compare the evaluated interface with a pattern, but also to show its weaknesses in order to help improve the usability (Agarval and Venkadesh 2002). Although Heuristic Evaluation (HE) is generally regarded as not the most effective method of usability assessment, it is the most often applied by usability experts with 93% reported popularity ratio (Rosenbaum, Rohn and Humburg 2000). Heuristic evaluation is a technic popularized by Nielsen and Molich as a cheap, fast, and easy way for interface inspection. Nielsen (1994) defined heuristic evaluation as a method for identifying usability problems in a user interface design examining the interface and judge its compliance with recognized usability principles (the "heuristics"). During a heuristic evaluation review, an evaluator uses a software or web site and assesses its usability against a set of principles or best practice guidelines. The main advantages of heuristic evaluation are a low price, ability to recognize problems and its intuitiveness. On the other hand, some weaknesses of the method can be found: the discovered problems are often either low-priority or "positive false" (Sears 1997). However according to Nielsen, during the heuristic evaluation major usability problems have a higher probability than minor problems to be found (Nielsen 1992).

Heuristic evaluation mostly bases on an assumption that the better usability is connected with meeting standards. Generally, the most popular set is ten heuristics by Nielsen and Molich, which were defined in 1990 and refined by Nielsen in 1994 on the basis of a factor analysis of 249 usability problems found in practical applications in various contexts (Nielsen and Molich 1990; Nielsen 1994). This set includes:

1. Visibility of system status (feedback for user)
2. Match between system and the real world (using users' language and follow real-world conventions)
3. User control and freedom
4. Consistency and standards (following conventions)
5. Error prevention
6. Recognition rather than recall (minimizing the user's memory load)
7. Flexibility and efficiency of use (personalization and accelerators for experienced users)
8. Aesthetic and minimalistic design
9. Error support (help users recognize, diagnose, and recover from errors)
10. Help and documentation (supportive, easy to use and search help system).

NASA (1997) proposed differentiation of the design guidelines according to their specificity level. General Design Principles are the most aggregate. They can be thought of as high-level design goals and present the same specificity level as Nielsen and Molich heuristics. Display Object Design Guidelines are more detailed, since they provide specific guidance for the design of objects on the display. The Interaction Guidelines are the most thoroughgoing and they cover design issues related to the interaction between the human and the software interface.

General Design Principles according to NASA are:

1. Directness (the interface should be direct in style),
2. Consistency (the interface should be consistent within a display, across displays within a system, and across systems that are to be used together)
3. Forgiveness (the interface should be forgiving mistakes and provide support for error recovery)
4. Feedback (visual and/or auditory feedback should accompany every user action)
5. Aesthetics (aesthetics should be considered in display design, as long as ease of use and functionality are not compromised).
6. Content and Navigation (a display should contain only the information that is relevant to the current task, in the proper format, at that point in time)
7. Organization (information within a display should be organized according to logic and accepted standard conventions).

The validated set of heuristics based mainly on a cognitive theory was formed by Jill Gerhardt-Powals (1996). This set concentrates on a reduction of cognitive overload during task realization and includes:

1. Automation of unwanted workload to eliminate unnecessary thinking and mental operations.
2. Uncertainty reduction (data displaying in a clear manner)
3. Data fusion (aggregating and summary)
4. Presentation of new information with meaningful aids to interpretation (familiar framework and terms)
5. Using names that are conceptually related to function (context-dependent and improving recall and recognition)
6. Grouping data in consistently meaningful ways
7. Limitation of data-driven tasks (a.o. appropriate use of color and graphics)
8. Displaying only that information needed by the user at a given time
9. Provide multiple coding of data when appropriate
10. Practice judicious redundancy.

Another set of general principles is defined in ISO 9421 series. The dialogue requirements described in part 110 established a framework of ergonomic principles (Oppermann and Reiterer 1997), so they play the same role as heuristics by Nielsen and Molich. The principles, and their definition in the ISO standard, are as follows (ISO 9241-110:2006; Travis 2007):

1. Suitability of the dialogue for the task and the user skill level
2. Dialogue self-descriptiveness (it is obvious to users which dialogue they are in and which actions can be taken)
3. Conformity with users' expectations (consistency with predictable contextual needs of the user and to commonly accepted conventions)

4.   Suitability for learning (dialogue supports and guides the user)

5.   Controllability (the pace and interaction sequence controlled by user)

6.   Error tolerance (achieved by means of damage control, error correction, or error management)

7.   Suitability for individualization (dialogue is capable of personalization).

The more precise requirements concentrated on user guidance (prompts, feedback, error support and help) are described in part 13 of ISO 9241 standard.

It is worth noticing that the general principles by NASA as well as cognitive heuristics by Gerhardt-Powals and guidelines formed in ISO 9241-110 standard tie in with Nielsen and Molich heuristics very well, as it is shown in Table 1.

Table 1 Comparison of different heuristic sets.

Nielsen & Molich	ISO 9241-110	NASA	Gerhardt-Powals
Visibility of system status	Dialogue self-descriptiveness	Feedback	Displaying only that information needed
Match between system and the real world		Organization	Practice judicious redundancy
User control and freedom	Suitability for the task and the user skill level Controllability		Automation of un-wanted workload Uncertainty reduction
Recognition rather than recall			Using names that are conceptually related to function
Consistency and standards	Conformity with user expectations	Consistency	Grouping data in consistently meaningful ways
Error prevention		Forgiveness	Provide multiple coding of data
Error support	Error tolerance		
Flexibility and efficiency of use	Suitability for individualization	Content and Navigation	Data fusion
Aesthetic and minimalistic design		Directness Aesthetics	Limitation of data-driven tasks
Help and documentation	Suitability for learning		Presentation of new information with meaningful aids

The main goal of all of these heuristics sets as well as standards is to help ensure consistency. They also provide a benchmark for evaluation. Nevertheless, standards help define good practice, especially for such a hard field to measure as usability.

All the presented heuristics sets however, require interpretation to be useful in particular interface evaluations (Oppermann and Reiterer 1997) according to the usability definition, which takes into account the context of use (ISO 9241-110:2006). A lot of other sets of guidelines and more thoroughgoing recommendations were formed on the basis of these data. Especially Nielsen and Molich heuristics were used to form guidelines and checklists, i.e. Pierotti's System Checklist (1995) which consists of 13 sections each with 3 to 50 questions, with 296 questions in total.

Hvannberg at al. (2007) compared the number and seriousness of usability problems found by applying two different sets of heuristics (Nielsen and Molich versus Gerhardt-Powals) and stated that there are no significant differences between results of the evaluation made by both sets of heuristics measured by their validity (percent of verified problems against the total number of problems identified, thoroughness and efficiency).

2 THE TOOL

The TECHA (Tool for E-Commerce Heuristic Assessment) structure was inspired by the ISO 9241 evaluator presented by Oppermann and Reiterer (1997). The components of object oriented requirements were assessed with consideration to ergonomic criteria described in the ISO 9241 standard, general principles by NASA, cognitive heuristics by Gerhardt-Powals as well as Nielsen and Molich heuristics. As it was shown in Table 1, all these heuristics sets have a common background, but in order to make the result analysis more clear, only the heuristics by Nielsen and Molich were used for presenting results.

The idea of the tool is to make assessment of e-commerce pages usability possible and to point their weaknesses focusing on either Nielsen and Molich heuristics or objects. The application is destined to be use by evaluators who know an investigated web site well (i.e. online shop owners or administrators) and possess some basic knowledge about usability, but still who are not usability experts.

A list of guidelines used in the tool was created on the basis of large literature reviews. For e-commerce web sites assessment the most helpful turned out to be the list of 247 web usability guidelines collected by David Travis (2009). The other supportive publication was the Guidelines For Designing User Interface Software made by Smith and Mosier (1986). Although the publication is over 25 years old, the collection 944 guidelines based on scores of research papers is still very usable. Another important source of the guidelines was The Research-Based Web Design & Usability Guidelines created by U.S. Department of Health and Human Services (HHS). All the 209 guidelines described in the publication are assessed against their relative importance and strength of evidence. The 'Relative Importance' ratings were revised based on an online survey in which 36 Web site professionals responded. The 'Strength of Evidence' ratings were revised by usability professionals, who rated each of the guidelines, and assigned 'Strength of Evidence' ratings. The raters all were very familiar the research literature, all had conducted their own studies, and there was a high level of agreement in their ratings

(Cronbach's alpha = .92). The list was reviewed with Pierotti's (1995) System Checklist and Connell's (2000) set of usability evaluation principles. The guidelines which were reduplicative, not validated or irrelevant to e-commerce web sites were eliminated. The final list of 136 guidelines was chosen to preliminary tests.

The first preliminary test was conducted on the set of 10 online shop representing different trade industries. Based on internet users' opinions, for this research stage the poor usability ecommerce web sites were chosen. The main goal of this research was to reduce number of guidelines by eliminating always true guidelines from a further analysis. Finally, the list of guidelines was reduced to 99. All the guidelines were assessed against their importance on the five point scale as the "relative importance" factor (HHS 2010) on the basis of literature review and expert evaluation. Each of the guidelines was assigned into the heuristics which requirements it met.

The second preliminary test was conducted as a "loud thinking" session with each of the five evaluators individually. The test allowed to eliminate misunderstandings, with a special attention being paid for familiarity of a language, phrases and terms to the evaluators. The other purpose of this test stage was choosing a proper scale for measuring a level of fulfilling a guideline by an evaluator. This problem was checked by A/B test with a three and five-level scale. All the investigated evaluators preferred three-level scale. The final version of the assessment tool consists of ten sheets. Three of them described general features of an e-commerce web site (the Layout, the Navigation, and Other categories). Seven sheets refer to objects which are present at each e-commerce web site like Home Page, Logo, Cart, Search engine, Login, Product page and Info site. The evaluator assesses each guideline separately using the three-level scale (0 – the investigated online shop does not comply with the guideline, 1 – the shop complies with the guideline partially, 2 – the guideline is complied totally). Blank spaces are allowed if the guideline does not regard the investigated web site or the evaluator does not have a clear opinion about fulfilling guideline. The level of usability of e-commerce site was assessed as a total number of points (Points – maximum 2 for each of the guideline), total number of points with taking into account the relative importance level (Weighted Points – from 2 to maximum 10 for each of the guideline), Results (the percent of Points in relation to maximum possible Points) and Weighted Results (the percent of Weighted Points in relation to maximum possible Weighted Points).

3   RESULTS

The research was carried out in January 2012. Each of the evaluator filled the tool assessing an online shop well known for themselves. The research was conducted on the set of two computers or a computer equipped with two displays, so an evaluator had a possibility to see assessed web site during the TECHA using. 18 online shops were assessed by 4 experts and 24 novices. A detailed instruction of using tool was given to an evaluator at the first window and in the last window one obtained the results of the evaluation both in graphic and text forms for object and heuristics dimensions separately. The average filling time was 48 minutes while minimum was 23 minutes and maximum 60 minutes.

Basic statistics of all the assessment metrics are presented in Table 2. The statistical significance of differences among all the assessment measures (Points, Weighted Points, Results and Weighted Results) was checked with Mann–Whitney U test because of abnormality of the analyzed variables. Since it was stated during the analysis that all the assessment measures are highly correlated (the least Spearman's rank correlation coefficient value was 0.86 and for Results and Weighted Results it was 0.94), further analysis was conducted for this measure only.

Table 2 Basic statistics of all the assessment metrics.

	Available max.	Mean	Min	Max	St. Dev.
Number of Answers	99	95.00	72.00	99.00	6.21
Points	198	143.65	87.00	173.00	23.35
Weighted Points	720	529.82	324.00	636.00	84.69
Results	100%	77.87%	53.72%	90.81%	10.26%
Weighted Results	100%	76.44%	45.65%	89.88%	11.85%

The mean values of Weighted Results for all the investigated ecommerce web sites are presented in Fig.1. separately for heuristics and objects. Among the objects, the highest and most consistent results regarded to guidelines connecting with a Homepage and a Product page. The lowest results were observed for an Info site and the least consistent observations regarded with Search engine object. According to the heuristics, the weakest results concerned the heuristics involving Error support, Error prevention and Help for the user.

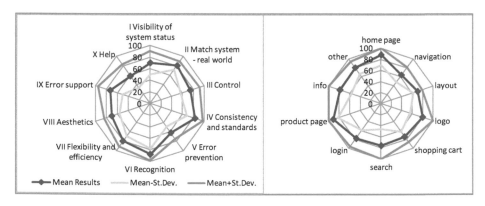

Figure 1 Mean Weighted Results of Heuristics and Object evaluation.

3. 1.   Comparison between novice and expert evaluation

The preliminary assumption regarding to TECHA using was that the evaluators assessment should be comparable whatever their knowledge level might be. This assumption was tested for each investigated online shop separately by Kendall's coefficient of concordance. Its mean value was relatively high (0.76 with minimum 0.58 and maximum 0.94 value; St.Dev. 0.12), which means the evaluators judged the investigated web sites similarly. The detailed analysis made with Mann–Whitney U test (Table 3) showed that the expert evaluation was comparable with the novice one for most objects and heuristics, although some differences are statistically significant, especially in the navigation field.

Table 3 Mean Weighted Results for expert and novice evaluation (1-3 column) and for large and small e-commerce web sites (4-6 column).

Object or heuristic	1 Expert	2 Novice	3 p	4 Large	5 Small	6 P
home page	78.33	91.67	0.070	86.25	88.89	0.326
navigation	45.08	73.30	<0.001	64.77	54.55	0.049
layout	70.56	70.37	0.741	71.67	66.42	0.031
Logo	78.70	75.00	0.741	88.54	76.54	0.280
shopping cart	70.49	77.95	0.032	78.84	69.56	<0.001
search	66.67	69.05	0.248	87.50	58.60	<0.001
Login	64.88	69.35	0.248	87.05	58.33	<0.001
product page	92.11	98.25	0.005	88.16	95.91	<0.001
Info	74.70	69.35	0.621	91.07	64.48	<0.001
Other	71.94	92.78	<0.001	79.58	78.70	0.238
I Visibility of system status	66.25	72.92	1.000	75.63	57.50	<0.001
II Match system - real world	74.11	85.27	<0.001	82.03	76.49	0.031
III Control	68.96	75.00	0.621	82.03	65.28	0.003
IV Consistency	75.55	78.55	0.138	86.27	74.59	0.169
V Error prevention	56.02	63.43	0.048	62.67	57.56	0.062
VI Recognition	80.24	88.81	0.008	89.46	81.43	<0.001
VII Flexibility and efficiency	74.05	81.67	0.187	81.96	75.24	0.095
VIII Aesthetics	69.14	75.93	0.322	70.83	69.14	0.491
IX Error support	54.90	82.35	0.001	83.09	60.78	<0.001
X Help	51.01	56.06	0.248	78.60	41.41	<0.001

Significant differences are bolded.

This differentiation can be observed in Fig. 2 especially for Error support heuristic. However, most of expert and novice evaluations were comparable, although the expert notes were slightly lower.

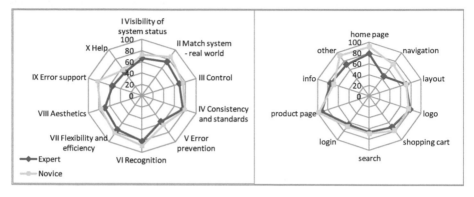

Figure 2 Heuristics oriented and object oriented comparison between expert and novice evaluators.

3.2   Comparison between large and small online shops

The comparison between investigated online shops divided into large and small groups was made. It was expected that large shops will better obtain the usability requirements described in TECHA. This hypothesis was checked by Mann–Whitney U test. The results are shown in Table 3. For all the objects and heuristics, large shops were assessed better than small ones (not all the differences were statistically significant) and the results were more consistent.

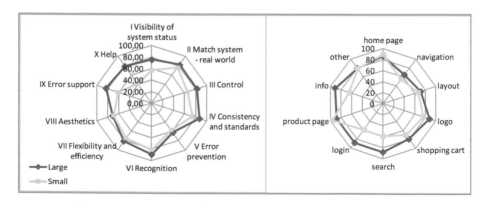

Figure 3 Heuristics oriented and object oriented comparison between large and small e-commerce web sites.

As it can be observed in Fig. 3., the weakest assessment measures concerned with navigation and info object. This result is compatible with Kulinski at al. (2011)

observations made on the basis of analysis the layout of 93 e-commerce web sites. In the investigation, small online shops reported high variety of an Info site, a Search engine and a Login function layout as well as a lack of standardization of vocabulary and terms.

4   CONCLUSIONS

Among seven main reasons of problems occurring during online shopping pointed by Polish internet users, all of them were connected with poor usability of web sites of online shops. Polish internet users recognized such factors as: lack or poor quality of product images (50%), lack of detailed information about products (38%), low quality search results (23%), lack of data safety information (22%), problems with finding shipping costs (19%), nonstandard order forms and procedures (13%) and problems with a login function (10%) (Dwornik at al. 2011). These results shows that even simple modifications of e-commerce web sites can cause significant reduction of e-commerce barrier for consumers.

Payne et. al (1990) found the crucial factor influencing the users' behavior, which is a prior knowledge possessed by user. Therefore, creating the web sites using the prior experience of the users with support of standards, help minimize the abandonment rate of shopping cart during online shopping. Kulinski at al. (2011) show the user's experience has a significant influence on focusing the preferences of layout the typical e-commerce shop elements like shopping cart, login form or main categories object. The investigated subjects were asked to create the best for use e-commerce web page layout of a single product. The more experienced users expressed better their preferences, it was easier for them to design the layout and the designed patterns were more consistent then the novice users.

The results consider with Chen and Macredie (2005) observations. Based on the large research review, the authors shown that design features as well as heuristics meeting play a crucial role in users' behavior and later satisfaction.

5   LIMITATIONS

The research sample was limited to Polish internet users, so the results may be not representative for other populations and cultures. However, the differentiation of Polish internet users and their online behaviors are similar to these observed in European Community ones (MMG 2011). Another important factor is relatively low number of evaluations of a single online shop. According to Hvang and Salvendy (2007) observations, the results obtained by different evaluators can vary significantly.

Next versions should include the guidelines connected with accessibility for impaired persons as well as for nonstandard web browsers and devices. To fulfill the function of heuristic evaluation (creating recommendations for usability improvement), the problems reporting form should be added.

It should be always taken into account that heuristic evaluation is only one of the method of usability testing, which should be applied prior to and in addition to user-testing, not instead of user-testing.

REFERENCES

Agarwal, R., V.Venkatesh. 2002. Assessing a firm's Web presence: A heuristic evaluation procedure for the measurement of usability. *Information Systems Research.* 13(2):168-186.

Chen, S.Y., R.D. Macredie. 2004. The assessment of usability of electronic shopping: A heuristic evaluation. *International Journal of Information Management.* 25:516-532.

Connell, I.W. 2000. Full Principles Set. Set of 30 usability evaluation principles compiled by the author from the HCI literature. Accessed April 4, 2011. http://www.cs.ucl.ac.uk/staff/ /i.connell/DocsPDF/PrinciplesSet.pdf.

Dwornik, B., B. Ratuszniak, B. Rumczyk, M. Rynkiewicz, B. Wawryszuk, P. Wilk. 2011. Raport e-commerce. Assessed January 12, 2012. http://interaktywnie.com/biznes/artykuly/raporty-interaktywnie-com/raport-interaktywnie-com-e-commerce-i-ranking-e-sklepow-22115.

Gerhardt-Powals, J. 1996. Cognitive engineering principles for enhancing human - computer performance. *International Journal of Human-Computer Interaction.* 8(2): 189-211.

Hwang, W., G. Salvendy. 2007. What Make Evaluators to Find More Usability Problems?: A Meta-analysis for Individual Detection Rates. In: *Human-Computer Interaction. Part I. HCII 2007.* Jacko J. (ed). 499-507.

Hvannberg, E.T., E. Lai-Chong Law, M.K. Lárusdóttir. 2007. Heuristic evaluation: Comparing ways of finding and reporting usability problems. *Interacting with Computers.* 19: 223-240.

ISO 9241-110:2006 Ergonomics of human-system interaction - Part 110: Dialogue principles.

Kuliński, M., K. Jach, R. Michalski, J. Grobelny. 2011. Modeling online buyers' preferences related to webpages layout: methodology and preliminary results. In: *Advances in applied digital human modeling.* Duffy V. G. (ed): CRC Press. 499-509.

U.S. Department of Health and Human Services (HHS). 2006 (first edition). Research-Based Web Design & Usability Guidelines. Assessed February 3, 2010. http://www.usability.gov/guidelines.

Lohse, G., P. Spiller. 1999. Internet retail store design: How the user interface influences traffic and sales. *Journal of Computer-Mediated Communication.* 5 (2). http://jcmc.indiana.edu/vol5/issue2/lohse.htm.

Krug, S. 2005. Don't Make Me Think. A Common Sense Approach to Web Usability, New Riders.

Miniwatts Marketing Group (MMG). 2011. Internet World Stats. Usage and Population Statistics. Assessed February 10, 2012. http://www.internetworldstats.com/stats4.htm.

Nielsen, J. , R. Molich. 1990. Heuristic evaluation of user interfaces. *Proceedings of CHI 90.* 249-256.

Nielsen, J. 1992. Finding Usability Problems Through Heuristic Evaluation. *Proc. ACM CHI'92 Conf.* (May 3-7). 373-380.

Nielsen, J. 1994. Enhancing the explanatory power of usability heuristics. *Proc. ACM CHI'94 Conf.* (Boston, MA, April 24-28). 152-158.

Nielsen, J. Ten Usability Heuristics. Assessed March 3, 2011. http://www.useit.com/papers/heuristic/heuristic_list.html.

National Aeronautics and Space Administration. 1997. Human-Computer Interface (HCI) Design Guide. Human Research Facility (HRF). (LS-71130).

Ouellet, M. 2011. Shopping Cart Abandonment Practices of the Internet Retailer 1000 Companies. Assessed November 12, 2011. http://www.shop.org/c/document_library/et_file?folderId=164&name=DLFE-936.pdf.

Oppermann, R., H. Reiterer. 1997. Software evaluation using the 9241 evaluator. *Behaviour & Information Technology*. 16 (4/5): 232-245.

Pierotti, D. 1995. Heuristic Evaluation - A System Checklist. Usability Analysis & Design. Xerox Corporation.

Rosenbaum, S., J.A. Rohn, J. Humburg. 2000. A Toolkit for Strategic Usability: Results from Workshops, Panels and Surveys. *CHI Letters*. 2(1): 337-344.

Smith, S.L, J.N. Mosier. 1986. Guidelines For Designing User Interface Software. Hanscom Air Force Base. MA: USAF Electronic Systems Division. Technical Report ESD-TR-86-278. http://www.hcibib.org/sam/index.html.

Travis, D. 2007. Usability Expert Reviews: Beyond Heuristic Evaluation. Assessed November 13, 2011. http://www.userfocus.co.uk/articles/expertreviews.html.

Travis, D. 2009. 247 web usability guidelines. Assessed May 20, 2011. http://www.userfocus.co.uk/resources/guidelines.html.

CHAPTER 49

Optimal Scroll Method to Browse Web Pages Using an Eye-gaze Input System

Atsuo MURATA, Kazuya Hayashi, Makoto Moriwaka and Takehito Hayami
Graduate School of Natural Science and Technology, Okayama University
Okayama, Japan
murata@iims.sys.okayama-u.ac.jp

ABSTRACT

To determine the optimal scroll method for an eye-gaze input system to browse Web pages, an empirical study was executed. Ten healthy undergraduate students participated in the experiment. Each participant executed a scroll-and-select task on the display consisted of two layers using an eye-tracking device. The task performance was compared between four scroll methods: a scroll-icon method and three auto-scroll methods. As a result, it was found that the improved auto-scroll method (quadratic and quadratic combination) with nonlinear relationship between the vertical scroll location and the scroll velocity was optimal from the viewpoints of the task completion time, the error rate and the subjective evaluation on usability.

Keywords: eye-gaze input system, scroll, web page, usability

1   INTRODUCTION

The technology for measuring a user's visual line of gaze in real time has been advancing. Appropriate human-computer interaction techniques that incorporate eye movements into a human-computer dialogue have been developed (Goldberger et.al., 1995, Huchinson et. al., 1989, Jacob, 1990, Jacob, 1991, Jacob, 1993a, Jacob 1993b, Jacob, 1994, Murata, 2006, Murata et.al., 2009a, Murata et.al., 2009b, Sibert et.al., 2000). These studies have found the advantage of eye-gaze input system. However, few studies except Murata, 2006 have examined the effectiveness of such systems with older adults. Murata, 2006 discussed the usability of an eye-gaze input system to aid interactions with computers for older adults. Systematically manipulating experimental conditions such as the movement distance, target size, and direction of movement, an eye-gaze input system was found to lead to faster pointing time as compared with mouse input especially for older adults. However, these studies cannot be applied to the real-world computer systems such as Internet Explorer (Murata et al., 2009a, Murata et al., 2009b).

However, many problems remain to be overcome so that the eye-gaze input systems can be put into practical use. The scroll method that changes the scroll velocity nonlinearly as a function of scroll location has not yet been proposed. It has also not been explored how the scroll location on Web browsers affects the performance. Although a few scroll methods have been proposed, these are not assumed to be used on Web pages. In this study, an attempt was made to determine empirically the optimal scroll method among the scroll methods (improved scroll-icon, auto-scroll, improved auto-scroll methods).

2   METHOD

2.1   Participants

Ten male undergraduate students aged from 21 to 23 years (average: 21.8 years) took part in the experiment.. The visual acuity of the participants in both young and older groups was matched and more than 20/20. They had no orthopedic or neurological diseases.

2.2   Apparatus

An eye-tracking device (EMR-AT VOXER, Mac Image Technology) was used to measure eye movement characteristics during the experimental task. The eye-tracker was connected with a personal computer (HP, DX5150MT) with a 15-inch (303mm x 231mm) CRT. The resolution was 1024 x 768 pixels. Another personal computer was also connected to the eye-tracker via a RS232C port to develop an eye-gaze input system. The line of gaze, via a RS232C port, is output to this computer with a sampling frequency of 60Hz. The illumination on the keyboard of a personal was about 200lx, and the mean brightness of 5 points (four edges and a center) on CRT was about 100cd/m2. The viewing distance was about 70 cm.

2.3   Scroll method

The following scroll methods were used: (1) auto-scroll method, (2) improved scroll-icon method, and (3) improved auto-scroll method with nonlinear relationship between vertical scroll location and scroll velocity. In the method (3) (improved auto-scroll method), two types of nonlinear relationship was used. One was the quadratic, and the other was the combination of two quadratic equations. The display of (1) auto-scroll method is shown in Figure 1. The fixation to the scroll area more than 0.1 s allowed the participant to scroll the display. The scroll velocity is given by

$$(0.0667y + 7) \text{ [pixel]} / 0.01\text{[s]} \qquad (1)$$

where y corresponds to the vertical coordinate in the display. The coordinates at the inner and the outer of the scroll area were set to $y=0$ and $y=120$, respectively. The minimum and the maximum velocities were 700 and 1500 pixel/s, respectively. The display of (2) improved scroll-icon method is explained in Figure 2. The fixation of more than 0.1 s to the scroll icon enabled the participant to scroll the display. The icons (1) and (4) corresponded to the fast scroll mode (1500pixel/s). The icons (2) and (3) were the slow scroll mode, the velocity of which corresponded to 1000pixel/s. Different from (1) auto-scroll method, this method allows for the wider use of the display. In Figure 3, the display of (3) improved auto-scroll method is shown together with the relationship between y and scroll velocity. In Figure 3, the relationship between y and scroll velocity for (1) auto-scroll method is also depicted. The scroll velocities are given by

$$(0.00056y^2 + 7) \text{ [pixel] / 0.01[s]} \qquad (2)$$
$$(0.00078y^2 + 7) \text{ [pixel] / 0.01[s]} \quad (0 \leqq y \leqq 80)$$
$$(-0.00188(y - 120)2 + 15) \text{ [pixel] / 0.01[s]} \quad (80 \leqq y \leqq 160) \qquad (3)$$

For both Eqs.(2) and (3), the minimum and the maximum velocities were 700 and 1500 pixel/s, respectively. The display for scroll when using a mouse in shown in Figure 4.

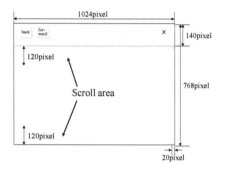

Figure 1 Display of auto-scroll method.

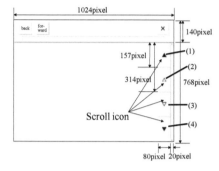

Figure 2 Display of scroll-icon method.

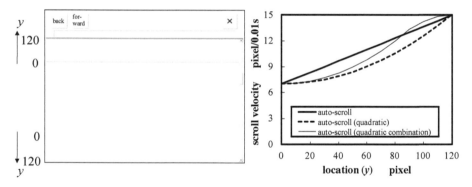

Figure 3   Relation between location *y* and scroll velocity in three types of auto-scroll methods.

2.4   Task

The task was to scroll and select a pre-specified menu item. The initial display before starting an experimental task is shown in Figure 5. The experimental display consisted of two layers. On the first layer, the participant was required to scroll the display and click the predetermined item such as the name of sightseeing site.

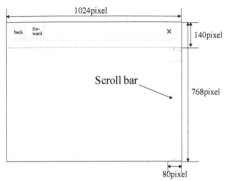

Figure 4 Display of scroll by mouse.

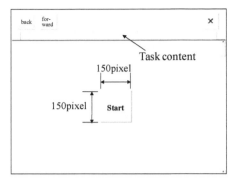

Figure 5 Display of scroll by mouse.

The display of the first layer (page) is depicted in Figure 6. On the second layer, the participant scrolled the display selected the menu to carry out the predetermined task such as the printout of the map of sightseeing site. The second layer (page) is demonstrated in Figure 7.

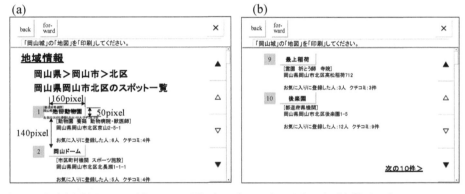

Figure 6 (a) Initial state of first page Display of experimental task. (b) State where page was scrolled until the bottom of first page.

Two scroll conditions were used: slow scroll and fast scroll. In the slow scroll condition (Task1), the participant was required to search for eight items from (3) to (10) (See Figure 7) on the first layer. The participant was required to search for four items from (7) to (10) in Figure 7 in the fast scroll condition (Task2). On the second layer, the participant was required to carry out a menu selection task on the area "A" shown in Figure 7 in the slow scroll condition (Task1). In the fast scroll condition (Task2), the participant was required to select a menu on the areas "B" or "C" in Figure 7 which required scrolling to a larger extent.

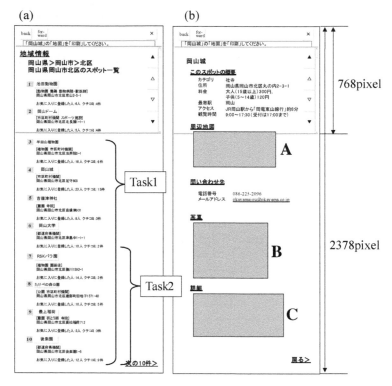

Figure 7 Whole page presentation of (a) first and (b) second page.

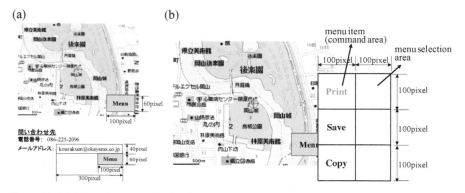

Figure 8 (a) Menu icon attached to image and E-mail address, (b) Menu items (print, save, and copy) (command and selection areas).

In this experiment, the menu was selected using I-QGSM (Improved-Quick Glance Selection Method) proposed by Murata et al., 2009b. In Figure 8, (a) menu icon attached to image and E-mail address and (b) menu items (print, save, and copy) (command and selection areas) are depicted.

2.5   Design and procedure

The scroll method and the scroll condition were within-subject experimental factors. As a control, the experimental task was also carried out using a traditional mouse. For each scroll condition, each participant carried out a pointing task 24 times using each of five scroll methods. The order of five scroll methods was randomized across the participants. The order of fast and slow scroll conditions was counterbalanced across the participants. One group performed the slow scroll condition first. The other group performed the fast scroll condition first. After all tasks were exhausted, the following psychological rating on usability was carried out using a 5-point scale.

The evaluation measures were the percentage correct recognition (error rate), the task completion time, and psychological rating on usability.

3   RESULTS

The task completion time, the percentage error, and the psychological rating of ease of operation and eye fatigue were compared among five scroll methods ((1) auto-scroll method, (2) improved scroll-icon method, (3) improved auto-scroll method (quadratic and combination of two quadratic equations), and mouse).

3.1   Task completion time

In Figure 9, the task completion time is plotted as a function of scroll method. In Figure 10, the task completion time is compared among scroll methods and between Task1 (slow scroll condition) and Task2 (fast scroll condition). The results of multiple comparison by Fisher's PLSD for task completion time is summarized in Table 1 ((1) Task1 (sloe scroll condition), (2) Task2 (fast scroll condition), and comparison between Task1 and Task2).

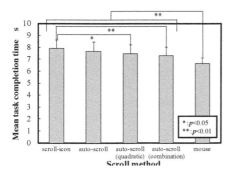

Figure 9 Mean task completion time as a function of scroll method.

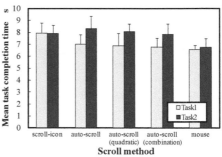

Figure 10   Mean task completion time as a function of scroll method   and task contents (task1 and task2).

Table 1 Results of multiple comparison by Fisher's PLSD for task completion time ((1) task1, (2) task2 and comparison between task1 and task2).

	scroll-icon	auto-scroll	auto-scroll (quadratic)	auto-scroll (combination)	mouse
scroll-icon					
auto-scroll	(1) *p*<0.01 (2) n.s.				
auto-scroll (quadratic)	(1) *p*<0.01 (2) n.s.	(1) n.s. (2) n.s.			
auto-scroll (combination)	(1) *p*<0.01 (2) n.s.	(1) n.s. (2) n.s.	(1) n.s. (2) n.s.		
mouse	(1) *p*<0.01 (2) *p*<0.01	(1) n.s. (2) *p*<0.01	(1) n.s. (2) *p*<0.01	(1) n.s. (2) *p*<0.01	

	scroll-icon	auto-scroll	auto-scroll (quadratic)	auto-scroll (combination)	mouse
Comparison between task1 and 2	n.s.	*p*<0.01	*p*<0.01	*p*<0.01	n.s.

The task completion time of the mouse was the shortest. The task completion time of the scroll-icon method was the longest. A one-way ANOVA carried out on the task completion time revealed a significant main effect of scroll method ($F_{(4,36)}$=9.740, p<0.01). Fisher's PLSD detected the significant differences between the following combinations: mouse and other scroll methods (p<0.01), scroll-icon and auto-scroll (quadratic) (p<0.05), and scroll-icon and auto-scroll (combination) (p<0.05).

A two-way (scroll method by scroll speed (fast and slow scroll)) ANOVA carried out on the task completion time revealed significant main effects of scroll method and scroll speed.

3.2 Error rate

Figure 11 shows the comparison of error rate among scroll methods. The error rate is shown as a function of scroll method and scroll condition in Figure 12. In Table 2, the results of multiple comparison by Fisher's PLSD for task completion time is summarized ((1) Task1 (sloe scroll condition), (2) Task2 (fast scroll condition), and comparison between Task1 and Task2). The error rate of the mouse was the lowest.

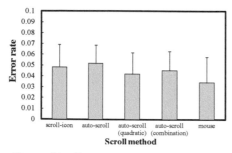

Figure 11 Error rate as a function of scroll method.

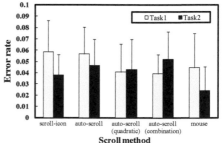

Figure 12 Error rate as a function of scroll method and task contents (task1 and task2).

Table 2 Results of multiple comparison by Fisher's PLSD for task completion time ((1) task1, (2) task2 and comparison between task1 and task2).

	scroll-icon	auto-scroll	auto-scroll (quadratic)	auto-scroll (combination)	mouse
scroll-icon					
auto-scroll	(1) n.s. (2) n.s.				
auto-scroll (quadratic)	(1) n.s. (2) n.s.	(1) n.s. (2) n.s.			
auto-scroll (combination)	(1) n.s. (2) n.s.	(1) n.s. (2) n.s.	(1) n.s. (2) n.s.		
mouse	(1) n.s. (2) n.s.	(1) n.s. (2) p<0.05	(1) n.s. (2) n.s.	(1) n.s. (2) p<0.05	

	scroll-icon	auto-scroll	auto-scroll (quadratic)	auto-scroll (combination)	mouse
Comparison between task1 and 2	p<0.05	n.s.	n.s.	n.s.	p<0.05

The error rate of the auto-scroll method was the highest.

A one-way ANOVA carried out on the error rate revealed no significant main effect of scroll method. As a result of a two-way (scroll method by scroll speed (fast and slow scroll)) ANOVA conducted on the error rate, a significant main effect of scroll speed ($F(1,18)$=6.720, p<0.05) was detected. A significant scroll method by scroll speed interaction ($F(4,36)$=2.796, p<0.01) was also detected.

3.3   Subjective rating on usability

In Figure 13, the score of ease of operation (usability) is plotted as a function of scroll method. The usability ranking was the following order: mouse, auto-scroll (combination), auto-scroll (quadratic), auto-scroll, and scroll-icon. A Kruskal-Wallis non-parametric test revealed a significant difference among scroll methods. As a result of further carrying out a Mann-Whitney U test, the following combinations were found to be statistically significant: mouse and other scroll methods (p<0.01), scroll-icon and other scroll methods (p<0.01).

470

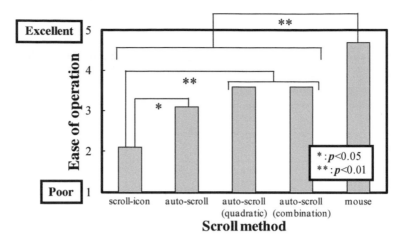

Figure 13 Subjective rating on ease of operation as a function of scroll method.

4 DISCUSSION

4.1 Task completion time

From the viewpoints of the task completion time, scroll method using an eye-gaze input system was inferior to a scroll using a mouse (See Figure 9). Contrary to this, it has been shown that the eye-gaze input system is faster than the mouse in a click operation (Murata, 2006, Murata et al., 2009a). It is necessary to realize a more effective eye-gaze input system by bearing such characteristics in mind.

From Figure 9, it is clear that the auto-scroll method (quadratic) was the fastest of all scroll methods using an eye-gaze input system. The scroll-icon method was the slowest. Task1 and Task2 corresponded to the condition under which scroll movements are short and that under which scroll movements are long. Analyzing the task completion time of Task1 and Task2 separately (See Figure 10) enables us to clarify the characteristics of each scroll method in more detail. For Task1 (slow scroll condition), the task completion time of the scroll-icon method was the longest, indicating that the scroll-icon method is not proper for short scroll movements. As the participant repeatedly scrolled while looking at the display and confirming it in Task1, the frequent eye movements occurred and inevitably led to longer task completion time. Concerning Task2, the task completion time of the scroll-icon method was the second shortest. This suggests that the scroll-icon method is usable for a scroll task where scroll movements are long and the participant need to scroll at a stretch to a larger extent.

While the task completion time did not differ between the mouse and the auto-scroll (combination) for Task1, the difference was enlarged for Task2. The reason can be inferred as follows. The scroll velocity when using a mouse is by far faster than that for the auto-scroll (combination). Therefore, when the participant recognizes that the target is located at the bottom of the display like Task2, the mouse enables us to scroll at a stretch until the bottom of the display for a shorter time.

From the viewpoint of velocity (speed), the auto-scroll methods with nonlinear relationship between the eye-gaze location and the scroll velocity was proper of all scroll methods using an eye-gaze input. However, it must be noted that fast scroll like a mouse is impossible, in particular, when longer scroll movements are necessary as Task2.

4.2   Error rate

While the auto-scroll method suffered from the most frequent error, the error rate of the auto-scroll method (quadratic) was the lowest of all scroll method using an eye-gaze input system. It is difficult for us to adjust to a slower scroll velocity with the auto-scroll method, because the relationship between the eye-gaze location and the scroll velocity is linear as in Figure 3. This certainly leads to the over-scrolling operation. Due to the nature of the eye camera measurement, the calibration error becomes dominant and the pointing accuracy consequently degrades. Such a property makes it difficult for the participant to point to a target under such a condition, leading to frequent errors. The nonlinear relationship between the eye-gaze location and the scroll velocity effectively worked to overcome such a problem related to the auto-scroll method.

The significant interaction of error rate between the scroll method and the scroll condition in Figure 12 is attributed to the following reason. While the error rate of both auto-scroll (quadratic) and auto-scroll (combination) methods was lower than that of the mouse for Task1, the error rate of the mouse was the lowest for Task2. The proposed auto-scroll methods were suitable for Task1 where scroll movements are short. The error rate of scroll-icon was the lowest of all scroll methods in case of Task2, because the scroll area placed not on the top and the bottom but on the right-edge of the display (See Fig.2) did not lead to the confusion between the scroll operation and pointing to a target.

4.3   Subjective rating on usability

The usability of a mouse was the highest. The usability of the auto-scroll (quadratic) and the auto-scroll (combination) methods was the highest of all of the four scroll methods, which indicates that the nonlinear relationship between the eye-gaze location and the scroll velocity also lead to higher psychological evaluation of usability. The usability of the scroll-icon was the lowest of all of the four scroll methods.

When executing a scroll operation using the scroll-icon method shown in Figure 3, the participant cannot help viewing the display using peripheral vision. The participant must confirm the scrolled display by moving the eye to it. Therefore, the eye movement distance becomes larger than other scroll methods, leading to the annoyance of the participants. For these reasons, the usability tended to be rated lower. Another reason can be interred as follows. The size of scroll area was determined so that it is larger than the minimum level by Murata et al., 2009b which assures the pointing accuracy and speed. However, the scroll area of the scroll-icon method is narrower than that of other scroll methods. Moreover, the scroll area is divided into four sub-areas according to the direction and the velocity of scroll. On the basis of the discussion above, the size of scroll area also seems to affect the usability rating.

4.4   General discussion

The scroll-icon method was found to be not proper for use in an eye-gaze input system on the basis of the results of the task completion time, the error rate, and the subjective rating on usability. It was found that the improved auto-scroll method (quadratic and quadratic combination) with nonlinear relationship between the vertical scroll location and the scroll velocity was optimal from the viewpoints of the task completion time, the error rate and the subjective evaluation on usability. Although the scrolling using the eye-gaze input system took longer than that using a mouse, the improved auto-scroll method with nonlinear relationship between the vertical scroll location and the scroll velocity led to a faster task completion and higher subjective rating on ease of operation. The result is indicative of the effectiveness of (3) improved auto-scroll method (quadratic and combination of two quadratic equations) when realizing a scroll function in an eye-gaze input system.

The condition to assure pointing accuracy and speed, the click method, drag & drop method, and the menu selection method have been investigated by Murata, 2006, Murata et al., 2009a and Murata et al., 2009b. These must be taken into account when putting an eye-gaze input system to practical use. When making use of the finding in this paper together with past findings above and designing an eye-gaze input system which is equipped with the scroll function, the following technical problems must be overcome: lower pricing of eye camera system, enlargement function of a target, auto-scroll function, proper input method of characters and drag & drop function.

REFERENCES

Goldberger,J.H. and Schryver, J.C., 1995. Eye-gaze determinationof user intent at the computer interface, In *Eye movement research*, eds. J.M.Finley, R.Walker, and R.W.Kentridge, Amsterdam: Elsevier Science B.V., 491-502.

Huchinson,T.E., White,K.P., Martin,W.N., Reichert,K.C., and Frey,L.A., 1989. Human-computer interaction using eye-gaze input, *IEEE Transaction on System, Man, and Cybernetics* 19: 1527-1534.

Jacob,R.J.K., 1993a. Eye-movement-based human-computer interaction techniques: Towards non-command interfaces. In *Advances in Human-Computer Interaction* 4, eds. Harston.H.R. and Hix.D,, Norwood, NJ:Ablex, 151-190.

Jacob, R.J.K., 1994. Eye tracking in advanced interface design. In *Advanced interface design and virtual environments*, eds. Baefield.W. and Furness.T., Oxford, UK: Oxford University Press, 212-231.

Jacob, R.J.K., 1991. The use of eye movements in human computer interaction techniques: What you look at is what you get, *ACM Transactions on Information Systems* 9: 152-169.

Jacob, R.J.K., 1990. What you look at is what you get: Eye-movement based interaction technique, *Proceedings of ACM CHI'90*: 11-18.

Jacob, R.J.K., 1993b. What you look at is what you get: Using eye movements as computer input, *Proceedings of Virtual Reality Systems '93*: 164-166.

Jacob, R.J.K., Sibert, L.E., Mcfarlanes, D.C., and Mullen, M.P., 1994. Integrality and separability of input devices, *ACM Transactions on Computer-Human Interaction*: 2-26.

Murata, A., 2006. Eye-gaze input versus mouse: cursor control as a function of age, *International Journal of Human-Computer Interaction* 21: 1-14.

Murata, A. and Moriwaka, M., 2009a. Basic Study for Development of Web Browser suitable for Eye-gaze Input System- Identification of Optimal Click Method -, *Proceedings of International Workshop on Computational Intelligence and Applications (IWCA) 2009*: 302-305.

Murata, A. and Moriwaka, M., 2009b. Effectiveness of the menu selection method for eye-gaze input system -Comparison between young and older adults-, *Proceedings of International Workshop on Computational Intelligence and Applications (IWCA) 2009*: 306-311.

Sibert, L.E. and Jacob, R.J.K., 2000. Evaluation of eye gaze interaction, *Proceedings of CHI2000*: 281-288.

CHAPTER 50

Usability Verification of Operation Method for Web Browsing on TV

Kazuyuki Matsubara, Atsuo Murata and Takehito Hayami

Graduate School of Natural Science and Technology, Okayama University
Okayama, Japan
matsubara@iims.sys.okayama-u.ac.jp

ABSTRACT

The usability of Web browsing was compared among four operation (input) methods (indirect cursor movement by a mouse, direct cursor movement by a Wii remote controller, focus movement by a TV remote controller, and hand-written character recognition by an iPod touch system). The task completion time of a mouse was the shortest of four operation methods, which is indicative of the effectiveness of a mouse under such environment.

Keywords: Usability, Web browsing, mouse, direct cursor movement, focus movement

1   INTRODUCTION

Although personal computers and mobile phones are popular among Internet terminals, the opportunity of to use TVs or TV game machines as Internet terminals is increasing more and more with the widespread of TVs that enable us to access Internet. There exist trends towards the use of mobile devices to control and manipulate objects such as displacing a cursor.

The usability of remote controller of Internet TVs is explored. The usability of Internet TVs using large display is discussed by Balakrishnan (2009) and Kahn et al. (2004). MaCallum et al. (2009) proposed an alternative to use mobile devices to control and manipulate objects. They proposed ARC (Absolute + Relative Cursor Pad) for interacting with large displays using a touch-panel of mobile phone. They showed that participants carried out a target acquisition task using the proposed

ARC faster without sacrificing accuracy. However, the proper operation method or device of the Web browser for Internet TVs has not been clarified until now. In other words, it has not been explored what type of device is proper for the Web browsing on Internet TVs. The proposal of the operation method that is suitable for the Web browsing on Internet TVs is necessary with the spread of Internet TVs.

The traditional operation method includes methods such as indirect cursor movement, direct cursor movement, or focus movement. This study made an attempt to verify the usability of Web browsing on TV among four operation(input) methods (indirect cursor movement by a mouse, direct cursor movement by a Wii remote controller, focus movement by a TV remote controller, and hand-written character recognition by an iPod touch system).

2   METHOD

2.1   Participants

Ten healthy males participated in the experiment. They daily used PC. All were male graduate and undergraduate students aged from 22 to 24 years. They hand no orthopaedic or neurological diseases.

2.2   Apparatus

The display size and the viewing distance were 819mm X 461mm and 2500mm, respectively. In the indirect cursor movement by a mouse, the resolution was fixed to 110 pixel/cm. The number of scroll lines was set to three, and the participant was allowed to use a scroll button. As for the direct cursor movement by Wii remote controller, the click and the scroll functions were substituted by "A" key and up and down of a cross key, respectively. In the focus movement by a TV remote controller, the click and the focus movement were carried out by an enter key and up and down of a cross key, respectively. Scroll started automatically when the focus disappeared from the display. The hand-written character recognition by an iPod touch system enabled the participant to select a menu item by the recognition of a handwritten number from 0 to 9 on iPod touch.

We used an A/D instrument PowerLab 8/30 and a bio-amplifier ML132 for EMG measurements. Surface EMG was recorded using A/D instrument silver/silver chloride surface electrodes (MLAWBT9), and sampled with a sampling frequency of 1 kHz.

The sketch of experimental setting and the devices used in the experiment are shown in Figure 1 and Figure 2, respectively.

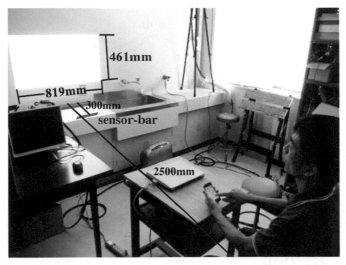

Figure 1 Outline of experimental setting.

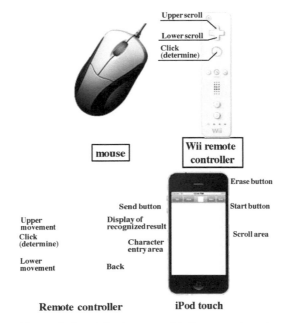

Figure 2 Input devices used in the experiment.

2.3   Design and procedure

The Web site used in the experiment is displayed in Figure 3. The experimental task was to search for the prefecture name and the city name which belonged to the

476

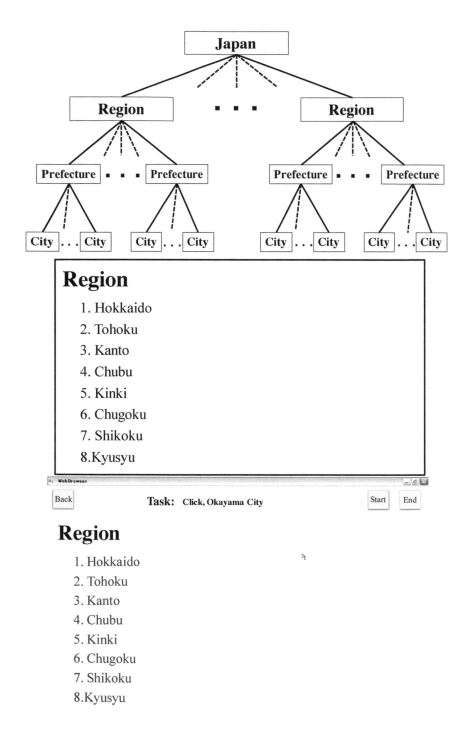

Figure 3   Web site used in the experiment.

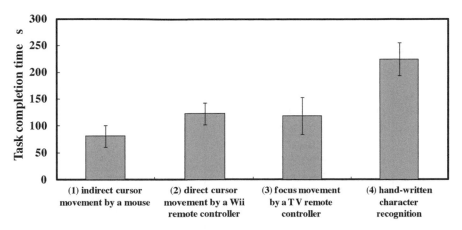

Figure 4 Task completion time as a function of operation method.

prefecture, and to click these using a pre-specified operation method (indirect cursor movement by a mouse, direct cursor movement by Wii remote controller, focus movement by a TV remote controller, and hand-written character recognition by an iPod touch system). In the hand-written character recognition by an iPod touch, the participant entered a one- or two-digit number using a character recognition system. The click and the scroll functions were necessary to carry out this task. Only when an error selection occurred, the participant was permitted to use (click) "Back" function. The main evaluation measures were the task completion time and the percentage correct of the task. The psychological rating on usability and workload to upper body was also conducted for each operation method. EMG of the trapezius muscle was measured while using each of four devices in order to assess physical workload for each operation method. The EMG measurements were converted into %MVC (Maximum Voluntary Contraction).

3   RESULTS

In Figure 4, the task completion time is plotted as a function of operation method ((1) indirect cursor movement by a mouse, (2) direct cursor movement by a Wii remote controller, (3) focus movement by a TV remote controller, and (4) hand-written character recognition by an iPod touch system). A one-way (operation method) ANOVA carried out on the task completion time revealed a significant main effect of operation method ($F(3,21)=69.83$, $p<0.01$). Multiple comparison by Fisher's PLSD revealed the following significant difference ($p<0.01$): (1) and (2), (1) and (3), (1) and (4), (2) and (4), and (3) and (4). The scroll time increased according to the following order: (1) indirect cursor movement by a mouse, (2) direct cursor movement by Wii remote controller, (4) hand-written character recognition by an iPod touch system, (3) focus movement by a TV remote controller.

Table 1 Frequency of errors compared among four operation methods.

	Cognitive errors	Operation errors
(1) indirect cursor movement by a mouse	0.3	1.5
(2) direct cursor movement by a Wii remote controller	0.2	67.4
(3) focus movement by a TV remote controller	0.3	7
(4) hand-written character recognition	0.1	3.8

The mean number of errors is compared among four operation methods in Table 1. The error trials were classified into the operation error and the error occurred during the cognitive process necessary for performing a search task. A one-way (operation method) ANOVA carried out on the number of errors revealed no significant main effect of operation method ($F(3,21)=1.13$). The frequency of operation errors was especially large for (2) direct cursor movement by Wii remote controller.

%MVC is plotted as a function of operation method in Figure 5. A one-way (operation method) ANOVA carried out on the number of errors revealed no significant main effect of operation method ($F(3,21)=10.96$, $p<0.01$). Multiple comparison by Fisher's PLSD revealed the following significant difference ($p<0.01$): (1) and (2), (2) and (3), and (2) and (4).

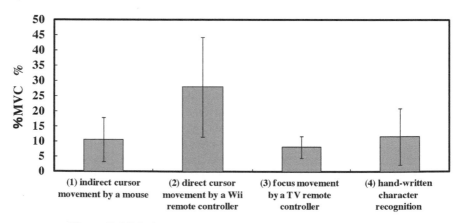

Figure 5 %MVC compared among operation methods.

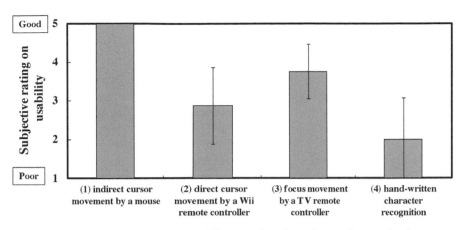

Figure 6 Subjective rating on usability as a function of operation method.

In Figure 6, the subjective rating on usability is compared among four input methods. Mann-Whitney non-parametric tests revealed the following

significant differences: (1) and (2), (1) and (3), (1) and (4), and (3) and (4). Figure 7 compares the subjective rating on workload to upper body among four input methods. As a result of Mann-Whitney non-parametric tests, the following significant differences were detected: (1) and (2), (1) and (3), and (1) and (4).

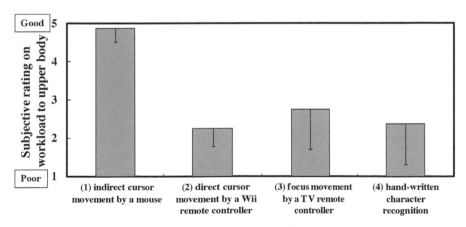

Figure 7 Subjective rating on workload to upper body as a function of operation method.

480

4   DISCUSSION

A mouse is generally not recommended to use for Web browsing on TV.

We hypothesized that the mouse is not proper for Web browsing on TV due to the small cursor size viewed from the distance of 2500 mm. In spite of this hypothesis, the subjective rating on suability and workload to upper body of (1) mouse was the best of four operation methods. This indicates the effectiveness of (1) mouse under such environment. In future work, it should be examined whether a similar result is obtained even for older adults.

The operation method (2) (direct cursor movement by Wii remote controller) suffered from the most frequent error trials. The reason is inferred as follows. The click of "A" key forced Wii remote controller to move, and this led to its recognition as a drug operation. In order to reduce such errors, the enlargement of character size and making the selection easier would be necessary. The higher recognition accuracy between click and drug operations should also be realized.

The focus movement by a TV remote controller (operation method (3)) unexpectedly led to shorter task completion time as compared with methods (2) and (4). In this operation method (3), the task completion time increased with the increase of the number of links. As for (4) hand-written character recognition by an iPod touch system, efforts must be exerted to reduce error

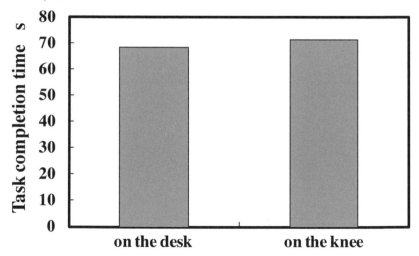

Figure 8   Task completion time compared between mouse operation with and without a desk.

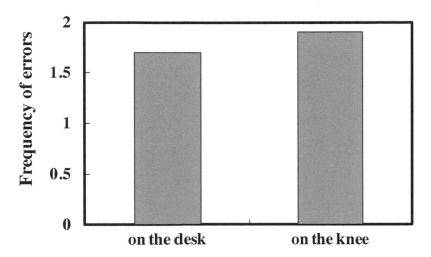

Figure 9 Frequency of errors compared between mouse operation with and without a desk.

trials to further extent.

Although the mouse was the most proper from all the evaluation measures and the most suitable for the Web browsing on Internet TVs, it is easily conceivable that one must use a mouse under environment where one cannot use a desk for operating mouse. Therefore, the following experiment was preliminary carried out to compare the task completion time and the frequency of errors between environment with and without a desk. Under environment without a desk, the participant was required to use a mouse on his knee. As in Figures 8 and 9, no significant differences of task completion time and frequency of errors were detected between two conditions. Thus, we can judge that the mouse is proper for such usage as the Web browsing on Internet TVs. As the mouse is generally not acceptable for older adults (Murata (2006)), the usability evaluation of the Web browsing on Internet TVs for older adults should be carried out in future research.

REFERENCES

Balakrishnan,X.B.R. 2009. Comparing usage of a large high-resolution display to single or dual desktop displays for daily work, *Proc. of CHI2009*: 1005-1014.

Dieen,J., Toussaint,H., Thissen,C. and Ven,A. 1993. Spectral analysis of erector spinae EMG during intermittent isometric fatiguing exercise, *Ergonomics* 36: 407-414, 1993.

Huang,H. and Lai,H. 2008. Factors influencing the usability of icons in the LCD touchscreen, *Display* 29: 339-344.

Kahn,A., Fitzmaurice,G., Almeida,D., Burtnyk,N. and Kurtenbach,G. 2004. A remote control interface for large displays, *Proc. of ACM Symposium on User Interface Software & Technology*: 127-136.

McCallum,D.C. and Irani,P. 2009. ARC-Pad: Absolute + relative cursor positioning for large displays with a mobile touchscreen, *Proc. of UIST'09*: 153-156. 2009.

Murata,A. 2006. Eye-gaze input versus mouse: cursor control as a function of age, *International Journal of Human-Computer Interaction* 21: 1-14, 2006.

CHAPTER 51

Evaluating WantEat: A Social Network of People and Objects

Alessandro Marcengo[1], Amon Rapp[2], Luca Console[2], Rossana Simeoni[1]

[1]Telecom Italia,
Torino, Italy
alessandro.marcengo@telecomitalia.it
rossana.simeoni@telecomitalia.it

[2]University of Torino,
Torino, Italy
amon.rapp@gmail.com
lconsole@di.unito.it

ABSTRACT

This paper describes the methodology adopted and the results obtained within the ergonomic design of the WantEat interactive system. The reference methodology for this project included different consolidated methods in the field of human factors, but also required a creative combination of different techniques in order to effectively evaluate the complexity of the system. Specially some aspects related to the evaluation of an ontology-based knowledge engine and the new concept of a social network of people and objects required to take a step further beyond a consolidated user centered approach.

Keywords: evaluation, user test, field test

1   INTRODUCTION

Some years ago we started the Piemonte Project based on some fundamental considerations shared by the partner of the projects (Telecom Italia, University of Turin – Department of Computer Science, University of Gastronomic Sciences in

Pollenzo and Slow Food). More or less we can resume it in this way. On one side the technological evolution, the miniaturization of electronic components and the increasing computing capabilities are enabling a gradual integration of content and digital intelligence into artifacts, goods, and supply chains under-supported until today by telecommunication functionality. On the other side the increasing simplification of advanced interfaces let to foresee a much more active role for the actors of productive sectors in their own digital communication activities. Transversally, it's clear that the success of enabling services today is increasingly influenced by a "positive" and "ethic" image of the same services and of the enabled application domain. The "sustainable future" has an enormous need of telecommunications and itself becomes a flag of high visibility and credibility.

According to these considerations in the Piemonte Project we chose to address the wine and food domain (as part of cultural and territorial heritage) partnering with best-in-class partners (like Slow Food) and finding the right balance with other techno-economic players in the digital supply chain. Nowadays are not available any "sites" or "portals" on a worldwide level for this domain: that was the start of WantEat product, the outcome of the Piemonte Project. The WantEat proposition is right this one, to fill this gap, also through the technology enablement of the territory. In this vision smart things can really play the role of gateways enhancing the mutual knowledge exchange between people and the heritage of a specific area. Physical things typical of our everyday reality, as human artifacts or even landscape elements, are able to "socialize" with people and other objects, telling about the relational world that surround them and sharing their relationships with other people or objects of the same area. Keeping track about what is going on to and around them, these "things" become able to tell stories and, through a direct interaction, they are also able to give people a significant and richer experience during they staying in the area. WantEat propose to address the theme of food not only as an economic matter but more on a cultural point of view, creating an interactive system with different entry points, each addressed to a specific actor in the wine and food domain. Starting from this large concept it was therefore developed an iPhone mobile application and a companion Web Site for general users, and a totally different web environment addressed instead to specific actors (producers, restaurateurs, etc.).

Developing these three system entry points and their interfaces, users have been involved since the early phases. The design of an interactive system as WantEat, made up of different applications, each addressing different types of users, led us to implement a highly structured evaluation plan. In order to collect useful data to the design process, we had to adopt different methods of assessment, putting out all the critical aspects of the system, as well as the unexpressed needs and desires of various user targets. The centrality of social networking features into the mobile application led us to balance laboratory assessment methods and field assessment. Indeed classic usability testing methods are not always sufficient to provide guidance to the designer of how the system will be used in everyday practice: too often laboratory tests are able to assess the usability of a system but fail to evaluate its real usefulness (Greenberg and Buxton, 2008). Through field assessment, instead, users can ecologically use the interactive system under test according with their own purposes and usage habits (Rogers, 2011). This is a focal point when you want to have feedbacks on the social aspects of an application. Iterative evaluations in the laboratory have been so alternated with field evaluation phases in which users used the application in a very similar way to that of an everyday context use.

Since it would take too long to treat the entire experimental plan applied to the three entry points (see Figure 1) of the system, we will focus in this paper on the strand that concerns only the mobile application, which was also the most widespread means of the WantEat system and the one with the most innovative interaction paradigm.

want	CONSUMERS		PRODUCERS	
Involved USERS	Mobile Client	Web App	Web Dashboard	
Requirements	12	Guidelines	Guidelines	Interviews
LabTest	12 + 26	User Test	Card Sorting	X
Requirements	24			Interviews
Field Test	684	«Salone del Gusto»	X	X
Requirements	12		Guidelines	Interviews
Field Game	157	«Cheese2011»	«Cheese2011»	X
Requirements	12			Interviews

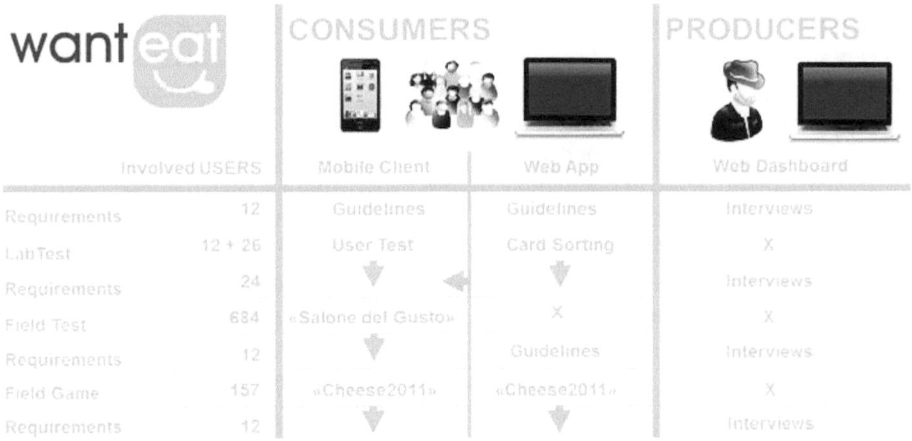

Figure 1 - The WantEat Evaluation Plan

Looking at a typical usage scenario it is possible for the consumer to simply frame the label (which must be loaded in the system by the producer) of any food product obtaining immediately not only information about the product, but also how this product is in relationship with other products, with the places where it is produced and distributed, with other people who like it. All these relationships contribute to create a social network, in which objects and people are strongly connected through associative logics (e.g. similarity, proximity, friendship, etc.). The user can freely explore these relationships, interact actively with the product, leaving comments, opinions and sharing them with other people in a social logic. The most innovative technology modules that enable this type of interaction are basically three. A visual search algorithm that recognizes the products through his label, a knowledge manager that manages the relationships between objects in the system and a social networking system that enables communication between all objects (both people and things). These aspects are blended by an innovative interaction paradigm in the mobile app arena called "The wheel" (PIEMONTE TEAM, 2011, 2012; Biamino et al., 2011).

2 THE HUMAN COMPUTER INTERFACE: LAB TESTING

As regard of the prototype interface the test was conducted through 5 tasks committing the user to find information and navigate the application, built in flash mock up and running on a touch screen tablet. Two additional tasks were defined to test some specific features of the application rendered through power point animated storyboards. Finally, a last task to evaluate the intuitiveness and effectiveness of the image recognition paradigm was performed on a prototype running on iPhone (see Figure 2). The questionnaires provided after each task have used 5-point self-anchoring scales to assess different dimensions (intuitiveness, clarity, ease of use, etc.). After the evaluation session a final questionnaire was given and a short interview conducted in order to gather feedback about the acceptability of the proposed service solution. The evaluation cycle has involved 12 members representing various possible targets. The variables considered for the composition of the sample were:

- "Age", which assumed a breakdown between population 18 to 35 years (young users) and population 36 to 60 years (old users)
- "Experience and openness to technology", which breakdown the user population in Hard users (daily using complex technological tools, having a smartphone touch) and Soft users (who do not own and do not daily use complex technological tools, and are not equipped with a touch smartphone).

Belonging to one of two categories was determined through a preliminary screening questionnaire. From the intersection of these two variables were formed 4 groups of 3 persons each. The tests were conducted individually.

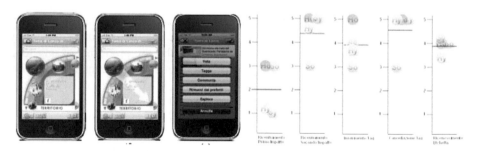

Figure 2 - Lab test: Human Computer Interface

Data analysis reveal that on average the soft users, especially in the age group 36-60 years, find more difficulties than hard user. This difficulty is reflected in a lower average rating on intuitiveness and clarity of some interface elements (average of 4.1 for intuitiveness and 4.3 for the clarity dimension for hard user category and 3.1 for intuitiveness and to 3.4 for the clarity dimension for the soft-user category). However, pleasantness and innovation impression on the application gather more favorable ratings among soft users (4.6 against 3.4 for hard users).

The hard users have proved on average to be more critical of the soft users in particular with regard to the interaction logic and information architecture. Obviously, the usefulness of the application was subordinated to the interest in the domain: that is why not all who have expressed a positive opinion about the application expressed themselves positively about a possible personal usage. The results led to the redesign of some elements of the interface and the implementation of a new high fidelity prototype for iPhone, which included all the insights obtained during the testing phase and also included new social features allowing the building process of a social network of users and objects.

3   KNOWLEDGE MANAGER: CARD SORTING

As we mentioned before we had in the early phase the problem to assess the appropriateness of the relationships (between objects, between people and between people and objects) proposed by the Knowledge Manager (KM). The Knowledge Manager is a component of the system that is responsible to calculate the similarities between objects (e.g. different cheeses) and then present those similarities to users according to their estimated needs.

To avoid the possible distortions introduced by the influence of the interface in the evaluation of the performances of the KM, we chose to do an assessment completely apart from the interface itself. We opted for the card sorting method, in order to both understand the most grouping criteria used by people, and have a direct impression by the users about the associations proposed by the KM. The findings were used to give a feedback to the development team about the validity and relevance of the actual performance of the Knowledge Manager. The card sorting session had the objective to discover the similarity criteria used by most users, bringing out their mental representations and having so a comparison yardstick about the similarity criteria and the relative weights currently employed by the Knowledge Manager. The evaluation session had the aim to find possible problems in the current performances of the KM.

It was therefore prepared an open card sorting where users had to classify some cards (35) representing different types of cheeses: users were able to pile them into as many groups as desired, without any constraints of size, giving them a label reflecting the similarity criterion used. At the end of the session participants were given a questionnaire in which were asked to:
- 1/ explain their classification criteria
- 2/ assess the similarities proposed by the Knowledge Manager
- 3/ guess the criteria used by the Knowledge Manager.

At the end of the sessions a brief group discussion brought out the points and judgments shared by all participants. This evaluation cycle involved 26 members representing various possible targets. The variable considered for the composition of the sample was "age", which assumed a breakdown between population 18 to 35 years (young users) and population 36 to 60 years (old users). All the participants shared an interest in the food and wine domain. Considering the age variable, were

488

formed 2 groups of 13 persons each. The tests were carried out collectively on sessions of 6-7 members (2 per group).

About the results of this evaluation in Figure 3 we can see the evidences as a dendrogram, a tree graph that shows the similarity matrix indicating the most common groupings made by users. From the data analysis it can be inferred that the "provenance" is among the most used criteria (1 of 2 user uses it). Other used criteria in classification are the "consistency" and the "maturing" of cheeses, while the "taste" (e.g. spicy, sweet, etc.) and the "type of use" (e.g. be grated, to melt, etc. or anyway related with use) are mostly used as secondary criteria. If the "provenance" is actually the most used criterion, the preferred granularity appears to be regional (cheese from Piedmont, Lombardy, etc.).

Figure 3 - Lab test: Knowledge Manager

The analysis of the quantitative data gave as result essentially an image of the preferred methods of categorization actually used by users. The "provenance" and cognitive/perceptive criteria (use, taste, appearance) are preferred over chemical and physical criteria. The structural criteria (type of composition, seasoning, milk and fat mass) are used only if disclosing a perceptive impact. The cheeses so grouped were compared with clusters produced by the Knowledge Manager to determine if the criteria implemented in the system actually were reflective of the classification ways of real users.

About the evaluation of existing Knowledge Manager performances, the average level of satisfaction about the similarities found by the KM itself stood below the neutral value (3). Users were unable to guess the criteria used by the Knowledge Manager: there was a general tendency among the users to project their own criteria in the system. The similarities proposed have not been seen as the result of stable criteria, but variable depending on the cheese, from more than 50% of users. The final group discussion revealed the difficulty of tracing at least one stable main criterion used by the system to provide all the similarities, not helping in this way users to form a mental model of the system.

4   MATURE PROTOTYPE: MASSIVE FIELD TEST

Results gathered about the interface of the mobile application and about the Knowledge Manager in laboratory test led to new requirements, then included in the development roadmap of the subsequent prototype. But we felt the need to gather data with a more ecological validity. The cooperation with a best-in-class partner such as Slow Food allowed us to bring the first prototype in a truly "natural" setting: the largest international fair of food and culture "Salone del Gusto" hosted every two years in Turin, Italy. Our intention in this case, after extensive quality testing in our laboratory, was to test the application with a very large number of people to gather feedback with a more quantitative character. Also we wanted to ensure that users could use the application in ways closer as possible to the real contexts of use of their daily lives.

4.1   Methodology

All participants in this massive trial were given the opportunity to freely try the WantEat application on iPhone loaned for a short period of 15-20 minutes. An interviewer presented, in a strictly standardized way, the application and the whole service, explaining the purpose and the basic functionalities. Participants were asked to frame the label of no matter what product exposed in one of the booth inside the fair: this action allowed the recognition of the object with its interactive and informative features. From here on the navigation could be carried out completely free.

After the test the interviewer submitted to each participant a structured interview, asking some personal data and their level of acceptance of the application, evaluated according to different dimensions (ease of use, understandability, pleasantness, usefulness) using self-anchoring scales with values from 1 to 4, where 1 meant not at all and 4 meant a lot. In the case that one of the ratings were negative (values from 1 to 2) it was asked more about the negative judgment (through a set of default answers that could be affirmative to more items per question). In the final part of the questionnaire it was investigated more deeply the usefulness of the whole service through direct and indirect questions concerning the possible desired usage contexts. Also few questions about future desired functionalities were asked in order to provide new directions in application improvement to the design team.

 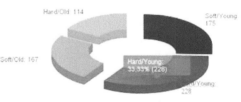

Figure 4 - Field Test at "Salone del Gusto"

The selected sample was a random sample. We wanted to gather impressions from 4 types of users. On the basis of our previous lab user tests we segmented the final sample in 4 sub-groups through the dimensions of "age" (Young vs. Old) and "attitude towards the new technologies of communication" (Hard vs. Soft). We called each group obtained crossing the two variables: Young/Hard Users, Old/Hard Users, Young/Soft Users, Old/Soft Users. The final sample was of 684 respondents, divided in 228 Young/Hard Users, 114 Old/hard Users, 175 Soft/young Users, 167 Soft/Old Users. All respondents were assumed to share a positive predisposition and interest towards the wine and food domain because the "Salone del Gusto" fair is dedicated exclusively to the food showing and tasting with an expensive admission fee.

4.2   Results

A first type of data processing was done with elementary statistics (average, mode, standard deviation, etc.) in order to deliver quick requirements to the design team to be included in the development roadmap. These aspects address usability and acceptability issues of the application.

For all the variables (ease of use, understandability, pleasantness, usefulness) the results were very good for all groups, in particular:

- Ease of use - 3,49 Young/Hard Users, 3,57 Old/Hard Users, 3,44 Soft/young Users, 3,24 Soft/Old Users with an average rating of 3,42 and a standard deviation of 0,68
- Understandability - 3,50 Young/Hard Users, 3,64 Old/Hard Users, 3,43 Soft/young Users, 3,32 Soft/Old Users with an average rating of 3,46 and a standard deviation of 0,67
- Pleasantness - 3,42 Young/Hard Users, 3,49 Old/Hard Users, 3,44 Soft/young Users, 3,38 Soft/Old Users with an average rating of 3,43 and a standard deviation of 0,67
- Usefulness   - 3,01 Young/Hard Users, 3,30 Old/Hard Users, 3,13 Soft/young Users, 3,09 Soft/Old Users with an average rating of 3,12 and a standard deviation of 0,84.

The result has been surprising good about a possible personal use of the application with an average of 76.46% of positive affirmations (with 66% Soft/Young Users expressing themselves with positive value that goes up to 90% in the case of Old/Hard Users). This value even raises when people were asked about a possible recommendation of the application to friends (86% for Soft/Young Users express themselves positively and even 94% was the result for Old/Hard Users). We then asked the users about new features they would like within the application even recognizing the poor predictive value of this kind of questions. Respondents (through a set of default answers that could be affirmative to more than one items) expressed themselves as follows: 70.17% would like to compare the prices of the products, 45.17% would compare the same class of products, 36.11% would be able to buy products through the application (e-commerce), while 35.09% would be able to book hotels and restaurants. Finally, only 24.85% of respondents would like to see videos about products.

All the results reveal a high rating particular for the category of Old/Hard Users (over 35 years and with a strong propensity to use new communication technologies). The higher age, which implies greater spending power and a more conscious lifestyle toward food quality, together with the habit to use the latest mobile phones and digital services, may explain the great success that the application has encountered. The application was, in any case, positively judged by the majority of the other types of users surveyed.

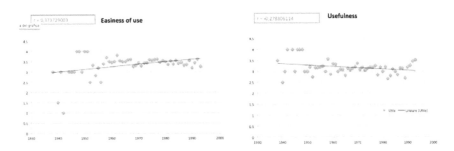

Figure 5 - Correlation Easiness of use – Age and Usefulness – Age

To complete the data analysis more advanced statistical processing were carried out later, so as to correlate different variables and highlight some trends. As an example we report the correlation between the "ease of use" and the "age" and the correlation between the "usefulness" and "age". As is clear from the graphs in the Figure 5, if the age decrease there is a trend towards increased perception of ease of use of the system, but the effect is reversed in the case of the correlation with the utility: as the "age" of the users decreases the system is perceived progressively less useful. This evidence is probably due to a lower interest related to food quality and his cultural aspects related to the youngest group of people.

5   CONCLUSIONS

Even if in this paper we just outlined the mobile client evaluation plan it easy to see how it was difficult to make meaningful evaluation due to the system complexity, the novelty and variety of technologies involved and the innovativeness of the concept. So it has been necessary to deploy a careful and personalized testing plan requiring a combination of methods and techniques. In particular after the "Salone del Gusto" fair we were not completely satisfied specially about the testing of the social network features of the system that, as it is clear, need to have an even small but running community to make sense to the users. So the following evaluation was carried out in another trial at another big fair ("Cheese 2011") that investigated new aspects and new issues right about the application of social network features, which in earlier stages was not possible to assess. In particular we built a specific context of use to stimulate the motivation to use those features. In this way it was possible to fill a gap that persists in the evaluation tests described in this paper, that is the lack of genuine engagement that users seems to show when they perform not self-motivated tasks. In essence, users even in the absence of specific requested tasks (like in a free field trial), if do not develop in a short time a mental model that match their motivations and objectives with the application possibility to satisfy these objectives, are not further motivated to explore and use the system in all its potentialities. To keep up with this problem we developed a new methodological approach that we propose to illustrate in a future work.

REFERENCES

Biamino, G., P. Grillo, I. Lombardi, F. Vernero, R. Simeoni, A. Marcengo and A. Rapp 2011 "The wheel" - an innovative visual model for interacting with a social web of things. In Proceedings of Visual Interfaces to the Social and Semantic Web (VISSW) 2011.

Greenberg, S. and B. Buxton 2008 Usability evaluation considered harmful (some of the time). In Proceedings of CHI 2008, ACM Press (2008), 111-120.

PIEMONTE TEAM, 2011 WantEat: interacting with "social" networks of intelligent things and people in the world of enogastronomy. In Proceedings of Workshop "Interacting with smart objects", held in conjunction with IUI 2011. Palo Alto, CA, USA (2011).

PIEMONTE TEAM, 2012 Interacting with a Social Web of Smart Objects for Enhancing Tourist Experiences. In Proceedings of eTourism Present and Future Services and Applications (ENTER2012).

Rogers, Y. 2011 Interaction Design Gone Wild: Striving for Wild Theory. *Interactions* 18, 4 (July + August 2011), 58-62.

CHAPTER 52

Assessment of Customer Satisfaction and Usability Evaluation of Social Networking Service

Ryang-Hee Kim, Kee-Sam Jeong, and Soon-Ei Bae
Dept. of Fashion Design, University of Hanseo,
360 Daegok-ri, Haemi-myeon,
Seosan-si, Chungcheongnam-do, ZIP: 356-706, South Korea,
Yanghee1003@naver.com

Young-Kuk Kwon, Kyung-Jun Min, Dong-Suck Son,
Seong-Hyun Park, Joong-Suck Choi,
Dept. of Safety Engineering, Seoul National University of Science and Technology,
KongNung 2-Dong, NoWon-Ku, Seoul, ZIP: 139-743, South Korea
safeman@seoultech.ac.kr

ABSTRACT

With the vision of ubiquitous computing, adjustable user interface as a new Social Networking Service (SNS, ex. 'Cyworld', 'Facebook', etc.) interaction paradigm suggests a novel way and direction using various personal network systems. Along with remarkable use of Smartphone, the types of SNS system in various domains have been developed, but there is lack of assessment for usability of various SNS interface types that are currently used, because everything has a potential to change, considering the fast growing ubiquitous computing environment of private SNS system. The purpose of this study was to assess user satisfaction and usability evaluation of Social Networking Service. The selected usability principles were classified systematically using statistical methods; the paired T-test and Principle Components Analysis (PCA) using Factor Analysis. A total of 203 valid questionnaires (two group: Youth group (N: 108, (male; N: 60, female; N: 48), Elders group (male; N: 45, female; N: 50) were received, and the data were estimated by using SPSS 18.0. The result of this test shows that 'Cyworld' which is called 'ssa-e', 'mini homepage', and 'Facebook' are most widely used on the SNS internet community system regardless of age. In addition, SNS service quality plays an important role in significantly moderating the influence of the-Age to use SNS system on technology acceptance in easy-fast way. As a result, 18 usability principles are categorized into 4 factors according to their correlations and we named them respectively, (1) User-Interaction, (2) User-Cognitive, (3) User–Performance, (4) User–Satisfaction, and several usability principles including utilization, convenience, amusement, use affect attitude, etc. show respectively differences depending on age and SNS types. Finally, we could suggest 18 customer satisfaction and usability questionnaires for SNS checklists system are produced on a 5-point Likert scale.

Keywords: Social Networking Service (SNS), Ubiquitous Computing, Customer Satisfaction, Usability Evaluation, the-Aged

1   INTRODUCTION

Social Networking Service (SNS) is a system which strengthens interpersonal relationships among people and allows them to form new relationships. In the past, SNS was only used through the internet, but recent days, because of the broad usage of smart phones, SNS is used more by people with Smartphone. Social Networking Service allows people to share their information on their portable internet, and helps communication among them. Since information shared by users on the SNS is concise and provides more reliability than the information obtained through search engines. As individual's expression has strengthened, high technological function of SNS is developing.

Also, SNS can be separated into many types in usability where functions that are commonly used for the public, and are provided for specialized services for specific groups of people. While online community services such as blogs, café, or groups share specific topics and provide closed services, SNS allows individuals to share their own interests with others. And, the functions of SNS are classified with one that is usually used for the most people and the others that focuses on special functions considering the characteristics of a few people.

Hence, this research was carried out statistically to find how people use Social Networking Service which is used representatively nowadays and which service is used the mostly. Also, the purpose of the research is to evaluate users' satisfaction and usability for SNS through the statistical research method depending on SNS types', and to propose the heuristic evaluation method for Aged-user Social Networking Service.

2   CONCEPTION OF SOCIAL NETWORKING SERVICE

"What is Social Networking Service (SNS)?" Social Networking Service, SNS is a system which strengthens interpersonal relationships among people and allows them to form new relationships. Several basic functions of SNS are post contents, share with users' friends, status updates, sharing photos and videos, and News Feed. Digital media convergence company, DMC Media has declared a report, targeted on 1300 people, on the user survey of SNS. On this survey, it was clear that different age groups preferred different types of SNS. While 44.6% of 20's preferred Cyworld, they showed little preference on blogs or café compared to other age groups. 24.9% of the age of 30 responded that they prefer Twitter, with 26.7% preferred Cyworld, showing a narrow margin, respectively. However, 40's preferred blogs (38.3%), and café (20%) which percentages are much higher than other age groups. Also, 21.7% of 40's preferred Twitter, which shows that most of Twitter users are in 30's~40's. 83.6% of total respondents use SNS and answered that they use SNS mainly for 'social networking', and 'exchange of information'. Twitter was favored for exchange of information (81.6%), Me2day for entertainment (60.8%), Cyworld for social networking (90.9%), and Facebook for business (26.7%). Through the result, SNS is classified with its services and utility values by the users (source: http://www.ddaily.co.kr/news/news_view.php?uid=66004).

A SNS website, My-space, had similar organization, but it started to differentiate from existing SNS with plenty funds and vast database. My-space assigned full controls to users of their own information, and the website itself was made into a 'space' where people can share their information. Not so long after, Facebook was created. This website was created by Harvard university student, Mark Zuckerberg and his friends, in order to promote interchange of information and opinions among students. Less than 3 months, Facebook successfully attracted 12 million users globally. Simple design and instant status updates functions made Facebook as standard type of SNS. Winced SNS market after the Cyworld in 2004, with successful advent of Social Networking Service which enables people sending and receiving news on the internet, is booming again. Especially, expert on website analysis, ranky.com analyzed recently becoming hot topic in the SNS companies, Twitter and Me2day, about users' using patterns. As a result, both two websites showed increasing user maintenance ratio; 5% of Twitter has increased up to 32%, and me2day, 26%. (Source: http://www.ukopia.com/ukoKoreaNews/?page_code=read&uid=129881&sid=16&sub=63).

Basically, important functions of Facebook can be separated into 3. Users can post and share contents on the Wall. By using Home page, and sharing menu on the top of profile, users can share their thoughts to others. And users can update their status and share photos and videos. The shared article appears as a new post, and on the news feed. A Minihompy service, provided by Cyworld, is a shortened version of 'Mini Homepage.' It is a one-person media which users can uniquely decorate their own Minihompy, and share their information. This is an important way of communication on the web. Users can leave their status or photos, mark important anniversaries, and set background music on the Minhompy. Twitter is a type of SNS which has unique function called 'Follow.' This function is much similar to other SNS's 'Friend Request' function, but the users are registered as 'Follower' unilaterally to a person who the users want to follow. Lastly, it has unique function called 'See collected comments'. When a user clicks a response for a comment, he or she can see not only the direct response to the comment, but replies made on those direct responses too.

Nearly half of Smartphone users responded that they use Smartphone to get in touch with their acquaintances. Trendmonitor made public that 45.9% of total respondents use text messages and SNS to form and maintain relationships, on survey of utilization of Smartphone. In other words, Smartphone has important role in expanding everyday social encounters into online webs. This survey, planned by Trendmonitor and progressed by Easysurvey, was based on 1000 adults who use Smartphone. Purposes other than interpersonal relations, Smartphone were mostly used for their technical functions. 18.9% of respondents used Smartphone for daily life and time management, and 15.3% used Smartphone for their high tech functions. Users who use Smartphone for high tech, showed interests when having conversations with others about latest technology and trends (source: http://www.zdnet.co.kr/news/news_view.asp?artice_id=20111026133819&type).

However, along with remarkable use of Smartphone, the use of SNS mobile internet users is being amplified recently. Also, there is no heuristic interface standard evaluation of SNS communication, and there is even good understanding of usability of various SNS interface types in the fast growing u-computing environment of private Social Networking Service system. Therefore, the objectives of this study were to assess user satisfaction and usability evaluation of Social Networking Service.

3   METHODOLOGY

The findings of previous studies show that 'Cyworld' which is called 'ssa-e', 'Mini-homepage', and 'Facebook' are used the most widely use on the SNS internet community system regardless of age (Palo Alto, 2010). We refined 18 usability principle items including Nielsen (1994)'s user interface checklist principles with five-score Likert scale (Table 1).

Table 1. Development of user interface prototype checklist for SNS by 5-point likert scale

Questionnaires	Very Not Important	Not Important	Average	Important	Very Important
	1	2	3	4	5
1. Must be easy to learn.					
2. Should be easy to use again.					
3. When learning, there should be helps to learn easily.					
4. System for on-screen menus and features the name, help should be clearly understood.					
5. Structure and how it is used should be simple.					
6. Should have elements that can be specified.					
7. It should be interworking with many devices (ex. Cell phone, tablet PC).					
8. There should be no errors or bugs					
9. System needs to be quick in prosecuting its functions.					
10. There should be a function to recover or cancel.					
11. Compared with previously used SNS, functions should be felt not much different.					
12. Should be interesting.					
13. There should be no problem to use even without specific knowledge.					
14. Need to feel affinity though it is first time using.					
15. There should be additional explanation on detailed feature.					
16. How satisfied are you for usage of this SNS?					
17. How satisfied are you for this SNS's visual aspect?					
18. Is this SNS easy to approach and convenient?					

Total of subjects were selected 203 people who have used both Facebook and Cyworld, and they were divided into two group: Youth group (N: 108, (male; N: 60, female; N: 48), Elders group (N: 95, (male; N: 45, female; N: 50)), valid questionnaires were received, and the data were collected the value of the dependent variables (user satisfaction and usability of SNS) according to the independent variables (SNS Type 1: Facebook, SNS Type 2: Cyworld (Minihompy)) between Youth group (N: 108), and Elders group (N: 95).

The evaluation and analysis of the results were statistically assessed by using the paired t-test with SPSS 18.0. The selected usability principles are classified systematically using statistical method, the paired T-test and Principle Components Analysis (PCA) using Factor Analysis.

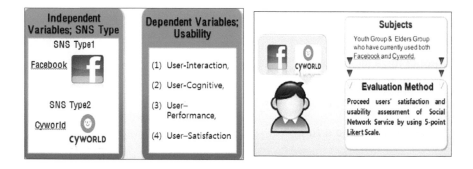

Figure 1 Design of survey protocol

4 RESULTS AND DISCUSSION

4.1 The Usage Results of Social Networking Service

The results of some general information about the survey respondents were as follows. As a questionnaires No. 3 ("How to connect SNS"), 162 people used Smart phones, 48 used Desktops, 30 used Laptop and 12 people used other methods to connect SNS (Figure 2), and 90 people answered to use SNS once a day, 60 said once in 2~3 days, 30 said once in a week, and 9 people used SNS more than twice in a day. As a questionnaires No. 6, 17 people said they use SNS at home, 11 people used at other places, 5 people used at school, and none of the respondents used SNS at work.

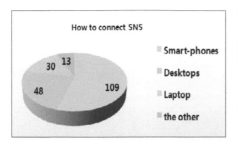
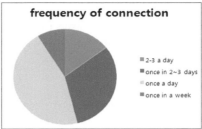

Figure 2 How to connect SNS (N: 203)

4.2   The Results of Assessment of User Satisfaction and Usability of Social Networking Service

To clarify the usability evaluation, usability metrics have been defined with factor analysis using SPSS 18.0. The selected usability principles are classified systematically using statistical method, Principle Components Analysis (PCA). Varimax rotated PCA was applied to the measured usability.

As the descriptive statistics of usability, all of the mean values of Youth group and Elders group are measured above 3.0 using 5-point Likert scales (Youth group; User-Interaction: 3.6485, User-Cognition: 4.3156, User-Performance: 3.7919, User–Satisfaction: 3.6463 and Elders group; User-Interaction: 3.2813, User-Cognition: 3.8111, User-Performance: 3.4310, User-Satisfaction: 3.4453). Figure 3~6 show the detail usability of SMGS1 and SMGS 2, and there are some significant differences in age for usability of SNS.

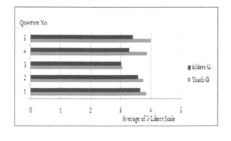
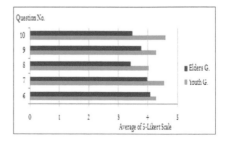

(A)   (B)

Figure 3 Mean values of User-Interaction (A), User- Cognition (B): Youth group (N=108), Elders group (N=95)

Users showed responses that overall understanding of how to use, menu, helps, and structures are important. From this result, it is clearly shown that SNS should be structured in easy way that users can understand and use SNS freely. It seems that users judged functional aspect of SNS is important. Especially, on function of the SNS should be interworking with various portable devices, and its restoring mistakes functions.

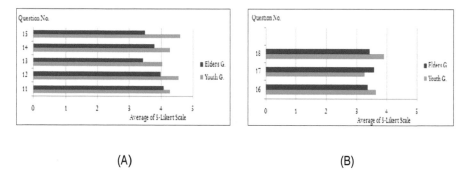

(A)   (B)

Figure 4 Mean values of User- Performance (A), User- Satisfaction (B): Youth group (N=108), Elders group (N=85)

Whereas satisfaction of using Facebook was appeared 3.63 out of 5, Cyworld's showed 3.36 which is a little lower than that of Facebook. 81.8% of respondents were satisfied with how to use Facebook; Cyworld, however, displayed 87.9% of users satisfied. Yet, Facebook was evaluated to have more user satisfaction than Cyworld, since Facebook had more 'satisfied' responses than 'normal' responses. For satisfaction on the visual aspect, Cyworld had average of 3.57 compare to 2.75 of Facebook; Cyworld had higher satisfaction on the visual aspect than Facebook. Also, in the percentage-wise, whereas 63.6% of users answered 'satisfied' with Facebook, 90.9% of Cyworld users answered 'satisfied' on the visual aspect. Facebook displayed average of 3.9 point in accessibility while Cyworld had average of 3.6; Facebook showed superiority in accessibility to Cyworld. For Facebook, 93.9% of users felt more than just satisfaction, and Cyworld, also 97% felt more than satisfaction, but Facebook had evenly distributed satisfaction on the scale.

The results of comparison with the usability between two groups showed in Table 2. There were significant differences between Youth and Elders groups in the comparison and analysis of the usability of SNS (p<.001***). So, users prefer SNS which has similar functions with other SNS, though first-time using. In terms of usability, Facebook was shown to average 4.12, and Cyworld, also, showed high mean of 3.8 points. For Facebook, 93.9% of users felt more than just satisfaction, and Cyworld, also 93.9% felt more than satisfaction.

Table 2 The analysis of user satisfation and usabilty evaluation of SNS between Youth and Elders groups (N=203)

	Differences of mean value between Youth(N=108) and Elders (N=95) groups	p-value
User-Interaction	-0.3672	<.001***
User-Cognition	-0.5045	<.014**
User-Performance	-0.3609	<.001***
User–Satisfaction	-0.2010	<.001***

1)* p<.05, ** p<.01 ,*** p<.001 (p-value for paired t-test))
2) Differences of mean value between Youth and Elders groups

5 CONCLUSIONS

Social Networking Service system has been used in various domains, and mobile internet users are being amplified recently. Because SNS system has changed along with the fast growing u-computing environment of private Social Networking Service system, it is necessary to enhance good understanding of usability assessment and users conscious of various SNS interface types

The purpose of this study was to assess user satisfaction and usability evaluation of Social Networking Service, and to develop and propose the heuristic evaluation method for the-Aged user Social Networking Service interface that is applied in well-pleasing way. We determined 18 usability principles among previous studies including Nielsen (1994)'s checklist terms of user interface main usability principles. The selected usability principles with five-score Likert scale are classified systematically using statistical method, T-test and Principle Components Analysis (PCA) using Factor Analysis. A total of 203(two group: Youth and Elders groups, female (N: 108), male (N: 95)), valid questionnaires were received, and the data were examined by using SPSS 18.0.

With the results of usability evaluation, 4 Factors of the checklist terms of user interface main usability principles were extracted. For User-Performance and User-Satisfaction, they showed that the users should be interested in using the SNS, and there should be no limitations on using SNS such as requirements of professional knowledge. Also, Facebook showed twice the users with 'Very Satisfied' compared to those of Cyworld. Overall result, Facebook seem to have better usability than Cyworld. Finally, we have researched <how to use, visual, applicability, and accessibility> of "Facebook" and "Cyworld". As a result, Facebook showed superiority on accessibility, applicability and the usage, whereas Cyworld has its strength on visual aspect.

The results of Principle Components Analysis (PCA) using Factor Analysis showed significant differences in two groups. We could suggest that the heuristic usability evaluation method for Social Networking Service interface developed in this research, can be used in well-pleasing way for any ages.

REFERENCES

E. M. Kim, D. H. Lee, Y. H. Lim, and K. I. Jeong, 2011, SNS myth of the revolution and the real, pp. 275-280, Seoul, Korea.: Nanam Press.

H. J. Yoo, 2010, Social Media PR Strategy Study on the Web using, *Journal of the E-Business Research Society*, Volume 11 No. 5, pp. 99-101.

ISO 9241-11, (1998), Ergonomic Requirements for Office Work with Visual Display Terminals (VDTs)-Part II: Guidance on usability, ISO, Geneva.

ISO 13207, (1999), Human-Centred Design Processes for Interactive Systems, ISO, Geneva.

K. B. Nam, 2010, Government guidelines on the organization of social media users, *Journal of the Korea Institute of Regional Informatization*, Volume 13 No. 3, pp. 41-63.

M. Fresier, S. Doota, 2010, E-revolution in social networks, Seoul, Korea.: Hanggan Press, pp. 20-22.

Nielsen, J., (1994b), Heuristic Evaluation, In, J. Nielsen and Mark, R. L. (Eds.), Usability Inspection Methods, pp. 25-62, New York: John Wiley and Sons.

S. M. Choi, 2009, Study on the Effects of SNS on the Social structure in the formation of social capital, *Journal of the Korean AcademicAassociation of Business Administration* 78: 13-22.

W. S. Jang, and Ji, Y. G., 2011, Usability Evaluation for Smartphone Augmented Reality Application User Interface, Thesis, University of Yonsei.

CHAPTER 53

Investigating Layout Preferences of Online Shop Buyers: A Case Study of a Purpose-built Software and Its Evolution

Marcin Kuliński, Katarzyna Jach

Institute of Organization and Management (I23)
Faculty of Computer Science and Management (W8)
Wrocław University of Technology
27 Wybrzeże Wyspiańskiego,
50-370 Wrocław, Poland

ABSTRACT

An evolution of software tools designed and built specifically for gathering and analyzing layout preferences data regarding e-commerce websites is presented in this chapter. The preferences were collected using an application called microSzu, whose initial functionality was to allow the participants to literally "draw" their preferred layouts of an imaginary web shop, as well as to visually present average preferences in the form of placement density tables. With the growing scale of the study being conducted, an ability to operate in multiple instances over the Internet and to write obtained data into a single remote database became a necessity. In turn, it allowed collecting a massive volume of preferences from student subjects (about 1000 participants in 4-year span). The sophisticated form of the user input data being two-dimensional, graphical representation of a preferable webpage layout resulted in the developing of a companion application, microVis, solely intended for data manipulation, filtering, analyzing and visualization. Not all the functionalities available at present were implemented from the beginning, because this software evolution process was driven by specific problems and needs identified gradually over time, for example a possibility to filter out inconsistent layout data or a means of measuring a statistical significance of preferred location differences between various user groups.

Keywords: web usability, e-commerce, programming, statistical analysis, HTML, JavaScript

1   INTRODUCTION

A usability study focused on online shops web page layout has been conducted between 2008 and 2012. During this period preferences related to the placement of 10 selected objects constituting an interface between user and that kind of a web page were collected from 961 student subjects with age ranging from 18 to 31 years and with different levels of online buying experience. Concurrently, some partial results from this research have been published (Kuliński and Jach, 2008, 2009, Kuliński et al., 2011) in a preliminary form. This chapter presents in details software tools designed and built specifically to facilitate the tasks involved in layout data gathering and analysis. The chronological perspective was used intentionally to emphasize the importance of problems and needs identified while the experiment continued for the process of improving and shaping up the software.

2   THE MICROSZU APPLICATION SOFTWARE

The idea of the microSzu application was inspired by numerous web usability studies focused on users' layout preferences. One such research in the context of a web page typical elements placement, namely a web page title, internal and external link groups, a homepage link, advertisement banners, and internal search engine was done by Bernard (2001a, 2001b), who used a board organized into a visible grid of numerous fields and a set of paper cards representing the elements under investigation, which were positioned within the grid by the subjects according to their expectations. The same method was used in Bernard's another study (Bernard, 2003), but the objects examined this time were related to the e-commerce web sites. A smilar research was conducted by Markum and Hall (2003), who used a 3 by 3 table and additionally performed some statistical analysis of the significance of differences in placement preferences gathered this way. Our conclusion after studying these publications was that using an appropriate software in an analogous study would give a possibility to carry it out at a larger scale and would facilitate the processes of data gathering and analyzing in a considerable way.

It was decided that for the sake of flexibility and ease of use from the researcher's perspective, the program should be implemented in the form of a web application, that is, being operational entirely from within a web browser. Static HTML and CSS (Cascading Style Sheets) code became the base of it, while all the computations and necessary GUI manipulations were carried out by JavaScript functions and event handlers with the help of DOM (Document Object Model) Level 2 (http://www.w3.org/DOM/) interface where applicable. The employed graphical user interface framework was fairly straightforward: the cards container situated on the left, the table grid where the cards can be placed at occupying the center of the window, and the controls in the form of buttons, pull-down lists, and textboxes at the bottom (see Figure 1). The size of all visible interface elements automatically adapted to the browser's window size, which helped in maintaining the program's portability.

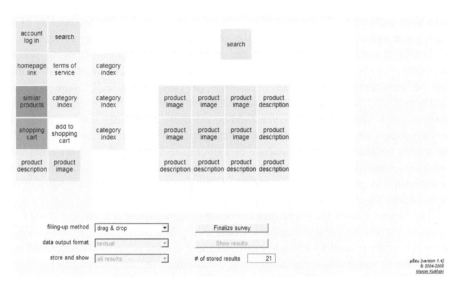

Figure 1 The main interface of the microSzu application showing an example of layout preferences filling-up process

The main principle of operation is to give an easy and convenient way of placing the cards selected from the container at a desired location on the table grid, thus allowing an examinee to construct his/her preferred layout in a short time. Two different modes of mouse manipulation have been implemented, called "drag & drop" and "pick & paint", respectively. In the first one a card can be selected from the container, literally moved over the table and then dropped at desired location, so it emulates physical activities linked with the real cards arrangement process, as used by Bernard (see above) and his followers (Shaikh and Lenz, 2006). The second one mimics the software paint applications' principle of work to some extent: a card picked up from the container with a mouse click stays surrounded by a rectangular frame and becomes a default "painting" tool, so user can quickly fill up a portion of the table with specified card simply by multiple mouse clicks. This mode speeds up the process of completing the task, but can be less intuitive to some individuals due to the lack of aforementioned analogy to the real world, as our observations suggest.

The presence of the controls is highly configurable, allowing to expose or hide the elements responsible for choosing the filling-up method, data output format, and so forth, so the GUI can be easily stripped down to the single button that begins and finishes a session for individual examinee, if necessary. Additionally, there is a possibility of presenting a short instruction manual at the right side of the window, which will guide users during an experiment.

Another optional component may be used for the purpose of collecting some additional information about the subjects, such as their age, gender, skill or experience specific to a study being conducted, or any other relevant data. It takes a form of a questionnaire displayed before the main microSzu interface and can by configured in a way that the primary experiment begins only if particular questions have been answered. That way it ensures that a certain level of understanding of the study's aim and/or its technical principles is achieved prior to actually entering it.

Displaying results collected from the subjects has been possible using either a graphical or textual form. As for the graphical one, a set of placement density tables can be generated, showing the percentage of every table cell usage for each card/object and across all participants. Moreover, table cells are displayed using scale of grays: the darker any given field is, the more frequently it was chosen as a preferred location for an object under investigation. The textual mode was suitable for storing obtained data in a database or a spreadsheet in case the experiments were conducted concurrently on several computers and if some sort of results aggregation or analysis was planned, as these specific tasks could not be done internally at that point of time (Kuliński and Michalski, 2005).

The format chosen for outputting collected information was a variant of the CSV (Comma Separated Values) standard, where every line of a text represents a single entry (i.e., data related to a single examinee) and consecutive fields separated by a semicolon, or some other special character, contain all the data related to that entry. Because there is no valid method of accessing the local filesystem for storage purposes from a JavaScript program, the only way to save the results was to manually select and copy them into the Notepad or similar text editor, and then store a resulting file. An example of the textual output generated by microSzu is shown in Figure 2. Although with the advent of the complementary microVis application this particular functionality became obsolete, the CSV format itself is still used for transferring data from one to the other application.

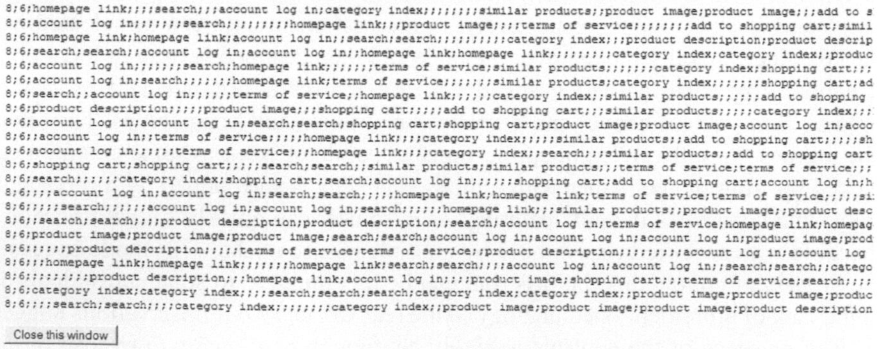

```
8;6;homepage link;;;;search;;;account log in;category index;;;;;;;;similar products;;product image;product image;;;add to s
8;6;account log in;;;;;;;search;;;;;;;;;homepage link;;;product image;;;;terms of service;;;;;;;;add to shopping cart;simil
8;6;homepage link;homepage link;account log in;;search;search;;;;;;;;;;category index;;;product description;product descrip
8;6;search;search;;account log in;account log in;;homepage link;homepage link;;;;;;;;;category index;category index;;produc
8;6;account log in;;;;;;;search;homepage link;;;;;;;terms of service;similar products;;;;;;;category index;shopping cart;;;
8;6;account log in;search;;;;;;;homepage link;terms of service;;;;;;;similar products;category index;;;;;;;shopping cart;ad
8;6;search;;account log in;;;;;;terms of service;;homepage link;;;;;;category index;;similar products;;;;;;add to shopping
8;6;product description;;;;;product image;;;shopping cart;;;;;add to shopping cart;;;similar products;;;;;category index;;;
8;6;account log in;account log in;search;search;shopping cart;shopping cart;product image;product image;account log in;acco
8;6;;account log in;;terms of service;;;;;homepage link;;;;category index;;;;;similar products;;add to shopping cart;;;;;sh
8;6;account log in;;;;;;terms of service;;homepage link;;;;category index;;search;;;similar products;;add to shopping cart
8;6;shopping cart;shopping cart;;;;;search;search;;similar products;similar products;;;terms of service;terms of service;;;
8;6;search;;;;;;category index;shopping cart;search;account log in;;;;;;shopping cart;add to shopping cart;account log in;h
8;6;;;;account log in;account log in;search;search;;;;;homepage link;homepage link;terms of service;terms of service;;;;;si
8;6;;;;;search;;;;;;account log in;account log in;search;;;;;;homepage link;;;similar products;;product image;;product desc
8;6;;search;search;;;;product description;product description;;search;account log in;terms of service;homepage link;homepag
8;6;product image;product image;product image;search;search;account log in;account log in;account log in;product image;prod
8;6;;;;;;product description;;;;;terms of service;terms of service;;product description;;;;;;;;;account log in;account log
8;6;;;homepage link;homepage link;;;;;;;homepage link;search;search;;;;account log in;account log in;;search;search;;catego
8;6;;;;;;;;;product description;;;homepage link;account log in;;;;product image;shopping cart;;;terms of service;search;;;;
8;6;category index;category index;;;;search;search;search;category index;category index;;product image;product image;produc
8;6;;;;search;search;;;;category index;;;;;;;;category index;;product image;product image;product image;product description
```


Figure 2 A sample of the textual output using CSV notation produced by microSzu

At a later stage of development the capability to automatically store the data on a remote web server via the Internet was added. It was done with the help of the PHP (PHP: Hypertext Preprocessor) server-side scripting language, which allowed us to interface with server's filesystem from the inside of a web browser running microSzu. As a side effect, it became possible to conduct an experiment with multiple computers and instances of the program operating concurrently, while interacting with a single database file. This particular feature turned out to be the key to efficiently perform a large scale research, as the example of our study proves.

3   THE MICROVIS APPLICATION SOFTWARE

The first version of the microVis program was conceived in 2008, just after the beginning of a massive web usability study involving students from Wrocław University of Technology and the Academy of Art and Design in Wrocław, Poland. The only functionality implemented at that time was opening and importing a single database file stored on a server with the use of the XMLHttpRequest() method, and then to visualize averaged users' layout preferences in a form of placement density tables (see Figure 3) – a manner that the original microSzu also used. The results could be then saved as a HTML file. Similar to microSzu, the application was running inside a web browser, negating right away the need of any form of software installation and allowing to invoke it across the Internet, as well as from a local filesystem. The entire user interface has been built around dynamically generated HTML content using JavaScript, CSS, and DOM Level 2.

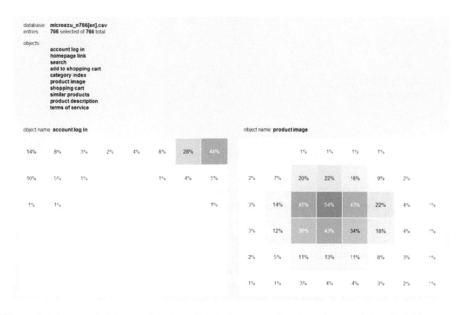

Figure 3 A fragment of the graphical results window presenting two placement density tables

508

Soon it became clear that some sort of data filtering and partitioning tool is needed, allowing for a visual comparison of placement density tables obtained from subjects at different age, with particular gender, online experience level, etc. The employed data subsets generation method is based on specific auxiliary information stored along with layout preferences, which can be collected from every subject prior to the main survey. Any given filter takes a special field (e.g., subject's age or gender) and all its corresponding values, from which the interesting ones can be selected – it may be considered as an logical OR operator between values chosen, or an union of the sets that every of them constitutes. If multiple filters are used, a data subset is collected using one or more logical AND operators – that is, it becomes an intersection of the sets generated from individual filters chosen. This method gives a great flexibility in selecting data for visualization and statistical analysis purposes. An example of selection rules usage is presented in Figure 4.

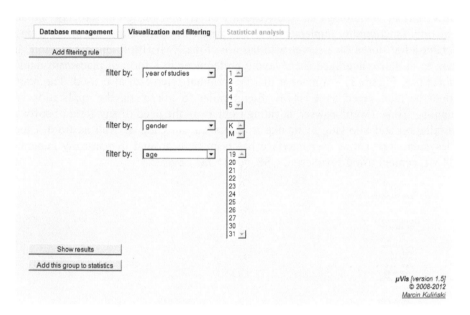

Figure 4 An example of multiple filtering rules used for results partitioning in the microVis program

Until mid-2009 our usability study collected entries from over 500 participants, thus allowing for some type of statistical analysis to be employed. Due to the nature of recorded preferences data only a non-parametric statistical test could be considered. The Pearson's chi-square test of independence formula was adopted to accommodate relatively large data structures (e.g., a graphical table of preferences composed of 8 columns and 6 rows, which was used in our study, results in both contingency tables needed for the calculations that can have as many as 2*48 cells each). It operates on data subsets generated by the filtering tool and allows for a statistical comparison of exactly two selected groups at once.

One of the formal requirements of the test is that all of the observed contingency table cells need to be populated adequately, that is, every cell should contain at least 5 observations assigned to it (Kirkman, 1996, Hill and Lewicki, 2007). To fulfill this requirement an iterative algorithm for automatic resizing of the observed contingency table and repopulating its cells by data rebinning was implemented. Generating the expected contingency table and comparing it with the observed one was the final step in obtaining the chi-square test value and the probability of significant similarity between groups tested. The only real obstacle here was to determine an exact p-value without the need of referring to a chi-square distribution table and without any sort of probability values interpolation. This one was accomplished with the help of the work done by Pezzullo, who is the author of an online scientific calculator written in JavaScript with built-in statistical distribution functions (Pezzullo, 2010). Its freely available source code was analyzed and functions related to the chi-square distribution were incorporated into microVis. An example of computational results from above-described software module using latest up to date data from the study is shown in Figure 5.

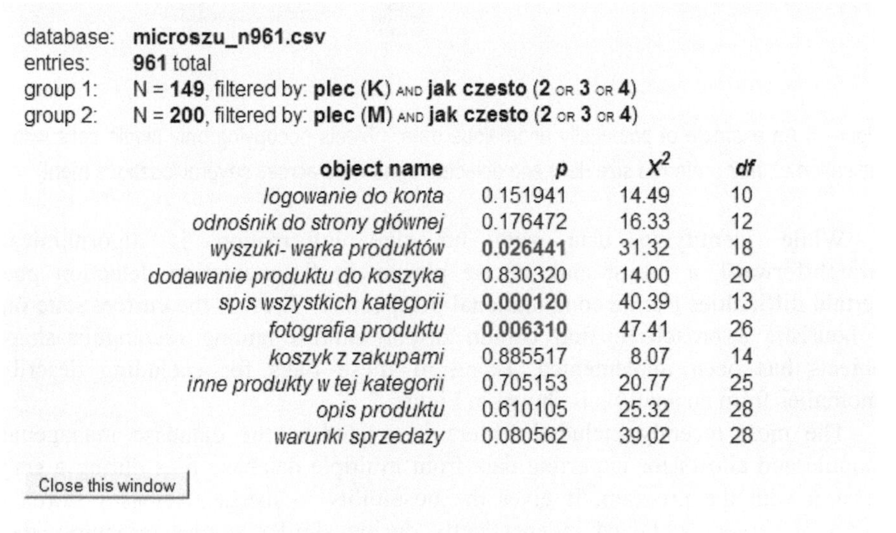

database: **microszu_n961.csv**
entries: **961** total
group 1: N = **149**, filtered by: **plec (K)** AND **jak czesto (2** OR **3** OR **4)**
group 2: N = **200**, filtered by: **plec (M)** AND **jak czesto (2** OR **3** OR **4)**

object name	p	x^2	df
logowanie do konta	0.151941	14.49	10
odnośnik do strony głównej	0.176472	16.33	12
wyszuki- warka produktów	**0.026441**	31.32	18
dodawanie produktu do koszyka	0.830320	14.00	20
spis wszystkich kategorii	**0.000120**	40.39	13
fotografia produktu	**0.006310**	47.41	26
koszyk z zakupami	0.885517	8.07	14
inne produkty w tej kategorii	0.705153	20.77	25
opis produktu	0.610105	25.32	28
warunki sprzedaży	0.080562	39.02	28

Close this window

Figure 5 The statistical results window showing a comparison of layout preferences collected from experienced male and female subjects. Statistically significant differences in objects' placement are marked bold and with red color

With the database constantly growing, the number of possible outliers, or some sort of anomalous entries, would also increase over time. In the context of the study described here, such anomalies may take a form of either a layout preferences where a subject positioned all of the interface objects using single cells only (i.e., sharing with us the information about objects' placement but not about their size) or an inconsistent one – with the very same object placed at two or more different locations, resulting in placement fragmentation – as depicted in Figure 6. However,

510

it is important to note that within other possible areas of interest, for example in a preferences study related to plant and facility layout, similar data arrangements could be considered perfectly normal and desirable. In order to maintain the software flexibility this kind of anomalous data needed to be identified and marked out, but not automatically excluded. To achieve this goal two special binary flags denoting both potentially abnormal conditions were incorporated into the existing filter system, thus leaving the freedom of choice to the researcher, who may or may not include marked entries into his/her analysis.

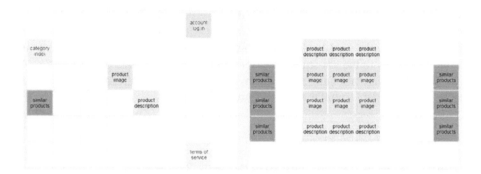

Figure 6 An example of potentially anomalous data: objects occupying only single cells with no indication of their preferred size (left) and objects fragmented across several locations (right)

While identifying data with no size information is algorithmically straightforward, a robust and precise placement fragmentation detection poses certain difficulties from a computational perspective. Thus, at the current state only a heuristic approach to find certain discontinuities among rectangular-shaped objects has been implemented. Using of these flags for excluding described anomalies from an analysis is shown in Figure 7.

The most recently included feature is related to the database management module and allows for importing data from multiple database files during a single session with the program. It gives the possibility to merge databases stored on different servers, obtained independently during similar studies or simply those saved manually during usage of microSzu in the offline mode. As a consequence, potential layout differences between data sets from various sources (e.g., preferred placements versus actual layouts) can be statistically tested for its significance. Again, a flexible way of applying this component was to introduce a special field which stores an unique ID number assigned to a particular database during its importing, that could be used within the scope of the filtering module. Most important properties of loaded databases are successively displayed within the database management panel, as an example in Figure 8 demonstrates.

It is important that some basic compatibility exists between data sets imported from different files, at least with regard to the list of objects used during a survey. As for the dimensions of tables used for placing those objects, if different sizes are detected during loading, an attempt to upscale the smaller ones to fit the largest table is being made, but it works only if the larger width and height are dividable by the smaller ones without a remainder (e.g., 4 by 3 and 8 by 6 tables are compatible, but 8 by 6 and 12 by 9 are not, thus can't be analyzed together).

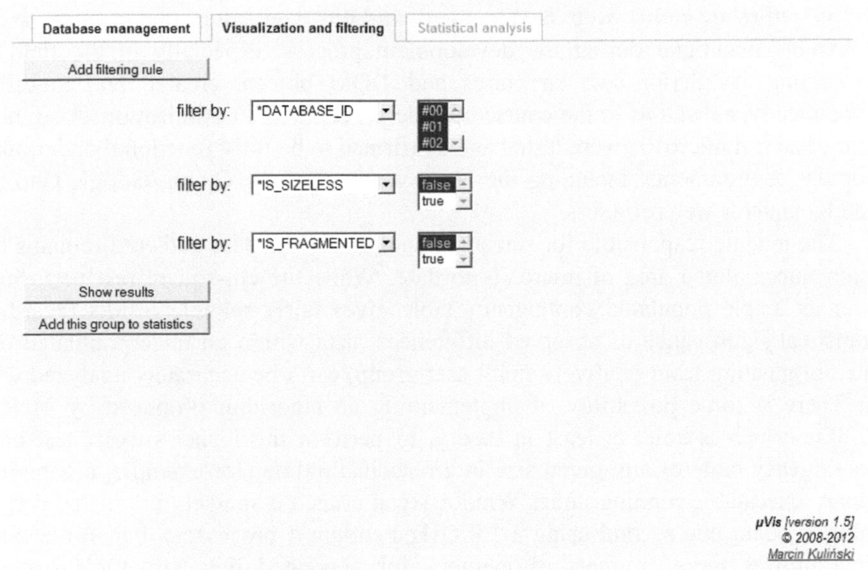

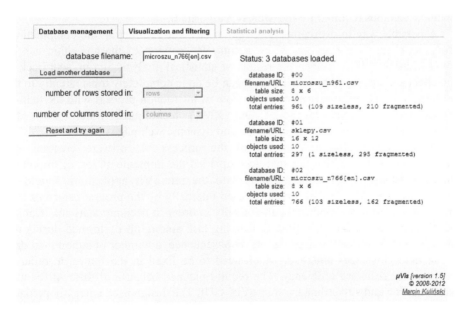

Figure 7 Filtering out all the inconsistent data from multiple database files imported into microVis

Figure 8 Loading multiple databases into microVis

4 CONCLUSIONS

The application software described in this chapter was created using the Firefox (http://www.mozilla-europe.org/pl/firefox/) web browser and the Firebug (http://getfirebug.com/) web development add-on. The latter one proved to be extremely beneficial during the development process, especially in the field of examining JavaScript data structures and DOM objects created and modified dynamically, as well as in the course of code profiling and optimization. Also, both microSzu and microVis were tested and confirmed to be fully functional under other popular environments, including the latest versions of the Opera, Google Chrome and Konqueror web browsers.

The module responsible for statistical analysis of the data gathered remains the main uncompleted area of microVis to date. While the chi-square test performed over an ample populated contingency table gives fairly reliable results regarding statistical significance of observed differences, data within an underpopulated one (i.e., originating from relatively small user group) can't be accurately analyzed with it. There is some possibility of implementing an algorithm proposed by McKay (2002), which is able, at least in theory, to perform the Fisher's exact test on a contingency table of any given size in a reasonable time (for example, a compiled binary executable running under Windows can crunch a sparsely populated 4 by 3 table in about one second using a 1.8 GHz Pentium 4 processor), but in the same time utilizes large amounts of memory for generated data structures storage. Consequently, there are concerns related to the nature of JavaScript itself, which currently remains the core of our software: being an interpreted language executed within a web browser may be sufficiently limiting factor in the context of speed and memory volumes required for the McKay's algorithm.

A possible way to overcome these limitations is to redirect all the computations from the client to the server, where a compiled version of the algorithm can be invoked using PHP, and then the results of statistical analysis can be sent back to the client and displayed by a browser. It can be done completely transparently to the user and without any impairment of the flow of interaction process with the help of AJAX (Asynchronous JavaScript and XML) techniques, which allow for background client-server communication and dynamic web page updates.

As for the evolutionary nature of the process of software creation and modification, some difficulties that accompanied the implementation of importing multiple database files simultaneously into the microVis application should be mentioned here. Certain data structures used internally by the program since its first versions needed to be modified quite heavily in order to accommodate this feature, which turned to be the most time consuming task among all performed during the last (to date) functionality upgrade. As a consequence, a number of suboptimal data organization instances emerged, which need to be fixed in the future, in order to maintain the software efficiency. The rebuild planned because of this will be also used to refine and streamline the microVis' GUI. The design used currently predates specific features implemented later, thus the interaction has become somewhat less intuitive and consistent, compared with previous versions of the program.

REFERENCES

Bernard, M. L. 2001. User expectations for the location of Web objects. *Proceedings of the CHI '01 Conference*, 171-172.

Bernard, M. L. 2001. Developing schemas for the location of common Web objects. *Proceedings of the Human Factors and Ergonomics Society 45th Annual Meeting*, 1161-1165.

Bernard, M. L. 2003. Examining user expectations for the location of common e-commerce web objects. *Proceedings of the Human Factors and Ergonomics Society 47th Annual Meeting*, 1356-1360.

Hill, T. and P. Lewicki. 2007. STATISTICS Methods and Applications. StatSoft, Tulsa, OK.

Kirkman, T. W. 1996. Statistics to Use. http://www.physics.csbsju.edu/stats/

Kuliński, M. and K. Jach. 2008. E-Commerce Websites Layout: Users' Preferences and Actual Location of Common Web Objects. In. Information Systems Architecture and Technology. Web Information Systems: Models, Concepts & Challenges, eds. L. Borzemski et al. Library of Informatics of University Level Schools, Wroclaw University of Technology.

Kuliński, M. and K. Jach. 2009. microVis: an application software for visualization and statistical analysis of layout preferences data. In. User Interface in Contemporary Ergonomics, ed. K. Hankiewicz. Publishing House of Poznan University of Technology.

Kuliński, M., K. Jach, R. Michalski, and J. Grobelny. 2011. Modeling online buyers' preferences related to webpages layout: methodology and preliminary results. In. Advances in Applied Digital Human Modeling, ed. V. G. Duffy. CRC Press, Taylor & Francis Group, Boca Raton, FL.

Kuliński, M. and R. Michalski. 2005. microSzu – a computer program for screen layout preferences analysis. In. Ergonomics and work safety in information community. Education and researches, eds. L. M. Pacholski, J. S. Marcinkowski, and W. M. Horst. Institute of Management Engineering, Poznan University of Technology.

Markum, J. and R. H. Hall. 2003. E-Commerce Web Objects: Importance and Expected Placement. Laboratory for Information Technology Evaluation Technical Report, Missouri University of Science and Technology. http://lite.mst.edu/documents/LITE-2003-02.pdf

McKay, I. C. 2002. An algorithm to apply Fisher's exact test to any two-dimensional contingency table. http://www.discourses.org.uk/statistics/fisher2.htm

Pezzullo, J. C. 2010. Web Pages that Perform Statistical Calculations. http://statpages.org/index.html

Shaikh, A. D. and K. Lenz. 2006. Where's the Search? Re-examining User Expectations of Web Objects. Usability News, 8(1), http://psychology.wichita.edu/surl/usabilitynews/81/webobjects.asp

Section IV

Miscellaneous

Aesthetic and Symbolic Aspects versus Usability: Evaluation of Daily Use Product - Lemon Squeezer

*Jamille Noretza de Lima Lanutti 1 Lívia Flávia de Albuquerque Campos 2
Douglas Daniel Pereira 3 Liara Mucio de Mattos 4 Elen Sayuri Inokuti 5
Luis Carlos Paschoarelli 6*

Unesp – Univ Estadual Paulista
Bauru, Brazil
jamille_lanutti@hotmail.com

ABSTRACT

Usability has been characterized as a requirement in the development of product design. Besides the physical aspects, the subjects' perception and their relationship with objects of daily use have been the focus of research in usability. In this sense, the rejection of a product by the user is not necessarily related to being complex or inefficient, but may be associated with factors and aesthetic significance. The objective of this study was to do a perceptual evaluation with objects of daily use, seeking to identify the relationship between product usage factors and factors related to the aesthetic and symbolic values that can be assigned to it. The procedures were based on ethical recommendations. Twelve subjects participated (average age 20,62, s.d. 1,66), that evaluated six different models of lemon manual through Semantic Differential Test composed of ten descriptors related to subjective aspects of the product (modern / classic, beautiful / ugly, common / unusual, essential / dispensable, elegant / inelegant, repulsive / attractive, funny / serious, known / unknown, rational / emotional, humble / sophisticated). The results were analyzed statistically using the nonparametric Wilcoxon test. This analysis demonstrated higher sensitivity ($p \leq$ 0,05) to the descriptors related to subjective aspects of the product (51,33% of cases) when compared to the descriptors of product use (20,66% of cases). However, significant differences ($p \leq 0.05$) with

518

these product use descriptors indicate that the shape and morphological structure of the products influence their perceived usability. As an example, two presses with completely opposite morphological characteristics showed significant results (p ≤ 0,05) opposites in all analyzed using descriptors. Thus, the results demonstrate that the Semantic Differential Test for descriptors related to the use of the product allows a significant perceptual assessment and other tests may contribute to the usability of product / system.

Keywords: Ergonomic Design, usability, perception, Semantic Differential, juicers.

1 INTRODUCTION

The development of product projects has evolved expressively in last years, with special interest in marketing requirements, where the users' needs and expectations stand out. In this sense, usability has become a fundamental requirement so technological products that satisfy consumers and users can be produced and commercialized.

Product usability, far beyond the efficiency and effectiveness (Tullis and Albert, 2008), should also consider the users perception. The relation between them and the everyday objects has become the focus of many researches in that field (Hassenzahl, 2004, Mahlke, Minge and Thüring, 2006, Ben-Bassat, Meyer and Tractinsky, 2006, Mahlke and Lindgaard, 2007, Mahlke and Thüring, 2007, Rafaeli and Yavetz, 2004).

In general, it is noted that users, already utilizing functional and easy-to-use products, tend to look for other attributes, such as emotional benefits (Buccini and Padovani, 2005).Those attributes are related to perception of the subjects, which establish links between people and objects, reinforcing the relation between the human being and the world (Barbosa, Menezes, Paschoarelli, and Alencar, 2008).

According to Hekkert (2006), the user experience is made up of effects that are caused by the interaction between a user and a product, including the degree to which all the senses are gratified (aesthetic experience), the meanings are assigned (symbolic experience) and the feelings and emotions are elicited (emotional experience). It is argued that "(...) the rejection of a product by the user is not necessarily linked to the fact of being complex or inefficient" (Medeiros and Okimoto, 2004), which allows to hypothesize that there are much more variables in the relationship between an object and a user, and the factors that characterize its usefulness. However, that statement seems incipient, since methods that aim to relate aesthetic and symbolic factors to the use of an object are still rare, despite being important for product design.

A tool that may be applied to analyze aesthetic and symbolic aspects of a product is the Semantic Differential (SD), which consists of the quantifying of user's perception through a Likert Scale (Tullis and Albert, 2008). The SD has been already applied with the objective of analyzing the semantic variation of products perception, though it isn't always related to the use of them. Therefore, it should be noted that such tool is related to one of the usability metrics reported by Tullis and Albert (2008), denominated as "auto-reported", that is, methods that seek to comprehend the product usability through questions realized directly to the user after use, obtaining subjective data.

This study had the objective of performing a perceptive assessment through the SD, during real activity, with everyday objects – lemon squeezer –, having the final purpose of identifying the validity of SD in face of the factors related to aesthetic and symbolic values, from the effective use of the product.

2 MATERIALS AND METHODS

The procedures were based on ethical recommendations (CNS, 1996; ABERGO, 2003). All subjects signed an informed consent statement previously approved by the local ethics committee (Of. 001/11 Feb 24, 2011/CEP-USC).

Twelve young adult subjects participated in the study, with average age of 20, 62 years old (s.d. 1,66), where 07 were female and 05 were male. All the participants were Design college students.

Six different manual (not electric) lemon squeezers were utilized (Figure 01). For the task, 1" (25,4mm) average diameter "limes" [*Citrusaurantifolia (Christm.) Swingle*] were used, previously cut in a uniform manner, besides the identification and evaluation protocols.

Figure 01 Citric fruits squeezers used in the study.

The semantic space for the SD protocol application was constituted by 10 pairs of bipolar descriptors, from adjectives described in manufacturers and e-commerce websites (Table 1). The distribution of adjectives in SD scale was random and the disposition of words, of which interpretation was ascending or descending, was alternated avoiding the conditioning of subjects to general affirmative or negative answers.

The SD scale utilized was composed by seven anchors arranged horizontally between criteria (adjectives), in a way that, the closer in position, the more concordance there would be in meaning with the presented criterion.

The fruit squeezers were arranged on a table, in a way the user had free access to each of them (Figure 02). It happened because the forms containing the pairs of descriptors and the letter of each corresponding juicer were randomized, so the users performed the test in different sequences.

520

Table 1 Words used in the DS scales.

Translated words	Original words used (Portuguese)
Modern / Classic	Moderno / Clássico
Beautiful / Ugly	Bonito / Feio
Common / Unusual	Comum / Inusitado
Essential / Dispensable	Essencial / Dispensável
Elegant / Inelegant	Elegante / Deselegante
Repulsive / Attractive	Repulsivo / Atrativo
Funny / Serious	Divertido / Sério
Known / Unknown	Conhecido / Desconhecido
Rational / Emotional	Racional / Emocional
Humble / Sophisticated	Humilde / Sofisticado

The subjects were oriented about the procedures. The task was to squeeze the lime juice, with the availability of more limes, in case of they would consider necessary to repeat the activity to certify the answer (Figure 02). After use, each participant was oriented to answer individually the SD test. The subjects performed the activity with all the six lemon squeezers (Figure 03). For data analysis, each anchor of SD protocol was evaluated from 1 to 7. The signed values were tabled and analyzed descriptively for the purpose of obtaining averages and standard deviation of each lemon squeezer. *Wilcoxon* test was applied, considering $p0,05$, to verify the statistic differences between them.

Figure 02 – The squeezers arrangement on the table.

Figure 03 – Interaction of participants with the products.

3. RESULTS

The results indicated that all the squeezers presented different averages for each analyzed criterion. However, when compared and submitted to *Wilcoxon* test, diverse and significant differences were found.

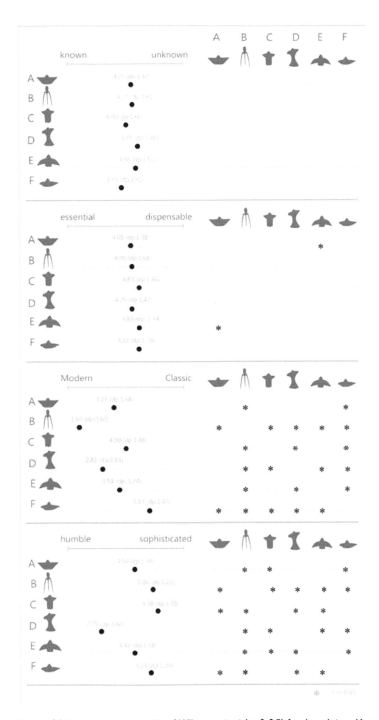

Figure 04 Averages and results of Wilcoxon test (p≤0,05) for descriptors Known/Unknown, Essential/Dispensable, Modern/Classic, Humble/Sophisticated.

522

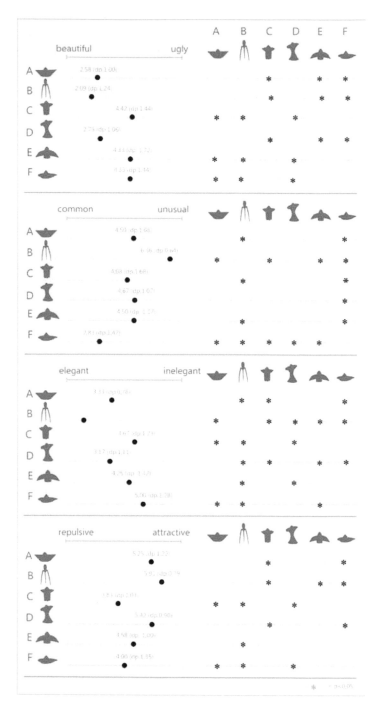

Figure 05 Averages and results of Wilcoxon test (p≤0.05) for descriptors Beautiful/Ugly, Common/Unusual, Elegant/Inelegant, Repulsive/ Attractive.

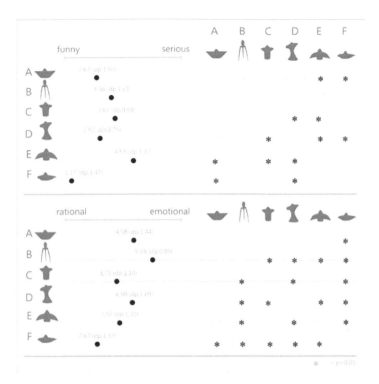

Figure 06 Averages and results of Wilcoxon test (p≤0.05) for descriptors Funny/Serious, Rational/Emotional.

The pair of descriptors Known/Unknown didn't present significant difference between any of the squeezers (Figure 04), though the most trivial one (F) had reached the average value nearer to the "known"; and one of those which presented innovative shape (D) had reached the average value nearer to the "unknown". It may be explained by the familiarity of the subject with the objects.

In a similar way, the pair of descriptors Essential/Dispensable presented significant difference only between lemon squeezers A and E (Figure 04). That result may be comprehended, once such descriptor standardizes the perception and the need for product use without the influence of subjective aspects on the perception.

The descriptor Modern/Classic presented significant difference between the most of the lemon squeezers (Figure 4), with B and F standing out, differentiated from the other five. B was characterized as the most modern (average 1,64) and F (average 5,17) as the most classic.

For pair of descriptors Humble/Sophisticated significant differences were detected too (Figure 4). The squeezer B also stands out, which was different form the other five. Besides the squeezers C,D, E and F, that were different from the other 4 squeezers. It may be affirmed that B, D and E are characterized as sophisticated – the greatest averages – and C and F are characterized as humble, with the smallest averages.

In the Beautiful/Ugly descriptor, all lemon squeezers were significantly different from the other three (Figure 05). A,B and D were characterized as beautiful – the smallest averages – and C, E and F as ugly – the greatest averages.

Common/Unusual is the pair of descriptor in which B and F are the most distant, both differentiating from the other five and one each other (Figure 05). F was considered common (average 2,83) and B unusual (average 6,36). The other squeezers had average values very approximate.

In the pair of descriptors Elegant/Inelegant, B stands out again, being the most elegant when significantly differentiated from the others (Figure 05). A and D may be classified as elegant too – smaller averages – and C, E and F as inelegant – the greatest averages.

The pair of descriptors Repulsive/Attractive presented relevant difference in B in relation to C,E and F; and C and F when related to A, B and D (Figure 05). C, E and F were considered as repulsive – the lowest averages – while A, B and D were considered as attractive.

In the results of descriptors pair Funny/Serious, the squeezers D and E were demonstrated as significantly different from the other three ones (Figure 06). A, D and F were perceived as funny – the lowest averages – and E considered as serious – the greatest average.

For pair of descriptors Rational/Emotional, F stands out, which differentiated significantly from the other five squeezers (average 2,67) as "rational" (Figure 06). B and D also stands out as "emotional", according to the subjects perception.

4. DISCUSSIONS AND CONCLUSIONS

The application of SD test during the effective use of product seems to be related to aesthetic and symbolic values of products. The presented study results demonstrate that the SD test allowed the capturing of subjects perception in a significant way in great part of the evaluated descriptors.

It was also observed that there was a strong antagonism between squeezers B (modern, sophisticated, beautiful, unusual, elegant, rational) and F (classic, humble, ugly, common, inelegant, repulsive, rational).

It's important to highlight that the squeezer F is very simple and common in shape and material with which it's produced. While squeezer B, commercially called Juicer Salif, designed by Philipe Starck in 1990 and recognized as a design piece, isn't commonly utilized by most people. The Juicer Salif also presents a peculiar shape, which justify the fact of being considered unusual. Silva and Nojima investigated the intuition of a product use, through its form and perceived that while 6 people recognized the product function, 16 didn't.

Also in this context, Russo and Moraes (2003), after conducting interviews with users, they emphasized the symbolic and aesthetic characteristics of the object – such as sophistication, elegance and beauty – to the detriment of the functional ones. Moreover, most of the subjects interviewed had the object exclusively as decorative piece.

The antagonism between B and F is characteristic of two lemon squeezers groups: one formed by B, D and also by A in some descriptors (perceived as modern, sophisticated, beautiful, elegant, attractive and emotional), and the other, formed by C, F and, in some descriptors, by E (perceived as classic, humble, ugly, inelegant, repulsive and rational).

In general, the SD test seemed sensitive to the evaluations in this study and can contribute expressively in analysis of usability, during the product design development.

Moreover, we can affirm that the SD test is demonstrated as a valid method in perceptive assessment of everyday objects with different shapes, when applied after the use of them. Thus, it is characterized as an expressive tool for evaluation of aesthetic and symbolic values contained in the objects, which may contribute as a method of product usability assessment.

ACKNOWLEDGMENTS

This study was supported by CNPq (Proc. 303138/2010-6) e FAPESP (Proc. 2011/04208-0).

REFERENCES

Abergo - *Code of Deontology of Certified Ergonomists*. Norm ERG BR 1002. Accessed December 20, 2009, http://www.abergo.org.br/arquivos/norma_ergbr_1002_deontologia /pdfdeontologia.pdf [In portuguese].

Barbosa, R.T. M.S. Menezes, L.C. Paschoarelli, and F. Alencar. "The logic design: think, create and feel." Paper presented at P&D 8th Brazilian Conference on Design Research and Development (UAM), São Paulo, 2008 [In portuguese].

Ben-Bassat, T., J. Meyer, and N. Tractinsky 2006. Economic and Subjective Measures of the Perceived Value of Aesthetics and Usability. *ACM Transactionson Computer-HumanInteraction*, 13(2), 210–234.

Buccini, M.; and S. Padivani. Methods for measuring emotions in design. Paper presented at 5th International Congress of Ergonomics and Usability of Interfaces, Rio de Janeiro, 2005 [In portuguese].

Espe, H., 1992. Symbolic qualities of watches. In: Susann Vihma(Eds.), Object and Images } Studies in Design and Advertising.University of Industrial Arts Helsinki UIAH, 124}131.

Hekkert, P. Design aesthetics: principles of pleasure in design. Psychology Science, 48, 2006. 157 – 172.

Hassenzahl, M. 2004. The interplay of beauty, goodness and usability. *Human-Computer Interaction*, 19(4), 319-349.

Hsu, S.H., M.C. Chuang, and C.C. Chang. 2000. A semantic differential study of designers' and users' product form perception. Int. J. Ind. Ergon. 25 (4), 375e391.

Mahlke, S., M. Minge, and M. Thüring, 2006.Measuring Multiple Components of Emotions in Interactive Contexts.*Proceedings of the CHI 2006*.Montréal, Québec, Canada.

Mahlke, S. and Lindgaard, G. 2007. Emotional Experiences and Quality Perceptions of Interactive Products. *Human-Computer Interaction HCII*, 164–173.

Mahlhe, S. and M. Thüring, 2007.Studying Antecedents of Emotional Experiences in Interactive Contexts.'*Proceedings of the CHI 2007: Emotion & Empathy,*' San Jose, CA, USA.

Medeiros, C. R. P. X., and M. L. L. R Okimoto. Perception of Aspects of Usability and Design in Product Acquisition: A Case Study in the Transport Sector.Paper presented at 6th Brazilian Conference on Design Research and Development, New York, 2004 [In portuguese].

Rafaeli, A. and V. Yavetz. 2004. Instrumentality, aesthetics and symbolism of physical artifacts as triggers of emotion.*Theoretical Issues in Ergonomics Science*, 5(1), 91–112.

Russo, B., and A. Moraes. The Lack of Usability in Design Icons. Paper presented at Designing Pleasurable Products, Pittsburgh. Proceedingof DPPI'03 - Pittsburgh - Pennsylvannia - USA, 2003.

Silva, C. M.A. and M.L.O. Okimoto, Considering the use of intuition in products: the case juicySalif. Paper presented at 11th Ergodesign, jointly with 11th USHIC (UFAM) Manaus, 2011 [In portuguese].

Tullis, T., W. Albert. 2008 *Measuring the User Experience: Collecting, Analysing, and Presenting Usability Metrics*. Burlington: Morgan Kaufmann.

CHAPTER 55

Aesthetics Sells: Mobile Phone Sales as a Function of Phone Attractiveness

Jeffrey M. Quinn, Ph.D.

Sprint
Overland Park, KS
jeffquinn5@gmail.com

ABSTRACT

Mobile phones are designed to have aesthetic appeal. Presumably, a phone's attractive appearance is one factor that induces consumers to purchase a particular device. The present research aimed to quantify the impact that the attractiveness of a phone's appearance has on consumers' purchase decisions. Participants rated the attractiveness of several phones. Mean attractiveness ratings for each phone were compared against actual sales figures from an internal Sprint database. A significant correlation emerged such that phones that received higher attractiveness ratings from research participants also tended to be purchased in higher volumes by consumers. Statistical analyses yielded a formula with which ratings of phone attractiveness can be used to estimate phone sales.

Keywords: aesthetics, attractiveness, sales, purchasing, mobile phone

1   INTRODUCTION

Appearance is a salient attribute which consumers use to evaluate a product. This is apparent in recent expert reviews of mobile phones, in which the aesthetic qualities of some designs garnered praise (e.g., "one gorgeous piece of hardware," "handsome and refined," "one of the sexiest pieces of hardware we've played with," Joire, 2011) and others failed to measure up (e.g., "ugly," design elements "look cheap and out of place," Miller, 2010). The more attractive-looking models likely have several advantages over their less attractive competitors. A product's attractive appearance captures consumers' attention, shapes initial impressions about positive product and brand attributes, and contributes to the user's enjoyment of the product (Bloch, 1995). Not surprisingly, attractiveness is thought to be a reason why consumers choose to purchase one brand over another (Bloch, Brunel, & Arnold, 2003; Hoegg & Alba, 2008; Norman, 2004).

Presumably, the relationship between product attractiveness and purchasing can be quantified. To the extent that attractiveness can be measured and the magnitude of its association with sales can be determined it should be possible to use this information to guide design-related decisions. For example, consumers' ratings of a new product design's attractiveness could be used to estimate its likely success upon launch. For this reason, it is important for designers and managers to better understand the attractiveness-sales relationship. The goal of this paper is to examine this relationship. It begins with a brief review of what is known from the published research in this area, which has focused on individuals' product preferences as a proxy for product purchase. In addition, it reports results from a study of attractiveness in the context of mobile phones, which uses sales data to determine the extent to which attractiveness and actual purchases covary.

1.2 Attractiveness and Product Preferences

Most of us, as consumers, have an intuitive understanding of the relationship between product attractiveness and purchasing decisions. We like attractive things and sometimes we purchase the products that we like. Research and theory are largely consistent with this view. According to Norman (2004), attractive-looking products elicit visceral, emotional responses from consumers. Early interactions with an attractive product may instigate positive emotional experiences such as excitement and pleasure from use, and long after purchase this positive affect could take the form of pride of ownership or pleasant memories associated with the object (Norman, 2004; Townsend & Shu, 2010). The affective impact of attractiveness on product evaluations takes hold quickly. Affect shapes people's initial impressions of whether or not they like a product, whereas other judgments (e.g., of a product's quality) can take longer to formulate (Page & Herr, 2002; Shiv & Fedorikhin, 1999). Affective responses to products can be strong and, under certain circumstances (e.g., low cognitive processing resources), difficult to override (Loewenstein, 1996; Metcalfe & Mischel, 1999; Shiv & Fedorikhin, 1999). As a result, "many a product is purchased on looks alone" (Norman, 2004, p. 69).

Empirical studies in this area typically have assessed participants' preferences for products that vary in attractiveness. For example, Yamamoto and Lambert (1994) examined attractiveness in the context of industrial products such as small gearmotors and solenoid valves. Employees of technical firms rated their preferences for different models of these products that varied in attractiveness and other attributes. Results indicated that attractiveness exerted a small influence on participants' preferences even though participants believed the appearance of these industrial products was not an important criterion for purchasing decisions. Attributes such as price and performance had a stronger effect on preferences, however, the attractiveness effect was "too strong in many cases to ignore" (p. 312). In another study, Townsend and Shu (2010) tested the impact of product attractiveness on hypothetical financial decisions. Specifically, participants viewed a financial document (i.e., a company's annual report) that was either high or low in attractiveness then evaluated the company described by the document (e.g., rated the likelihood with which they would invest in the company). Participants (who, in one study, were experienced investors) believed that such decisions should be made on the basis of a company's financial performance and not the extent to which its annual report looked pretty. However, their ratings varied as a function of attractiveness with attractive annual reports receiving more favorable company ratings than unattractive annual reports. Indeed, the authors concluded that "aesthetic elements impact experienced investors' likelihood of investing as much as financial factors such as year-to-year change in profit or revenue" (p. 457).

Studies of the impact of product attractiveness on research participants' preferences provide invaluable insights. However, preferences, purchase intentions, and other evaluations may not always be equivalent to consumers' actual purchasing behavior (Eagly & Chaiken, 1993; Ji & Wood, 2007). That is, although an individual may express a preference for a product or an intention to purchase the product, this doesn't necessarily ensure that the individual will complete the behavior of purchasing the product. Failure to behave according to reported intentions may occur for a variety of reasons (e.g., intentions change before the opportunity for purchase arises; cost or other constraints prevent purchase despite intentions). Thus, it is important to demonstrate a relationship between attractiveness and actual product sales to bolster findings from research focusing on product preferences. This is the objective of the present research.

1.3   The Present Research

This study investigated product attractiveness in the context of mobile phones, which often are designed to have aesthetic appeal. In contrast, the studies by Yamamoto and Lambert (1994) and Townsend and Shu (2010) examined products for which attractiveness was thought to be relatively unimportant. Given that attractiveness influenced product choices even in these extreme conditions, it seems likely that attractiveness also will be one factor that drives consumers' choice of phones.

Participants rated their perception of the attractiveness of several phones. They also rated the extent to which they would consider purchasing each phone. This measure approximates the product choice or preference measures typically used as a dependent measure in the marketing literature, and it allowed me to try to replicate the finding that consumers' preferences favor attractive versus unattractive products. The present research also included an objective measure of phone sales collected from Sprint's internal records. This makes it possible to test a second hypothesis, i.e., that phone attractiveness correlates significantly with actual phone sales. To my knowledge, this study provides the first test of the attractiveness-sales relationship.

Table 1 Phone characteristics and ratings

Phone	Months since launch	Design/ Form factor	Smart-phone	Touch-screen	# of raters	Phone attractiveness		Interest in purchase	
						M	SD	M	SD
Phone 1	8	Bar	Yes	No	254	4.60	1.39	2.23	0.77
Phone 2	4	Bar	Yes	No	219	4.41	1.44	2.36	0.72
Phone 3	5	Slider	Yes	Yes	244	5.62	1.21	1.71	0.70
Phone 4	38	Flip	No	No	235	2.47	1.34	2.80	0.47
Phone 5	28	Flip	No	No	294	2.01	1.21	2.86	0.41
Phone 6	4	Flip	No	No	317	4.18	1.85	2.41	0.74
Phone 7	6	Slider	Yes	Yes	295	5.51	1.17	1.79	0.71
Phone 8	3	Flip	No	No	243	3.00	1.57	2.65	0.64
Phone 9	2	Bar	Yes	Yes	232	5.20	1.31	1.96	0.74
Phone 10	20	Bar	No	No	284	4.15	1.55	2.47	0.67
Phone 11	1	Flip	No	No	247	3.39	1.64	2.55	0.71
Phone 12	4	Flip	No	No	297	2.86	1.53	2.65	0.66
Phone 13	3	Bar	Yes	Yes	312	4.95	1.52	2.04	0.78
Phone 14	25	Bar	Yes	No	222	4.81	1.40	2.29	0.73
Phone 15	13	Bar	Yes	No	209	4.75	1.48	2.28	0.75
Phone 16	7	Bar	Yes	Yes	246	5.47	1.41	1.72	0.75
Phone 17	15	Bar	Yes	Yes	275	5.25	1.34	1.84	0.74
Phone 18	5	Bar	Yes	Yes	322	4.82	1.46	2.16	0.74
Phone 19	18	Flip	No	No	276	3.83	1.66	2.58	0.66
Phone 20	27	Flip	No	No	261	2.57	1.50	2.73	0.57
Phone 21	12	Flip	No	No	260	3.14	1.61	2.65	0.61
Phone 22	15	Slider	No	No	325	4.10	1.40	2.52	0.66
Phone 23	10	Flip	No	No	291	3.58	1.61	2.57	0.64
Phone 24	14	Flip	No	No	324	4.13	1.68	2.42	0.72
Phone 25	15	Bar	Yes	No	230	4.50	1.53	2.34	0.72
Phone 26	12	Flip	No	No	305	4.01	1.76	2.44	0.72
Phone 27	16	Flip	No	No	264	3.04	1.62	2.58	0.65
Phone 28	14	Bar	Yes	Yes	241	4.62	1.34	2.23	0.75
Phone 29	16	Slider	Yes	Yes	295	5.41	1.26	1.80	0.77
Phone 30	17	Slider	No	No	246	4.07	1.70	2.48	0.74
Phone 31	8	Slider	No	No	253	4.10	1.60	2.44	0.71
Phone 32	28	Flip	No	No	232	3.59	1.57	2.54	0.68
Phone 33	7	Slider	No	No	249	4.06	1.53	2.55	0.65
Phone 34	10	Slider	No	Yes	246	4.61	1.45	2.24	0.69
Phone 35	8	Slider	No	No	317	4.88	1.43	2.14	0.79
Phone 36	3	Flip	Yes	No	237	4.77	1.59	2.14	0.77
Phone 37	3	Slider	Yes	Yes	257	5.31	1.22	1.85	0.73

2   METHOD

2.2   Participants

Participants were 900 Sprint customers (56% women, 44% men) in the United States. They ranged in age from 16 to 85 (median = 40.00, M = 41.95, SD = 14.59) and each received as compensation a $5 credit on an upcoming monthly bill.

2.3   Stimuli

The set of stimuli participants rated included 37 different models of Sprint phones. Participants viewed digital photos of phones on a computer screen as part of an online survey. At the time the research was conducted, all 37 phones were available for purchase in stores or online. As can be seen in Table 1, the phones varied on a number of characteristics.

2.4   Procedure

Data were collected using an online survey. Upon arriving at the survey website, participants received instructions stating that the purpose of the study was to ask their opinion about the appearance of several mobile phones. They were told they would view digital photos of several phones and that they should base their ratings of each phone only on its appearance, not their perceptions of other qualities of the phones (e.g., cost, presence of desired applications). Participants were instructed to respond quickly and to focus on their first impressions of each phone.

After reading the instructions, participants viewed and rated phones. The survey was programmed such that 12 phones were randomly selected from the set of 37 phones and the selected phones were shown to the participant in a random sequence.

Participants viewed each phone model one at a time. Pictures of the phone to be rated appeared at the top of the computer screen above several survey questions. For each phone, participants first rated the attractiveness of the phone's appearance then reported whether they had ever owned this model of phone and whether they would consider purchasing this model of phone. After rating 12 phones participants reported their age and sex before the survey concluded.

2.5   Measures

Phone attractiveness. Participants rated their perception of the attractiveness of a phone's appearance with a 7-item measure. Using four 7-point scales, participants rated the phone as: (a) attractive versus unattractive, (b) beautiful versus ugly, (c) eye-catching versus plain, and (d) interesting versus boring. Also, using a scale from 1 (*strongly disagree*) to 7 (*strongly agree*), participants rated their agreement with three statements: (a) I like the way this phone looks, (b) this phone is exciting, and (c) this phone is cool. The set of seven items possessed high internal consistency (Cronbach's $\alpha = .91$). The mean rating across these seven items served as the measure of attractiveness. Higher values indicated greater attractiveness.

Phone ownership. For each phone, participants indicated whether they had ever owned the same model of phone shown on the screen. Instructions specified, "for example, if you are rating a BlackBerry Curve, select 'yes' only if you've owned a BlackBerry Curve (any color). Select 'no' if you haven't owned a BlackBerry Curve, even if you've owned some other model of BlackBerry phone." Participants selected either *yes* or *no*. Ratings of phones that participants had owned in the past were omitted from statistical analyses. This resulted in removal of 944 phone ratings (9% of the original total of 10,800 ratings).

Interest in purchasing the phone. For each phone, participants responded to the question, "If you were looking for a new phone to purchase, would you consider buying the phone that you are rating now?" They selected from three responses: (1) Yes, I would consider buying this phone, (2) Maybe, I might consider buying this phone, or (3) No, I would not consider buying this phone. Lower values indicated greater interest in purchasing the phone.

Actual phone sales. Monthly sales data were gathered for each phone from an internal Sprint database, not from survey responses. These data showed the number of phones for each phone model that were sold to consumers during the month in which the survey data were collected and in each of the five preceding months. Analyses were performed using the average monthly sales for this 6-month period.

Phone price. Purchase price for the 37 phones came from internal Sprint records of point of sale price disregarding mail-in rebates. Mean price was calculated for each phone over the 6-month period for which phone sales data were available. Price ranged from $20 to $342 (median = $120, M = $144.75, SD = 82.22).

3   RESULTS

Table 1 summarizes participants' responses to survey questions pertaining to attractiveness and interest in phone purchase. Participants perceived some phones to be highly attractive (highest rating = 5.62) and others to be unattractive (lowest rating = 2.01; M = 4.21, SD = 0.93). Attractiveness was associated with the number of months since a phone launched such that participants perceived newer phones to be more attractive than older phones, $r(35) = -.48$, $p < .01$. Plausibly, this indicates that novelty plays a role in perceived phone attractiveness and that attractiveness declines over a phone's lifespan. This is consistent with research showing that prolonged exposure can cause an object's appeal to "gradually succumb to a 'tedium' factor" (Berlyne, 1970, p. 281; see also Bornstein, 1989; Frijda, 1988).

Participant-rated interest in purchasing these phones also varied quite a bit (M=2.32, SD = 0.31, greatest interest = 1.71, least interest = 2.86). The percentage of participants who would consider purchasing the phone that they rated ranged from 43% for the most well-liked phone to 3% for the least popular model.

3.1   Phone Attractivenss and Interest in Phone Purchase

The design of the present research resulted in a hierarchical data structure with up to 12 phone ratings nested within participants. Multilevel modeling was used to account for the non-independence of the phone ratings. This analytic approach resembles regression but allows parameters (e.g., slopes, intercepts) to vary across individuals (for a detailed description see Singer & Willett, 2003; Kenny, Kashy, & Bolger, 1998). Analyses were conducted with the Mixed Models function of the SPSS 19.0 statistical program (IBM, 2010) following procedures described by Singer and Willett (2003).

A multilevel model was constructed to predict participants' interest in purchasing a phone as a function of their ratings of that phone's attractiveness. This analysis produced a statistically significant effect, $b = -0.30$, $SE = .004$, $t(3610.25) = -79.06$, $p < .001$, such that participants showed greater interest in purchasing phones that were high (versus low) in attractiveness. Each one unit increase in perceived attractiveness was associated with a 0.30-point increase in the rater's likelihood of considering a phone for purchase.

3.2   Phone Attractivenss and Phone Sales

Mean attractiveness ratings were calculated for each of the 37 phones using participants' survey responses. The correlation between the phone-level attractiveness ratings and phone sales was statistically significant, $r(35) = .45$, $p < .01$. This relationship remained significant after omitting data for two phones that had very high sales (>3 SD above the mean sales) and could be considered outliers, $r(33) = .41$, $p < .05$.

Sales were higher for attractive phones than for less attractive phones even despite the fact that purchase price tended to be higher for attractive phones than for less attractive phones in this sample, $r(33) = .64$, $p < .001$. Interestingly, price was unrelated to sales, $r(33)= .20$, ns.

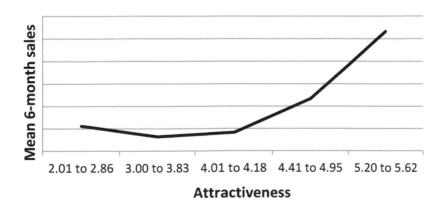

Figure 1 Sales as a function of attractiveness

3.3   How Attractive Must a Phone Be to Achieve High Sales?

Figure 1 shows mean sales values at various levels of attractiveness. Although actual sales values have been omitted to protect proprietary information, the general pattern of the attractiveness-sales relationship may provide guidance regarding which level of attractiveness can be expected to generate desired sales. Generally, the phones that received attractiveness ratings from 2.01 to 4.18 achieved the lowest sales. Sales more than doubled for phones rated between 4.41 and 4.95. For the phones rated highest in attractiveness (5.20 to 5.62) sales more than doubled again compared to phones that scored between 4.41 and 4.95.

Although the analyses described previously focused on linear relationships between attractiveness and sales the pattern in Figure 1 suggests a possible curvilinear relationship, which was tested using stepwise regression. Both the linear and quadratic components of the attractiveness effect significantly predicted sales (see Table 2). The significant change in R^2 indicates that the inclusion of the quadratic component improves the model's ability to explain variance in sales. Results of this analysis provide an equation which can be used to estimate sales using consumers' ratings of a phone's attractiveness:

$$Sales = X - (128{,}007.03*A) + (19{,}668.97*A^2) + error.$$

In this equation, A represents attractiveness ratings and X represents the intercept of the model (omitted to obscure the proprietary sales data).

4   DISCUSSION

The results of this study showed that phone attractiveness is associated with two measures related to purchasing. Consistent with past studies from the marketing literature, participants were more likely to report that they would consider purchasing a phone if they perceived the phone to be high (versus low) in attractiveness. Preferences and purchase intentions are not necessarily equivalent to actual purchases. Indeed, in this sample the correlation between sales and the mean interest in purchase (aggregated at the phone level) was only moderate in magnitude, $r(33) = -.45$, $p < .01$. Thus, it is important to demonstrate that attractiveness also is associated with purchasing behavior and the present research is likely the first study to do so. Mean attractiveness ratings collected from participants were significantly associated with phone sales.

The attractiveness-sales relationship was not strictly linear in this research. Sales were flat for phones that scored around or below the midpoint of the attractiveness scale. Above this level sales climbed steeply. It is possible that these benchmarks could help answer the question, 'what is a good score on the attractiveness measure?' However, an important caveat is that the changes in sales associated with these attractiveness values will not apply for all categories of products. For example, whereas attractiveness had a moderate effect on sales for mobile phones

Table 1 Stepwise regression of sales as a function of attractiveness

Step	Significant predictors	b	β	R^2	Change in R^2
1	Attractiveness, linear	28315.74**	0.45	.20	
2	Attractiveness, linear	-128007.03***	-2.04	.31	.10*
	Attractiveness, quadratic	19668.97*	2.52		

Note. *p < .05; **p < .01; ***p < .10. b = unstandardized regression coefficient; β = standardized regression coefficient.

(R^2 = .31), the relationship between attractiveness and sales is likely much smaller for gearmotors and solenoid valves (Yamamoto & Lambert, 1994). For these kinds of products the difference in sales for models rated high versus low in attractiveness could be much more modest than those observed for attractive versus unattractive mobile phones.

4.1   Implications for Designers and Managers

Understanding the relationship between attractiveness and sales within a particular industry offers several opportunities to improve product design-related decisions. Prior to the launch of a product, it should be possible to measure the extent to which potential consumers perceive a design to be attractive and then to use attractiveness ratings to predict future sales. This approach could be used to determine which of several competing designs should be adopted, based on estimated future sales. Similarly, product attractiveness could be measured at several points over time to determine the rate at which attractiveness is changing. Such information could help forecast sales and determine the appropriate number of devices to maintain in inventory. Finally, it is possible to use estimated sales to determine the return on investment of design work that focuses on improving the attractiveness of a product. Changes in consumer ratings of attractiveness before and after a redesign could be used to calculate changes in expected sales and revenue associated with design improvements.

The results of this research demonstrated the importance of product attractiveness to consumers. Used in the manner described above, measurements of product attractiveness could be equally important to designers and managers to support strategic decisions pertaining to product design.

ACKNOWLEDGMENTS

Thank you to Jason Ward, Tuan Tran, and Clyde Heppner for their contributions at various stages of this research.

REFERENCES

Berlyne, D. E. (1970). Novelty, complexity, and hedonic value. *Perception & Psychophysics, 8*, 279-286.

Bloch, P. H. (1995). Seeking the ideal form: Product design and consumer response. *Journal of Marketing, 59*, 16-29.

Bloch, P. H., Brunel, F. F., & Arnold, T. J. (2003). Individual differences in the centrality of visual product aesthetics: Concept and measurement. *Journal of Consumer Research, 29*, 551-565.

Bornstein, R. F. (1989). Exposure and affect: Overview and meta-analysis of research, 1968-1987. *Psychological Bulletin, 106*, 265-289.

Eagly, A. H., & Chaiken, S. (1993). *The psychology of attitudes*. Fort Worth, TX: Harcourt Brace Jovanovich.

Frijda, N. H. (1988). The laws of emotion. *American Psychologist, 43*, 349-358.

Hoegg, J. & Alba, J. W. (2008). A role for aesthetics in consumer psychology. In C. P. Haugvedt, P. M. Herr, & F. R. Kardes (Eds.), *Handbook of consumer psychology* (pp. 733-754). New York: Taylor & Francis.

IBM. (2010). SPSS for windows (Version 19.0) [computer software]. Chicago: IBM.

Ji, M. F., & Wood, W. (2007). Purchase and consumption habits: Not necessarily what you intend. *Journal of Consumer Psychology, 17*, 261-276.

Joire, M. (April 11, 2011). Nokia E7 review. Retrieved from www.engadget.com/2011/01/11/nokia-e7-review/

Kenny, D. A., Kashy, D. A., & Bolger, N. (1998). Data analysis in social psychology. In D. T. Gilbert, S. T. Fiske, & G. Lindzey (Eds.), *The handbook of social psychology* (4th ed., pp. 233-265). New York: McGraw-Hill.

Loewenstien, G. (1996). Out of control: Visceral influences on behavior. *Organizational Behavior and Human Decision Processes, 65*, 272-292.

Metcalfe, J., & Mischel, W. (1999). A hot/cool-system analysis of delay of gratification: Dynamics of willpower. *Psychological Review, 106*, 3-19.

Miller, P. (November 5, 2010). T-Mobile myTouch 4G review. Retrieved from www.engadget.com/2010/11/05/t-mobile-mytouch-4g-review/

Norman, D. A. (2004). *Emotional design: Why we love (or hate) everyday things*. New York: Basic Books.

Page, C., & Herr, P. M. (2002). An investigation of the processes by which product design and brand strength interact to determine initial affect and quality judgments. *Journal of Consumer Psychology, 12*, 133-147.

Shiv, B., & Fedorikhin, A. (1999). Heart and mind in conflict: The interplay of affect and cognition in consumer decision making. *Journal of Consumer Research, 26*, 278-292.

Singer, J. D., & Willett, J. B. (2003). *Applied longitudinal data analysis: Modeling change and event occurrence*. New York: Oxford University Press.

Townsend, C., & Shu, S. B. (2010). When and how aesthetics influences financial decisions. *Journal of Consumer Psychology, 20*, 452-458.

Yamamoto, M., & Lambert, D. R. (1994). The impact of product aesthetics on the evaluation of industrial products. *Journal of Product Innovation Management, 11*, 309-324.

CHAPTER 56

Evaluation of the Users' Emotional Interactions in the Use of Washing Machine

Alexandre Barros Neves, Maria Lúcia L. R. Okimoto

Universidade Federal do Paraná
Curitiba - PR, Brazil
alexandre.b.neves@electrolux.com.br, lucia.demec@ufpr.br

ABSTRACT

The growing awareness with regards to the role emotion plays on the choice and use of products has been leading both companies and universities to a deepening in the study of the relationship between users and products. Knowing the changes in user's perception that happen during different moments of this relationship may contribute with relevant information, so that designers while creating new products, seek to guarantee emotional consistency throughout product's life time. In this context, this research aimed at identifying the changes that happen in the relationship between users and a washing machine in four different moments of the interaction, over a period of six months. The results of the evaluation were plotted in a three dimensional graphic showing the evolution of the perception users had about the product. Users were also asked to highlight the area of the product that influenced their evaluations for each adjective. This activity was done twice. A comparison between these two assessments rendered a clear visual map of the items that most influenced the perception of the participants. The numeric data gathered with the questionnaires was also evaluated using a number of statistic tests. Significant variations on users' perceptions about the product, throughout the different levels of interaction evaluated were identified. This fact points to the utility of such method during the appraisal of the human-product relationships. It also reinforces the need for instruments able to measure these relations during time, in order to improve the attachment between users and products for a longer period.

Keywords: design and emotion, home appliances, semantic differential.

1   INTRODUCTION

Researches conducted in the field of design and emotion have been evaluating and measuring the emotions objects evoke in individuals. Many studies use pictures of products as stimulus for the assessment (DESMET, 2003, SILVA et al. 2008; MCDONAGH et al. 2002; JORDAN, 2002; MONDRAGON et al. 2005). Several authors such as Norman, Hekkert, Nagamashi, Helander, etc. contributed to leverage the research in the subject, and inspire us in the search of some deeper questions about the emotions caused by the products in regards to temporal aspects, considering the dynamic changes of emotion in the time line.

Woolley (2003) links the different types of pleasure generated by the product at each stage in time, promoting different types of interaction. The author notes that even for products with greater durability, they are easily replaced and discarded if do not provide pleasure during use anymore.

Bonapace (2002) also proposes that the perception individuals have of a given object varies according to the time and intensity of interaction they are exposed to the product. When examining an object for the first time, one unconsciously summarizes sensations communicated by this object as being pleasurable, irrelevant, or repulsive. After a period of time, during which the person receives different stimuli form the product, the assessment becomes more precise. However, she points that before being able to perform a fair judgment based on the real attributes of the object, the person has already made an appraisal based on the initial pleasure elicited by the object, remaining strongly subjected to this first sensation.

Hassenzahl (2007) identifies two dimensions to classify the perception individuals have of interactive products: pragmatic and hedonic. The first are related to what the author calls 'do-goals' and are related to the actions the user needs to realize and which the object were made for. The 'be-goals' constitute the hedonic dimension and refer to the human needs that go beyond the instrumental features of the product. The hedonic/pragmatic model assumes that the experience alters the users' perception of a given object, where the hedonic dimension tends to decrease, while the pragmatic on rises.

The research presented here sought to investigate the aspects of how emotions change over time in the individual-object interaction. The objective of this study is to identify the changes in emotional relationship applied in different stages of interaction of an appliance.

2   METHOD

This study, conducted in collaboration with a major appliances manufacturer, analyzed four different moments of the interaction between users and a washing machine. These moments were: a) emotional image of the product category, without any visual stimulus, b) presentation of a visual stimulus of a new product, c) first physical contact with the product and d) after six months of frequent use of the product. The experiment was carried out with 10 individuals that took part on the field test of a new model of washing machine. The field test is a procedure used by the manufacturer with the purpose of evaluating pre-production products and identifying possible problems before they go in production. All participants of the field test are real potential consumers of the product, i.e. they all belong to the target group defined for the product being tested, in terms of gender, age, income, etc.

Self reporting questionnaires containing 30 semantic differential scales were applied on each phase in order to assess the responses to emotional stimuli generated by the product. The semantic differential method, proposed initially by Osgood, Suci and Tannenbaun (1957), is based on pairs of opposite adjectives in a scale, where participants choose a point in between. A ten steps method based on four different sources of information was used in order to the select and classify the adjectives used in the questionnaires, as described by Neves and Okimoto (2010). These adjectives were divided into three dimensions, according to Jordan (2002, p.6): functionality, usability and pleasure.

3   POSITIVE ASPECTS MAPPING

On phases 3 and 4, the participants were asked to mark on a picture of the washing machine which part of the product most influenced their evaluation for each pair of adjective. The results of this exercise, shown below, gave some valuable insights for a better understanding on the way users perceived the different components of the tested product. As this evaluation was done on phase 3 - immediately after the first physical contact with the product- and repeated on phase 4 -after six months of product use- it was possible to analyze and compare the changes on the perception, affected by a deeper experience in using the washing machine. For this evaluation only the pairs of adjectives that received a positive value were considered, given they were the great majority of the indications. The data gathered was then classified in the three dimensions evaluated in the research: functionality, usability and pleasure.

Figure 1 presents the comparison of the positive marks for the *functionality* dimension between phases 3 and 4. For this dimension, a great concentration of markings is observed on the right hand side of the control panel. In this area a number of washing options buttons are located, with emphasis to the 'Economy' function, that renders the process of reusing the water much easier. Despite the slight reduction on the number of positive indications for this region in phase 4, the concentration is still quite big, indicating satisfaction with regards to the expectation related to these features.

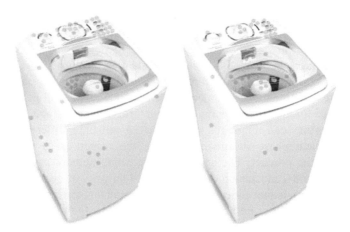

Figure 1 Map of the regions of the washing machine that elicited positive appraisal for the adjectives that compose the dimension *functionality* on phases 3 (left) and 4 (right).

For the *usability* dimension, one of the great changes in the perception is related to the central knob area that, while keeping a good number of indications on phase 4, had a significant reduction when compared to the third phase. Another important variation happened on the lid handle region that received several indications on phase 3 and almost none on phase 4. However, this data should be considered along with the results of the *pleasure* dimension, where exactly the opposite happened. It is also important to observe that the number of marks on the dispensers areas (detergent, bleach and softener) was consistent between the two moments assessed. This fact indicates that they were well perceived by the users during their first physical contact with the product and this perception was kept during time. The *usability* dimension marks are shown in picture 2.

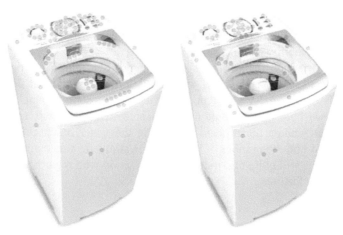

Figure 2 Map of the regions of the washing machine that elicited positive appraisal for the adjectives that compose the dimension *usability* on phases 3 (left) and 4 (right).

Several participants evaluated positively the graphic elements on the glass lid when considered the adjectives that composed the *pleasure* dimension on both phases. These marks are shown in picture 3. The indications on the control panel in general did not change much, with just a slight improvement on the central knob area on phase 4. On the other hand, the consumables dispensers that were very well evaluated on phase 3, had no marks on phase 4. This can be explained by the fact that these components are seen as essentially functional ones, mainly after a frequent use of the product, avoiding any association with an adjective related to *pleasure*. As mentioned above, on the *pleasure* dimension the amount of indications on the lid handle area grew substantially between phases 3 and 4, while the opposite happened on the *usability* dimension. This may denote a migration of what was initially perceived as usable in the first contact to a pleasurable element after the real use if the washing machine.

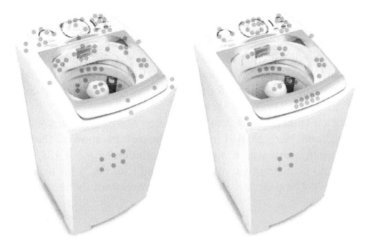

Figure 3 Map of the regions of the washing machine that elicited positive appraisal for the adjectives that compose the dimension *pleasure* on phases 3 (left) and 4 (right).

4   Adjectives Comprehension

In order to verify the consistency of interpretation of the adjectives by the participants, a synonym table was used. For each of the 60 adjectives presented on the semantic differential questionnaires, a number of synonyms were displayed. The participants were then asked to choose which of them had the closest meaning to the original adjective, according to their own interpretation. Through this exercise, it was possible to identify eventual interpretation discrepancies users could have regarding the adjectives.

The result of this test showed that for 49 out of the 60 adjectives (82%), at least 50% of the respondents chose the same synonym. Besides, none of the 11 adjectives that had a synonym selected by less than half of the participants had its opposite in scale with less than 50% of indications. Due to the fact that in the differential semantic scales the adjectives are presented in pairs with their opposites, the understanding of one of them, helps the interpretation of the other. It is relevant to mention also that for 17 adjectives (28%) 75% or more of the respondents selected the same synonym.

Considering that in some cases up to 6 different synonymous were presented to each adjective, and that some of them had extremely similar meanings, these results can be understood an indication that the participants had a very consistent comprehension of the adjectives presented on the semantic differential scales.

5 STATISTIC ANALYSIS OF THE RESULTS

The evaluations made by the participants and registered on the semantic differential questionnaires were converted in numeric data by measuring the positions marked on the linear scales. The values attributed to the adjectives were grouped by each of the 3 dimensions and summed for each participant, originating a set of 30 (10 participants x 3 dimensions) numbers for each of the 4 phases.

The symmetry of the distribution of this set of data was evaluated based on the values of *curtosis* and *skewness* and the Shapiro-Wilk test was applied. The results of these evaluations pointed to a trend for a normal distribution of the data. Based on this information, both ANOVA and t-test were applied.

For each dimension (functionality, usability and pleasure) the oneway ANOVA test was performed, considering all 4 phases. The results showed that for the functionality dimension, there was a significant variation ($F_{3,35}$=4,423, p<0,05), while for the others, no significant variation was found (Sig.=0,26 for usability and Sig.=0,33 for pleasure). The picture 4 shows the comparison of the means for each dimension in each phase.

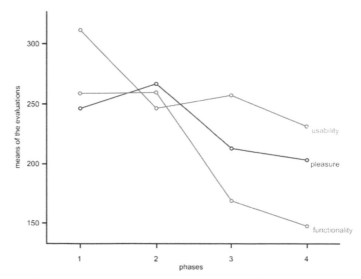

Figure 4 Means of the evaluations of each dimension during the different phases.

The paired t-test was performed comparing the phases in pairs. The aim of this analysis was to verify the statistic significance of the variations of the results in each dimension between one phase and its successive.

For the functionality dimension, the t-test showed a significant variation between phases 2 and 3. In this case, the null hypothesis can be rejected as $t(9)=2.664$, $p<0,05$. This variation is very relevant in the context of this research because it represents the change in the participants' perception between the first visual stimulus of the washing machine and the first physical contact with the object. It also shows that the assessment of the perception an individual has of an object, based only on visual stimulus may be biased. The t-test didn't reveal any other significant variations for the functionality dimension on the other intervals.

The variations occurred for the usability dimension were quite different from what happened with the functionality dimension. The t-test showed a significant variation between the participants expectation of a washing machine (phase 1) and their appraisal of the picture of the product (phase 2). Between these two phases, for the usability dimension, the variation was significant, once $t(9)=2.409$, $p<0,05$. However, between phases 2 and 3 and between phases 3 and 4, no significant variations were observed. For the pleasure dimension, no significance was perceived in any of the intervals.

An evaluation of the correlation among the dimensions was also performed in order to verify the influence one dimension would have on the other. The values attributed individually by each participant to the adjectives that composed each dimension were summed. This value then was subtracted from the correspondent value of the predecessor phase. Thus, for example, there was the value attributed by the participant 3 for the functionality dimension at phase 2. From this value was subtracted the amount given by the same participant (3) for the same dimension (functionality) on the previous phase (1). The resulting number corresponds to the variation on the perception of this specific participant, and can be positive in the case of an improvement of the perception, or negative on the other hand.

The next step was the verification of the correlation among the dimensions, taken in pairs. Thus, the difference between functionality dimension in phases 2 and in phase 1 had its correlation tested against the usability dimension under the same conditions. The process was repeated between functionality and pleasure and between usability and pleasure for all phases. The result is shown in picture 5.

It can be observed in picture 5 a big change between the first group (difference between phases 1 and 2) and the other 2. This may have been a consequence of the fact that in phase 1 the participants had no stimulus, i.e. only the conceptual model of the product category was assessed. When the picture of the washing machine was presented, the correspondence to what they expected varied in unevenly among the dimensions. Despite the not very high correlation of this first interval, on the others, it was always bigger than 0.66, indicating that one dimension influences and is influenced by the others.

544

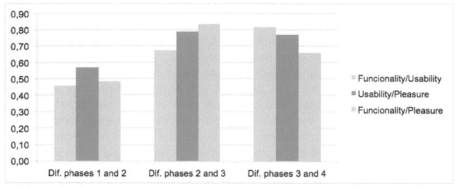

Figure 5 Correlation between the dimensions considering the differences between the phases

6   DISCUSSION

The changes identified on the perception individuals have of a product, as presented above, are important indications that a fair assessment of the affective relationship between an individual and a product, should consider a number of different moments of the interaction.

It is relevant also to point that the results suggest that the classification of the perception based on the three dimensions proposed by Jordan (2002) is more effective than the one proposed by Hassenzahl (2007). This fact is supported by the uneven variation between the functionality and usability dimensions observed during the four phases of the study. The adjectives that compose these dimensions would be classifies under the pragmatic dimension.

The results also suggest that the use of pictures as the only stimulus on the evaluation process of the affective perception individuals have of an object, is not effective. As shown in picture 4, on phase 2 (when participants are stimulated by a picture of the product), the evaluations of the three dimensions converge. This behavior is unique during this phase when compared to the others. The emphasis here is on the pleasure dimension, which has its higher result on this phase.

REFERENCES

BONAPACE, L. Linking product properties to pleasure: the sensorial quality assessment method - SEQUAM. In: GREEN, W. S; JORDAN, P. W. (ed.) **Pleasure with Products: Beyond Usability**. London: Taylor and Francis, 2002. p.189-217.

DESMET, P.M.A., & HEKKERT, P. (2002). **The basis of product emotions**. In: W. Green and P. Jordan (Eds.), Pleasure with Products, beyond usability (60-68). London: Taylor & Francis.

DAMAZIO, V. M. ; LIMA, J. ; MEYER, G. C. . . In: Mont´Alvão, C; Damazio, V. (Org.). **Design, ergonomia e emoção** 1a ed. Rio de Janeiro: MAUAD X, 2008, p1.

DE MASI, D. **O Ócio Criativo**. Rio de Janeiro: Sextante, 2000.

DESMET, P.. **A multilayered model of product emotions**. The Design Journal, p.1-13, 2003.

HASSENZAHL, M. The hedonic/pragmatic model of user experience. In: Law, E. et al (ed.). **Towards a UX manifesto.** Lancaster: Mause, 2007

JORDAN, P.W. **Designing pleasure products**. London: Taylor and Francis, 2002.

MCDONAGH et al. 2002; **Design and emotion**. London: Taylor and Francis, 2004.

MONDRAGON, S.; Company, P.; Vergara, M.. Semantic Diferential applied to the evaluation of machine tool design. **International Journal of Industrial Ergonomics**, v.35, n.11, p.1021-1029, nov. 2005.

NEVES, A. B. ; OKIMOTO, M. L. R. **Usability: Timeline and emotions**. In: David B. Kaber; Guy Boy. (Org.). Advances in Cognitive Ergonomics. 1 ed. Boca Raton: CRC Press / Taylor & Francis Group, 2010, v. 1, p. 317-327.

OSGOOD, C.E.; SUCI, C.J. TANNENBAUM, P.H. **The measurement of meaning**. Urbana: University of Illinois Press, 1957.

SILVA, T. B. P. et al. Teste de personalidade dos produtos. In: Congresso Brasileiro de Pesquisa e Desenvolvimento em Design, 8, 2008. **Anais**... São Paulo: AEND/Brasil, 2008. p. 3038-3043.

WOOLLEY, M. Designing Pleasurable Products And Interfaces .In: International Conference on Designing Pleasurable Products and Interfaces. **Proceedings.** Pittsburgh, PA, USA, 2003.

CHAPTER 57

Juicy Salif – Affective or Functional Design

Julio van der Linden, Andre Lacerda, Renata Porto,
Liliane Basso, Mariana Seferin

Federal University of Rio Grande do Sul
Porto Alegre, Brazil
julio.linden@ufrgs.br

ABSTRACT

The paper deals issues of functionality, usability and enjoyment of users in interaction with artifacts of symbolic value. For this end, was proposed an experiment to graduates of Design utilizing the squeezer Juicy Salif. Through observation of users and subsequent textual descriptions of them, the views were evaluated in two stages (before use and after use) and a record of experience in user memory. The idea of observation is to understand how the product performs under natural conditions without imposing boundary constraints that would arise with a set evaluation protocol. From the material written by eleven students, textual descriptions were analyzed which describes individual experiences of pre-use, post-use and how it formed the record in memory. In the reports of the experiences of pre-use were identified terms that relate the squeezer to a formal and aesthetic attributes, with compromised functionality which the form invalidates the function. The descriptions after use indicate the artifact is fragile, light, unstable, easily flowing and functional. After 14 days, the students described the artifact as being fragile, modern, unstable and strong aesthetic appeal. Are new concepts not previously mentioned and this would explain why when a memory is recorded, we keep more than just the object instance, but we create a meaningful context for the object itself. The responses suggest that objects are always seen in a context of things, situations and users, including the observer himself . So when we recall an object, that is not ordinary as a general concept, we use the episodic memory, which leads us to associate meanings inherent to the context. This explains why we did not associate the Juicy Salif squeezer directly to their real function, rather, we connect to other previous contexts of meanings.

Keywords: emotional design, design and memory, Juicy Salif

1   MEMORY, PRE-CONCEPTIONS AND PRODUCT EVALUATION

From a conducted experiment with the *Juicy Salif* lemon squeezer, the following hypothesis was formulated: What kind of meaning a product has before and after his use, and what is the register in the user memories. Based on the experiences performed with the lemon squeezer we also discussed the affirmation made by Donald Norman in the first chapter of his book "Attractive things work better". From the undeniable attractive appearance of the lemon squeezer designed by Philippe Starck and his average functioning when compared with any other lemon squeezer, causing the product to be remembered in the category of different and fun products, and ended up does not fitting into a routine of a kitchen, when we need the function of squeezing lemons.

When we remember a certain moment, we assign values in order to summarize a few characteristics, classifying the meanings that reflect that experience (GLADWELL, 2008). When we remember a product, we also use these characteristics to rescue the experience gained with it. However, even when we have no direct experience, we build an anticipation of the event (DESMET E HEKKERT, 2002). Mills (2009) explains that first rule for understanding the human condition is that men live in second-hand worlds, in other words, we usually generate a pre-conception even before a direct experience. When evaluate an object which we did not have a previous experience, we appropriate the associative experiences in an instinctive and self-preservation character (FIALHO, 2001). If there is a pre-perception there is also a register of this, according to Iida (2008), not all perceptions are registered and only some of them are transferred to long-term memory (IIDA, 2008, p.260). But what usually happens is that we have a larger register of experiences memories, than imaginative or preconceptions memories.

> The semantic system serves on the one hand to represent the relations between representations of different modality-specific systems (e.g., word form of the word "shoe," picture form of the object "shoe") in a corresponding concept (concept of SHOE), and it serves on the other hand to represent connections among concepts in so called propositions. In episodic memory, typical connections are predicates to objects such as "the shoe is dirty." (Zimmer, 2001)

It explains why the register of the memory keep more than just the object instance, but create a meaningful context for the object itself. So when we recall an object, that is not ordinary as a general concept, we use the episodic memory, which leads us to associate meanings inherent to the context. As well as this explains why we do not associate the *Juicy Salif* lemon squeezer directly to its function, but to connect to other contexts of meaning. If we ask someone to imagine a comfortable chair, this person will probably visualize a particular product in your mind, but if we ask for this person to imagine a beautiful chair, the product model will probably be another. This is because the preferences of a person on matters of comfort and aesthetic aspects are separate in your mind. The preference for a comfortable chair will probably emerge from the experience and commitment that this person had, while beliefs about the beauty of a chair will rise even without direct experience. What we show is that there are two different languages involved in the same artifact. "Experience shows us that the images as well as words acquires different meanings depending on the use we make and the context that we situate" (CALVERA 2007, p.127). In a visual language, the explicit references in the image of the object, are not given, but understood by the circumstances of its use and dynamic connection established between them or the user. Thus, passing to consider the meaning of the products not only in the context in which its forms are used, but as these make some sense (KRIPPENDORFF, 1989).

Currently is increasing research in the field of Design and Emotion, where we find a new human factors adding more value to products. Perceived quality is also linked to the affect that the product generates in the individual as claimed by Jordan (2000, p.5), "when people get used to having something, they soon begin to look for something else." In the hierarchy of consumer needs created by Jordan, we have the first level the functionality, usability in the second and the third is the pleasure which emphasized the importance of "products offer something extra", the products are seen as "living objects" with whom people have relationships. Gladwell (2008) states that this tendency to think of giving character to the objects is called in psychology, the fundamental attribution error. Humans fail to overestimate the importance of fundamental character traits and underestimating the importance of the situation and context. The fundamental error, also argues Gladwell (2008), makes the world more simple and understandable. Another feature is that the "products can be viewed in terms of its appeal" (DESMET AND HEKKERT, 2002). Thus, products such as aesthetic objects would be attractive to anyone who has an aesthetic, relationship or social attitude. The aesthetic attitude is reflected when one chooses a product due to your own taste. The attitude of relationship would be one that denotes a sentimental bond with another person. While the social attitude denotes the collective taste, which can be well characterized by fashion. Patrick Jordan researched the "pleasure" to better understand the relationship between people and products. And within this context of pleasure with products, found three distinct types of benefits: practical related to the tasks performed by the products, their practical functions; emotional, related to the positive mood changes due to the use of products, satisfaction; hedonic, this associates the sensory and aesthetic pleasure, in the use of products (JORDAN, 2000).

2   MORE AN EXPERIENCE THAN AN EXPERIMENT

The first experiment did not have originally the purpose of evaluating systematically the *Juicy Salif*, but instead understand and get familiar with it through the use. In a discipline of the Master's degree course in Design at UFRGS, called Theory of Design, one of the participants brought an exemplar of the squeezer to be used during the class, with the intention of discussing its meaning based on the affirmation made by Philippe Starck: "this squeezer is made to start the conversation" (LLOYD; SNELDERS, 2003). However, when the squeezer was presented to the group, came the idea of systematically record their use, and others were invited to expand the number of observations of use, and balance the distribution related to gender. Thus, in addition to the discipline group (teacher and seven students - six men and two women) had joined four other women (undergraduate and graduate students, and the secretary of the program). The use of the *Juicy Salif* was made by the twelve participants (six men and six women), recorded in digital video. Each participant squeezed half a lemon according to their method of operation without any orientation. The gestures had small variations, but all chose to stabilize the squeezer with one hand, while the lemon was squeezed with the other hand (Figure 1).

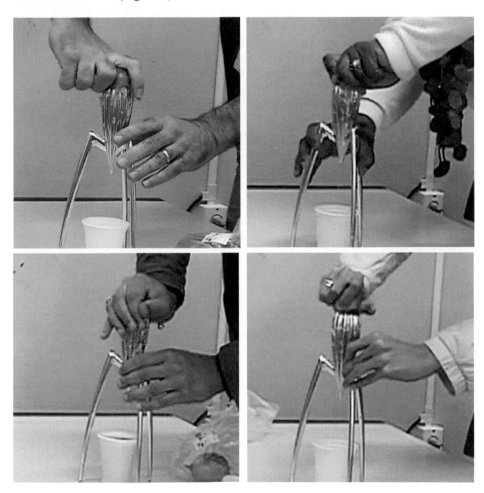

Figure 1 Illustration of how different subjects used the *Juicy Salif*

It may be observed in Table 1, that the time to perform the activity ranged between 1 minute and 45 seconds for the first subject, to 10 seconds in the case of the last one. The variation in the duration of activity has several explanations: subject 1 presented the proposal for the group, and get to long with many comments; subject 2, a woman, said that the lemon was not ripe, so hard to squeeze (used half a lemon used by the subject 1); subject 10, also woman, commented that she was very weak; and finally, subject 12, a man, claimed to have much experience in squeezing lemons and used another variety of lemons, which seemed to be easier to squeeze. Between these extremes, the others carried out the activity, between 21 and 44 seconds (average: 32.3 seconds). There was no assessment of the residue of each lemon, so it impossible to say that everyone had similar results. The use of such data has the unique purpose of demonstrating that the group's overall performance was satisfactory, allowing state, based on the video analysis, that the *Juicy Salif* worked as squeezer. As an opened user experience, the comments made during use and the opinions of those who were watching may have influenced the performance of each one, but this was not registered.

Table 1 Time to develop the task and subjects' comments

Subject	Gender	Duration	Observations
1	M	00:01:45	He squeezed slowly, making comments
2	F	00:01:13	She commented that the lemon was not ripe
3	M	00:00:37	He prepared the lemon before squeezing
4	F	00:00:37	
5	M	00:00:36	
6	F	00:00:44	She argued that she did not know squeeze
7	M	00:00:21	
8	F	00:00:23	
9	M	00:00:28	
10	F	00:01:22	She commented that she was weak
11	F	00:00:27	She used another variety of lemons
12	M	00:00:10	He used another variety of lemons

As the proposal was only the collective observation of the use, was not possible to record the opinions of all subjects. Just a few comments issued at the time were using the squeezer. Subject 1, who initially had restrictions on the *Juicy Salif*, remarked with surprise, "It's working ... look guys, I think it's working." Another positive comment was made by subject 8 who said, "but it is very easy ...". During use, the negative comments were subject 6, who said "it does not makes much easier for inexperienced users" and subject 12, for whom the *Juicy Salif* "It's not very practical." Dialogues occurred during use of the squeezer lemon, as when the subject 1 was a finishing their use:

- I did not find functional ... This whole mess ... For women it is not functional, to the man who is sloppy in the kitchen [can be] I thought that it is not practical (subject 12)
- But it is not so bad (subject 1)
- Not bad, but I prefer to squeeze with hands (subject 12)

After the conclusion of this experience of use, the issue was discussed during class, but there was no immediate advance. A few weeks later, the topic returned in a conversation in which participated three of the authors of this paper which also had participated of the experiment. One of the authors said that the *Juicy Salif* had worked effectively to squeeze lemons, but that it had left no emotional connections. His vision, based on memory of the use, led him to defend the *Juicy Salif* as squeezer and not as objects of symbolic desire, unlike the dominant discourse. This led to two decisions which gave a concrete basis for this work: return to the participants of the experiment to check your memory of the use, and perform a controlled experiment with undergraduate students in design to explore systematically the effect of memory in the evaluation of this product.

For the first case, messages to the others participants of this experiment were sent by electronic mail. Despite efforts to involve the subjects, only four of the nine answered. Table 2 summarizes the memory of these subjects. Subject 2 was a woman who had difficulty to squeeze the lemon and took to complete the task. She had a good impression of the *Juicy Salif,* but she did not consider it a functional product. The other woman, subject 8, considered the squeezer not just a symbolic object, in her opinion it works effectively, but it has ergonomic problems in its structure. Differently, the two men who responded, subjects 3 and 9, reported positive experiences..

Table 2 Summary of subjects' memory of the first experiment

Subject	Positive	Negative
2	It's beautiful, durable and hygienic	It doesn't fulfill its functions
3	It appeared non-functional	
	By using, it showed that's effective	
8	It isn't only symbolic	It isn't ergonomically appropriate
	It fulfill its functions	
9	It provokes the desire to use	

Although with few responses, the answers reflect the two dominant perspectives on this experience. The negative can be exemplified by the following comment, made by subject 2:

> About its use, I found problems in the direction of force, as it has three support eventually bend when the force is not perfectly vertical. In my opinion this squeezer, it's an ornament, because squeezers "simpler" and less expensive, satisfactorily fulfill the purpose, while it does not.

While the positive view, can be exemplified by the narrative of subject 3:

> Before using the squeezer, we can doubt the good functioning of the product, however it is effective! If we want to criticize it, we can mention his height (which may not work well on countertops too high or very low people), their polish (which may seem strange to an object made to deal with acid fruits), or its base with three support points (which may not be ideal to apply force and motion while squeezing a lemon). But the juicer works. However, the hypothesis that the use test proved, even more definitively, is that the *Juicy Salif* is a conversation starter. Even without electric and been absolutely silent, their use generates a lot of noise, transforming an ordinary share in the spotlight.

This view reduces the weight of the functionalist view that requires the product to maximum technical efficiency, even despite the loss of their emotional and symbolic content. Considers that a product for personal use, not offering risk to their users, can set aside the maximum functionality in terms of other attributes. In this sense it is relevant to mention that the observation of the use of the *Juicy Salif* has shown that it works properly as a squeezer, all participants performed the task, it just does not offer a filtering function for the juice.

3   THE EXPERIMENT THAT WAS EFFECTIVELY ONE

The experiment to explore systematically the effect of memory on *Juicy Salif* assessment was conducted with design undergraduate students in the end of the course. Previous the experiment, it was verified that all subject knew the *Juicy Salif,* at least as an object of theoretical discussion. Initially, they were invited to observe the squeezer and write their views concerning the object. Following, they were invited to squeeze half a lemon (as was done at the first experiment). At the end, all were invited to write their impressions after use. Two weeks later, a message was sent by email requesting to comment what they remembered about the experience.

Several reports previous of the use presented the oddity with the product and lack of communication of its function, as well as the surprise when was perceived that it works. The lack of stability was frequently reported both as a perception after use as a memory about the product. A case concerning subject 4 summarizes emphatically the experience of the students who had an opinion about Juicy Salif without user experience. In this student's speech is relevant to point how he frankly presented the prejudice against Phillipe Starck.

> (PRIOR TO USE)
> The *Juicy Salif* is an object of great visual and aesthetics impact. It has modern lines, but at the same time, gives an impression of an object from the last century, probably by its external metallic characteristic and smoothness of some curves. Do not communicate well its function, giving strangeness to the user and does not give me some assurance that their use will be really pleasurable. It is an object oriented to art and decoration much more than the design.
> (AFTER USE)
> The use, to some surprise, was satisfactory. Fulfills its function well, but it doesn't have great stability. So it is still distrust me.

(MEMORY)
I remember having tested the stability of the juicer and that worked well with most peers. I have a certain prejudice against Starck, but the product proved to be very usable, unlike my initial expectation.

The experiment included 11 students and 8 responded to that message, but two of them used only keywords. Thus, the analysis was based on the opinion of only six who participated effectively of the experiment. The summary of the opinions presented in Table 3 allows to observe some changes based on usage.

Table 3 Summary of reviews and of subjects memory of the second experiment

Subject	Before	After	Memory
1	It has aesthetic impact *It looks pretty bad*	*It looks fragile* *It is not functional*	*It is fragile*
2	It has aesthetic impact	It is useful It is easy to wash	It has aesthetic impact It is functional It is easy to use
4	It has aesthetic impact *It does not inform the function*	It is functional *It is unstable*	It is functional It is easy to use
6	It has aesthetic impact *It seems difficult to use* *It seems very dirty*	It is functional *It is unstable* It drains well the juice	It has aesthetic impact It is attractive It instigates use *It is unstable* It drains well the juice
7	It has aesthetic impact *It does not inform the function*	*It is unstable* It drains well the juice *It looks fragile* It is functional	*It is unstable* It drains well the juice
8	*It does not inform the function* It has aesthetic impact	*It is unstable* It drains well the juice	It is functional It is easy to use

It should be noted that the negative impressions (marked in *italic*) tend to diminish or disappear with the use of experience and memory. In particular, the perception of the lack of stability, including commentaries by two of them that do not answered the request about what they remembered about the experience. That does not mean that perception no longer exists, but its importance is smaller with time.

Moreover, the positive perceptions tend to be maintained or even enhanced with time. In reference to its aesthetic appeal, consolidated with the phrase "has aesthetic impact", an exception was observed. All the subjects analyzed and even other three that are not part of the analysis described the *Juicy Salif* as impressive or distinct. However, in the group that reported the memory on the *Juicy Salif,* only two have mentioned this aspect:

- subject 2 said "The juicer, although quite different from ordinary juicers, is efficient, practical and easy to use.
- subject 6 said "The juicer used in class had a very interesting way, which was distracting and urged its use"

Those are two different views: one considers that form does not affect ther performance and the other considers the squeezer's form as a motivating factor to use.

4   CONCLUSIONS

The concept of semantic networks, in the case of pre-trial operation of an artifact, says that human memory is associative (FIALHO, 2001). Ulrich Neisser noted that people do not realize pure forms, unrelated objects, or things like meaning. The difference between what is and what it means an object cannot be demonstrated, as in perceptual data are in the origin. The responses suggest that objects are always seen in the context of things, situations and users, including the observer himself. (IIDA, 2008)

It not appropriate to say that a product with an aesthetic appeal, or as it is called the Juicy Salif lemon squeezer, with a fun appeal, that causes a strangeness or surprise, are low-functioning, or that the function becomes secondary. This dilemma between the use and the image of a product is something that permeates cognitive issues that are not present only in the lemon squeezer by Philippe Starck. The Aeron chair from Herman Miller, suffered from the lack of user acceptance, in the tests prior to its release. Although designed to give the best possible comfort for the user, visually the chair, not pass this: "People looked at the structure, which seemed made of wire and wondered if the chair would stand them and then looked at the net and thought if it could be comfortable "(GLADWELL, 2005). Herman Miller noted consumers looking for chairs, that choosing an office chair is basically driven by the assumed status, and this is also what will convey comfort to that user.

Studies suggest that the squeezer does not work properly (RUSSO and MORAES, 2003). If the object is deliberately isolated of the environment and its functions are observed only with parameters such as how fast the task is done, for example, we may qualify the *Juicy Salif* lemon squeezer, as inappropriate. But what we saw when we propose to several users who make use of the way they thought most appropriate was that the product worked and somewhat amused users. And surprisingly, it demonstrated to be a lemon juice extractor even more efficient than others in some items, such as the ease of cleaning.

The user experience has a role in learning about the product and redefining their image. It is important to remember that most people involved in this experiment were students of design. Apart from their prejudice with respect to the *Juicy Salif*, they brought to the experiment their vision about design and their methods of product evaluation. Another profile would give different results, probably.

REFERENCES

CALVERA, A. De lo bello de las cosas: Materiales para una estética del diseño. Barcelona: G. Gili, 2007.

DESMET, P.M.A., HEKKERT, P.P.M. The Basis of Product Emotion. In: GREEN, William S.; JORDAN, Patrick W. Pleasure with Products: Beyond Usability. London: Taylor & Francis, 2002.

FIALHO, F. A. P. Ciências da cognição. Florianópolis: Insular, 2001.

GLADWELL, M. Outliers: The Story of Success. Boston: Little, Brown, 2008.

GLADWELL, M. Blink: a decisão num piscar de olhos. Rio de Janeiro: Rocco, 2005.

GREEN, William S.; JORDAN, Patrick W. Pleasure with Products: Beyond Usability. London: Taylor & Francis, 2002.

IIDA, I. Ergonomia: projeto e produção. São Paulo: E. Blucher, 2005.

JORDAN, P. An Introduction to Usability. London: Taylor & Francis, 2002.

JORDAN, P. W. Designing Pleasurable Products: an introduction to the new human factors. London: Taylor and Francis, 2000.

KRIPPENDORFF, K. On the Essential Contexts of Artifacts or on the Proposition That "Design Is Making Sense (Of Things)". Design Issues, Vol. 5, No. 2, 1989.

MILLS, C. W. Sobre o artesanato intelectual e outros ensaios. Rio de Janeiro: J. Zahar, 2009.

LLOYD, P.A., SNELDERs, H.M.J.J. What Was Philippe Starck Thinking of? Design Studies, 24, pp 237-253, 2003.

RUSSO, B. and MORAES, A. de. The Lack of Usability in Design Icons An Affective Case Study About Juicy Salif.Pproceedings of the DPPI03 pp 146-147, 2003.

ZIMMER, H. D. et al. Memory for action. New York: Oxford University Press, 2001.

CHAPTER 58

Ergonomics and Emotional Values in Design Process: the Case Study Daciano da Costa

Ana Moreira da Silva

CIAUD, Technical University of Lisbon
Portugal
anamoreiradasilva@gmail.com

ABSTRACT

This paper, which stems from a developing research project, aims to disseminate among the international scientific community the thought of Daciano da Costa (1930-2005) given the importance he conferred to Emotional Values in the Design Process, throughout his teaching career and his professional practice as a reference personage in the Design of the twentieth century in Portugal. We focus on the important role played by Emotional Values within the conceptual Design process, through the case study Daciano da Costa. What Daciano brought to the practice and teaching was a modernization of processes, a new perspective on the emerging themes of design, like Emotional Values. Daciano believed that designing was providing a service. This task was understood as the building of a relationship with users. The main issue is that emotions have a crucial role in the human ability to understand the world and how they learn new things. This concept developed by Donald Norman (2004) in his book *Emotional Design*, was already in Daciano's mind as a designer and as a professor.

Daciano da Costa was responsible for numerous interior and installation design projects of outstanding design quality from the early 60's onwards. His work made him the most important figure of a first generation of designers in Portugal.

Keywords: Ergonomics, Emotional Values, Daciano da Costa, Design Process.

1   INTRODUCTION

Emotional Design is both the title of a book by Donald Norman (2004) and the concept it represents. The main issue is that emotions have a crucial role in the human being and to the affinity the user feels for an object, due to the formation of an emotional connection with the object.

"User friendliness, human based technology, human factors – these terms are all synonyms for the basic notion that if we want the best out of people their technology and environment should be fashioned to fit their wishes, expectations and abilities." (Oborne, 1995)

"Advances in our understanding of emotion and affect have implications for the science of design. (...) Good design means that beauty and usability are in balance." (Norman, 2002)

Those concepts were already in Daciano's mind as a designer and as a professor.

Daciano da Costa brought to the practice and teaching, during the XX century in Portugal, a modernization of processes, a new perspective on the emerging themes of design, like Emotional Values.

2   ERGONOMICS AND EMOTIONAL VALUES IN THE DESIGN PROCESS OF DACIANO DA COSTA

We cannot speak about Portuguese Design without referring to Daciano da Costa. He was one of the pioneers of industrial design in Portugal. For more then four decades he conceived furniture and interior design, still used satisfactory nowadays.

Daciano da Costa stated that "man is the real protagonist of space" (1998, p.26). He was concerned with the design of spaces in which people carry out work or fruition.

According to Cushman and Rosenberg (1991) product design is the process of creating newest and better products for people to use. Ergonomics is responsible for the product usability focusing in the comfort, efficiency and safety.

In the design for everyday life situations, the focus of ergonomics is the human being.

For Daciano da Costa (1998), ergonomics contributes for the design and evaluation of work systems and products, in a way to ensure that the working environment must be designed to fit people's thoughts, wishes and abilities.

As Oborne (1998) outlines, a major role of ergonomics is to identify design issues which involve the human component of the work system.

According to Daciano objects and spaces have not only useful functions but also symbolic functions by means of fruition by the users. These functions are subjective, like aesthetics, affective, enjoyment or pleasure, but they represent an added value that will appeal the user due to an emotional connection with the object. (1998, p.50)

In the interiors he designed, such as the reading rooms at the archives of the Lisbon National Library and the Calouste Gulbenkian Foundation Library, the comfortable armchairs in distinguished surroundings create an atmosphere in the relation between the architecture and its furnishings, where users set and remain pushing them to the limits of their capacity for study and concentration, experiencing the emotional values of such spaces specially created for this purposes. We manage to deduce in his projects a clear *modus operandi* from a sensory analysis. The 'physical' dimensions and the 'hidden' dimensions became fundamental concepts. It would not simply be a case of distributing areas and uses; in the design of the interior spaces, the physical dimensions of the spaces and their metric relations would have to be closely aligned with the sensory dimensions of the space, which are measured by the quality of the environments (lighting quality, colour effect, furnishing comfort, etc.) humanizing this spaces under the emotional values point of view.

Daciano da Costa played a pioneer role on Design's theory fundaments in Portugal and an important pedagogic role as a teacher.

Daciano created and implemented the Design Graduation Degree Plan, established in 1991 at the Faculty of Architecture, Technical University of Lisbon, being its coordinator since then until 2003.

As a professor when conceiving, developing and supervising the exercises he set, the student's recognition of the importance of emotional values were one of his main goals.

Daciano da Costa believed that designing was providing a service. This task was understood as the building of a relationship with users. This is why he liked to paraphrase Orson Welles, saying "in this age of supermarkets, you can always count on your friendly neighbourhood grocer." Daciano aspired to be a "friendly grocer" in the design world (Martins and Spencer, 2009). For him, objects and spaces have not only useful functions but also symbolic functions by means of fruition by the users and they represent an added value due to a formation of an emotional connection between the user and the object.

To exemplify those issues in Daciano da Costa's work we present the following figures showing a clear and direct demonstration of the importance he conferred to ergonomics and emotional values in the design conceptual process, since the first ideas until the evaluation and implementation of the several solutions there is a direct and constant relationship with the human figure, a connection with the user.

Figure 1. Lisbon National Library (1965-1968). (Costa 2001, p.62)

Figure 2. Calouste Gulbenkian Foundation Library (1966-1969). (Costa 2001, p.132)

Figure 3. University of Lisbon - Rectory interiors drawing (1960-1961). (Costa 2001, p.112)

Figure 4 – Lisbon City Council. Councillors' bench sketches (1997). (Costa 2001, p.194)

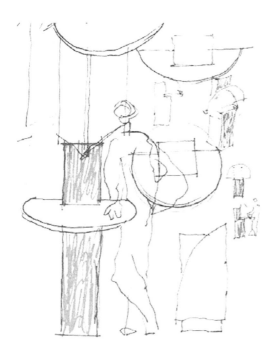

Figure 5. Furniture line Metropolis sketches (1988). (Costa 2001, p.276)

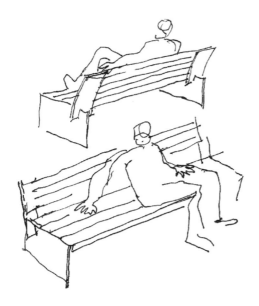

Figure 6. Urban design. Garden benches sketches (1994). (Costa 2001, p.233)

3   CONCLUSIONS

The person-centered view of ergonomics is so important today as it was years ago to Daciano da Costa, although, with the passage of time, more material is now available to support the argument that whatever happens within the system it is the individual human being who is the "prime actor". The work of Daciano da Costa, in his teaching career and in his professional practice, can be considered as an innovative one, bringing a new perspective in those times, given the importance he conferred to Ergonomics and Emotional Values in the Design Process.

ACKNOWLEDGMENTS

The author would like to acknowledge the support given by CIAUD – Research Center in Architecture, Urbanism and Design, Portugal, and FCT – Foundation for the Science and Technology, Portugal.

REFERENCES

Costa, D. 1998. *Design e Mal-Estar*, Centro Português de Design, Lisboa.
Costa, D. 2001. *Catálogo da Exposição Daciano da Costa Designer*, Fundação Calouste Gulbenkian, Lisboa, pp. 62, 112, 132, 194, 233, 276.
Cushman, W. H. and Rosenberg, D. 1991. *Human Factors in Product Design*, Elsevier, New York.
Martins, J. P. and Spencer, J. 2009. *Atelier Daciano da Costa*, Neves, J. M. (ed.), True Team Publishing and Design, Cascais.
Norman, D. 2002. *Emotion and design: Attractive things work better*, Interactions Magazine, ix (4) pp. 36-42.
Norman, D. 2004. *Emotional Design: Why We Love (or Hate) Everyday Things*, Basic Books, New York.
Oborne, D. 1995. *Ergonomics at Work – Human Factors in Design and Development*, John Wiley & Sons, New York.

CHAPTER 59

Function versus Emotion in a Wheelchair Design

Paulo Costa, Fernando M. Silva, Carlos Figueiredo

Polytechnic Institute of Guarda, Portugal
UDI - Research Unit for Inland Development
CIAUD – Research Centre in Architecture, Urban Planning and Design,
TU Lisbon,
Lisbon, 1349-055, PORTUGAL
paulocosta12@hotmail.com, fms.fautl@gmail.com

ABSTRACT

This article analyzes the various construction parameters for a wheelchair by a group of users and one of nonusers. It was used a conceptual approach where usability is quantified and parameterized, resulting in a functional analysis of the object. For an emotional approach were used 14 emotions based on surveys, which tried to understand how individual differences, such as education level, age, gender or be active professionally may influence the choice of a standard wheelchair. Based on the results and some background qualitative information selection, criteria were established for wheelchairs due to the individual, where the functional approach intersect with an affective construction of the use experience.

Keywords: function, emotion, wheelchair

1   INDIVIDUAL DIFFERENCES IN THE PERCEPTION OF AN OBJECT

Subjects respond differently to various products. This finding has led to a demand for a development of methods to understand the exact relationship between individual differences and choices made in selecting products. Studies carried out by Kim and Lee (2005) or by Desmet, Hekkert & Jacobs (2000) revealed a relationship between the cultural and emotional responses in the choice of objects. However, there is still a clear relationship between cause and effect which may be used in the development and design of products in a generalized form. The very definition of culture, whether in relation to a nation or an individual raises questions, especially if it is necessary quantification. Cultural levels can be identified (Xiang-Liang He, 1992; Leong & Clark, 2003), but how each of these levels can be quantified and its influence on the user's relationship with the product is yet to be determined.

In this study, and in order to somehow overcome this problem, it will not be mentioned the term culture, just the education level of the individual as a selection criteria. The term usability (Norman, 2002) in its dimensions of efficiency, effectiveness and ease of use has made easier the understanding of the relationship between the user and the product, but only as a theoretical concept. This is due to the need of for each product type be used specific analysis parameters in accordance with the product. Assessing the usability of a wheelchair cannot meet the same electrical parameters of a fryer or a television, so we must also think about the diversification of products under review. Some specific parameters for the wheelchair will also be analyzed with the aim of defining quantitatively assessing that users do the same.

The three types of experiences with possible products for a user in Hekkert (2006) are the aesthetic pleasure, attribution of meaning and emotional response. In the assessment of a wheelchair, all this three types of assessment are used, but differently, depending on whether is a case of a user or a non-user. For aesthetic pleasure, the user will make a more positive assessment of such an object than a non-user as will be shown.

We will not make a theoretical assessment of whether the answer is cognitive (Crilly, Moultrie and Clarkson (2004)) or if there are cognitive processes or emotions at this level (Norman, 2004). The attribution of meaning is already regarded as a cognitive process, by consensus. The associations that can be made on a wheelchair will influence the overall assessment of a wheelchair, as a subject of exclusion for non-users or technical assistance for users. These combinations and cognitive processes play an important role in the positive or negative observation of the wheelchair.

The emotional experience is considered by Frijda (1986) to be functional, since it defines our position on the object or event. Emotion is the result of a cognitive process, often unconscious and automatic (Desmet & Hekkert, 2007). All these processes can theoretically be explained, but in practical cases is extremely difficult to separate the two of them. Studies involving perceptions of affective experiences or emotions always require a definition of the meaning of the emotion in question. This definition can be understood differently from user to user, requiring a more thorough study of user-product experience.

2   CONCEPTUAL PARAMETERS OF A WHEELCHAIR

In order to classify a wheelchair for their usability some interviews were conducted to wheelchair users and studied the ANSI / RESNA standards. There were defined nine key parameters to consider: stability, comfort, storage, control and maneuverability, weight, price, stiffness, shape and color.

A sampling of 114 interviews were made to users of wheelchairs to find out the degree of importance of each of the parameters and the relation between them. The results are shown in Table 1.

Table 1. Analysis of statistical significance of average differences according to the ranking of factors

	Scoring for the total of sample (N=114)			
	M	DP		p
1 - Stability	4,58	,593	1=2; 1=3; 1=4; 1≠5	n.s.
2 - Comfort	4,53	,583	2=3; 2=4; 2=5	n.s.
3 - Storage	4,48	,875	3=4; 3=5	n.s.
4 - Control	4,48	,655	4=5	n.s.
5 - Weight	4,40	,817	5≠6	<.05
6 - Price	4,04	,954	6≠7	<.05
7 - Stiffness	3,85	,952	7≠8	<.05
8 - Form	3,39	1,149	8≠9	<.05
9 - Color	2,20	1,082		

Table 1 was obtained from a study verifying the differences to a number of parameters (one sample t-test) in order to determine whether the ranking factors had some significance. It could be concluded that there are five groups with statistically significant differences, being the first of stability, comfort, storage, weight and control. These parameters do not differ in a statistically significant way and can be included in the first group. It represents the most important group in the selection of parameters by the users. The following groups are, in sequence, always with statistically significant differences: price, stiffness, shape and color.

There were defined some variables such as gender, age and duration of use of a wheelchair to users in order to find statistically significant associations with the set parameters. These results are shown in Table 2 and 3, which have been considered according previous findings in several classes of variables.

Table 2. *Spearman* correlation coefficient between the variables gender, age and duration of use of a wheelchair, and the degree of importance attributed to factors weight, price control, stability and comfort

	Parameters considered in a wheelchair				
	Weight	Price	Control	Stability	Confort
Sex	.167	.202*	.163	.245**	.292**
Age	.038	.072	.042	.061	.006
Use Time	.129	- .067	.083	.127	.089

* p<.05 ** p<.01 ***p<.001

Table 3. *Spearman* correlation coefficient between the variables gender, age and duration of use of a wheelchair, and the degree of importance assigned to the factors of color, shape, stiffness and storage

	Parameters considered in a wheelchair			
	Color	Form	Stiffness	Storage
Sex	- .038	- .019	.110	.187*
Age	- .067	- .069	.043	- .061
Time of use	.018	.055	.074	- .042

* p<.05 ** p<.01 ***p<.001

The *Spearman* correlation coefficient allows to investigate the association between gender, age and duration of use and the parameters to which users respond.

Given the results of Tables 2 and 3, it can be seen that there are only statistically significant associations between the gender variable and the importance attributed to the factors price (ρ = .202, p <.05), stability (ρ = .245; p <.01), comfort (ρ = .292, p <.01) and storage (ρ = .187, p <.05). However the value of these correlations appears to be low. These statistically significant correlations indicate a tendency to be the female to give more importance to the parameters mentioned above.

3   THE WHEELCHAIR AS AN EMOTIONAL OBJECT

The emotional response of the individual to a wheelchair is translated by his connection to objects and society, conditioned by numerous factors that with time define the individual, emotionally and cognitively. The result of such an analysis can provide some results that are statistically significant for the study survey, supplemented by a qualitative analysis.

In the preparation of the inquiry were used six models of wheelchairs, four of them the most used by the 114 users group (model 2-5 of Table 4) and the other two because they present new and unusual aesthetic construction solutions and mobility. The users responded to the survey where was necessary to sort, in a Likert scale from 1 to 7, the intensity of each one of the 14 emotions presented for each wheelchair. The emotions chosen for this process were the same used by Desmet (2002) in his method PrEmo (desire, fascination, satisfaction, disappointment, dissatisfaction, contempt, unpleasant surprise, boredom, pleasant surprise, disgust, indignation, inspiration, amusement and admiration). All emotions were went along with a description of their meanings.

In order to understand the results obtained, it is necessary to identify all the types of commercial wheelchairs used in the inquiry, that can be observed in the Table 4.

Table 4. Models of wheelchairs used in the investigation

Wheelchair	Trade name
1	Marvel
2	Cyclone (Future)
3	Invacare (Küschall Champion)
4	Invacare (Action 3)
5	Invacare (Küschall/light weight/foldable)
6	Able to Enjoy

In an initial analysis were used the variables age, gender, education level, active employment status and activities outside the professional context.

For the surveys 89 completed answers were collected from wheelchair's users and 96 from non-users.

The analysis will follow the theoretical conclusions of Frijda (1986), Roseman (1991), Scherer (1988), Smith & Ellsworth (1985) and Lazarus (1991) in the relation between the emotions product appraisal and product concern (Desmet, 2002). All findings are based on statistical processing validated in accordance with the results obtained and confirmed by qualitative data from interviews.

The conclusions can be summarized as follows:

1 - Active wheelchairs users with activities outside the professional context appear to be more emotionally able to demonstrate positive or negative emotions in relation to the models presented.

2 - Users do not generally show a greater tendency to evaluate negatively the wheelchairs as non-users generally do.

3 - The non-users have a greater number of statistically significant differences in the evaluation of the emotions given by subgroups then the users, which indicates that there is less uniformity in the emotions of non-users before the assessment of wheelchairs.

4 – Both users and non-users react positively to new and different aesthetics (in general, without analyzing subgroups).

5 - The older the people are the less importance are given to aesthetics and innovative solutions.

6 - The higher the education level is, the more value are given to aesthetics and innovative solutions.

7 - The females do not appreciate novelty aesthetics so much as the males, especially when it could jeopardize stability and security.

8 - Professionally active users give great importance to the functional aspects, including stability, storage and weight. These are essential parameters so they can be autonomous in their daily commute.

4   FUNCTIONAL PARAMETERS AND EMOTIONAL EVALUATION

After analyzing the two types of evaluation we can conclude that the parameters presented in the conceptual analysis are perceived differently according to the variables defined for emotional analysis. Despite the stability factor beeing chosen as the most important factor in the selecting of a wheelchair, it can be neglected due to a more aesthetically pleasing performance by some user groups, including those of males under 30 years. This result only became apparent and significant after the emotional analysis of the set of six wheelchairs.

Not always a good usability can impose itself above aesthetic factors in the choosing of a wheelchair, even when the security of the individual is less assured due to a lower stability.

In order to facilitate the understanding of the choices made by users in order of on their age, gender, education level, employment status or activities outside the professional context before the selection of a wheelchair, figures 1-5 were drawn:

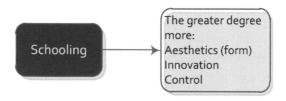

Figure 1 Options for the variable schooling

The higher level of education of users leads to the aesthetic factor or some innovation factors are considered positive, making that aspects associated with traditional wheelchairs are more easily rejected.

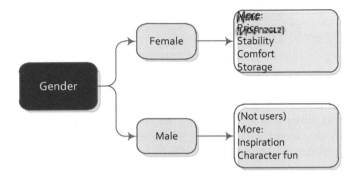

Figure 2 Options for the variable Gender

The variable gender, through the emotional analysis, confirmed the choice of wheelchairs with greater stability, comfort or storage by females. This difference of attitudes decreases with the advancing age of the user.

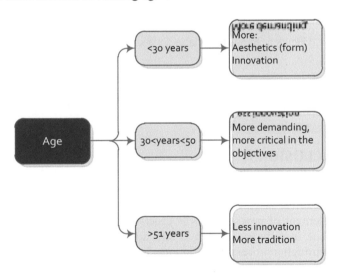

Figure 3 Options for the variable Age

As the age of the users increase they begin to prefer more traditional wheelchairs where stability and comfort tend to be more assured than proposals for new models.

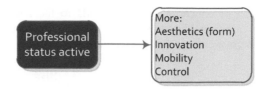

Figure 4 Options for variable employment status

This factor is one of the most demanding to check the assumptions of usability defined in Table 1, by the need for constant use of the wheelchair in outdoor environments, often with their own car. The combination of parameters usability are also usually associated with a strong aesthetics component. It is one of the most demanding groups in choosing a wheelchair.

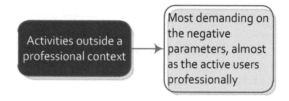

Figure 5 Options for variable activities outsider a professional context

This group presents an evaluation near the verified with active employment status users, and is even more demanding in particular aspects that can aesthetic or functional.

5 CONCLUSIONS

Throughout this study was always underlying the importance of the influence of Habitus (Bourdieu, 2001) in the relation between individuals and objects. Culture as an living element of new attitudes is crucial to the humanization of society. The exclusion due to difference, when is still not viewed as a disability, can only be countered through a process that promotes individual and social creation of identity through culture, the only really distinguishing element in society.

The choice for the traditional that is seen in older people in relation to wheelchairs may be a reflection of an attitude of resistance and protection in relation to the processing speed of the postmodern world of "liquid modernity" (Bauman, 2000) and "corrosion of character" (Sennett, 1998). The exposure of individuals to a changing world does not allow an adaptation of identity and an investment in individual differences, which need to be changed by greater exposure to cultural events and a lower absorption of predominantly visual culture.

The approach of conceptual and expressive aspects are complementary and can only be understood if considered as part of a whole. In the case of wheelchairs both approaches were important to draw some conclusions about the preference of users and non-users of most solutions of wheelchairs commercially proposals.

REFERENCES

Bauman, Zygmunt. Liquid Modernity. Polity Press. 2000

Bourdieu, P..(2001). Razões Práticas Sobre a Teoria da Acção. Celta Editora

Crilly, N., Moultrie, J. and Clarkson, P.J.. (2004). Seeing things: consumer response to the visual domain in product design. Design Studies, 25, (pp.547-577).

Desmet, P.M.A. (2002). Designing Emotions. Delft (NL): Delft University of Technology. BRP Publishers

Desmet, P.M.A., Hekkert, P. and Jacobs, J.J. (2000). When a car makes you smile: development and application of an instrument to mesure product emotions. In S.J. Hoch and R.J. Meyer (Eds.), Advances in Consumer Research, volume 27, (pp. 111-117). Provo, UT: Association for Consumer Research.

Desmet, Pieter and Hekkert, Paul. (2007). Framework of product experience. International Journal of Design, 1(1), (pp.13-23).

Frijda, N. H.. (1986). The emotions. Cambridge: Cambridge University Press

He, X.L. (1992). The worship of Chinese gods of nature (in Chinese). Shunghi:San-Lian book store.

Hekkert, P.. (2006). Design aesthetics: Principles of pleasure in product design. Psychology Science, 48(2), (pp.157-172).

Kim, J. H. and Lee, K.P. (2005). Cultural difference and mobile phone interface design: icon recognition according to level of abstraction. In: Proceedings of the seventh International Conference on Human Computer Interaction with Mobile Devices and Services. Austria: University of Salzburg.

Lazarus, R.S. (1991). Emotion and Adaptation. Oxford: Oxford University Press

Leong, B. C. H. and Clarck, H. (2003). Culture-based knowledge towards new design thinking and practice: a dialogue. Design Issues, vol.19(3), (pp. 48-58)

Norman, D. A.. (2002). The Design of Everyday Things. New York: Basic Books.

Norman, Donald A.. (2004). Emotional design - Why we love (or hate) everyday things. Basic Books.

Roseman, I. J. (1991). Appraisal determinants of discrete emotions. Cognition and emotion, 5, (pp. 161-200)

Scherer, K.R. (1988). Criteria for emotion-antecedent appraisal: a review. In V. Hamilton, G. H. Bower, & N.H. Frijda (Eds.), Cognitive perspectives on emotion and motivation: vol.44. Nato ASI series D: Behavioural and social sciences. (pp. 89-126). Dordrecht, Netherlands: Kluwer

Sennett, Richard. (1998). The corrosion of character - The personal consequences of work in the new capitalism. W.W. Norton & Company, New York, London.

Smith, C.A. and Ellworth, P.C. (1985). Patterns of cognitive appraisal in emotion. Journal of Personality and Social Psychology. 48, (pp. 223-269).

CHAPTER 60

Materials Emotional Profile from the Point of View of Graduation Design Students in Portugal

Pedro Oliveira[1,2], Fernando Moreira da Silva[2,3]

[1] ISEC – Instituto Superior de Educação e Ciências
[2] CIAUD – Centro de Investigação em Arquitectura, Urbanismo e Design
[3] FAUTL – Faculdade de Arquitectura da Universidade Técnica de Lisboa
Lisboa, PORTUGAL
ppoliveira@gmail.com; fms.fautl@gmail.com

ABSTRACT

The study subjects of this research were design students in graduation and post-graduation courses in Portugal. Students were tested in order to comprehend their intuitive perceptions upon symbolic, emotional, and functional/performance aspects of materials. The aim of this research is to determine if student's vision upon materials is too apart from current material possibilities. It was used a methodology of Research Survey among a representative sample of the student population. The results show a profound misunderstanding of the characteristics and capabilities of the most innovative materials, mainly metallic ones. This observation is transversal to all respondents, and is verifiable regardless of the study level, specific nature of the design courses, age or sex. The results achieved are considered to be of the utmost importance in what regards design education improvements has they provide an updated framework upon the prejudices student held on most known materials. They point out to a need of more holistic material teaching in graduation design courses in Portugal.

Keywords: material, metal, symbolism, emotion, identity, education, design.

CONTEXT

The design of emotional aspects may have more impact on the success of a product than the practical aspects or their strict functionality (Norman, 1998). Everything we do has both a cognitive and an affective dimension. The cognitive dimension assigns meaning at things, the affective ads value.

In this context, the project of materials primary qualities must be considered as a major design task as it is a key aspect in the design of more emotionally assertive products. The organoleptic material qualities play a decisive role in the seductive power of the objects they embody.

If one does not consider the emotional / cultural dimension spread by materials image, is led to a strictly functional decision. That is why materials selection should not be made by software but by designers able to perceived and use the semantic significant associated to their image.

In order to confirm the mental representation that higher education design students en Portugal have towards main material families, it has been conducted a survey through student groups. The survey was divided into two parts: the first one involved only ISEC – Instituto Superior de Educação e Ciências – students (which inquired about all the main material families) and a second survey, more complex, involving students from several universities, and that showed samples of materials.

INTUITIVE ASSOCIATIONS HELD TOWARDS MAIN MATERIAL FAMILIES

The first part of the survey research study was made in June 2008 at Design students from ISEC. The number of valid inquiries corresponded to 76% of students within the 1st graduation year, 45% at 2nd year, and 88% at 3rd year. The universe of respondents was divided equally between males and females. The subjects were inquired without showing any material sample, since the specific features of these could endanger the trustiness of each material family aprioristic representation.

In order to draw materials emotional profile it was suggested to pinpoint the proximity of each material to pair of adjectives able to describe materials (binomials), which were extracted from Ashby and Johnson (2002, p.77), as they are the most frequently used in design magazines to characterize materials. The average of the results allowed the definition, for each material family, of a specific emotional profile (Figure 1).

The survey results, analyzed in terms of total response frequency and gender differences, led to indict some dissonance with everyday material reality and some common assumptions. Metals, for example, were mostly associated with the terms "Hard," "Durable," "Futurist" and "Aggressive", defined as "Male", "Aggressive" and "Serious", especially by respondents of the male gender. Respondents of the female gender, although sharing much the same opinion, are less affirmative in their associations pinpointing. There was also a disagreement between genders regarding the characterization of binomials "Classic / Trendy" and "Extravagant / Restrained", with respondents of the female gender to characterize it as being "Trendy" and as "Extravagant" and male respondents to consider it as "Classic".

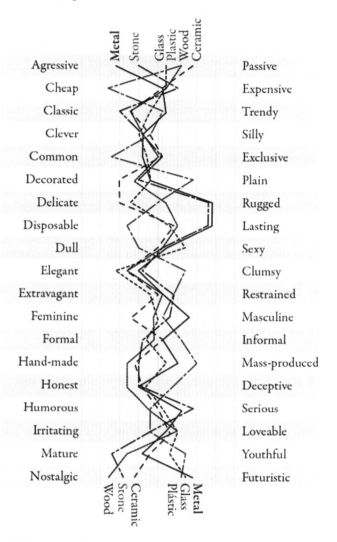

Figure 1: Results of the survey with respect to associations with the materials: Wood, Ceramics, Glass, Metal, Plastic, Stone compared with different pairs of binomials.

Glass was associated with the adjectives "Elegant", "Sexy", "Delicate" and "Futuristic".

Plastic was connoted very differently from other families. It was definitely associated with the adjectives "Cheap", "Common" "Plain", "Disposable", "Dull" "Clumsy", "Informal", "Mass-produced", "Deceptive" and "Youthful".

Stone was considered the most "Expensive" of all materials and also "Classic", "Serious", "Mature" and "Formal" and only surpassed by Metal in the association to the terms "Rugged" and "Lasting".

Wood, more than any other family, was associated to the adjectives "Handmade", "Honest" and "Nostalgic".

Ceramic, by other hand, was predominantly associated to the adjectives "Passive", "Decorated", "Delicate" and "Feminine".

The results indicate that emotional stereotypes held by students upon material families are derived from the characteristics of everyday life applications. The fact that innovative materials characteristics are not yet spread, make it possible for an outdated image of materials to outweighed them.

Aprioristic images of materials are expected to change in response to consistent and lasting everyday observations, but they will always tend to be outdated.

Furthermore it must be considered that a material family characterization may not reveal the complete diversity within each of its single members. This is why it was drawn another survey, specifically upon a material family – metals.

INTUITIVE ASSOCIATIONS HELD TOWARDS THE MOST REPRESENTATIVE METALS

In order to draw the individual associations aprioristically held towards main metals member was made a second survey, between November 2010 and June 2011, at higher education design students in Portugal. To obtain a correct sampling of the population it was defined a tolerance error of 5% and a 90% confidence level. It resulted from here a recommended sample size of at least 263 students (the total number of students in design courses in Portugal was 8.618 – data from 2008/09 collected by CPD). The sample was then stratified by study level (graduation, master, doctorate), by geographic area (north, centre, and south) and by gender of respondents.

The selection criteria for the selection of students institution origin consider the number of students enrolled during academic year 2008/09 in the following territorial divisions of Portugal: North (33.4% of students enrolled equivalent to 5 of the 15 design schools there held), Centre (63.5%, equivalent to 10 of 15 of the design schools); South and Islands (3.1%, equivalent to one school). Thus, from the North it has been chosen ESAD-Matosinhos, ESDGT-UA, EST-IPCA, FBAUP, FE-UBI; from Center ESAD-FRESS, ESAD-IPL, ESART-IPCB, ESTT-IPT, FAUTL, FBAUL, IADE, ISEC, Lusófona and Lusíada; from the South and Islands it was chosen ESE-UALG. Consistent with the same logic of proportionate representation, the survey planned to be distributed to a minimum of 223 students in Bachelor design courses (84.7% of the 263 required), 37 students in Master in design degree courses (13.9%) and 4 students in PhD design courses (1.4%).

The results of the survey (Figure 2) point out to a significant coincidence in the correlation between heavier materials like Steel, Iron and Bronze and their hardness qualities – which is consistent with the theory that the traditional metal family identity introduced a close correlation between weight and strength.

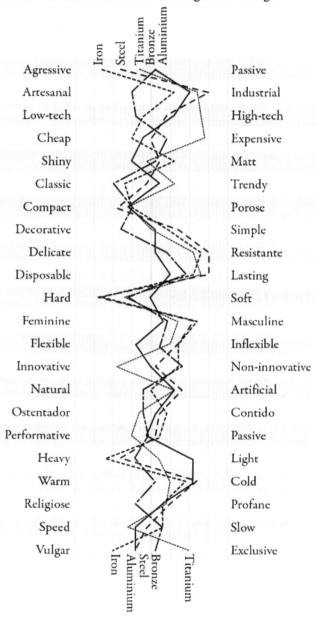

Figure 2: Associations made with metal Steel, Aluminium, Bronze, Iron and Titanium compared with different pairs of binomials.

On the other hand it was verifiable a tendency to associate the newest metals like Aluminum and Titanium, also the lighter ones, to a presumption of less hardness (even though, at least in the case that Titanium is evidently a mislead assumption).

Most recent metals (like Aluminum and Titanium) are associated to "Speed", "Performative" and "High-tech", while the oldest metals (Iron, Steel) are linked to other concepts: "Heavy", Aggressive" and "Masculine".

The analysis of the various responses brings out a striking difference between the image of the recent past main metals – Iron and Steel –, and the latest metals – Aluminium especially, but also the Titanium. The first ones perpetuate an image of strength, masculinity, aggressiveness, toughness, and a proportionality between weight and resistance, which prevailed for centuries as the identity of the entire family of metals- The second one's – Aluminium and Titanium – are, on the other hand, mainly associated to an image of lightness, high-technology, an unbind of the correlation weight / resistance, and a masculine identity best illustrated by the concept of speed as performance.

It was observed that, in general, most metals remain tightly bound to the idea of hardness and durability (as having potential to ensure longevity to the objects they embody). The exception to the longevity association is detected towards metals that oxidize – a characteristic of Western culture that continues to dominate in the present time. There was notice, however, a significant difference between the aging presumption of earliest metals – Gold, Silver, Iron, Copper – and the latest one's – Aluminium and Titanium. To the earliest it is assumed that the aging process is a result of many factors, such as oxidation, loss of gloss, color change, scratches, ecc.; To the second group, the aging evidences seem to be only focused in scratch evidences – which allows us to demonstrate that also regarding aging expectations, the new metals draw a difference from the others.

CONCLUSIONS

Throughout design history a poor attention has been paid to scientific research on materials primary characteristics. But, in order to fully design a product one must comprehend the materials physical and mechanical characteristics as well as its sensory qualities, symbolism and cultural values – their emotional profile. In order to sustain an assertive design one must be aware of its one prejudices towards material characteristics/possibilities. This is why it was drawn a survey to be submitted at design students that allowed the definition of individual and family material (predjudice) profiles. The results evidence that most materials are expected to perform bellow their present possibilities and points to the need to increase sensory and symbols content in the discipline of materials in Design courses.

ACKNOWLEDGMENTS

The authors would like to acknowledge the contribution of all schools involved in the study that authorized students to respond to the survey and also to the professors that patiently allow the use of time of their classes.

REFERENCES

Ashby, M. and K. Johnson, *Materials and Design – The Art and Science of Material Selection in Product Design*, Oxford, Elsevier Butterworth-Heinemann, 2002.
Norman, D. *The Design of Everyday Things*. Cambridge: MIT Press, 1998.

CHAPTER 61

Ergonomics as an Important Factor of Change in Women's Clothing during "Belle Époque" Period

Maria Heloisa Albuquerque

Faculdade de Arquitectura da Universidade Técnica
Lisboa, Portugal
heloisafalbuquerque@gmail.com

ABSTRACT

Throughout times, attempts were made to the welfare of human beings on behalf of the concepts of beauty of various ages. Fashion shaped the male and female body over centuries. And finally at the end of the 19th century ergonomic principles were a turning point during the Belle Époque.

Keywords: Belle Époque; silhouette; feminists; fashion

1. ERGONOMICS AS AN IMPORTANT FACTOR OF CHANGE IN WOMEN'S CLOTHING AT "BELLE ÉPOQUE" PERIOD

Ergonomics is the scientific subject related to the understanding of the human being's interactions with other elements of a system, and also is the profession that enforces theory, principles, data and methods to project, in order to optimize well-being and overall performance of a system.

Theory, data and methods to project, ***in order to optimize well-being and overall performance of a system.***

No doubt that nowadays an object, a machine, a chair is (almost) always (almost) ergonomic.

There is nothing that humans use that interferes more with their well-being than the clothes they wear! And what we said about the constant presence of ergonomic concerns in relation to design, can we say today in relation to fashion design? Is Ergonomics a principle enshrined in everything we use, whether clothes or accessories?

1.2 COSTUME BEFORE ERGONOMICS

The truth is that when we see pictures of old costumes we grieve to think how painful it would be to use it in everyday life. The same thing happens when we read that the Japanese girls' feet were bandaged at birth, so they did not grow, since it was considered to be more beautiful when ladies had little feet.

Figure 1- Vest (century. XV) for malformed bodies

Throughout times, attempts were made to the welfare of human beings on behalf of the concepts of beauty of various ages. The male and female body has been shaped over centuries as was fashion. Women's breasts were being squeezed, flattened or more stressed according to fashion. Like the male silhouette was framed as it was considered a handsome man with broad shoulders, prominent breasts and waist, so did the modifications of the female silhouette throughout history.

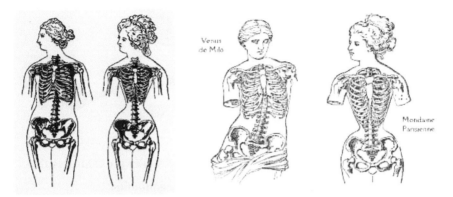

Figure 2 – Effect of the corset on the human body

These modifications of the human body were made using artefacts, some of which seemed more like real instruments of physical torture, only reported by doctors as harmful to health in the nineteenth century.

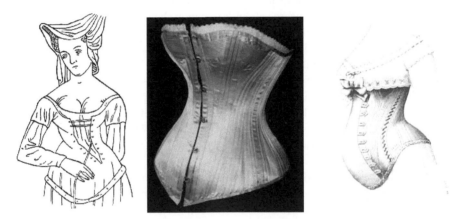

Figure 3 – Corset of the 16th century and 17th and a corset for a pregnant lady

Despite the warning, it was common practice to use the corset during the nineteenth century, whether to separate the breasts by a complicated system of fins at the time of the Empire, or during Restoration when the waist goes back into its right place and it was fashionable to have a marked and thin waist, maintaining the corset as king until the silhouette in S of the Belle Époque. Once again the comfort bows in front of fashion.

But designers of underwear were not the only ones to sin in regard to ergonomics. If we look at the history of footwear, we can notice that the shape of the foot was not the first concern in the design of the shoes through the ages. For centuries, footwear has always been far from the principles of good design and ergonomics in particular.

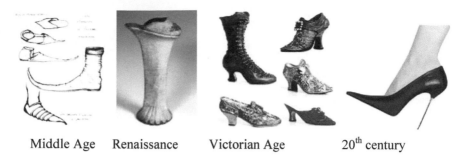

| Middle Age | Renaissance | Victorian Age | 20th century |

**Figure 4 - Format and bones of the foot / shoes over the centuries that
contradict the principles of ergonomics**

1.3 THE BEGINNING OF FASHION PREOCCUPATIONS WITH WELL-BEING AND HEALTH

Despite all these attacks on health made in the name of beauty and fashion, ergonomics is an important factor in the change of female costume in the Belle Époque, and the concerns about it began in the middle of the century.

In fact, in the second half of the nineteenth century, apart from doctors and hygienists, others connected to the female world and the fashion in particular, were concerned with the adaptation of costume to the female body, thus indicating ergonomic concerns. Also in the mid-nineteenth century several voices were raised against the injustice of the situation of women and their rights.

Amelie Bloomer (1818-1894) and Elizabeth Smith Miller (1822-1911), supporters of female Rights, who had participated on National Women's Rights Convention and were considered the first real feminists, were the first people who fought for women's rights.

Elizabeth Miller founded the Rational Dress Society in 1881 seen as responsible for the creation of the bloomers, a kind of Turkish trousers, as they became famous with Amelie Bloomer, and hence often considered her own creation.

Amelie Bloomer owned a feminist magazine, The Lily, where following an article published by her husband Dexter Bloomer, who was also owner and director of a weekly newspaper, The Seneca County Courier, she advocates simplification of female costume for the practice of sport. This costume consists of baggy pants tight at the ankle and a wide skirt to the knee, that only really had success in the late 80's and 90's, since yet many women continued to play sports such as tennis, badminton or croquet with their clothes of the afternoon, despite the fact that bloomers had been modified to adapt to fashion (Figure 4b).

a   b   d

Figure 5--a- Amelie Bloomer using bloomers; b Cyclists using the composed

of jacket and bloomers in the Belle Époque; *Tailleur* **1890s**

Rational dress, commonly called the suit (tailor in English) with skirt, was much easier to use when compared to other costumes. The style was simple, usable by all classes and found a huge success in the Belle Époque, only since it was simple enough to be worn by working women, while being and elegant enough to be used as a business suit for the ladies of the upper classes (Figure 4c). It was also adapted to cycling and other sports, as seen in figure 4, by only replacing the skirt for pants.

The problem of rationality of female costume was also the subject of study for artists and intellectuals, such as the ones from The Aesthetic Dress Movement integrated into the Aesthetic Movement a movement influenced by the Pre - Raphaelites.; Examples are: William Morris, Oscar Wilde and Arthur Liberty,

The Aesthetic dress was a protest against the fashion of the time which forced women to achieve a certain silhouette by means of corsets and poufs. Only a small number of women involved in artistic and literary world, adopted these dresses inspired by the Middle Age. Nevertheless this group had a major importance, as it was one more group opposing to fashion's impositions.

Figure 6 – Aesthetic dress

Also Charles Frederick Worth, the father of haute couture, was concerned with the rationalization of the suit and the adjustment to this new life of women in the second half of the nineteenth century. His creations of the '70s revolutionized the female silhouette. Still in the 60s he created one costume, consisting of a tunic to the knees and a long skirt, of oriental inspiration. Later he set aside the skirts of large circular wheel, achieved through the crinoline, passing the volume to the back and making large tails fashionable. He redefines the feminine silhouette and the concept of elegance, avoiding the excess of adornments.

Figure 7– From crinoline to bustle

In fact the feminine silhouette changed during the nineteenth century due to modifications and improvements made in underwear. From 1860 until the silhouette S of Belle Époque, the artifacts that formed the silhouette kept adapting to the demanding of fashion. Under the direction of Charles Worth, from the mid 70's, the silhouette acquires a more natural shape, achieved with the use of corsets (corsets / bustier) tight back and tournures and petticoats that concentrated the volume on the back.

In the Eighties the silhouette suffers a major transformation due to the tournure and corsets with fillings in the stomach area and very tight at the waist, giving origin to what is usually referred to as the pigeon craw. The feminine silhouette had the shape of S, the wavy line of Art Nouveau.

The 90s saw again changes to the feminine silhouette. The English call the silhouette of this decade, hourglass figure, since the wide sleeves that were used, the very narrow waist and bell-shaped skirt, made the silhouette a reminiscent of, in fact, an hourglass. It was this silhouette that marked the '90s, one of the most emblematic decades of the Belle Époque. It was at the end of this century, that skirt and jacket became common, whether a short jacket or a three-quarters jacket, but always with very voluminous sleeves in the area of the shoulders and with a very marked waist. In fact two visible features in all of these costumes, whether they were prom dresses, riding costumes or any other sport's costume.

Figure 8 – Costumes of Belle Époque

This figure was fashionable for all ages, and as the image of the right of figure 7 shows, also for boys. At least until they were three years old, clothes of girls and boys were not different. To children and adolescents the waist was also tightened and the shoulders were made to stand out with big balloon sleeves.

Fashion was not sorry for the discomfort that these garments represented and for how it hindered the children's day-to-day play. Just as fashion did not pity women; although there was already the notion that it was necessary to take health into account in a time when women were violently squeezed by the corsets that up to the end of the century had to be tightened by a third person.

The problems that corsets caused to those who used them lead to them progressively becoming softer, more flexible, whether by using different materials or by the way of tying them less tight.

These exaggerations have disappeared, partly because fashion imposed so. Paul Poiret inspired by the Ballet Russes who acted for the first time in Paris in 1909, still in the Belle Époque, revolutionizes the female dress.

Models, colors and prints are inspired by the traditional costumes of the East. Just like hats have been replaced by large flaps turbans, the feminine silhouette suffers a radical change, it becomes longilineal for which the turbans also contribute, Poiret abandoned the corset which continued to be used to construct the silhouette of Belle Époque, which in a certain way will coexist with this more radical one.

Figure 9 – Illustrations of Paul Poiret's dresses

2. CONCLUSIONS

As we have seen, the principles of ergonomics in the mid-nineteenth century begin to drive, though shy and with slow changes, the female silhouette. The progressive elimination of the heavy and cumbersome frames worn under the petticoats and corsets that began to be progressively designed thinking of the body shape, are ergonomic concerns of the nineteenth century.

Indeed the problematic of ergonomics was not yet a concern, at least one known by that name, but the concerns of design in order to optimize human well-being, were undoubtedly a guideline for the various fashion designers (then not yet referred to as such) and the struggles of feminists, intellectuals and artists who have made the defense of women's well being their flag.

As mentioned above ergonomics in the design of consumer products was already a concern in the Belle Époque, although it became an issue in previous decades by doctors, intellectuals and fashion designers.

REFERENCES

ALVIM, Mª Helena Villas-Boas (2005) Do Tempo e da moda, Museu Nacional do Traje, Lisboa

BAILLEUX, Nathalie, REMAURY, Bruno(1998) Modes & Vêtements, Gallimard, Paris

BAUDOT, François (2006) Poiret, Assouline Publishing, New York

BLUM, Stella, Ed (1985) Fashion and Costumes from Godey's Lady's Book, Dover Publications, Inc, New York

BOUCHER, François, (1995) Histoire du Costume en Occident de l'Antiquité à nos Jours, Flammarion, Paris

Catalogue de l'Éxposition Sous l'Émpire des Crinolines, Musée Galliera

FUKAI, Akiko dirc (2010) Moda, Taschen, Köln

LAVER, James (2002) Costume and Fashion, A Concise History, Thames & Hudson, London

NERET, Gilles (2001) Dessous, Taschen, Köln

PERROT, Phillipe (1994) Fashioning the Bourgeoisie, Princeton University Press, New Jersey

SIMON, Marie (1998) Les Dessous, Les Carnets de la mode, Editions du Chêne,

WILLETT, C, CUNNINGTON, P (1992) The History of Underclothes, Dover Publications, Inc, New York

CHAPTER 62

Cars Interior Design Customization as an Ergonomics Major Factor

Paulo Dinis, Fernando Moreira da Silva

CIAUD, Technical University of Lisbon,
Portugal
paudini@gmail.com, fms.fautl@gmail.com

ABSTRACT

This article refers to an ongoing investigation that has as main objective the identification, development and implementation of new concepts of products tailored to individual niche vehicles.

The Portuguese car components industry has adapted to new paradigms of production and is focused in particular on the needs of segment market.

The redesigned product development and / or custom integrators to the brands can be seen as an economic opportunity for designers and project-oriented and business.

Coupled with the customization of the components related issues arise with the interior in the car and the resulting adjustments to the needs of different users, genders, ages, etc..

Flexibility, innovation and responsiveness of small and medium-sized Portuguese components, allow integration of new projects in the area of mobility, incubation and testing of new segments, taking into account the future needs of cities and their users.

Keywords: Portuguese car components industry, ergonomics, car customization, car design individual segment.

1   THE INDUSTRY OF PORTUGUESE CAR COMPONENTS

The car is now an inescapable presence in people's lives, having a key role in economic activity and organization of society itself.

In the last two decades we have witnessed the globalization of markets and information. The brands of the world need to operate in a global market and design centers tend to be conceptually oriented to the demands and culture of the market they target.

Today we talk about design developed for the European, Asian or American, meaning a style guide for each of the markets, regardless of country of origin of the mark (Marcelino, 2008).

The Portuguese sector of car industry is characterized by its mainstreaming and has assumed an overwhelming importance in the recognition and development of Portuguese industry, reflecting a positive trade balance. This business sector has to restructure itself in a progressive manner, in order to monitor and respond effectively to new challenges imposed by internationalization and globalization (AFIA, 2008).

In the Portuguese car history, there are few significant examples of a complete car production, as well as few automotive products and brands that hold their own. In most cases, the design or improvement of a product is defined previously by integrating brand, leaving no room for companies to create their own products (Marcelino, 2008).

According to António Castro Guerra (2007), Portugal has shown significant growth in the export of technology and higher value products with growing domestic know-how. The country has sent to the big brands the ability to adapt to new paradigms of production car with a flexible supply chain and the potential for production oriented market niches.

The use of research and development also identified opportunities for the design sector. (Marcelino, 2008).

We need factories and small workshops. (Papanek, 1995, p.70).

According to Kunde (2010), small and medium enterprises should invest in a component parallel market, where they could compete with the big car brands through customized production. Large companies have difficulties in changing their mode of production to respond to a small order from a customer. This is where small and medium sized businesses can take advantage of getting quickly adjust their equipment, production lines and assembly to be able to respond to customer needs. The speed and efficiency in a short period of time requires these companies to seek quick, efficient, creative and profitable, generating new products not only for that customer as well as for future projects.

If the products produced result in unexpected success, the company must negotiate with another company in the same sector distinct geographical area. Thus, the products are produced and marketed without travel costs and the aggravated range of companies will remain. Another advantage for the sharing of projects characterized by the ability to adapt to local conditions of use (gender, environment, function, culture, etc..) (Papanek, 1995).

Portugal offers good conditions to serve as a living laboratory for experiencing new concepts of production and mobility. The small size of the territory joins the propensity of the Portuguese for innovation, bringing together the ingredients to make Portugal one area of testing and experimentation. (Pinto, 2011). According to Victor Papanek (1995), young people from around the world spend millions of dollars annually to personalize their cars. This expression of "vernacular like" and popular culture reinforces the paradigm that car manufacturers and their designers are unimaginative and that cars suffer from this monotony.

The average individual anywhere in the world is better informed and more aware of their needs than any designer (...) is quite obvious that the creative needs of most people are best met through collaboration between users and designers. (Papanek, 1995, p.220).

For Gonçalo Quadros (2011), young people were instrumental in major changes in society due to his irreverence and his ability to think "outside the box". The future users require that the projects result in diversified products, which offer differentiated and varied forms of mobility - car, motorcycle, bicycle, other. Future generations are demanding and creative, and daring the rule for new proposals (Borroni-Bird, 2011).

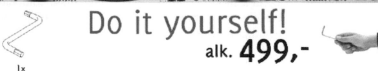

Figure 1 The New Do-It-Yourself VW Golf 2010, accessed 28 Fev 2012,
< http://www.nerdnirvana.org/2010/03/01/the-new-do-it-yourself-vw-golf/ >.

The concept- "do it yourself" has been growing steadily for economic reasons in terms of product assembly at the factory and transport with its reduction and optimization of space.

The consumer / user has been forced to become an owner / builder for the variety of products you buy a kit or small parts to finish assembly. However, this requirement produces side effects beneficial to the user, since it enriches the client learning experience and becomes easier to understand the functioning of the subject / apparatus. On the human level, this practice produces a feeling of satisfaction and self-esteem on participation in the construction of your product.

Build from kits allow people to improvise, devise workarounds and become more inventive and creative in adapting the car to their own needs, (...) (Papanek, 1995, p. 223)

In the 60 policy guidance on imports in the automotive sector was marked by the so-called "Law of the assembly," which restricted the import of vehicles (CBU) [1] "fully constructed, "and liberalized imports of vehicles (CKD) [2] "kits for assembling vehicles".

This legal framework has remained almost unchanged until 1972. However, the second law is assumed as a solution for continuity, establishing minimum percentage of incorporation of the components disassembled vehicles national imported.

As a result of this policy has seen the proliferation of several assembly plants nationally through the operations of foreign direct investment or licensing contracts. The manufacture of motor vehicles in Portugal destined mainly for the domestic market by the end of the 70 (Vale, 1999).

Based on these statements about the industry, economy and consumption Portuguese and the evolution of how the products are offered for sale now - fully assembled, in kit form to be completed by the user; and kit to be assembled by a company - we can expect a continuous increase in the number, scale and complexity of products sold, but also allow the growth of small and medium enterprises with production of niche market and their customization as well as provide distribution services, installation and maintenance.

---

[1] *Completely built up*

[2] *Completely knock down*

2   CAR MARKET SEGMENTATION

The economists Carlos Miguel Coutinho and White (2001), the car market has a high level of competitiveness and dynamism compared to other market sectors, demonstrating the production capacity, quality, technology and advertising. The automotive industry currently faces many challenges as:
- Surplus production;
- Harmonization of prices across Europe;
- Reduction of pollutant emissions;
- Slaughter and recycling of end-of-life;
- New automobile distribution strategies;
- Use and dissemination of new communication technologies.

Manufacturers tend to predict the demand for vehicles in terms of volumes, types and models, versions or options. This type of product exists since the beginning of the automotive industry.

The segmentation of products is made by identification of subsets with characteristics similar to each other and are distinguished from other subsets. Among the different types of criteria to define the segment of each passenger car, you can distinguish them by characteristics such as dimensions, powertrain, body, function, price.

The segmentation of the body of a passenger car is distinguished by:
- Two doors, cabrio, coupe, sedan;
- 3 ports: sedan, station wagon (van), hatchback;
- 4 ports: sedan;
- 5 ports: hatcback, station wagon (van), MPV, SUV.

The segmentation of the customers are grouped by:
- Private customers, fleet operators, state and local authorities, public utilities, members of the diplomatic and consular corps, small and medium-sized domestic companies, multinational companies and international organizations;

The main target of customization is to produce a large variety of products or services so that nearly everyone finds exactly what they want at a reasonable price. (Kunde, 2010).

Figure 2 Mini, 2010, acessed 02 Fev 2012,
< http://www.aprancheta.com/tag/customizacao/page/2/ >

In the case of the automotive sector is already possible to order via internet the brand, model, color, interiors, equipment, accessories and other components of the car according to the needs and tastes of each user.

Figure 3 Fiat 500, acessed 27 Oct 2009
< http://cinquecento.fiat.com.br/ >

3   INTERIOR CAR

The ergonomics of the product focuses on the interface of an object or machine to Man, where there are requirements of the Human Being as a user of a product or service elements such as security, adaptability, practicality, robustness, suitability and comfort (Marcelino, 2008).

According Larica (2003), the car is no longer understood merely as a means of locomotion and also became part of the home and work, due to the increase in distance and time people spend in cars. For this it is necessary to consider several factors in the project such as comfort, safety and ergonomics.

The appearance of a new aesthetic consisting of ecological and environmental considerations will be unpredictable in terms of shape, color, texture, and range, while exciting remarkably since, unlike all the new styles the last one hundred and twenty years handler will not be a restatement of what belongs to the past (Papanek, 1995, p.272).

In the last decade the major car brands have invested in technological innovation within the vehicle in response to the changing needs of drivers. The car became an extension of their homes and new consumers also became more critical, demanding, informed and multiple choice (Larica, 2003).

Munchies (2003) lists some key items to create a layout for the interior of the car:

- Modeling of interfaces between man and the automobile;
- Visual aspects (shape, color and texture);
- Materials (adequacy and security);
- Banks of the vehicle (comfort);
- Positioning of the whole wheel, pedals and quadrant commands (handling);
- Visibility (security);
- Sound insulation (comfort);
- Free internal space (comfort and safety);
- Entry and exit of the vehicle (circulation);
- Storage spaces (multiple use);
- Global factors (society, culture, economy, etc.).

The factors to guide the design of the interior of the car is not just for the anthropometric and ergonomic aspects of the human body, but also by the ability of the signals, the preparation of the recognition of situations and the decisions that volunteer to share. The vehicle has to reassure the user, confidence, ensuring that the components behave according to the specifications of its operation.

Papanek says that designers and users should communicate more and allow people to participate in seeking solutions to their problems. As an example of this practice, Papanek, identifies a target group in need of services consumed by industry and designers, the elderly, whose share in world population has increased considerably in recent years.

For this group of users, in the case of cars leads to an optimization to seek the feasibility of driving the vehicles due to the difficult conditions of access to cities, the number of vehicles and the ability to drive is increasingly reduced with advancing age (Borroni-Bird, 2011).

4 EXPECTED RESULTS

It is anticipated that this project would enable information to cross to explore opportunities for attracting investment to the Portuguese automobile industry in the area of automotive customization. The rapprochement between the designer and the user will provide a faster response, effective and aware of user needs. It is intended to highlight the importance of design and ergonomics national automotive industry, focused on developing niche projects such as the redesign, exterior and interior customization and testing of new segments, centering your target audience on the future users: new generation of young users vs. new generation of elderly users. In the second phase, the exploration of strategies for the incubation of projects in Portugal, will test solutions for international markets in the area of individual mobility, production of individual vehicles and production of niche vehicles in kits with the incorporation of custom components Portuguese.

REFERENCES

AFIA (2008), *Crise do Sector Automóvel*. Associação de Fabricantes para a Indústria Automóvel, Porto.

Borroni-Bird, C 2011, "O Futuro visto pelo futuro", *Jornal de Negócios,* (suplemento), 26 de Maio de 2011, pp. 2-3.

Coutinho, C & Branco M C 2001, "Segmentação do Mercado Automóvel", *Anuário da Economia Portuguesa*, pp. 200-212, acessd 15 May 2010.
< http://markzone.files.wordpress.com/2007/03/segmentacaomercadoautomovel.pdf>.

Guerra, A C (1990), *Formas e Determinantes do Envolvimento Externo das Empresas: Internacionalização da Indústria Automóvel e Integração da Indústria Portuguesa na Indústria Automóvel Mundial*. Dissertação de Doutoramento em Economia, ISEG-UTL, Lisboa.

Kunde, W G 2010, *Gestão da Produção e Qualidade*, accessed 22 Fev 2012,
<http://portal.pr.sebrae.com.br/blogs/posts/gestaoproducao?c=1100>.

Larica, N (2003), *Design de Transportes: Arte em Função da Mobilidade*. 2AB Editora. Rio de Janeiro.

Marcelino, R (2008), *Gestão do Design – Sector Automóvel*. IAPMEI, Lisboa.

Papanek, V (1995), *Arquitectura e Design – Ecologia e Ética*. Edições 70, Lisboa.

Pinto, F (2011) "O Futuro visto pelo futuro", *Jornal de Negócios,* 26 de Maio 2011, pp. 2-8.

Quadros, G 2011, "O Futuro visto pelo futuro", *Jornal de Negócios,* (suplemento), 26 de Maio de 2011, p. 8.

Vale, M 1999, Geografia da industry Automóvel num Contexto de Globalização – Imbricação Espacial do Sistema AutoEuropa. Dissertação de Doutoramento em Geografia, Universidade de Lisboa, Lisboa.

CHAPTER 63

Color in Urban Furniture: A Methodology for Urban Mapping and Wayshowing

Margarida Gamito 1, Fernando Moreira da Silva 2

Faculty of Architecture, Technical University of Lisbon; CIAUD – Research Centre in Architecture, Urban Planning and Design
Lisbon, Portugal
margamito@gmail.com

ABSTRACT

This paper presents a new methodology which is an important part of the contributions of the research project for a PhD in Design. The aim of the research was to define and emphasize the importance of a pertinent color application to urban furniture, arguing that it may contribute for a better visualization of these elements and, consequently, to ameliorate its use. The study also addresses the issue of the color application to urban furniture in the city environment, mainly for city areas identification and how its use may increment the orientation within the city, becoming color a mapping and wayshowing factor. The development and implementation of the new methodology will allow the determination, with a higher scientific approach and rigor, the color planning to be applied to urban furniture in each quarter, or urban area, of a city.

Keywords: color, urban furniture, methodology, color mapping, wayshowing

1   INTRODUCTION

Orientation within the cities, the problem of wayshowing, is not always easy to solve, independently from the individual locomotion mode. Occasional visitors and, even, the inhabitants often have difficulties to locate places, or find their way, in cities which architecture is more or less similar in their different quarters, and where there is a lack of obvious reference points.

Colour as a mean for wayshowing has been successfully employed, though punctually, in architecture and urban areas. Its criteria and widespread application appears to be a way to solve successfully the orientation problem.

For this purpose, this research has sought to apply color to urban furniture in a way that originates a system which will function simultaneously as an identification factor for the different city quarters and as an orientation factor for its inhabitants and visitors. In parallel, colour application to urban furniture will also become an inclusivity factor, by incrementing these elements visibility.

Figure 1   View of a Place of the Baixa quarter, with and without colour application to urban furniture.

2   FUNDAMENTATION

Human beings show psychologic and physiologic reactions to color that make it an important factor for information, communication and understanding the environment. In reference to this, Michael Lancaster (1996:8) states that: *The functions of color are to attract attention, to impart information, to aid deception and to stimulate the emotions.*

These connotations have already been considered on color applications to architecture, examples of which are the chromatic plans for the cities of Turin and Barcelona, and Jean-Philippe Lenclos, Michael Lancaster or Tom Porter projects, among others.

However, these concerns are rarely considered in relation to urban furniture, despite multiple warnings about their lack of visibility made by various authors:

"Color – architecture – cities – colorful cities – color on the urban scene – how does it all fit together? Is a colorful urban backdrop enough, will more color really change our living and working environment? This is the most basic question of all. To put it differently: Is it possible to raise a city's visual accessibility, the quality of experience and orientation, without merely underlining its character as a huge, conglomerate consumer object?" (Machnow & Reuss 1976:21).

For example, in 2009, the City of Austin Design Commission developed very specific and detailed urban design guidelines for Austin, but unfortunately they didn't mention once the importance of color planning or use. And sometimes when this is mentioned in Urban Planning, we never find it related with the urban equipment.

We may consider that urban furniture, although not with this designation, is contemporary to the existence of villages. People always needed benches to rest, being that the church marches or a stone on the edge of a path; railings were always used to prevent access; and, on streets and fairs, there always have been trading booths. When the populations increased and become organize, also the implantation of these elements started to become systematized, despite the urban furniture concept only has institutionalized from the mid-20th century (Serra 1998). Nevertheless the choice of its color or form only rarely obeys to a logic thought.

The urban furniture choice exceeds aesthetic beauty, or the simple wish to decorate the city, it must accomplish its functionality requirements in order to fulfill the population necessities, facilitating their lives and giving them comfort. So, to assure its functionality, urban furniture must protect the health and well-being of the city inhabitants; facilitate the accessibility and use to people with visual or motor difficulties; reinforce the local identity, representing a formal *family* that is coherent and values the surroundings (Águas 2003). However, while recognizing its necessity, the urban furniture functional possibilities have not been used to their fullest extent.

In fact, in order to be used, urban furniture must be clearly seen and, therefore, it must stand out from the environment. As a reinforcement to the necessity of more visibility to urban furniture, we must consider that the city population is constituted by a wide variety of people, with different visual acuities and limitations and, also, by a high percentage of older people. Insofar as people grow older, their ability to see small details decreases and eyes have a crescent difficulty of adaptation to sudden changes of light or a quick change in focus. Bearing in mind the visual limited population, only a small percentage is unable to see any color and the main part is able to distinguish luminosity differences (Lindemann et al 2004). Therefore, to have better visibility conditions, under an inclusive design perspective, urban furniture must present a good chromatic and luminosity contrast. Considering this, Per Mollerup (2005:161) states that "color can be seen from longer distances than other graphic elements" and that "in signage differentiation is the first and foremost role of color".

Cities are, frequently, a complex assortment of streets and buildings that may show an almost monotone similarity or, in opposite, can be extremely diversified. In consequence, orientation within the cities, the wayshowing problem, is not always easy to solve. Occasional visitors and, even, the inhabitants often have difficulties to locate places, or find their way, in cities which architecture is more or less similar in their different quarters, and where there are a lack obvious reference points. Concerning this issue, Charles Hilgenhurst wrote:

"Today we are the strangers in our towns. We do not know and cannot see how things work. Our support systems... are remote.

The information supplied in the environment is largely irrelevant to our immediate purposes or to an understanding of the world in which we live." (Hilgenhurst *apud* Berger 2005:18)

Along with the cities development, complex traffic and transport systems networks emerged. Those caused difficulties to the city visitors' orientation and aroused the necessity to create tools that could help guiding people and, at the same time, renew their identification with the city (Berger, 2005: 10). Gallen Minah (2005: 401), also states that the contemporary city development originates a great diversity and complexity in architecture that *fights* against visibility. Color bypasses its function as an element for definition and unification, and becomes a visual characteristic within the chaos and complexity of the visual field.

Concerning the orientation within the city, and the identification of its different zones, we may consider city maps that differentiate them through the use of different colors. However, on the urban space those colors don't show up, and there is no concern in establishing the correspondence to a real use on this space. The ideal would be to identify the city quarters by specific colors which may differentiate them and, as well, stand out the different urban furniture elements, such as: dustbins, benches, telephone boxes, bus stops, street lamps, bollards, etc.

Golledge (1999:1) explains the process of drawing cognitive mapps by the manipulation of selective information which, despite the existence of generalized information (maps, written descriptions, etc.), allows each person to choose the references that will help in their route marking out. He also explains that the *wayfinding* process depends on legibility, the ease with which a path becomes known through a pattern of nominations (indications?). Moreover according to Golledge:

"[…] For successful travel, it is necessary to be able to identify origin and destination, to determine turn angles, to identify segment lengths and directions of movement, to recognize on route and distant landmarks, and to embed the route to be taken in some larger reference frame" (p7).

Kevin Lynch (1960), in his study of three American cities, seeks to point out directions through the streets network, that in new towns may be protruding elements, but which are not in most of the European cities. He also proposes orientation through specific buildings (the corner street shop or a striking building) which in big cities, where architecture tends to become uniform, does not provide great guidance data. On the other hand, these data could only be used by the quarters inhabitants, or usual visitors, as they will not be recognized by whom uses the route for the first time.

Per Mollerup (2005:17) refers buildings, in several cities, which are so characteristics that become real signaling milestones. However, not every cities has buildings such as the *Eiffel Tower* or the *Centre Pompidou* in Paris, the New York *Empire State Building*, the *Sydney's Opera House*, or even the *Guggenheim Museums*, so they require other systems in order to facilitate the orientation of their visitors.

Gallen Minah (2005:401) explains that most cities which managed to control the visual order are *compact historical cities* where it already existed a hierarchy defined among their spatial elements. Modern cities are generally more dispersed and their architectonic elements are fragmented and autonomous. Because their hierarchy is less clear, order and harmony become difficult experiences from the pedestrian point of view. The development of the contemporary city originates great diversities and complexities in their architecture which compete for visibility. Color exceeds its function of definition and unification element and becomes a visual characteristic in the midst of the visual field complexity and chaos.

"Color is one of the repetitive visual elements that define the formal, spatial, and material phenomena in the city. One experiences color in a city through its combination with, and definition of, architectural elements in the visual field" (Minah 2005:401-402).

City maps indicate its different zones, or quarters, through colors, but these colors don't show up in the physical space, i.e., the concern to match these colors with a real use on its urban space doesn't exist. The ideal would be to identify the different city zones by means of colors that could differentiate them and, at the same time, called the attention to the urban furniture elements, such as: garbage bins, garden benches, bus stops, telephone booths, etc.

So, the application of this new methodology of color planning to urban furniture, may originate a system which will function simultaneously as an identification factor for the different city quarters and as an orientation factor for its inhabitants and visitors. In parallel, color application to urban furniture will also become an inclusivity factor, by incrementing these elements visibility and use.

3   METHODOLOGY FOR COLOR PLANNING

The research process is focused in Lisbon, establishing as result a color plan that can be applied to urban furniture at the different zones of the city. A Case Study methodology was used, including three different cases: three quarters of the city of Lisbon with their own particular specificities. The first one, *Baixa*, CBD city area, is the very heart of the city and in 2004 was a candidate to the world heritage; the second, *Campo de Ourique,* is a traditional quarter, both commercial and residential; and the third one, *Parque das Nações*, is a recent quarter which is still under development.

During the research process we acknowledge that the existent methodologies of support to data recording and creation of Chromatic Plans, which are generally linked to architecture, were not sufficiently adequate to reach the objectives of the present study, neither the expected and desired results.

Consequently, a new methodology was developed, using an extensive direct observation, with the use of mechanical devices, including photographic mapping of both urban furniture and signage, in order to evaluate their visibility and legibility, as well as their color applications. In each quarter, and to facilitate the study, a sample route was defined, including the main streets and places and, also, some secondary ones, with the intention of encompassing the quarter most representative

Figure 2 Material samples from buildings, pavements elements with relative permanence, such as transports, and vegetation.

zones, those with specific characteristics. Along with the chosen route, an exhaustive record of all the environmental colors was made, including material samples not only from the buildings, but also from pavements, vegetation and any additional elements present with a relative permanence in the urban environment (the non permanent colors) that must be taken into account for the spatial chromatic readings, which were then classified using the Natural Color System (NCS). These collections were completed by photographs of the environment elements and panoramic views from the different blocks, using as well urban plans, architectural elevations and sections of the selected paths, which act as elements of the environment color components.

All these records were methodically indexed in forms and maps, previously designed and tested, which allowed creating a data base guided by scientific rigor, in order to determine a chromatic palette for each quarter and, consequently, to establish a coherent chromatic plan that may be applied to urban furniture.

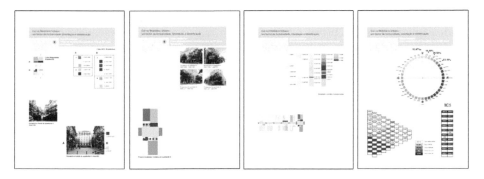

Figure 3 1 and 2: Block chromatic identification form ; 3 and 4: Street chromatic identification form — summary of present colors.

Each "Chromatic identification form" for every street, or place, contains:
1. Identification of the city zone (quarter), street or place, and block;
2. Map of the street, or place, with the block marked;

3. Photographic record of every color applied to architecture and the correspondent NCS notations;
4. Chromatic record and NCS notations of the other present elements;
5. Photographs of each block different views, completing the systematic record of the street, or place, chromatic environment;
6. Records from the existing relation between the buildings height and the streets width, in order to evaluate the quantity of day light and the sky color preponderance, in addition to the tarmac color and quantity, as well as the green elements present;
7. Summary of all present colors, proportionally represented, and their position on the NCS circle and triangle.

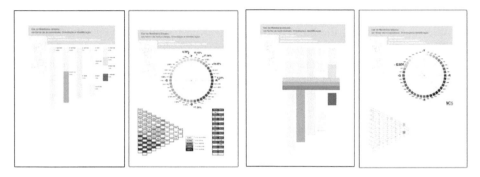

Figure 3   Quarter chromatic identification forms: 1 and 2 – dominant colors; 3 and 4: colors proposal for urban furniture.

In order to guarantee the scientific rigor on each quarter chromatic plan determination, we considered the dominant colors, proportionally represented, choosing colors to the urban furniture which may establish an adequate chromatic and luminosity contrast with the dominant colors and, also, respect the traditions, culture, identity and history of the quarter.

The urban furniture chromatic plan, that will be different for every quarter, must stand out from the environment, contributing for a better legibility and identification of these elements and, in the same way, will become a city's area identification element which may be used in different supports and, this way, facilitate the orientation and wayfinding within the city.

2   CONCLUSIONS

The main starting points of this project where that, in most cities, there was a problem of wayshowing and the urban furniture lack of visibility. Therefore, the research sought to create a chromatic planning system which would join together the orientation within the city, the identification of its different zones and would promote the inclusivity of its inhabitants. To achieve these aims it became necessary to develop a new methodology that would include the record of the entire city quarters environmental colors, in order to establish their urban furniture colors.

The new methodology and the chosen chromatic plans were tested by three focus groups, and their intervention resulted in their validation.

Consequently, we may consider that the development and implementation of the new methodology will allow the determination, with a higher scientific approach and rigor, the color planning to be applied to urban furniture in each quarter, or urban area, of a city.

ACKNOWLEDGMENTS

The authors would like to acknowledge FCT – Foundation for Science and Technology, Scholarship SFRH/BD/41096/2007; and CIAUD – Research Centre in Architecture, Urban Planning and Design, from the FAUTL, Lisbon

REFERENCES

Águas, S. 2003. *Urban furniture Design, a Multidisciplinary Approach to Design Sustainable Urban Furniture*. M.Sc Dissertation. University of Salford. Salford.
Berger, C. 2005. *Wayfinding – Designing and Implementing Graphic Navigational Systems*. Switzerland: Rotovision SA.
Lindemann et al (Eds.). 2004. *Regulated Agent-based Social Systems*. Germany: Springer.
Machnow, H. and Reuss, W. 1976. *Farbe im Stadtbild*. Berlin: Abakon Verlagsgesellschaft mbH.
Minah, G. 2005. *Memory Constellations: Urban Color and Place Legibility from a Pedestrian View*. AIC Color 05 – 10th congress of the International Color Association.
Mollerup, P. *Wayshowing*. Baden: Lars Müller Publishers.

CHAPTER 64

Perceptive and Ergonomics concerns in Corporate Visual Identity

Daniel Raposo Martins, Fernando Moreira da Silva

CIAUD – Research Center in Architecture, Urban Planning and Design
TU of Lisbon, Portugal
draposo@ipcb.pt
fms.fautl@gmail.com

ABSTRACT

This article is an explanation based on the bibliography of the specialties of communication design, Corporate Visual Identity, ergonomics and visual perception of graphic signs, with the intention of increasing communicational efficiency in designers' work.

Based on the assumptions of perceptual efficiency, ergonomics and symbolic, graphic brands are contextualized as strategic elements of Corporate Visual Identity.

The globalized knowledge society for the information that surrounds the Internet on a global scale, the cultural changes and technological developments appear to facilitate that the design of symbols creates problems in a semantic, perceptual and ergonomic dimension. In particular, new digital media seem to foster graphic solutions with a strong mesmerizing power, but moving at a pace that the evolution of human cognitive system is unable to follow.

Keywords: perception, ergonomics, graphic brands, corporate visual identity

1   A GAP BETWEEN VISUAL ERGONOMICS AND EMPTY AESTHETICS

Much of the literature devoted to design, places the Man as the centre of all the projectual activities of the designer. From this perspective, design starts to meet a human need contributing to improve life quality in an eco and sustainable while conciliating with commercial or market issues. However, the history of communication design and other visual arts is marked by stylistic variations to which images and brands representing corporations cannot be dissociated.

In order to make themselves known, even to compete with other and establish an emotional connection with audiences, brands seek notoriety through the combination of signs of Visual Identity conveyed by various graphic means, sonorous, digital or video (Perez, 2005).

According to Gasch (1991) the design of a graphic brand implies some perceptual requirements such as legibility (the possibility to read at a considerable speed and distance), memorability (enable recognition, being expressive and impactful for easy memorization) and flexibility (the possibility to reproduce in various sizes, shapes and graphics production processes), as well as semantic, such as the graphic expression (suitable according to company values), bimedia meanings (relationship between graphic style, image and text) and formal validity (adequacy to the codes of time and culture).

According to Chaves and Bellucia (2003) the role of graphic brands is to identify, differentiate and contribute to the reputation of the represented entity, although reality shows evidence that graphic expression does not always follow these assumptions, given that it changes according to the style of the author. For this reason, it's not difficult to collect graphic brands whose symbols are formally complex, overly detailed, somewhat contrasting or ambiguous.

2 DESIGN AS CULTURAL INTERFACE FOCUSED ON THE HUMAN BEING

The origins of the term "communication" come from the Latin communicatio, communi which, in turn, comes from commune, meaning "common good", "public good", "participation", "sharing" and "to have in common" .So, briefly, in a general perspective, "communication is a form of interaction between two separate beings, and the support of interaction is the exchange of information" (Castro, 2007, p.26). Communicating is to share or to have in common, as a transmitter and receiver, a piece of information.

In this study, objects of communication design are meant as strategic tools designed to a purpose, interfaces between sender and receiver (appropriate to their shared inter-subjective codes). To that purpose, the designer Providencia (2000, p.16) considers that through design, communication design, it creates artefacts of communication, that is to say, instruments that promote the decoding of the message by recipients.

According to Frascara (2008, pp. 23–24), "visual design communication deals with the construction of visual messages in order to drive knowledge, attitudes and behaviour of people" and the task of the communication designer consists in the interpretation and representation of messages according to a program, "a work that is beyond cosmetics, which is related to the planning and structuring of communications, with its production and evaluation".

Authors such as Joan Costa (2008) reported that the definition of Corporate Visual Identity (CVI) is particularly complex, since visual language does not have a repertoire of universal, unequivocal signs as it occurs with writing. Moreover, as stated by Acaso (2006, p.27) visual language is "the oldest semi-structured communication system known... and the one with the most universal character".

And as with the visual language in general, Corporate Visual Identity depends on successful integration of signs into a coherent system whose value is created by the whole instead of all the parties "the message - that is, the interpretation created by the public - is a cognitive/emotional/operative integral unit, which can only be divided into several components with the purpose of studying its structure" (Frascara, 2006, p.75), for example, as it happened with the Gestalt Theory.

For Acaso (2006), the characteristic that most distinguishes visual language from the others is its resemblance to reality, and the great variety of ways to represent it. The designer decides which level of approximation to reality and gender he is interested in incorporating to the visual message.

In summary, the effectiveness of Corporate Visual Identity depends on a number of signs, inter-subjective and shared by sender and receiver.

The verification of the peculiarities of visual communication, emphasize the need for studies that meet communication design's point of view, that is to say, contemplate the perspective of "message builders" and its receptors (Frascara, 2008, p.93).

The social responsibility of the designer, Alexandre Wollner (Stolarski, 2005) advocates the need to scale signs of identity based on the location and context of use. Calculating the position and size of a sign on a building is a plausible example that must be projected so that it is noticeable and readable from a passer-by's point of view without contributing excessively to visual pollution or constitute an obstacle (Plácido da Silva, Nakata, Paschoarelli, and Raposo, 2010). To that purpose the conceptual description of the design department of communication HFM (where he studied) describes that "the experiences in the department aim to establish a clear possible coordination between visual messages and its purpose. For this, methods must be developed, adding knowledge through the theory of perception and meaning" (Wollner, 2003, p.83).

3   DRAWING ON A HUMAN SCALE

As described by Costa (2011, p.131) "drawing a brand is to give visible form to an idea, which is that way conveyed. "To that definition of the author it might be added that perceiving that one graphic sign is seeing it, identifying it and correctly understand their meaning, so that "designing for the eyes is designing for the brain" (Costa, p.12).

For Arnheim (1965), the perception of shapes begins with the recognition of the most evident structures, both the limits and outlines of their skeletons. Also, Gestalt psychologists describe how the human eye prefers to establish their structural connections as simple as possible (Gomes Filho, 2003).

The sequence of cognition presented by Wheeler (2003, p.7) begins in the recognition of shape, the semantic evocation created by colour and finally by denotative content. The brain takes more time to process language than identifying shapes.

Gomes Filho (2010, p.161) states that visual perception depends essentially on the "capacity, facility and quickness" of the decode, which depend on the graphic shape and cultural knowledge of the receiver. However, this dependence on the recipient's culture is reduced when it is about natural and symbolic signs. Thus, the functioning of the graphic sign depends on whether its design is systemic, the relationship between the sign, the context and the user, and the appropriate relationship between the various graphical resources (tone, texture, shape, position, orientation, size, proportion and movement).

The eye does not work like a photography that captures everything. Aicher (2004, p.140) argues that human behaviour is especially prone to saving efforts, leading him to simplify and economize and interesting only by what contrasts and stands out - "We only see what it has meaning for us, make a selection (...), filter and simplify the redundant material." The author wishes to emphasize that the interest requires a definition of levels of importance to ensure concentration.

The filter is symbolic and cognitive. Neumeier (2006, p.34) considers that "differentiation occurs through the human cognitive system, where the brain acts as a filter that protects the vast amount of irrelevant information that surrounds people every day." For the author, visual cognition involves the Gestalt laws but also aesthetics to the level of differentiation.

The eye's tendency in retaining a summary of the envisioned shape explains why the generalization of a graphic brand should be as simple as possible so that it can be quickly perceived and memorized. Thus, through a scientific research on the eye and the entire system of visual perception, would be expected that design would evidence concerns or adjustments to the human eye. However, the rapid and growing development of cities and companies has led to an excess of graphic communication elements in constant competition, upsetting people and creating what is known as visual pollution (Gomes Filho, 2003; Plácido da Silva, Nakata, Paschoarelli and Raposo, 2010).

According to Plácido da Silva, Nakata, Paschoarelli and Raposo (2010) for financial reasons or lack of appreciation towards design, most SMEs do not invest completely in a Visual Identity project. Whenever it is urgent to communicate, "a non-specialized service is hired, to charge cheaper, culminating in an excess of harmony errors both in shapes, typography and colours". Therefore it is not uncommon for errors to arise within the hierarchy of graphic signs, lack of coordination or identity, as well as scaling, contrast and framing in the context problems.

Considering the perspective of visual and cognitive ergonomics, Gomes Filho (2010) believes that one of the primary objectives of graphic signs is to ensure the function, contributing to the people's comfort and safety. In this sense, the sign must have in mind stereotypes (contradicting or reinforcing), the location and scale display in accordance to the best conditions for its viewing so that the person does not have to undertake a major effort.

The perception of the meaning in graphic forms is dependent on the experiences, culture and filters of each individual (Arnheim, 1965), but derives largely from the evolution of the human species. In this regard, Frutiger (2005, p.18) illustrates that, contrary to what occurs with other animals, the human eye is conditioned at a cognitive and symbolic level because the Man moves in a horizontal plane and because, usually, danger zones came from the sides rather than top. "This millenary effort, Man's genetic legacy, led to the fact that our visual field is much wider in the horizontal dimension than vertical."

For Frutiger (2005), Man tends to be located as an active vertical element towards a passive horizontal plan (real or mentally established by the observer). Based on this principle, individuals tend to judge all graphic shapes by comparison to the horizontal plan. Considering how the most dynamic and active shapes contrast with the horizontal plan, on which are based those seen as more inert, stable and discrete.

Seeing the graphical shape, the observer draws parallels with the real world, assessing the signs towards a horizontal and vertical plan. For this reason, Dondis (1976) establishes symbolic relations with geometric shapes in which the square means boredom, honesty, righteousness, the triangle is action, conflict and tension, while the circle corresponds to the cosmos, is continuity, protection and warmth. According to Perez (2005), angular shapes are associated with masculinity, toughness, stability and conflict, while the circular are soft, dynamic and feminine.

Everything is learned and created according to our reality. And in this perspective nothing is new, it just changes according to the way you look at that reality. Artificial graphic shapes are created having as a model the human proportions, experience and culture, which assume in the two-dimensional plan some of the principles of reality, such as gravity (Bruni and Krebs, 1999).

By nature, graphic shapes are bound to the two-dimensional plan. However, over time, both large and small shapes, light and dark, overlapping or in perspective, are examples of the different graphic techniques used to suggest three-dimensionality, depth and hierarchy.

If in its essence the graphic sign only suggests volume, it might result tempting to give it three-dimensionality, which can lead to ambiguous solutions. Three-dimensionality implies attention to a greater number of issues that are not only symbolic, but especially perceptive such as the viewing angle, depth, light and shadow or optical illusions (Frutiger, 2005, p.63).

While the two-dimensional graphic sign is admittedly artificial, the efficiency of the three-dimensional representation requires a greater similarity to the fact that surrounds the Man (Jacobson, 1999).

The very notion of symmetry is no stranger to the contemplation of the human body and objects in the world in which we live. For Frutiger (2005, p.22), "certainly we feel very safe or quiet when we see a geometric construction figure, while not ignoring that its interior may contain asymmetrically arranged elements for functional reasons." A permanent tension between the internal and external elements could be accepted.

Graphic signs and human behaviour have relationships with these principles, as it is proved by the symmetry of the Latin alphabet vowels, but also in the Phoenician writing, which is part of the history of our alphabet (Costa and Raposo, 2008).

Being acquired in the Western society to observe the signs from left to right, this is the result of established conventions throughout history.

However, asymmetry can be used as a means to obtain contrast, since graphical shapes eventually end up being more dynamic and unbalanced.

Spivey (2005) relies on art history to show how cultural factors influence the representation of graphic shapes, which although related to real ones may be more geometric or organic. For this author, Greek civilization was the first to master drawing, painting and sculpture techniques, allowing them to accurately represent real, which eventually resulted too human and hardly fascinating. The reality does not satisfy and the search for graphic shapes with greater power of fascination led to the choice of formal exaggeration, to the superhuman, the choice of the Greek, as well as previous and subsequent civilizations.

The investigations of the neuroscientist V.S. Ramachandran indicate that the graphic and symbolic accentuation of shapes and the unreal largely contribute to the power of fascination in the brains of individuals (Spivey, 2005). Costa and Moles (Moles and Janiszewski, 1990) explain that in this context, fascination refers to expressiveness, to the power of the graphic shapes to attract and retain the attention of the eye. A shape that not only attracts, but also retains the human eye.

4   CONCLUSIONS

The specificity and social character of the discipline of communication design, give it a degree of considerable importance because of its contribution to a more intelligible world, to increase the quality of life, provide information and dissemination of culture.

In the manifesto, "First Things First", published by 22 signatories in Design, the Architects' Journal, the SIA Journal, Ark, Modern Publicity, The Guardian, in April 1964, updated and republished in 2000 in Émigré and Eye magazines, sought to establish itself as an awakening of conscience. An inversion of priorities to value the utility, durability and visual messages of democracy and less consumerism agenda.

In 1997, in the Mexican magazine DX, Joan Costa also published his manifesto "For the design of the XXI century", republished in 2000, proposing the appointment of design and designers to improve the quality of life of individuals. Communication design is contextualized in the Age of Information and Service Culture, assuming an effective role in improving the quality of life, referring to the communication of functional assets, aesthetic and cultural that guarantees a contribution as a service and useful knowledge to the society.

These principles seem to be consistent with the idea that communication design works for the eyes and the brain.

For the communication design to play a truly responsible and focused on the user role it is only possible to meet the expectations and limitations of people. To improve people's quality of life it is necessary to contribute to resting the eye, drawing signs of identity that are not dubious, properly sized for the environment, viewing context, culture and support in question.

Thus, in most cases, it is essential to consider the principles of simplicity, visibility, identification, semantics, technical and cultural constraints. Draw for the eyes as the brain perceives, without complicating graphic shapes, without limitation regarding fashions, trends or personal tastes.

As stated by Fascioni and Vieira (2001), in order to quickly grasp their audience, companies adopt colours, sounds and visual effects, often poorly selected and unhelpful. This pursuit of media coverage may result in sensationalism and kitsch.

Montesinos and Hurtuna (2004) believe that graphic signs involve the articulation of its symbolic and graphic aspects in order to meet ergonomic and perceptive questions. According to the authors, in each job, the designer needs to establish a hierarchy among the symbolic and graphic dimensions, which are not incompatible.

The evolution of graphic signs accompanied human history, developing with ways of thinking, culture and economy. Graphically, the brand is dependent on the human know-how, trends, technical and technological advances, but also on cultural developments and tastes.

Driven by strong convictions, in 1924, sociologist Otto Neurath began a project to develop a universal graphical language named ISOTYPE - International System of Typographic Picture Education, but only in 1928 the first symbols were designed. In collaboration with the Gestalt theory, ISOTYPE would greatly influence design communication.

According to Adrian Frutiger (2002, p.86), the 1939-1945 war would have stopped "any creative impulse in Europe," but out of the conflict, Switzerland followed its course in these areas. In the Art schools of Basel and Zürich, and in the context of the International Typographic Style design was developed coupled with a clear rejection of constructivism. Steiner in Zürich and Hoffmann in Basel, would have been the founders of a new direction in the field of graphic design, in which Eidenbenz, Falle, Piali and other reformed poster conception and Emil Ruder, the typographic design. In "symbols", figurative representation was abandoned in favour of simplification or graphic synthesis where the drawing would be limited to white-black contrast or to shape-background. The linear signs would gain greater formal relationship, with a constant thickness and proportion (Frutiger, 2002).

The industrial expansion of the Post-War gave a new impetus to the market of products and services, enabling the emergence of big companies and to the growth of the reputation of designers and design offices dedicated to corporate visual identity such as Paul Rand, Lester Beall, Saul Bass, Lippincott & Margulles, Chermayeff & Geismar, among others...

The complexity and size of the companies and world events from the fifties, sixties, and the HFG German School in Ulm, headed by Otl Aicher, contributed to an extensive and systematic organization controlling all visual applications, so it has become a following example worldwide and continue on projects such as Lufthansa German Airlines in 1962.

With the introduction of the personal computer (1980) the emergence of Desktop Publishing was made possible (1985) and the slow assembling Letraset and photocomposition began to be substituted by PostScript and various software that allowed various combinations of fonts in different bodies, images and layouts (Clair and Busic-Snyder, 2009).

The emergence of the first personal computers was a pretext to the appearance of printers that such as monitors supported only low resolutions, forcing the creation of letters that remained legible in these conditions (Gaudêncio, 2004).

Regarded as pioneers of digital graphic design, April Greiman, Rudy VanderLans, John Hersey and Zuzana Licko began a project-oriented approach in which design takes dual advantage of computers: as a tool and as a means of expression, particularly by allusion to pixels. Other designers such as Neville Brody, Matthew Carter and David Carson followed their footsteps (Meggs, 2000).

Interestingly, as computer technology was perfected and allowed greater accuracy and legibility in design, designers experienced in seemingly opposite directions. On the other hand, as stated by Fascioni and Vieira (2001, p.7), since the late '50s, "perhaps by the excess of information and functions," the services and electronic equipment or gadgets industry has adopted a corporate style approaching kitsch.

Presented to the public in 1981, since its creation MTV realized the potential of the television, using a three-dimensional logo with great graphical variation. This was followed by the flexibility of customization offered by digital printing and the Internet, which enabled the emergence of brands created in the aesthetics of Digital Art, Web 2.0 or translucent bright or metallic effects.

It will be undeniable that the new digital media eliminated restrictions on the graphic reproduction and new ways to create meaning to enhance the interaction between brands and people. A relationship where the metamorphic brands may be the latest innovation.

The use of metamorphic graphic signs is found especially in brands aimed at a very broad audience and especially associated with such technologies as Swisscom, AOL, Optimus, Oi, Carat, or Google, but also cities like NY, São Paulo, and in Portugal, Guimarães, the Casa da Música and the bookstore Wook.

The graphic brands NY and AOL are clear examples of the absence of a structure or common constant skeleton to facilitate the perception of the identity sign. The AOL logo gets to be replaced by images capable of conveying feelings or sensations, a sort of image-brand that presents new challenges for research dedicated to ergonomics and perception of graphic signs of identity.

Interestingly, there has been a restyling of the identity signs of major brands, which tend to go towards the formal complexity, namely by the introduction of close to reality three-dimensionality. For example, the automotive industry has been adapting to paper a graphical version of the symbol that resembles what a car traditionally consists in its physical composition.

612

ACKNOWLEDGMENTS

The authors acknowledge the CIAUD – Research Center in Architecture, Urban Planning and Design and to Polytechnic Institute of Castelo Branco.

REFERENCES

Acaso, M., 2006. *El lenguaje visual*. Barcelona: Ediciones Paidós Ibérica S.A.

Aicher, O., 2004. *Tipografía*. Vanència: Campgràfic

Arnheim, R., 1965. *Art and Visual Perception: A psychology of the Creative Eye*. Berkeley and Losa Angeles: University of Califórnia Press.

Bruni, D. and Krebs, M., 1999. *Norm: Einführung/Introduction*. Zürich: Norm

Chaves, N. and Belluccia, R., 2003. *La marca Corporativa: Gestión y diseño de símbolos y logotipos*. Buenos Aires: Paidós

Clair, K. and Busic-Snyder, C., 2005. *Manual de Tipografia: A história, a técnica e a arte*. 2nd ed. São Paulo: Aertmed Editora / Bookman

Costa, J., 2008. *En torno a los 60 años de la Ciencias de las Comunicaciones*. Lição magistral. Barcelona: Universitat Abat Oliba CEU

Costa, J., 2011. *Design para os Olhos: Marca, Cor, Identidade, Sinalética*. Lisboa: Dinalivro

Costa, J. and Raposo, D., 2010. *A rebelião dos signos. A alma da letra*. Lisboa: Dinalivro

Dondis, D., 1976. *La Sintaxis de la Imagen*. Barcelona: Editorial Gustavo Gili

Fascioni, L. C. and Vieira, M. H., 2001. *O kitsch na comunicação visual das empresas de base tecnológica*. In: 15º Simpósio Nacional de Geometria Descritiva e Desenho Técnico. São Paulo: Anais do Graphica

Filho, J. G., 2005. *Ergonomia do objeto: sistema técnico de leitura ergonômica*. 2nd ed. São Paulo: Escrituras

Filho, J. G., 2003. *Gestalt do objeto: sistema de leitura visual da forma*. 5th ed. São Paulo: Escrituras

Frascara, J., 2006. *El Diseño de Comunicacion*. Buenos Aires: Ediciones Infinito

Frascara, J., 2008. *Diseño gráfico para la gente: Comunicaciones de masa y cambio social*. Buenos Aires: Ediciones Infinito

Frutiger, A., 2002. *En torno de la tipografía*. Barcelona: Editorial Gustavo Gilli

Frutiger, A., 2005. *Signos, Símbolos, Marcas, Señales: Elementos, morfología, representación, significación*. 5th ed. Barcelona: Editorial Gustavo Gilli

Gasch, M. ed., 1991. *Curso Práctico de Desenho por computador*. Madrid: Ediciones Génesis

Jacobson, R. ed., 1999. *Information Design*. Cambridge: MIT Press

Meggs, P. B., 2000. *Historia del Diseño Gráfico*. 3rd ed. Santa Fé: McGRAW-HILL Interamericana Editores

Moles, A. and Janiszewski, L. ed., 1990. *Grafismo Funcional*. Enciclopedia del diseño. Barcelona: Ediciones CEAC

Montesinos, J. L. M. and Hurtuna, M. M., 2004. *Manual de tipografia: del plomo a la era digital*. Valencia: Campgràfic Editors

Neumeier, M., 2006. *The Brand Gap: How to bridge the distance between business strategy and design*. Berkley: AIGA

Perez, C., 2005. *Signos da Marca: Expressividade e Sensorialidade*. São Paulo: Pioneira Thomson Learning

Pinto e Castro, J., 2007. *Comunicação de Marketing*. 2nd ed. Lisboa: Edições Sílabo

Plácido da Silva, J., Nakata, M., Paschoarelli, L., Raposo, D., 2010. *A contribuição do projeto de identidade corporativa na redução da poluição visual nas cidades*, Revista Convergências, nº5 [online] Available at: http://convergencias.esart.ipcb.pt/artigo/74 [Accessed 28 November 2011].

Providência, F., 2000. *Design de Comunicação/Gráfico*. In: Directorio de Design 1999-2000. Lisboa: Centro Português de Design

How Art Made The World: How humans made art and art made us human. 2005. [DVD Video] Spivey, N. London: BBC

Stolarski, A. ed., 2005. *Alexandre Wollner e a formação do design moderno no Brasil*. São Paulo: Cosac & Naify

Wheeler, A., 2003. *Designing Brand Identity: A complete guide to creating, building, and Maintaining Strong Brands*. New Jersey: John Wiley & Sons. Inc.

Wollner, A., 2003. *Design Visual 50 anos*. São Paulo: Cosac & Naify

CHAPTER 65

Touristic Information: The wayfinding and Signage Systems Contribution for an Inclusive Design

João Neves, Fernando Moreira da Silva
CIAUD – Research Centre in Architecture, Urban Planning and Design,
TU Lisbon,
Lisbon, 1349-055, PORTUGAL

ABSTRACT

Based on previous studies related with this investigation, it was found that the sign systems for tourist information lacked uniformity and were developed empirically, without recourse to methodological procedures or models that could give formal unity, forming ergonomically unsuitable systems, less inclusive and hindering the mobility of users.

The survey data seem to reinforce the importance of exploratory research in the area of tourism signage and wayfinding, particularly in methodological and ergonomic terms in order to address the needs of organizations and designers in the application of models able to facilitate and enhance the quality of their projects, particularly in development of inclusive signage systems.

It is purpose of the present work the study of signs systems for tourist information, specifically the definition of a model that can be applied on the development of signage systems and that promotes the design process, production and implementation by professionals and the decoding and code perception by the user.

This research works in the field of design and focuses on the area of tourism signage and wayfinding.

Keywords: Wayfinding, Signage Systems, Pictograms, Ergonomics, Touristic information

INTRODUCTION

The growth in recent decades in tourism activities, coupled with an increasingly global world-wide, provided a more or less general abolition of physical borders, linguistic and even cultural, facilitating the free movement of people and goods, enhancing trade, industry, recreation and other activities related to tourism.

The mobility and greater affluence of people to certain places or attractives, raised the need to target these people in an unknown space and communicate basic messages in a universal language, expressed through images in order to facilitate the understanding and reduction of written messages in any language. To Massironi (1983, p. 118), this type of images (pictograms) help on orientation in stations, airports, hotels, services, but also currently find on maps, tourist guides, multimedia applications, among others, and for which the requirements of export markets and circulation can not predict the use of one language or the confusion of many languages at once.

The displacement within the tourism activities is made many times in unknown places, raising the need to seize new rules, which will then be formalized through signs that facilitate access or movement to certain places.

The growth and evolution of cities, the complexity of transportation routes, trade relations and communications become essential in the signaling environment, necessary for the safe use of urban facilities, providing business and exchange of knowledge and ideas (Velho, 2007, p. 12).

Tourism, as an activity which involves displacement or mobility of visitors, generates various needs, whether at the level of tourism resources, services, or offer. It is on the tourist information a huge contribution to the mobility, the quality of accommodation and the provision of tourist services and, ultimately, to develop more inclusive places and territories. Tourist information is presented in various forms, from tourist maps, books, leaflets, guides, panels, advertising panels, multimedia applications, web sites, signage, tourist signs, among other forms of communication.

WAYFINDING DESIGN

According to Britto (2006), the creation and transmission of a message about a given product / service or equipment rental is a process that triggers the connection between supply (product / service / equipment) and demand (actual or potential tourists) and ensures their complete satisfaction. Such signals are integral components of a directory, system, or a signage system directory, specially designed for many different situations, may or may not be referenced by systems or official directories of tourist signs, especially on issues of internal signage, when the freedom of establishment may be maintained (Britto, 2006).

The tourist signs or symbols can not have a dubious character, or be based on a certain code of restricted access to some users. It is also important for the interpretation of such signs the physical environment where they are (World Tourism Organization, 2003, p. 6).

The tourist guidance cue is the communication by means of a set of panels, implanted successively along a established route, with a ordered written messages, pictograms and arrows. This set is used to inform about the existence of tourist attractions and other references, to tell which are the best access routes, and over these, distance to be traveled to reach the desired location.

It is also important in this context to define a concept applied to the systems of signs and its relationship with its surroundings: Wayfinding. It is a word that has been used to identify the theme of spatial orientation and 'navigation', especially in urban areas. It is an important area for design, for architecture and ergonomics that is not limited to the design of pictograms and signs, but everything that concerns human interaction with the space (Arthur, Passini, 1992) and formulation adapted spaces to the user, from the viewpoint of ergonomics visual cognitive and anthropometry, in order to planning and design more inclusive spaces.

Several definitions for the word wayfinding are known, as a methodology for organizing indicators to guide people to their destination (Beneicke; Biesek, Brandon, 2003). Wayfinding can also be an orientation process that uses spatial information and the environment (natural, urban or built).

Wayfinding can be considered as a method for providing consistent information in a clear and obvious way to guide a person to their destination. This information may include maps and signs, clear clues for architectural and interior design facilities or through the use of standard color and texture. Advanced systems of Wayfinding may also be effective systems of information that support organizational identity and branding strategies (Hablamos Juntos, s.d.).

Wayfinding is the organization and communication of our dynamic relationship with the space and environment. Wayfinding design aims to provide the user: (1) determine its location in a certain environment, (2) determine their destination, and (3) develop a plan that will take their from its initial location to their destination. The creation of wayfinding systems should include: (1) identification and marking of spaces, (2) public spaces, and (3) connection and organization of spaces with architectural features and graphics (Center for Inclusive Design and Environmental Access, s.d.).

Wayfinding can be described as the process of using the spatial information and the environment to find our path in an building environment. Wayfinding can also be defined from the viewpoint of the designer and the customer, seeking to establish or improve the function of a particular environment (Kelly Brandon Design, s.d.).

Wayfinding design is the creation of resources and information systems about space and environment, to guide people. It may also be considered as the process of organizing spatial information to help users find their way. Wayfinding should not be considered an activity other than or different from the traditional "design cue 'but in a larger extent and more inclusive to assess all the environmental issues that affect our ability to find our way (Kelly Brandon Design, s.d.) .

A wayfinding system including brands, signs, maps and directional devices that tell us where we are, where we want to go and how to get there. An effective wayfinding system can add an important dimension to the image of a museum, a transit system, an airport, an office building or an entire city. It can be designed as an aid to understanding which provides information and guidance for people in a clear, appropriate and user-friendly, to help find your way through and out of an environment (Wyman, 2004.)

All projects of wayfinding have a common factor . Are these in a large scale or small-scale, long-term or short term, for public or commercial spaces, for new visitors or employees, indoor or outdoor, pedestrian or car drivers, all projects of wayfinding intended to be used by people. This means that all projects of wayfinding will have to take into account human perception and human psychology (Mijksenaar, 2009-2011).

Thus, the projects of wayfinding includes a combination of various articles, which should combine in order to provide project development design user-centered, more inclusive, usable and, most of all, noticeable and mobility helpers. Signage systems incorporate wider range of wayfinding systems and contribute to the transformation of visual signs in signaling information messages.

SIGN, CODE AND SYSTEM

Signs, as words, need to be understood, to be organized in a speech or text, else they can became not understandable as an whole. Therefore, the signs relation in the same network must be understood so that one can comprehend their meaning. Blending in groups of related signs, having in mind its utilization rules, one is in a presence of a code. Otherwise, a code is a system of signs with relations and meanings. (Raposo, 2008, p. 12).

In communication, according to Aicher and Krampan (1995, p. 9), there are elements from two main groups that interrelate: Those from a fundamental group of signs; Those from a crucial message group admitted from the signs. Code is what one calls to the coordination of these two main groups.

It is the code that establishes that a certain sign has certain meaning. Meaning is not natural when one looks at a sign. The signs whose meaning is determined by a code care for an apprenticeship of its meaning (Fidalgo, 2005).

Therefore if, by code we refer to a system of signs with relation and meaning, it is important to deepen the definition of system, wich can be grouped in diversely ie. building signage, companies, traffic sign, theme parks, organizations, etc.

Beni (2001) defines system as the set of parts that interact in order to achieve a given goal, according to a plan or principle, logically ordered and sufficiently coherent to describe and explain the functioning of the whole. For Britto (2006), systems are part of a whole, coordinated among themselves and that they function as organized structure.

Heskett (2005, p. 145), defines system as a set of interrelated elements, interacting entities or independent that form, or one may consider them to form, a collective entity. The objective of a system is to provide clear information on the consequences of choosing a route or a particular direction, but leaving the users decide exactly where they want to go.

A system requires principles, rules and procedures to ensure a harmonious interaction and ordered in the interrelation of ideas with forms. This means having systematically qualities of thought, from which it implies methodical, logical and certain procedures (Heskett, 2005, p.145). The author also adds that the objective of a system is to provide clear information on the consequences of choosing a route or a certain direction, thus leaving the users to decide exactly where they want to go.

Each artifact unit (signal) helps to form a whole (the system), that is, the signs (artifacts built by man) are not individually designed, but taking into account the collective entity which unites them. The signal (unit belonging to a whole) is then a physical object with different meanings and with unique features that make it, on one hand different from the rest and on the other still relating to the system. Being a sign a physical object, with a self-image and to which it is conventionally assigned a meaning, then we are faced with a sign (Neves, 2006, p. 178).

The signage systems are composed by independent elements - the signs - which convey certain information or an obligation of a action and that interrelate with the function of communicating messages with meaning (through code). Signage systems, to communicate messages, involve the use of pictograms, which are not more than simplified figurative signs that represent things and objects from the surrounding (Costa, 1998, p. 219).

The pictographic systems discussed in this work are understood as signage elements, signals or information interrelated, making use of simplified figurative signs that represent things and objects of the environment (pictograms). Simply put, it is understood by pictographic system the set of signage elements (signs, signals or information) that relates to form a code and that involve the use of pictograms.

The highest affluence of people to certain places such as airports, commercial areas, events, public services, tourism installations, etc., led to the need to guide these persons in unknown places and to communicate basic messages with a language understandable by everyone.

CONCLUSIONS

How to promote and guide the flow of visitors in a given territory, contributing to their knowledge and at the same time articulate the need for guidance, information, signaling and potentiate the inclusion and mobility of citizens? The answer may lie in a restricted set of uniform, standardized, recognized symbols by many users, with application potential in various contexts.

For this it is essential to create mechanisms to develop complementary systems that do not require the seizure of new codes and to facilitate displacement and access to tourists and travelers, generating visual codes to supplement the expressive limitations of images and text as code for use in tourist messages incorporating a new universal sign language, inclusive and instantaneous. It is also necessary to incorporate into projects for tourist information a methodology that covers all the steps involved in the process and sufficiently adaptable to various types of projects and that will generate more efficient systems from the point of view of communication, ergonomics, functional and even aesthetic.

A signage system for tourist information has more qualitative and perceptive potential applying a methodology in its development? Taking as starting point for investigating this issue, another associate and that this study sought to answer relates to the possibility of defining a methodology applicable in the whole process of systems development for tourist information signs.

In this sense, it was understood very important the development and application of a methodology for developing systems for tourist information, recognizing the vital importance of the contribution of wayfinding and sign systems design for a more inclusive and mobility users.

REFERENCES

AICHER, Otl e Krampen – Sistemas de signos en la comunicación visual. 4.ª ed. México: Gustavo Gili, 1995.

ARTHUR, Paul; PASSINI, Romedi - Wayfinding: People, Signs, and Architecture. New York: McGraw-Hill, 1992. ISBN: 0-07-551016-2.

BENEICKE, Alice; BIESEK, Jack; BRANDON, Kelley - Wayfinding and Signage in Library Design. Libris Design Project. [Em linha]. (2003), p. 1-20. [Consultado 19 Dez. 2005]. Disponível na Internet: <http://librisdesign.org/docs/WayfindingSignage.pdf>.

BENI, Mário Carlos - Análise Estrutural do Turismo. São Paulo: Editora SENAC, 2001.

CARNEIRO, R.J.B. - Sinalização turística: diretórios e sistemas nacionais e internacionais. São Paulo: ECA/USP, 2001. 206 f. Programa de Pós-Graduação em Ciências da Comunicação. Dissertação de Mestrado.

Center for Inclusive Design and Environmental Access - Universal Design New York: 4.1c Wayfinding. [Em linha]. (s. d.). [Consultado 17 Nov. 2009]. Disponível na Internet: <http://www.ap.buffalo.edu/idea/udny/section4-1c.htm>.

COSTA, Joan – La esquemática: Visualizar la información. 1.ª ed. Barcelona: Paidós, 1998. ISBN 84-493-0611-6.

FIDALGO, António – Sinais e Signos: aproximação aos conceitos de signo e de semiótica. [Em linha]. (2005). [Consultado em 20 de Abril de 2005]. Disponível na Internet:

620

<http://ubista.ubi.pt/~comum/fidalgo-sinais-signos.html>. Universidade da Beira Interior.

Hablamos Juntos - Universal Symbols for Health Care. [Em linha]. (s. d.). [Consultado 17 Nov. 2009]. Disponível na Internet: <http://www.hablamosjuntos.org/signage/default.index.asp>.

HESKETT, John – El diseño en la vida cotidiana. 1.ª ed. Barcelona: Gustavo Gili, 2005. ISBN 84-252-1981-7.

BRITTO, Janaina - Sistema de sinalização turística: a importância da sinalização turística para o desenvolvimento sustentável do Turismo. Revista de Estudos Turísticos, Edição n.º 24. [Em linha]. (2006). [Consultado 2 Maio 2011]. Disponível na Internet: <http://www.etur.com.br/conteudocompleto.asp?IDConteudo=2887>. ISSN 1809-6468.

Kelly Brandon Design - Wayfinding. [Em linha]. (s. d.). [Consultado 17 Nov. 2009]. Disponível na Internet: <http://www.kellybrandondesign.com/IGDWayfinding.html>.

MASSIRONI, Manfredo – Ver pelo desenho: aspectos técnicos, cognitivos, comunicativos. 1.ª ed. Lisboa: Edições 70, 1983

MIJKSENAAR, Paul - What is wayfinding? [Em linha]. (2009-2011). [Consultado 7 Jan. 2009]. Disponível na Internet: <http://www.mijksenaar.com/content/18-wayfinding.html>.

NEVES, João – O sistema de sinalização vertical em Portugal. Aveiro: Departamento de Comunicação e Artes da Universidade de Aveiro, 2006. Dissertação de Mestrado em Design, materiais e Gestão do Produto.

Organização Mundial do Turismo – Sinais e símbolos turísticos: Guia ilustrado e descritivo. 1.ª Ed. São Paulo: Roca, 2003. ISBN 85-7241-450-9.

RAPOSO, Daniel – Design de identidade e imagem corporativa. 1.ª ed. Castelo Branco: Edições IPCB, 2008. ISBN: 978-989-8196-07-1.

VELHO, Ana Lucia de Oliveira Leite - O Design de Sinalização no Brasil: a introdução de novos conceitos de 1970 a 2000. Rio de Janeiro: Pontifícia Universidade Católica do Rio de Janeiro, 2007. Dissertação de Mestrado em Artes e Design.

VELHO, Ana Lucia de Oliveira Leite - O Design de Sinalização no Brasil: a introdução de novos conceitos de 1970 a 2000. Rio de Janeiro: Pontifícia Universidade Católica do Rio de Janeiro, 2007. Dissertação de Mestrado em Artes e Design.

WYMAN, Lance - Wayfinding Systems. Webesteem magazine, n.º 9. [Em linha]. (2004). [Consultado 17 Nov. 2009]. Disponível na Internet: <http://art.webesteem.pl/9/wyman_en.php>.

CHAPTER 66

Ergonomic Garment Design for a Seamless Instrumented Swimsuit

Gianni Montagna 1, Hélder Carvalho 2,

André Catarino 2, Fernando Moreira da Silva 1,

Heloísa Albuquerque1

1Faculty of Architecture,
Technical University of Lisbon
Portugal
2 University of Minho,
Guimarães
Portugal

g.montagna@gmail.com

ABSTRACT

In the relationship of the human body with the environment, there is no doubt that garments are the most used medium and the most widespread way of communication. In sports garments and in technical apparel, the relationship with the user is even closer, in order to obtain better results and an advantage that could derive from the best combination between the user and the study object.

In this paper, the study of different factors affecting athlete's swimming performance are analyzed. The factors related to textile substrate, the shape and measure of the swimsuit, as well as the construction methods are taken into account for an optimal design.

Keywords: Ergonomics, Smart Garments, Human Factors

1   STATE OF THE ART

The development of sports garments or garments for a specific use, usually made for technical and functional garments, has been a central matter in the last decade, with the application of new materials and the development of new markets.

In sports, generally, specifically in swimming, a lot of developments have been made in the last few years by institutional research and in particular by commercial brands like Speedo, Arena, Tyr and others, with the aim to overturn new sports records and gain global market in this field.

The implication of new technologies, the development of new materials and the application of new finishing to the textiles, are some of the developments used by the different suppliers for the new garments that gain new records at every competition, as could be seen in the last Olympic Games.

After the last 2008 Olympics of Beijing (China) the international governing body of swimming, diving, water polo, synchronized swimming and open water swimming (FINA) implemented new rules for high performance competitions on swimming and prohibited the use of any kind of equipment beyond the swimsuit, with specific instructions on the parts of the body that have to be visible or what could be covered by the suit. Since a few hundredths of seconds could make the difference between a gold or a silver medal (Montagna, 2009), any single way to increase swimming performance is auspicious and desirable.

The need to control and organize any output of the body in order to increase swimming performance could be an effective way to perform at best. Recognized critical problems of the swimmer are described in scientific literature: hydrodynamic resistance, also called drag problem, and the production of high levels of blood lactate during training are just some of the problems associated to swimming that are also usually described as a disturbing for better athletic response.

2   METHODOLOGY (DESIGN CONTEXT)

The basic and simple relationship form/function that has always been used in Design and that represents the primary source of needs of the user and needs of the designer is applied here as a basic instrument to plan and try to solve some of the most important questions related to the design object.

Principles of co-design are shared in the development of this project when the interaction between user and designer is developed on a daily basis, interacts with the product and reacts during the process on two levels- on the creative and adaptive plan and on the user´s tests and feedback.

As considered by the Executive Council of the International Ergonomics Association (IEA) – "Ergonomics (or human factors) is the scientific discipline concerned with the understanding of the interactions among human and other elements of a system, and the profession that apply theoretical principles, data and methods to Design in order to optimize human well-being and overall system performance". Nowadays, as said by Yamada & Price, previous values as functionality, reliability and cost are complemented by others such as comfort, satisfaction or usability. (Yamada & Price, 1991).

During the last years, different approaches to functional and smart garments have been proposed. The methodology proposed by Jane McCann (McCann 2005) and defined by the author as "critical path" is a route that considers to operate the appropriate technology to give response to the users´ needs. However, the focus on end users in terms of co-design and user cooperation is very poor and the end user is not a player in the development process. Also the theoretical methodology formulation described in Ariyatum's work (Ariyatum 2005), proposes the following core issue arguments: 1. user requirements, 2. electronics and 3. clothing. The analysis puts more focus on the methodology system than on the user.

In this work, a new methodology has been proposed with the aim to have the final user taking part of the research group and participating in the different stages of the project development. So, a new design methodology for smart and performance wear has been created and applied to the design project. The new methodology is based on a new conceptual layered system - the UCLM. The aim of the UCLM methodology is to take into account the different levels of the user's needs whilst helping to organize and plan the different tasks of the working group, in order to achieve the best result for the final user, no matter the kind of user, and always in a user centred perspective (Montagna, 2011).

3 THE GARMENT DESIGN AND CONSTRUCTION

3.1 Definition of requirements and basic swimsuit design

The identification of the users' needs has been made by a questionnaire directed to trainers of high competition swimming athletes. As a result, the most recommended items have been taken into account and applied to the swimsuit.

Comfort of the suit is probably the "must" issue requested by the athletes and the trainers, and its translation to the suit has been made on different, but integrated levels.

Taking into account trainers and users' needs, the research group identified and drove specific attention to areas and issues that were considered as most problematic and critical for the development of the study object. Some of the most important issues covered by the research group will be described next.

Even if not for high competition, but only for training, body suit with sensing functions to monitor physical and biomedical variables, could be very useful in order to gather, organize, review and elaborate information on the user's performance and health status.

The need to integrate sensors for biomedical and biomechanical signal monitoring soon led the team to decide to use a double layered swimsuit. The inner layer would hold all the sensors, whilst the outer layer would provide adequate protection and isolation of the inner layer.

The sensors used can be classified in two types. First, electrodes based on electrically conductive yarns, fully integrated into the fabric using jacquard knitting techniques, are used to measure electrocardiographical and electromiographical sensors. Second, conventional off-the-shelf sensors are integrated to measure variables such as body temperature, accelerations and pressure on hands and feet.

The insertion of the smart system into the swimsuit didn't change its ergonomics properties; therefore it is not a vital issue from an ergonomics point of view and shall not be examined in depth.

3.2   The Textile Substrate

The textile substrate is the primary element of identification and construction of the swimsuit. Swimmers and trainers agree on the fact that the textile substrate has to be firm and malleable at the same time, so as to easily permit the right body movements, adapt to the body and permit a quick recovery of the fibers and substrate to its original form.

Different fibers have been tested and used to develop different textile knit structures. Polyamide and polypropylene-based textile yarns have been used to produce textile swatches in three different textile structures: jersey, single piquet and double piquet, which have been tested with different forces on a dynamometer. The result of the produced knitted tubes revealed that extension forces increase at the same time that knit loop length decreases (Montagna, 2009).

With the use of seamless technology, different tubes have been produced. In each machine rotation, 8 different knitting systems works at the same time, producing a structure fed with 8 textile yarns and 8 elastane yarns at any machine rotation.

Code	Raw Material	Lu (cm)	Force (N) @ 435 mm		
			Jersey	Single Pique	Double Pique
AA		0.3	35.7	170	186.6
AB	PA 78 / EL 78	0.28	45.7	307	311.2
AC		0.24	59.3	495	486.5
BA		0.3	23.3	140	137.7
BB	PA 78 / EL 44	0.28	26	285.6	260.8
BC		0.24	34	518	489.5
CA		0.3	57.7	810	862
CB	PP 90 / EL 78	0.28	53	335.2	---
CC		0.24	81.3	---	---

Figure 1. Measured force of fabric tube when stretched to 435 mm distance between jaws. Different fibres were tested in different textile structures and settings (Montagna, 2009)

Using different raw materials, structures and machine parameters, it was possible to conclude that the factors that seem to be the most important are structure and loop length. Single pique and a 0.24 cm loop length (Lu) produced an interesting weft knitted fabric.

An important factor is textile recovery after stretching and deformation. Cyclic tests revealed that deformed textiles recover its primary form slowly, even when in combination with elastane yarns.

3.3   Measurements of the Human Body

Body scanning of the athlete for whom the swimsuit is designed has been made in order to be able to reproduce her body shape. With the aim to reproduce bidimensionally (2D) the form of a tridimensional (3D) body, the need of a rigorous measurement of the entire body is associated to the need to understand the posture and the structural characteristics of the athlete. Even if the swimsuit is made of a textile knitted fabric, that offers specific comfort and augmented stretch, the importance to be able to reproduce the athlete's posture is important to permit the swimmer to develop her training maximizing self-confidence and reducing body energy wastage.

The user taking part of the research project has been scanned with swimming suits of different commercial brands and a comparative analysis was implemented.

A control scan with basic underwear has been made.. The resulting 3D images have been sectioned every 30 mm and each contour cut was measured.

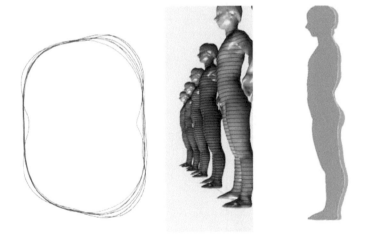

Figure 2. 2a) Comparative analysis of contour perimeter with different swimsuits. 2b) Body scans with 30mm distance cuts for measurements. 2c). Body modification analysis caused by different swimsuits.

3D body scan analysis allowed the research group to work on the swimmer body structure in different moments. During the development phase it is necessary to link body scanned measured points to the same point in 2D pattern making design, and establish the right connection and relation between body form and measurements.

3.4 The Swimsuit Patternmaking

The patternmaking methodology used to develop the study object needs to have a new vision and a different approach. A renovated and a broader vision on patternmaking *modus operandi* is necessary with the aim to combine 2D and 3D patternmaking methods.

The seamless technology used for the production of the textile knit fabric offers the capability of a full circular jacquard knitting machine, combined with design and production software modules. The garment design module presents the possibility to create and design different garment forms and different knit structures in order to produce a garment with a minimum number of cuts. The garment design is made on a cylindrical module that imitates the machine cylinder. No measurements are possible inside the design module and any measure has to be calculated by the numbers of needles. The fact that the seamless machine does not accept metrical measurements has actually revealed to be a great handicap. To be able to draw a proper garment pattern with the desired form and with the expected body measurements of the swimmer, a 2D pattern has to be made and associated to the 3D cylinder produced by the seamless machine.

Figure 3. Design and programming modules of Merz jacquard circular knitting machine

The Lectra Systemes *Modaris* software by has been used to develop the 2D pattern for the swimsuit. Developed from a basic leotard basic block, the pattern has been adapted to the body form of the swimmer by the application of the body measurements from the 3D body scan. Basic leotard pattern has been marked with sectional cuts as performed on the 3D swimmer image, and the measurements were applied to the exact corresponding points. Body measurements and 2D/3D matching points have been marked on the 2D swimsuit pattern on both vertical and horizontal axes. Fabric shrinkage will be applied in X and in Y axes depending on the textile knitting section dimension and characteristics.

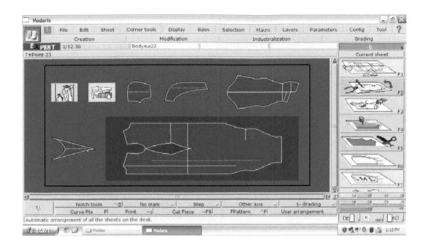

Figure 4. 2D pattern design of the swimsuit made by Lectra Modaris

3.5 The Swimsuit Construction

The swimsuit construction is an integral part of the garment itself. The evaluation of a garment from a technical point of view could not separate patternmaking and garment construction, because they have to be seen as two parts depending on each other. Garment construction is completely dependent on all variables present into the study object, including needs from users and researchers and also technical impediments and adaptations.

Garment construction is a result between a 2D garment pattern and a 3D knitted cylinder, in which the advantage of having few cuts around the knit tube is actually a big handicap in terms of fitting of the swimsuit. It is difficult to mold the knit tube without cuts, and without generating a leftover fabric on the waist line.

The application of cuts to the tube reveals some problems on the textile substrate structure due to its construction made by polyamide fibers associated with elastane. During cut and zip sewing on the swimsuit center back, sliding of elastane yarns is a problem that could affect the capacity of the substrate to compress the body properly.

Different kinds of sewing stitches were tested on the substrate. Several sewing stitches and seam constructions were tested, finally, flat seams with three-thread covering stitches were adopted. In combination with polyester threads, the seams presented good resistance and elasticity of about 70%.

3.6 Prototype Test and Correction

Prototypes of the swimsuit have been made in different stages. Since the beginning, different parts of the swimsuit have been developed with the aim to insert sensors for data acquisition.

A full swimsuit prototype has been developed as nearly as all requirements could be defined in terms of textile substrate, user and research group requirements.

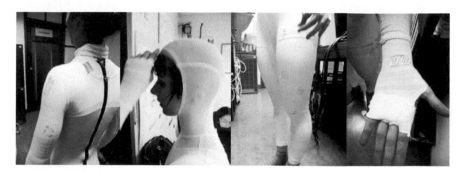

Figure 5. Pictures of the first full prototype of the sensored swimsuit.

Some difficulties has been encountered during prototype making, due to the specific material and also to the small dimension of the garment. Garment made 1:1size, due to a body compression on the swimmer body is cutted out at 67% of real size.

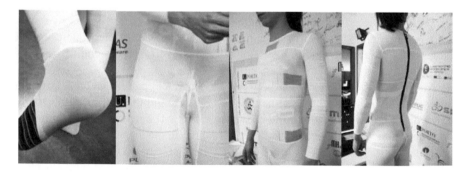

Figure 6. Pictures of the second full prototype of the swimsuit.

Corrections have been made to the first prototype and a second one has been proposed. Since the body suit has the correct shape from the beginning no significant modifications have been made in terms of patternmaking. Changes on some elements of the swimsuit were shown to be necessary. Changes have been made to the end of the leg and a half sock has been added. The hood, has been removed to give the swimmer best neck and head movement comfort.

4 RESULT DISCUSSION

The results obtained on garment construction seem to be very interesting. Tests have been made by the swimmer and the feedback has been very good. The suit does not limit any movement and works apparently as a second skin on the swimmer body. The swimsuit adapts to the swimmer body, applying a uniform sense of compression, permitting the textile to adapt to the movements with minimum wrinkles and with no textile leftover, not causing any discomfort.

5 CONCLUSIONS

The development of a swimsuit with the characteristics exposed in this paper is a task that requires the application of multiple technics and methods. The scientific methods are constantly tested by the multiple variables that from the textile substrate to the users' body adaptation are constantly changing and adapting to different conditions.

The use of seamless technology for the production of this kind of garment is a great advantage in terms of body adaptation and freedom of swimmer movements. Seamless technology allows a construction of a jacquard knit textile substrate with different parameters of stiffness and adaptability, permitting a great choice of diverse textile structures and its characteristics.

The combination of 2D patternmaking and 3D cylinder design, even if not easy to integrate, enables a high level of adaptation to the users' anthropometrics needs and permit the development of a high compatible and adaptable garment for swimming, giving a high level of freedom of movement and mobility.

ACKNOWLEDGMENTS

The authors wish to thank FCT, which is funding this research through project number PTDC/EEA-ELC/70803/2006. Acknowledgments also to Human Solutions for the body scanning of the swimmer and YKK for special swimming zip supplier. Special thanks to the Research Centre for Architecture, Urbanism and Design (CIAUD) .

REFERENCES

ARIYATUM, B. & HOLLAND, R. 2005. A Strategic Approach to New Product Development in Smart Clothing. PhD Theses, Brunel University.
IEA, INTERNATIONAL ERGONOMICS ASSOCIATION'S EXECUTIVE COUNCIL-. 2000. IEA Definitions of Ergonomics. In: KARWOWSKI, W. (ed.) International Encyclopedia of Ergonomics and Human Factors. London and New York: Taylor and Francis.

MCCANN, J., HURFORD, R. & MARTIN, A. 2005. A Design Process for the Development of Innovative Smart Clothing that Addresses End-User Needs from Technical, Functional, Aesthetic and Cultural View Points. Proceedings of the Ninth IEEE International Symposium on Wearable Computers. IEEE Computer Society.

MONTAGNA, G., CARVALHO, H., CATARINO, A. & MOREIRA DA SILVA, F. Year. Metodologia User Centered Design para Vestuário Funcional e Inteligente. In: CIPED 6 Congresso Internacional de Pesquisa e Investigação em Design, 2011 Lisbon, Portugal. Ciped 6.

MONTAGNA, G., CATARINO, A., CARVALHO, H. & ROCHA, A. Year. Study and Optimization of Swimming Performace in Swimsuit Designed with Seamless Technology. In: Autex 2009, 2009 Irmir, Turkey.

WEGGE, K. P. & ZIMMERMANN, D. 2007. Accessibility, usability, safety, ergonomics: concepts, models, and differences. Proceedings of the 4th international conference on Universal access in human computer interaction: coping with diversity. Beijing, China: Springer-Verlag.

YAMADA, S. & PRICE, H. E. Year. The Human Technology Project in Japan. In: Human Factor Society, 35th Annual Meeting, 1991 San Francisco, California, USA. p. 1194 - 1198.

CHAPTER 67

Impact of Background Music on Visual Search Performance

Ruifeng Yu, Yuchen Cheng

Department of Industrial Engineering, Tsinghua University
Beijing, China
Email address: yurf@tsinghua.edu.cn

ABSTRACT

The purpose of the experiment is to investigate the influence of background music on visual search performance. A 2(tempo: fast, slow) X 2(loudness: high, low) within-subject design was used in addition to a control condition. Twenty University students participated in the experiment. The data of search time and correct rate were obtained for indicating the related performance. The results obtained here showed that there were significant differences among 5 different conditions. Background music loudness had significant impact on visual search performance. No significant differences for visual performance were found across tempo. Music with low loudness acted better for enhancing visual search performance.

Keywords: Visual search; Environmental distractions; Background music

1   INTRODUCTION

Since 9.11 terrorist attacks event in the United States, many nations paid more attention on the public security. Almost every nation began to put a lot of resources to improve and enhance the public security level. As an important part of public security, the public transportation security is of significance. Security check usually locates in the area of entrance for a public transportation. The passengers are required to put the luggage with them onto the convey of the X-ray security-checking equipment. The staffs visually search the X-ray profile scanned by the equipment on a visual display terminal. According to the color and shape of the image, they identify the dangerous objects. In a sense, the performance of X-ray viewers directly decides the security level of passengers in transportation systems. The above scenarios commonly exist in metro, airport, subway and other public transportation systems. Moreover, most of the security systems are open and the staffs directly expose to the background noise. Then, a question regarding the impact of background noise on the visual search performance occurs. In addition, some subway stations try to broadcast music in order to improve the noisy environment and relax. Is this measure effective? Does it affect the performance of X-ray viewers working in the security systems?

There were numerous researches on the effects of music on a person's performance. By conducting the experiments, the scholars investigated the effects of music in the areas of memory, writing, reading, treadmill exercise, etc. Schlittmeier and Hellbruck (2009) studied the impacts of continuous music, intermittent music and continuous noise on conversation and working memory. The results indicated that the intermittent music broadcasted in a noisy environment of open-plan office had a deteriorative effect for sequential memory. However, the deteriorative effect was not significant for continuous music in a noisy environment. Ransdell and Gilroy (2001) studied the effect of background on writing performance. 45 participants took part in a between-subject experiment consisting of 3 scenarios, no music, music without lyrics and music with lyrics. The obtained data revealed that even the music without lyrics also used many working memory in addition to interrupting writing. Kallinen（2002）studied the disturbing effect of music tempo on reading. 60 participants joined in a between-subject experiment including 3 scenarios, i.e., no music, fast tempo music, and slow tempo music. The obtained results revealed that a male generally preferred slow tempo music while a female preferred no music condition. The participants' reading efficiency of slow tempo group was much lower than that of fast tempo one. Edworthy and Waring（2006）did research on the impacts of music tempo and volume on the performance of treadmill exercise. 30 participants joined in a within-subject experiment consisting of 4 scenarios. The results indicated that the music, especially fast tempo music, had positive effect on enhancing the performance.

This study investigates the effects of two important factors of music (i.e., tempo, volume) on the performance of visual search task (i.e., search time, correct rate). The obtained results can benefit the design and optimization of environment for visual search tasks (such as security check, etc.) in public transportation systems.

2. EXPERIMENT

2.1 Design

A 2(tempo: fast, slow) X 2(loudness: high, low) within-subject design was used in addition to a control condition. Based on the actual noise level in a subway (i.e., 72 dB) as well as the influence of music volume on noise environment (Schlittmeier and Hellbruck, 2009), the high and low music volume were set as 77dB (i.e., 72+5 dB), 67dB(i.e., 72-5 dB), respectively. Fast tempo and slow tempo were 150 bpm and 70 bpm, respectively. The dependent variables were the search time and correct rate of the participants.

Twenty college students, 10 male and 10 female, voluntarily participated in this experiment whose ages ranged from 20 to 23. The mean of their ages was 22.2. Each of them has normal near foveal acuity and none of them has any former experience in visual search tasks.

The software used in this experiment was an Adobe Air-based program developed on the Adobe Flex platform. It allows the users to select the type of search target (single or multiple Landolt C Rings with gaps oriented in the up, down, left or right directions), number of search targets (1 or 2), angular velocity and the longest search time limit for a single search task. Also it records the time taken for the observers to locate the targets. Each stimulus image had only one search target. The two types of search targets used were a Landolt C Ring with a gap on the left or right; the background character used was a closed circular ring with the same line width as the Landolt C Ring. The target and background character were shown in Figure 1.

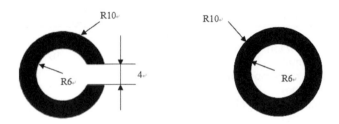

Figure 1 dimensions of target and background characters

The program was run on a Lenovo ThinkPad X201i notebook computer, with the resolution set as 1280 x 720 pixels and a refresh frequency of 60 Hz. The computer was connected to a 47-inch liquid crystal display via a VGA cable. The participants controlled the tasks with the use of a mouse connected to the notebook computer. They sat on a chair with adjustable height and at a distance of 800 mm directly in front of the liquid crystal display. Adjustment was made to ensure that the participant's eyes were at the same level as the center of the screen.

2.2 Procedure

The experimenter selected the type of target and then gave a demonstration on interacting with the facility and interfaces. The participants were required to remember the target type and location in each trial. The participants practiced searching for the target under the guidance of the experimenter. The participant was then asked to carry out a series of practice searches that were of an identical format to those used in the experiment. The practice session lasted for about 10 minutes. After the practice session ended, the actual experiment began. Each participant completed one session which consisted of six visual search tasks. There was a break for 5 minutes between two different music conditions.

When the participants found the target, they would click the left mouse button and a response screen would replace the stimulus screen. In the response screen, the participants were asked to identify the type of target they saw and the location within the screen. Then they clicked the 'Confirm' icon and the program returned to the starting image and a new search task began. The search time was recorded by the program automatically.

3. RESULTS AND DISCUSSION

Totally 4000 data were collected from 20 participants. The search frequency, error frequency and error rate were shown in Table 1. In general, the error frequency of search task increased in the condition of music in contrast to that in the condition of no music.

Table 1 statistics of error frequency for search task

Condition	Search frequency	Error frequency	Error rate
noise	800	10	1.25%
noise + music(slow tempo, high volume)	800	19	2.37%
noise + music(slow tempo, low volume)	800	15	1.88%
noise + music(fast tempo, high volume)	800	18	2.25%
noise + music(fast tempo, low volume)	800	24	3.00 %
total	4000	86	2.15%

Under the different conditions for combination of noise and music, the statistics of search times were shown in Table 2.

Table 2 statistics of search time for different conditions

Condition	Mean	SD	Min	Max	Median
noise	11.017	11.645	0.745	99.451	7.659
noise + music(slow tempo, high volume)	10.135	12.361	0.695	95.299	6.227
noise + music(slow tempo, low volume)	8.374	9.330	0.761	103.205	5.505
noise + music(fast tempo, high volume)	9.560	9.872	0.742	71.843	6.643
noise + music(fast tempo, low volume)	8.480	9.364	0.634	112.996	5.250

From Table 2, we can see that the mean and median of search time in the condition of music adjustment for background noise were smaller than those in the condition of pure noise. It was revealed that music had influence in improving the noisy environment and the visual search performance. In addition, the music of low volume was more conducive for the performance enhancement than the one of high colume. All data in five conditions conformed to exponential distribution. It indicated that the participants used random search strategy. In a random search, each fixation is randomly distributed within the stimulus image and so certain locations may be repeatedly fixated upon while certain other locations may never be fixated. The test results revealed that there was significant difference in visual search performance among 5 conditions (Pearson $\chi^2 = 90.056$, $p = 0.000$). The post hoc results between pure noise condition and other four music ones were shown in Table 3.

Table 3 partial post hoc test results for search performance

Condition	Pearson χ^2	p
noise vs noise + music(slow tempo, high volume)	16.280	0.092
noise vs noise + music(slow tempo, low volume)	53.273	0.000
noise vs noise + music(fast tempo, high volume)	16.910	0.076
noise vs noise + music(fast tempo, low volume)	43.841	0.000

The significant improvement for the performance could be seen in the music conditions of slow tempo and low volume as well as fast tempo and low volume, comparing with the conditions of two high volumes. It seemed that the volume might be a key factor or force in affecting the effect of music adjustment in noisy environment. The effect of tempo of music on search performance was shown in Table 4. There was no significant difference between fast and slow tempos.

Table 4 effects of volume and tempo on search performance

Condition	Pearson χ^2	p
High volume vs low volume	27.643	0.002
Fast tempo vs slow tempo	5.814	0.831

4. CONCLUSION

The study examined the impacts of volume and tempo of background music on visual search performance. Background music loudness had significantly positive impact on visual search performance. No significant differences for visual performance were found across tempo. Music with low loudness acted better for enhancing visual search performance.

ACKNOWLEDGMENTS

This work was fully supported by a grant from National Natural Science Foundation of China (Project No.71071085).

REFERENCES

Schlittmeier, S.J. and J. Hellbrück. 2009. Background music as noise abatement in open-plan offices: A laboratory study on performance effects and subjective preferences. *Applied Cognitive Psychology*. 23(5): 684-697

Ransdell, S. E. and L. Gilroy. 2001. The effects of background music on word processed writing. *Division of Psychology*. Lauderdale: Florida Atlantic University.

Kallinen, K. 2002. Reading news from a pocket computer in a distracting environment: effects of the tempo of background music. *Computers in Human Behavior*. 18(5): 537-551

Edworthy, J. and H.Waring. 2006. The effects of music tempo and loudness level on treadmill exercise. *Ergonomics*, 49(15): 1597 – 1610

Author Index

T - #0297 - 071024 - C656 - 234/156/29 - PB - 9780367381080 - Gloss Lamination